AF179598

ENDOHEDRAL FULLERENES

From Fundamentals to Applications

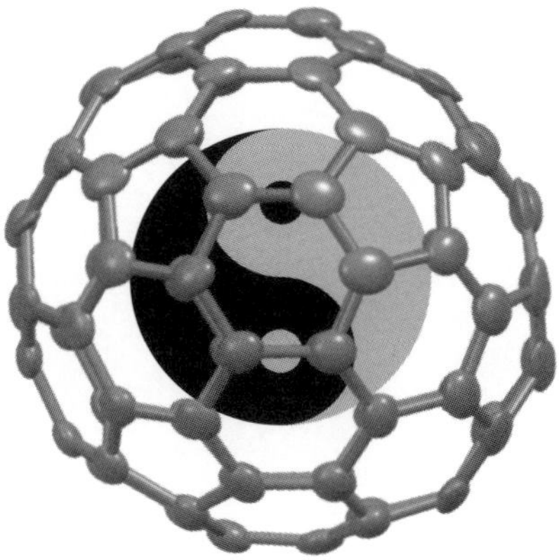

ENDOHEDRAL FULLERENES

From Fundamentals to Applications

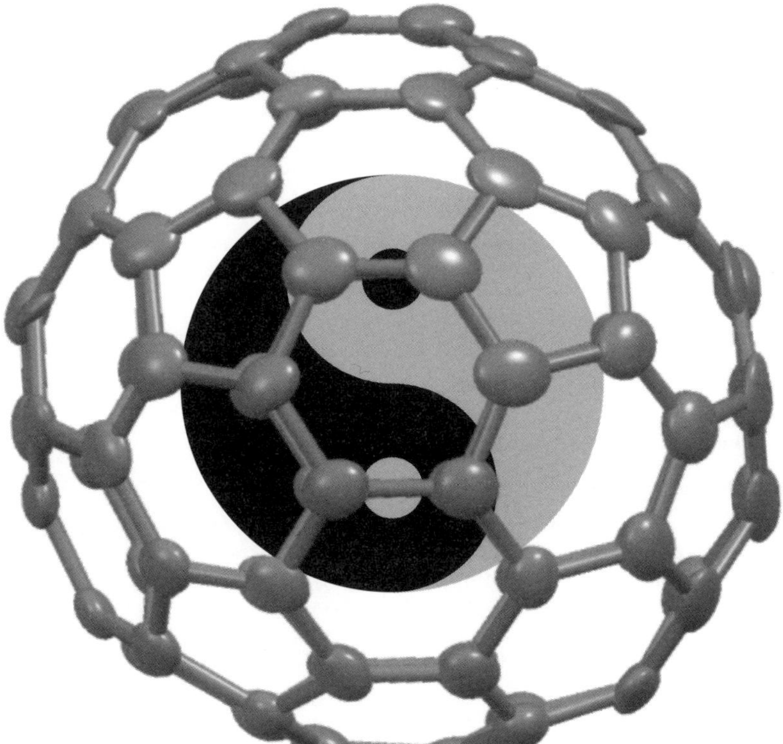

Editors

Shangfeng Yang

University of Science and Technology of China, China

Chun-Ru Wang

Chinese Academy of Sciences, China

World Scientific

NEW JERSEY • LONDON • SINGAPORE • BEIJING • SHANGHAI • HONG KONG • TAIPEI • CHENNAI

Published by

World Scientific Publishing Co. Pte. Ltd.
5 Toh Tuck Link, Singapore 596224
USA office: 27 Warren Street, Suite 401-402, Hackensack, NJ 07601
UK office: 57 Shelton Street, Covent Garden, London WC2H 9HE

British Library Cataloguing-in-Publication Data
A catalogue record for this book is available from the British Library.

ENDOHEDRAL FULLERENES
From Fundamentals to Applications

Copyright © 2014 by World Scientific Publishing Co. Pte. Ltd.

All rights reserved. This book, or parts thereof, may not be reproduced in any form or by any means, electronic or mechanical, including photocopying, recording or any information storage and retrieval system now known or to be invented, without written permission from the publisher.

For photocopying of material in this volume, please pay a copying fee through the Copyright Clearance Center, Inc., 222 Rosewood Drive, Danvers, MA 01923, USA. In this case permission to photocopy is not required from the publisher.

ISBN 978-981-4489-83-6

Typeset by Stallion Press
Email: enquiries@stallionpress.com

Printed in Singapore by Mainland Press Pte Ltd.

We dedicate this monograph to the memory of Lothar Dunsch.

Lothar Dunsch (1948–2013) studied chemistry at the TU Bergakademie Freiberg, Germany, and received his diploma there in chemistry (1972) and his Ph.D. (1973) in electrochemistry. In 1974, he turned to the Institute of Solid State Research of the Academy of Sciences in Dresden and moved in 1978 to the Institute of Polymer Technology before he joined the IFW Dresden in 1992, heading the Department of Electrochemistry and Conducting Polymers and the Center of Spectroelectrochemistry. After his habilitation in 1996, he taught at the TU Dresden, where he was a Professor in Electrochemistry of Organic Systems. His research interests were focused on fullerenes, conducting polymers and oligomers as well as the different methods of in situ spectroelectrochemistry.

Lothar Dunsch was among the first German scientists who started studying fullerenes in early 1990s. The synthesis, characterization, and electrochemical and spectroelectrochemical studies of endohedral fullerenes were among his main research interests over the last two decades. His heritage includes more than 400 published papers, of which 115 are related to the field of endohedral fullerenes. Lothar Dunsch was among those scientists whose contributions shaped the field of endohedral fullerenes. Besides research, Lothar Dunsch was also a passionate writer who, among others science articles and books, published a series of biographies on famous early chemists. Lothar Dunsch's knowledge, energy, and passion will be sorrowfully missed.

Foreword

The last book on endohedral fullerenes, edited by Takeshi Akasaka and Shigeru Nagase, appeared more than 10 years ago. This new collection of articles covers the progress made in the meantime, highlighting the many new developments in this rich field. Over the past years our colleagues from China have joined the endo-field, and the growing importance of their work is exemplified by the editors Shangfeng Yang and Chunru Wang, by the authors of most of the contributions, and finally by the expressive cover design of this book.

A historical overview is given by one of the grand masters in the field, Hisanori Shinohara. It may be worthwhile to remember that endohedral fullerene research originated when the late Richard Smalley attempted to prove that C_{60} exhibits a closed-cage structure, as he and his co-workers had suggested in their famous *Nature* article. Smalley succeeded in enclosing lanthanum inside C_{60}. However, all this research took place on the scale of molecular beams, i.e. with minute amounts of material. The real game started when bulk amounts of endohedral fullerenes could be produced. For quite a while, it was a matter of dispute whether the metal atom in such compounds is located inside or outside the cage. This was resolved when X-ray diffraction gave the final proof of the endohedral nature.

Following the course of events, the rare gas endohedrals were the second type of endo-species to be investigated. Martin Saunders opened this field when he detected traces of the buffer gas helium inside fullerenes produced by an arc-discharge reactor. He then succeeded in improving the endo yield considerably. One of his co-workers, R. James Cross reports in this volume about exciting features of these compounds.

That atomic nitrogen could be enclosed by C_{60} came as a big surprise. This fascinating compound is still waiting for applications. The spin of the nitrogen atom may be useful for nano-scale computer memories and quantum computing, as outlined in one of the articles.

Nitrogen seems to be also important in another respect, namely in forming endohedral metal-nitrides. This field of research was opened by Harry Dorn, who discovered a procedure to produce $Sc_3N@C_{80}$ in surprisingly high yield, considering the complexity of this compound. Here Dorn and co-workers report on biomedical applications of metal-nitride fullerenes which are promising and may also become successful commercially.

For quite a while, fullerene- and endofullerene compounds were discussed as components in solar cells — as outlined in a chapter of this book. Research on solar cells is a highly competitive field. Should fullerenes prove superior, another market for large scale fullerene application would open.

Intrigued by the discovery of Dorn, many groups in the US, Japan, Germany, and recently also in China, have started to prepare endos of the type $M_3N@C_{2n}$. Substitutions in the metal components M and changes of the size of the fullerene cage were successfully achieved. Various state-of-the-art techniques were applied to characterize the clusters and to study their properties, in particular their chemical functionalization and electrochemistry. Along with high level quantum calculations, these efforts are aiming at a better understanding of charge transfer and geometry of the inner composite and outside cage. A considerable number of contributions are devoted to these issues.

Not only nitrides, but also carbides and oxides in combination with metals have been encapsulated and characterised successfully. Details can be found in several chapters of this book. The diversity of enclosed structures is really becoming breathtaking. One wonders what kind of endo-structure will emerge next.

What in my opinion has remained a problem from the beginning to the present day are the sometimes extremely low yields by which the desired endo-compound can be produced, and the laborious and time-consuming separation processes necessary to isolate and characterize the compound of interest. Various tricks have been developed to facilitate separation but the yield problem seems to remain. Up-scaling the production may be a method to obtain more of the endo material, but it may also be necessary to develop entirely different production approaches, for example by searching for low temperature processes which involve incomplete combustion.

To me, one of the highlights of endo research is the discovery by Wang and co-workers that one can overcome the restrictions imposed on the cage-structure by the isolated pentagon rule. With a suitable inside species one thus may be able to obtain fullerene cages smaller than that of C_{60}. How far down in cage size one can go remains to be seen. In any event, one aspect of fullerene research seems to become important again: Endo-fullerenes may serve as "pseudo-elements", meaning that these compounds form another periodic table in which the entries are not elements but endo-fullerenes. Solids built up from such pseudo-atoms may show interesting and exciting features, for example superconductivity or extreme hardness, just to name a couple. Nevertheless, to produce such solids and study their properties is, I fear, still a long way to go.

Unfortunately, there is also a rather sad message to announce: Lothar Dunsch, one of the leading researchers in the endo-fullerene field, passed away in November 2013. With his untimely death the community loses an admirable friend and colleague. This book is dedicated to his memory.

Wolfgang Krätschmer Heidelberg, December 2013

Contents

CHAPTER 1

The Early Days of Metallofullerene Research

*Hisanori Shinohara**

1.1. Fullerenes with Metal Atom Encapsulated: Discovery

Endohedral metallofullerenes are an interesting class of fullerenes because electron transfer from the encaged metal atom to the carbon cage has been known to occur, and this oftentimes (dramatically) alters the electronic and magnetic properties of the fullerenes.

In this chapter, I will briefly describe the history and evolution of metallofullerene research in the early days, where a series of exciting and novel results were obtained on daily basis.

The first idea (and also an experimental trial of the synthesis) of metallofullerenes was presented by the Sussex University–Rice University joint research team headed by Harry Kroto and Rick Smalley in 1985. A week after the first experimental observation of the so-called "magic number" soccerball-shaped C_{60} in a laser-vaporized cluster beam mass spectrum by Kroto *et al.*,[1] the same research group also found a magic number feature due to LaC_{60} in a mass spectrum prepared by laser vaporization of a $LaCl_2$-impregnated graphite rod.[2]

* *Department of Chemistry and Institute for Advanced Research, Nagoya University, Nagoya 464-8602, Japan.* Email: noris@nagoya-u.jp

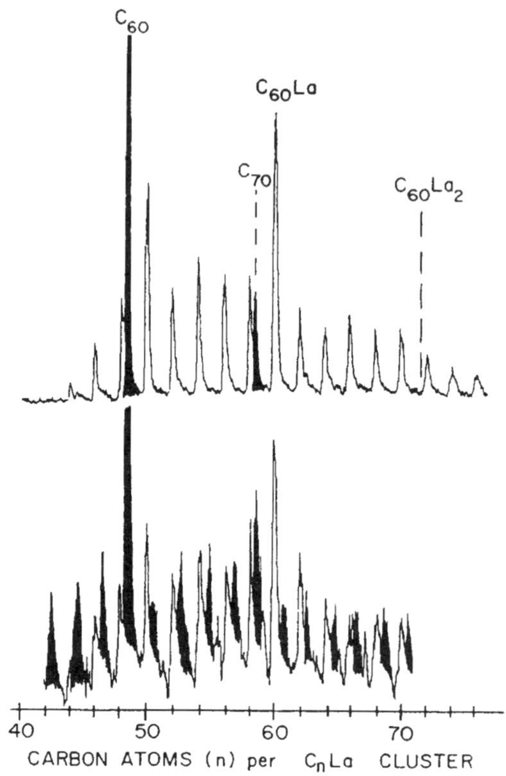

Fig. 1 Laser-vaporization supersonic cluster-beam Time-of-Flight mass spectrum of various lathanum-carbon clusters. LaC_{60} is seen as an enhanced (magic number) peak.

They observed a series of C_n^+ and LaC_n^+ ion species with LaC^+_{60} as a magic number ion in the mass spectrum (Fig. 1) and concluded that a La atom might be encaged within the (then hypothetical) soccerball-shaped C_{60}. This was obviously the first experimental proposal of the so-called "endohedral metallofullerenes" concept. They first tried Fe with no success and then found that La was the correct atom for encapsulation within fullerenes. Interestingly, even today Fe has not been encapsulated by fullerenes because of a substantial lack of electron transfer from Fe to fullerenes. Instead, Fe has been known as one of the best metal catalysts in synthesis of single-wall carbon nanotubes.[3]

Further circumstantial (but not direct) evidence that metal atoms may be encaged in C_{60} was also reported by the Rice group, showing that LaC_{60}^+

ions did not react with H_2, O_2, NO and NH_3,[4]. This suggests that reactive metal atoms are protected from the surrounding gases and are indeed trapped inside the C_{60} cage.

The first direct evidence of the soccerball (truncated icosahedron) C_{60} was amply demonstrated in 1990 by a historical experiment done by Kraetschmer, Huffman and co-workers. They succeeded in producing macroscopic quantities of soccerball-shaped C_{60} by using resistive heating of graphite rods under a He atmosphere.[5,6] The resistive heating method was then superseded by the so-called contact arc discharge method developed by the Rice group[7] since the arc discharge method can produce fullerenes order of magnitudes larger than that by resistive heating. Since then, this arc discharge method has become a standard method for fullerene synthesis.

1.2. Macroscopic Synthesis of Metallofullerenes

The first production of macroscopic quantities of endohedral metallofullerenes were also reported by the Rice group.[8] They used the high-temperature laser vaporization method, originally developed for C_{60} synthesis, of La_2O_3/graphite composite rods and the contact arc discharge technique to produce various sizes of La-metallofullerenes. Contrary to the previous expectation, only the La@C_{82} fullerene survived in solvent and was extractable by toluene even though La@C_{60} and La@C_{70} were also seen in the mass spectra of the sublimed film from raw soot. In other words, the major La-metallofullerenes with air stability was La@C_{82}, and La@C_{60} and La@C_{70} were somehow unstable in air and in solvents.

I would like to point out here in passing that the symbol @ is conventionally used to indicate that atoms listed to the left of the @ symbol are encaged in the fullerenes. For example, a C_{60}-encaged metal species (M) is then written as M@C_{60}.[8] The corresponding IUPAC nomenclature is, however, different from this conventional M@C_{60} representation. It is recommended by IUPAC that La@C_{82} should be called [82] fullerene-*incar*-lanthanum and be written *i*LaC_{82}.[9] However, throughout this chapter the conventional M@C_{2n} description is used for endohedral metallofullerenes for brevity, unless otherwise noted.

The speciality of the La@C_{82} fullerene was soon confirmed by Whetten and co-workers.[10] However, they also observed that at relatively high

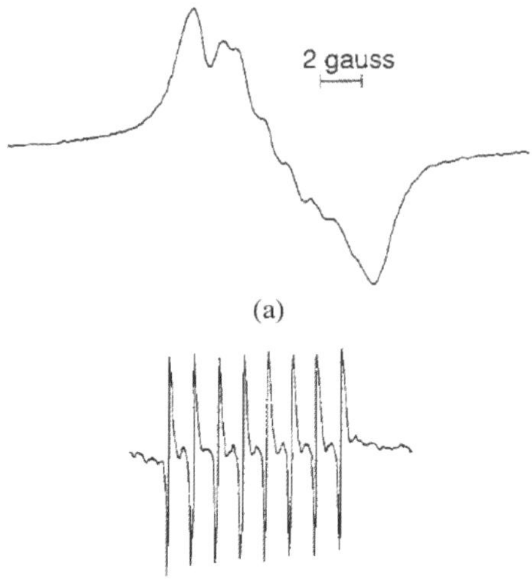

Fig. 2. EPR spectrum (9.112 GHz) at ambient temperature of (a) solid degassed toluene extract (dried) resulting from arc burning of a composite graphite/La_2O_3 rod and (b) a degassed solution of the dried extract in 1,1,2,2-tetrachloroethane[12].

loading ratios of La_2O_3 in composite rods, a di-lanthanofullerene, $La_2@C_{80}$, was also produced by the resistive-heating method and found to be another solvent-extractable major lanthanofullerene.[10,11]

The first important information on the electronic structure of $La@C_{82}$ was provided by the IBM Almaden research group. The charge state of the encaged La atom was studied by Johnson *et al.*[12] using electron paramagnetic resonance (EPR). The EPR hyperfine splitting (hfs) analysis of $La@C_{82}$ revealed that the La atom is in the +3 charge state and that the formal charge state of $La@C_{82}$ should be written as $La^{3+}@C_{82}^{3-}$: three outer electrons of La are transferring to the C_{82} cage.[13]

Several other research groups extended their works to endohedral yttrium compounds. The Rice–Minnesota University[14] and Nagoya University[15] research groups also reported solvent-extractable $Y@C_{82}$ and $Y@C_{82}$ fullerenes and observed the EPR hfs of $Y@C_{82}$. From the hfs analyses, both groups concluded that the charge state of the Y atom is also +3 and that a similar intra-fullerene electron transfer is taking place in $Y@C_{82}$ as in $La@C_{82}$. These results were also confirmed by the Naval

Research Laboratory (NRL) group.[16] In addition, they also reported the production of mixed di-metallofullerenes like (LaY)@C_{80}. McElvany[17] reported the production of a series of yttrium fullerenes, Y_m@C_n including Y@C_{82}, by direct laser vaporization of samples containing graphite, yttrium oxide and fullerenes in the gas phase.

Scandium metallofullerenes were also produced in macroscopic quantity and solvent-extracted by Shinohara *et al.*[18] and Yannoni *et al.*[19] Both research groups found that the Sc fullerenes exist in extracts as a variety of species (mono-, di-, tri-and even tetra-scandium fullerenes), typically as Sc@C_{82}, Sc_2@C_{74}, Sc_2@C_{82}, Sc_2@C_{84}, Sc_3@C_{82} and Sc_4@C_{82}. It was found that Sc_3@C_{82} was also an ESR-active species whereas di-and tetra-scandium fullerenes like Sc_2@C_{84} and Sc_4@C_{82} were ESR-silent. (Later, some of these Sc fullerenes were found to be the so-called "carbide metallofullerenes,"[20] such as Sc_2C_2@C_{82} and Sc_3C_2@C_{80}.)

The formation of lanthanide metallofullerenes R@C_{82} (R = Ce, Pr, Nd, Sm, Eu, Gd, Tb, Dy, Ho, Er, Yb and Lu) was also reported by the UCLA[21] and SRI international[22] groups. These metallofullerenes were also based on the C_{82} fullerene. In addition to group 3 (Sc, Y, La) and the lanthanide metallofullerenes, group 2 metal atoms (Ca, Sr, Ba) were also found to form endohedral metallofullerenes, and have been produced and isolated in mg quantity.[23–28] Notably, these metal atoms have been encaged not only by C_{82} and C_{84} but also by such higher fullerenes as C_{72}, C_{74} and C_{80}.

One big mystery concomitant with the metallofullerene research is that M@C_{60} (M=metal) metallofullerenes (even if they have been found in soot co-existing with other metallofullerenes as described above) have not been solvent extracted, separated and purified. One of the main causes of this is due to a high reactivity of M^{3+}@C_{60}^{3-}, which is predicted to possess triply degenerate half-filled open shell electronic structures.[29] Metallofullerenes based on C_{60} are, therefore, unstable in air and in normal fullerene solvents such as toluene and carbon disulfide.

In the ensuing years, however, C_{60}-based metallofullerenes such as Ca@C_{60}[30–34] were synthesized. The Ca@C_{60} and U@C_{60} fullerenes are unique metallofullerenes in which Ca and U atoms are encaged by C_{60}, and are quite different from group 3 and lanthanide, M@C_{82} type, metallofullerenes. An *ab initio* SCF Hartree–Fock calculation indicates that the Ca ion in Ca@C_{60} is displaced by 0.7 Å from the centre and that the

Table 1 A 'Bucky Periodic Table' Showing the Elements Which Have Been Reported to Form Endohedral Metallofullerenes and Isolated as Purified Forms

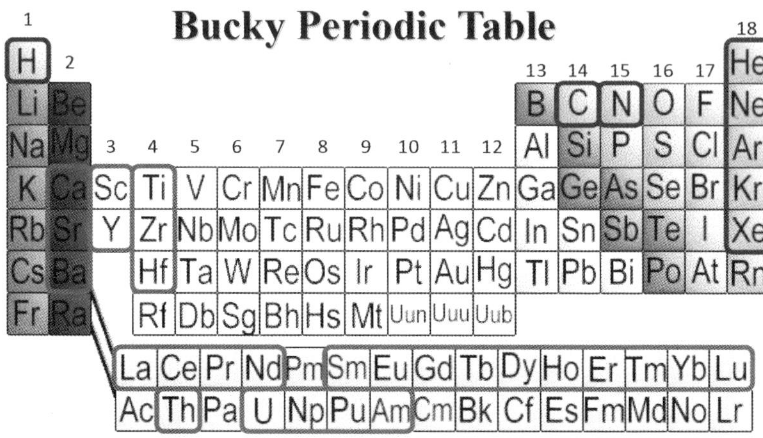

electronic charge of Ca is 2+.[31,35] A similar theoretical prediction has been made on Sc@C_{60} by Scuseria and co-workers.[36] The metal atoms which have been reported to form endohedral metallofullerenes are shown in Table 1.

1.3. Separation, Purification and Isolation

As usually employed in the separation of hollow fullerenes,[37,38] liquid chromatography (LC) is also the main purification technique for metallofullerenes. LC has been frequently and traditionally used in separation chemistry. One of the most powerful LC techniques is high-performance LC (HPLC) which allows separation of fullerenes according to their molecular weight, size, shape or other parameters.[39–43] The HPLC technique can even allow us to separate structural isomers of various matallofullerenes.[29]

The purification of endohedral metallofullerenes via HPLC had been difficult, mainly because the content of metallofullerenes in raw soot is normally very limited and, furthermore, the solubility in normal HPLC solvents is generally lower than that of various empty higher fullerenes. It took almost two years for metallofullerenes to be completely isolated by the HPLC method[44,45] after the first extraction of La@C_{82} by the Rice group.[8] Following these first isolations of metallofullerenes, isolation with different HPLC columns was also reported.[46,47]

The success of the purification/isolation was a breakthrough for further characterization of the endohedral metallofullerenes. Scandium metallofullerenes are, in particular, interesting in terms of separation and purification, because scandium fullerenes appear as mono-, di-, tri- and tetra-scandium fullerenes with several structural isomers which can be separated completely by HPLC. The scandium fullerenes, such as $Sc@C_{82}$, $Sc_2C_2@C_{82}$ and $Sc_3C_2@C_{80}$, were separated and isolated from various hollow (C_{60}–C_{110}) fullerenes by the so-called two-stage HPLC method.[44,,45,48] The two-stage HPLC method uses two complementary HPLC columns which have different types of fullerene adsorption mechanisms and realizes a complete separation of the metallofullerenes.

The two-stage HPLC method was first successfully applied to the isolation of several di-scandium fullerenes including $Sc_2@C_{74}$, $Sc_2@C_{82}$ and $Sc_2C_2@C_{82}$,[44] as shown in Fig. 3. To simplify the separation, an automated HPLC separation on some endohedral metallofullerenes was reported by Stevenson.[49,50] It has been found[51,52] that most of the mono-metallofullerenes, $M@C_{82}$, have at least two types of structural isomers (conventionally denoted as isomers I and II), which can be separated by the two-stage HPLC technique.

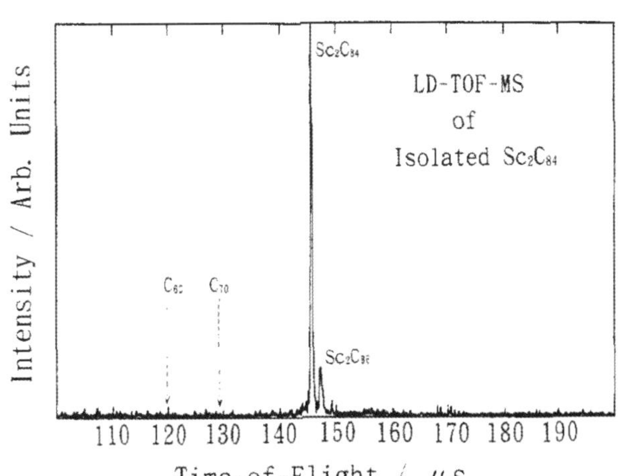

Fig. 3 The first mass spectrum of isolated metallofullerenes: $Sc_2@C_{84}$ (which is currently identified as $Sc_2C_2@C_{82}$).

Up until fairly recently, the separation and purification of metallofullerenes by these HPLC techniques has been a time-consuming process in metallofullerene studies. However, the recent breakthrough in the non-HPLC rapid separation[53] of various metallofullerenes from empty fullerenes using the $TiCl_4$ Lewis acid method has made it possible to overcome this difficulty and to ensure a large scale of purification of metallofullerens such as gram quantity.

1.4. Endohedral Structure of Metallofullerenes: *Endohedral or Exohedral?*

Since the first studies on production and solvent extraction of metallofullerenes such as La/Y/Sc@C_{82}, there had been considerable controversy as to whether or not the metal atom is really trapped inside the fullerene cage.[8]

The stability of endohedral metallofullerenes was studied in the gas phase by laser photofragmentation for La@C_{82} and Sc_2@C_{84},[55,56] collisional fragmentation with atomic and molecular targets for La@C_{82} and Gd@C_{82}[57] and fragmentation induced by surface impact for La_2@C_{80},[11] La@C_{82}, La_2@C_{100},[58] La@C_{60},[59] Ce@C_{82} and Ce_2@C_{100},[60] Y@C_{82} and Ca@C_{84}.[61] Although the most extensive fragmentation was observed in the laser photofragmentation, the general tendency of the fragmentation induced by the three excitations was found to be similar: the main fragments from La@C_{82} were C_2-loss species such as La@C_{80}, La@C_{78}, La@C_{76}, etc., and the empty C_{82} fragment was not observed. This result was interpreted as being due to the endohedral nature of La@C_{82} since exohedral La(C_{60})[62] and Fe(C_{60})[63] prepared by gas phase reactions gave C_{60} as the main product upon collisional fragmentation against rare-gas targets.

However, in the solid state, the evidence for the endohedral nature of the metallofullerenes had not been convincing up until 1995. For instance, two contradictory results were reported for extended X-ray absorption fine structure (EXAFS) experiments on an unpurified extract of Y@C_{82} (i.e. a mixture of Y@C_{82} and empty fullerenes). Soderholm[64] reported that the yttrium atom is exohedrally attached from the outside to the C_{82} cage, whereas Park[65] reported an endohedral nature of Y@C_{82}: the nearest-neighbour C–Y distances obtained were 0.253 ± 0.002 and 0.24 nm, respectively. Kikuchi[66] performed an EXAFS experiment on a purified La@C_{82} powder material and

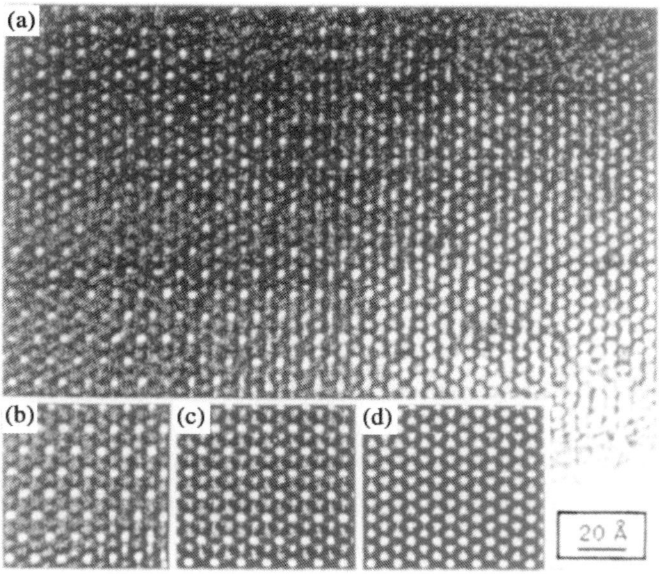

Fig. 4 (a) High-resolution TEM image of a $Sc_2@C_{84}$ crystal, taken along the [0001] direction. Inset (b) is a Fourier-filtered image that brings out the periodicities present in the original image. Insets (c) and (d) are simulated images of 67 Å thick $Sc_2@C_{84}$ (currently identified as $Sc_2c_2@C_{82}$) and C_{84} crystals, respectively.

reported that the nearest-and next-to-nearest-neighbour C–La distances are 0.247 ± 0.002 and 0.294 ± 0.007 nm, respectively.

Most of the major experimental evidence suggested, however, the endohedral nature of the metallofullerenes. The IBM Almaden group reported a high-resolution transmission electron microscopy (HRTEM) experiment on a purified $Sc_2@C_{84}$ material which suggests that the two scandium atoms are encapsulated in the C_{84} cage (see Fig. 4).[67] A similar evidence on endohedral nature based on HRTEM images was reported on $Gd@C_{82}$ by Tanaka.[68]

The UCLA group reported a high-energy collision experiment on $La_2@C_{80}$ against silicon surfaces and found that no collision fragments such as La atoms and C_{80} were observed, also suggesting an endohedral structure of $La_2@C_{80}$.[11] Similar surface-induced dissociation experiments were done on $La@C_{82}$ and $La@C_{60}$[59] against a self-assembled monolayer film and on $Y@C_{82}$, $Ca@C_{82}$ and $Ca@C_{84}$ against solid (silicon and gold) surfaces and SAM films,[61] all of which strongly suggested the endohedral nature of these metallofullerenes.

The Tohoku–Nagoya University joint group reported a series of ultra-high vacuum scanning tunnelling microscopy (UHV-STM) studies on $Sc_2@C_{84}$ and $Y@C_{82}$ adsorbed on silicon and copper clean surfaces, respectively.[69–71] All of the obtained STM images showed a spherical shape, which strongly suggests that the metal atoms are encapsulated in the fullerene cages. Gimzewski also studied $Sc_2@C_{84}$ molecules deposited from a CS_2 solution onto Au(110) by STM and obtained some internal structure on the top part of the images.[72] Although the above experimental results strongly suggest the endohedral nature of the metallofullerenes, the final confirmation of the endohedral nature and detailed endohedral structures of the metallofullerenes was obtained by synchrotron X-ray diffraction measurements on purified powder samples.[73]

1.5. Confirmation on Endohedral Structures by X-ray Diffraction

The experimental evidence described above including EXAFS[65,66] and HRTEM[67] suggested that the metal atoms are inside the fullerenes. Theoretical calculations also indicated that such endohedral metallofullerenes are stable.[31,36,74–84] However, the first conclusive experimental evidence on the endohedral nature of a metallofullerene, $Y@C_{82}$, was obtained by a synchrotron X-ray diffraction study. The result indicated that the yttrium atom is indeed encapsulated within the C_{82} fullerene and is strongly bound to the carbon cage.[85]

The X-ray powder data were analyzed in an iterative way of combination of Rietveld analysis[86] and the maximum entropy method (MEM).[87,88] The MEM can produce an election density distribution map from a set of X-ray structure factors without using any structural model. By the MEM analysis,[89,90] the R_I become as low as 1.5% for $Y@C_{82}$.

To visualize the endohedral nature of the $Y@C_{82}$, the MEM electron density distribution of $Y@C_{82}$ is shown in Fig. 5. There exists a high-density area just inside the C_{82} cage. The density maximum at the interior of the C_{82} cage corresponds to the yttrium atom, indicating the endohedral structure of the metallofullerene. It was also found that the cage structure

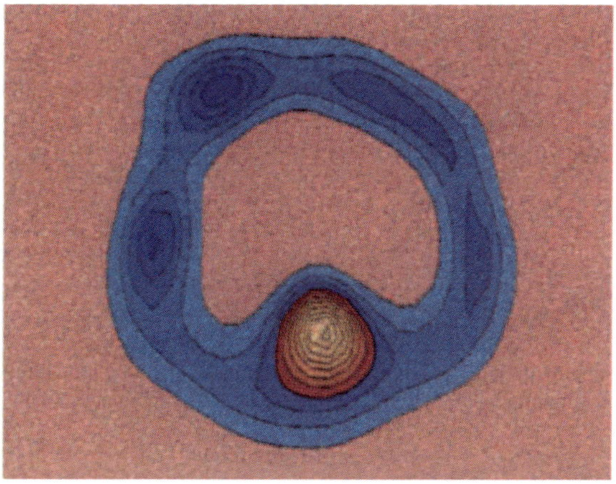

Fig. 5 The first clear and direct confirmation of the endohedral nature of $Y@C_{82}$. The MEM electron density distribution of $Y@C_{82}$ for the (001) section.

of $Y@C_{82}$ differs from that of the hollow C_{82} fullerene. There are many local maxima along the cage in $Y@C_{82}$, whereas electron densities of the C_{82} cage are relatively uniform. This suggests that in $Y@C_{82}$ the rotation of the C_{82} cage is very limited around a certain axis even at room temperature, while that in C_{82} is almost free.

Importantly, the MEM election density map further reveals that the yttrium atom does not reside at the centre of the C_{82} cage but is very close to the carbon cage, as suggested theoretically.[36,80,82,83,84,91] The ESR[14,15] and theoretical[84,92] studies suggest the presence of a strong charge transfer interaction between the Y^{3+} ion and the C^{3-} cage which may cause the aspherical electron density distribution of atoms. The Y–C distance calculated from the MEM map is 2.9(3) Å which is slightly longer than a theoretical prediction of 2.55–2.65 Å.[84] The X-ray study also reveals that the $Y@C_{82}$ molecules are aligned along the [001] direction in a head-to-tail (...$Y@C_{82}$ ...$Y@C_{82}$ ...$Y@C_{82}$ ...) order in the crystal, indicating the presence of a strong dipole–dipole and charge transfer interactions among the $Y@C_{82}$ fullerenes.

Ever since this confirmation of the endohedral nature of the metallofullerene, a large number of X-day diffraction experiments have been reported, which include single-crystal X-day diffractions.[93]

1.6. A Future Challenge and Dream: Superconductive Metallofullerenes?

One of the most exciting and enticing topics on metallofullerenes is whether or not a superconductive metallofullerene could exist as a high-temperature fullerene superconductor such as alkaline-doped C_{60} (K_3C_{60}, Rb_3C_{60}, $RbCs_2C_{60}$, Cs_3C_{60}, etc.).

Trivalent $M^{3+}@C_{60}^{3-}$ (M = typically trivalent lanthanoid metal atoms such as La, Y, Ce, Er, Gd) metallofullerenes might exhibit superconductive behavior in low temperatures, because, as described in Section 1.2, the electronic structure of $M^{3+}@C_{60}^{3-}$ is exactly the same as that of the superconductive $K^{3+}C_{60}^{3-}$, where the triply degenerate half-filled LUMO should play a crucial role in the emergence of superconductivity.

To do this, all we have to do first is to try to obtain purifed $M@C_{60}$ materials which have not been experimentally available so far due to their high reactivity towards other fullerenes and amorphous carbons present in the original arc-processed soot. However, it is well known that the amount of $M@C_{60}$ is generally much higher than that of the conventional $M@C_{82}$ in original raw soot, so that the future purification of $M@C_{60}$ is, I believe, highly probable. This is definitely worthwhile to change.

REFERENCES

1. Kroto H, Heath JR, O'Brien SC, *et al.* (1985) C_{60}: Buckminsterfullerene. *Nature* **318:** 162–163
2. Heath J R, O'Brien S C, Zhang Q, *et al.* (1985) Lanthanum Complexes of Spheroidal Carbon Shells. *J Am Chem Soc* **107:** 7779–7780.
3. Iijima S, Ichihashi T. (1993) Single-shell carbon nanotubes of 1-nm diameter. *Nature* **363:** 603–605.
4. Weiss FD, Elkind JL, O'Brien S C, *et al.* (1988) Photophysics of metal complexes of spheroidal carbon shells. *J Am Chem Soc* **110:** 4464–4465.
5. Krätschmer W, Lamb LD, Fostiropoulos K, *et al.* (1990) Solid C_{60}: a New Form of Carbon. *Nature* **347:** 354–358.
6. Kraetschmer W, Fostiropoulos K, Huffman DR. (1990) The infrared and ultraviolet absorption spectra of laboratory-produced carbon dust: evidence for the presence of the C_{60} molecule. *Chem Phys Lett* **170:** 167–170.
7. Haufler RE, Chai Y, Chibante LPF, *et al.* (1991) *Cluster-Assembled Materials*, ed RSAverback, J Bernhoc and D L Nelson (Pittsburgh: Materials Research Society) **206:** 627–637.

8. Chai Y, Guo T, Jin CM, *et al.* (1991) Fullerenes with Metals Inside. *J Phys Chem* **95:** 7564–7568.
9. Godly EW, Taylor R. (1997) Nomenclatůre and terminology of fullerenes: A preliminary study. *Pure Appl Chem* **69:** 1411–1434
10. Alvarez MM, Gillan EG, Holczer K, *et al.* (1991) $La_2@C_{80}$: A Soluble Dimetallofullerene. *J Phys Chem* **95:** 10561–10563
11. Yeretzian C, Hansen K, Alvarez MM, *et al.* (1992) Collisional probes and possible structures of La_2C_{80}. *Chem Phys Lett* **196:** 337–342
12. Johnson RD, de Vries MS, Salem J, *et al.* (1992) Electron paramagnetic resonance studies of lanthanum-containing C_{82}. *Nature* **355:** 239–240.
13. Bethune D S, Johnson RD, Salem JR, *et al.* (1993) Atoms in carbon cages: the structure and properties of endohedral fullerenes. *Nature* **366:** 123–128
14. Weaver JH, Chai Y, Kroll GH, *et al.* (1992) XPS probes of carbon-caged metals. *Chem Phys Lett* **190:** 460–464.
15. Shinohara H, Sato H, Saito Y, *et al.* (1992) Mass spectroscopic and ESR characterization of soluble yttrium-containing metallofullerenes YC_{82} and Y2C82. *J Phys Chem* **96:** 3571–3573
16. Ross MM, Nelson HH, Callahan JH, *et al.* (1992) Production and characterization of metallofullerenes. *J Phys Chem* **96:** 5231–5234.
17. McElvany SW. (1992) Production of endohedral yttrium-fullerene cations by direct laser vaporization. *J Phys Chem* **96:** 4935–4937.
18. Shinohara H, Sato H, Ohchochi M, *et al.* (1992) Encapsulation of a scandium trimer in C_{82}. *Nature* **357:** 52–54.
19. Yannoni CS, Hoinkis M, de Vries MS, *et al.* (1992) Scandium clusters in fullerene cages. *Science* **256:** 1191–1992.
20. Wang CR, Kai T, Yomiyama T, *et al.* (2001) A Scandium Carbide Endohedral Metallofullerene: $(Sc_2C_2)@C_{84}$. *Angew Chem Int Ed* **40:** 397–399.
21. Gillan E, Yeretzian C, Min K S, *et al.* (1992) Endohedral Rare-Earth Fullerene Complexes. *J Phys Chem* **96:** 6869–6871
22. Moro L, Ruoff RS, Becker CH, *et al.* (1993) Studies of Metallofullerene Primary Soots by Laser and Thermal Desorption Mass Spectrometry. *J Phys Chem* **97:** 6801–6805.
23. Xu Z, Nakane T, Shinohara H. (1996) Production and Isolation of $Ca@C_{82}$ (I-IV) and $Ca@C_{84}$ (I,II) Metallofullerenes. *J Am Chem Soc* **118:** 11309–11310.
24. Dennis TJS, Shinohara H. (1997) Production and isolation of endohedral strontium- and barium-based mono-metallofullerenes: $Sr/Ba@C_{82}$ and $Sr/Ba@C_{84}$. *Chem Phys Lett* **278:** 107–110
25. Dennis TJS, Shinohara H. (1997) Fullerenes: Recent Advances in the Chemistry and Physics of Fullerenes and Related Materials.**4:** ed.

26. Dennis TJS, Kai T, Tomiyama T, *et al.* (1998) Isolation and characterisation of the two major isomers of [84]fullerene (C_{84}).*Chem Commun* 619–620.
27. Wan SM, Zhang HW, Tso TSC, *et al.* (1997) *Fullerenes: Recent Advances in the Chemistry and Physics of Fullerenes and Related Materials*, ed K Kadish and R Ruoff (Pennington: Electrochemical Society) **4:** 490–506.
28. Wan SM, Zhang HW, Nakane T, *et al.* (1998) Production, Isolation, and Electronic Properties of Missing Fullerenes: Ca@C_{72} and Ca@C_{74}. *J Am Chem Soc* **120:** 6806–6807.
29. Shinohara H. (2000) Endohedral metallofullerenes. *Rep Prog Phys* **63:** 843–892.
30. Guo T, Diener MD, Chai Y, *et al.* (1992) Uranium Stabilization of C_{28}: A Tetravalent Fullerene. *Science* **257:** 1661–1664.
31. Wang LS, Alford JM, Chai Y, *et al.* (1993) The electronic structure of Ca@C_{60}. *Chem Phys Lett* **207:** 354–359.
32. Wang LS, Alford JM, Chai Y, *et al.* (1993) Photoelectron spectroscopy and electronic structure of Ca@C_{60}. *Z Phys* D **26:** 297–299.
33. Wang Y, Tomanek D, Ruoff R. (1993) Stability of M@C_{60} endohedral complexes. *Chem Phys Lett* **208:** 79–85.
34. Kubozono Y, Ohta T, Hayashibara T, *et al.* (1995) Preparation and extraction of Ca@C_{60}. *Chem Lett* **24:** 457–458.
35. Scuseria GE. (1992) Comparison of coupled-cluster results with a hybrid of Hartree–Fock and density functional theory. *J Chem Phys* **97:** 7528–7530.
36. Guo T, Odom GK, Scuseria GE. (1994) Electronic Structure of Sc@C_{60}: An ab Initio Theoretical Study. *J Phys Chem* **98:** 7745–7747.
37. Taylor R, Hare JP, Abdul-Sada AK, *et al.* (1990) Isolation, Separation and Characterisation of the Fullerenes C_{60} and C_{70}: The Third Form of Carbon. *J Chem Soc Chem Commun* 1423–1425.
38. Scrivens WA, Bedworth PV, Tour MJ. (1992) Purification of Cram Quantities of C_{60}. A New Inexpensive and facile method. *J Am Chem Soc* **114:** 7917–7919.
39. Kikuchi K, Nakahara N, Honda M, *et al.* (1991) Separation, Detection, and UV/Visible Absorption Spectra of Fullerenes; C_{76}, C_{78}, and C_{84}.*Chem Lett* **20:** 1607–1610.
40. Kikuchi K, Nakahara N, Wakabayashi T, *et al.* (1992) Isolation and identification of fullerene family: C_{76}, C_{78}, C_{82}, C_{84}, C_{90} and C_{96}. *Chem Phys Lett* **188:** 177–180.
41. Klute RC, Dorn HC ,Mc Nair HM. (1992) HPLC Separation of Higher (C_{84+}) Fullerenes. *J Chromatogr Sci* **30:** 438–442.

42. Meier MS, Selegue JP. (1992) Efficient Preparative Separation of C_{60} and C_{70}. Gel Permeation Chromatography of Fullerenes Using 100% Toluene as Mobile Phase. *J Organ Chem* **57:** 1924–1926.
43. Jinno K, Saito Y. (1996) Separation of fullerenes by liquid chromatography — molecular recognition mechanism in liquid chromatographic separation. *Adv Chromatogr* **36:** 65–125.
44. Shinohara H, Yamaguchi H, Hayashi N, *et al.* (1993) Isolation and Spectroscopic Properties of $Sc_2@C_{74}$, $Sc_2@C_{82}$, and $Sc_2@C_{84}$. *J Phys Chem* **97:** 4259–4261.
45. Kikuchi K, Suzuki S, Nakao Y, *et al.* (1993) Isolation and characterization of the metallofullerene LaC_{82}. *Chem Phys Lett* **216:** 67–71.
46. Yamamoto K, Funasaka H, Takahashi T, *et al.* (1994) Isolation of an ESR-Active Metallofullerene of $La@C_{82}$. *J Phys Chem* **98:** 2008–2011.
47. Savina M, Martin G, Xiao J, *et al.* (1994) *Fullerenes: Recent Advances in the Chemistry and Physics of Fullerenes and Related Materials* ed K Kadish and R Ruoff (Pennington: Electrochemical Society) **1:**1309–1319.
48. Shinohara H, Inakuma M, Hayashi N, *et al.* (1994) Spectroscopic Properties of Isolated $Sc_3@C_{82}$ Metallofullerene. *J Phys Chem* **98:** 8597–8599.
49. Stevenson S, Dorn HC, Burbank P, *et al.* (1994) Automated HPLC Separation of Endohedral Metallofullerene Sc@C2n and Y@C2n Fractions. *Anal Chem* **66:** 2675–2679.
50. Stevenson S, Dorn HC, Burbank P, *et al.* (1994) Isolation and Monitoring of the Endohedral Metallofullerenes $Y@C_{82}$ and $Sc3@C_{82}$: On-Line Chromatographic Separation with EPR Detection. *Anal Chem* **66:** 2680–2685.
51. Yamamoto K, Funasaka H, Takahashi T, *et al.* (1994) Isolation and Characterization of an ESR-Active $La@C_{82}$ Isomer. *J Phys Chem* **98:** 12831–12833.
52. Inakuma M, OhnoM, Shinohara H. (1995) *Fullerenes: Recent Advances in the Chemistry and Physics of Fullerenes and Related Materials,* ed K Kadish and R Ruoff (Pennington: Electrochemical Society) **2:** 330–342.
53. Akiyama K, Hamano T, Nakanishi Y, *et al.* (2012) Non-HPLC Rapid Separation of Metallofullerenes and Empty Cages with $TiCl_4$ Lewis Acid. *J Am Chem Soc* **134:** 9762–9767.
54. Wang Z, Nakanishi Y, Noda S, *et al.* (2012) The Origin and Mechanism of Non-HPLC Purification of Metallofullerenes with $TiCl_4$. *J Phys Chem C* **116:** 25563–25567.
55. Wakabayashi T, Shiromaru H, Suzuki S, *et al.* (1996) C_2-Loss fragmentation of higher fullerenes and metallofullerenes. *Surf Rev Lett* **3:** 793–798.

56. Suzuki S, Kojima Y, Shiromaru H, *et al.* (1997) Photoionization/fragmentation of endohedral fullerenes. *Z Phys D* **40:** 410–413.
57. Lorents DC, Yu DH, Brink C, *et al.* (1995) Collisional fragmentation of endohedral fullerenes. *Chem Phys Lett* **236:** 141–149
58. Beck RD, Weis P, Rockenberger J *et al.* (1996) Fragmentation of fullerenes and metallofullerenes by surface impact. *Surf Rev Lett* **3:** 881–885.
59. Kimura T, Sugai T, Shinohara H. (1999) Surface-induced dissociation of lanthanum metallofullerenes on a fluorinated self-assembled monolayer film. *Chem Phys Lett* **304:** 211–216.
60. Beck RD, Weis P, Rockenberger J *et al.* (1996) Delayed ionization of fullerenes and fullerene derivatives upon laser desorption and surface collision. *Surf Rev Lett* **3:** 771–775.
61. Kimura T, Sugai T, Shinohara H. (1999) Surface-induced fragmentation of higher fullerenes and endohedral metallofullerenes. *J Chem Phys* **110:** 9681–9687.
62. Huang Y, Freiser BS. (1991) Externally bound metal ion complexes of buckminsterfullerene, MC_{60}^{+}, in the gas phase. *J Am Chem Soc* **131:** 9418–9419
63. Roth LM, Huang Y, Schwedler JT, *et al.* (1991) Evidence for an Externally Bound Fe^{+}-Buckminsterfullerene Complex, Few^{+}, in the Gas Phase. *J Am Chem Soc* **113:** 6298–6299.
64. Soderholm L, Wurz P, Lykke KR, *et al.* (1992) An EXAFS Study of the Metallofullerene YC_{82}: Is the Yttrium Inside the Cage? *J Phys Chem* **96:** 7153–7156.
65. Park CH, Wells BO, DiCarlo J, *et al.* (1993) Structural information on Y ions in C_{82} from EXAFS experiments. *Chem Phys Lett* **213:** 196–201.
66. Kikuchi K, Nakao Y, Achiba Y, *et al.* (1994) *Fullerenes: Recent Advances in the Chemistry and Physics of Fullerenes and Related Materials,* ed K Kadish, R Ruoff (Pennington: Electrochemical Society) **1:**1300–1308.
67. Beyers R, Kiang CH, Johnson RD, *et al.* (1994) Preparation and structure of crystals of the metallofullerene Sc_2@C_{84}. *Nature,* **370:** 196–199.
68. Tanaka N, Honda Y, Kawahara M, *et al.* (1996) High-resolution electron microscopy of Gd@C_{82} metalfullerenes grown on MgO(001) surfaces. *Thin Solid Films* **281–282:** 613–617.
69. Wang XD, Hashizume T, Xue Q, *et al.* (1993) Geometry of metallofullerenes adsorbed on the Si(100)2x1 surface studied by scanning tunneling microscopy. *Chem Phys Lett* **216:** 409–412.
70. Shinohara H, Hayashi N, Sato H, *et al.* (1993) Direct STM Imaging of Spherical Endohedral Sc2@C_{84} Bullerenes. *J Phys Chem* **97:** 13438–13440
71. Sakurai T, Wang XD, Xue QK, *et al.* (1996) Scanning Tunneling Microscopy Study Of Fullerenes. *Prog Surf Sci* **51:** 263–408.

72. Gimzewski JK. (1996) *The Chemical Physics of Fullerenes 10 (and 5) Years Later* ed W Andreoni (Dordrecht: Kluwer) **316:** 117–136.
73. Takata M, Umeda B, Nishibori E, *et al.* (1995) Confirmation by X-ray diffraction of the endohedral nature of the metallofullerene Y@C_{82}. *Nature* **377:** 46–49
74. Ros´enA, Waestberg B. (1988) First-Principle Calculations of the Ionization Potentials and Electron Affinities of the Spheroidal Molecules C_{60} and LaC_{60}. *J Am Chem Soc* **110:** 8701–8703.
75. Ros´enA, Waestberg B. (1989) Electronic structure of spheroidal metal containing carbon shells: study of the LaC_{60} and C_{60} clusters and their ions within the local density approximation. *Z Phys* D **12:** 387–390
76. Cioslowski J, Fleishcmann ED. (1991) Endohedral complexes: Atoms and ions inside the C_{60} cage. *J Chem Phys* **94:** 3730–3734.
77. Chang A HH, Ermler WC, Pitzer RM. (1991) The ground and excited states of C_{60}M and C_{60}M+ (M = O, F, K, Ca, Mn, Cs, Sa, La,Eu,U). *J Chem Phys* **94:** 5004–5010.
78. Manolopoulos DE, Fowler PW. (1991) Structural proposals for endohedral metal-fullerene complexes. *Chem Phys Lett.* **187:** 1–7.
79. Manolopoulos D E, Fowler PW. (1992) Hypothetical lsomerisations of LaC_{82}. *J Chem Soc Faraday Trans* **88:** 1225–1226.
80. Laasonen K, AndreoniW, Parrinello M. (1992) Structural and Electronic Properties of La@C_{82}. *Science* **258:** 1916–1918.
81. Saito S, Sawada S. (1992) Growth mechanism and geometry of La@C_{82}. *Chem Phys Lett* **198:** 466–471.
82. Andreoni W, Curioni A. (1996) Freedom and Constraints of a Metal Atom Encapsulated in Fullerene Cages. *Phys Rev Lett* **77:** 834–837.
83. Andreoni W, Curioni A. (1996) *The Chemical Physics of Fullerenes 10 (and 5) Years Later* ed W Andreoni (Dordrecht: Kluwer) 183–196.
84. Nagase S, Kobayashi K. (1993) Metallofullerenes MC82 (M = Sc, Y, and La). A theoretical study of the electronic and structural aspects. *Chem Phys Lett* **314:** 57–63.
85. Takata M, Umeda B, Nishibori E, *et al.* (1995) Confirmation by X-ray diffraction of the endohedral nature of the metallofullerene Y@C_{82}. *Nature* **377:** 46–49
86. Rietveld HM. (1969) A profile refinement method for nuclear and magnetic structures. *J Appl Crystallogr* **2:** 65–71.
87. Collins DM. (1982) Electron density images from imperfect data by iterative entropy maximization. *Nature* **298:** 49–51.
88. Bricogne G. (1988) Bayesian statistical theory of the phase problem. I. A. multichannel maximum-entropy formlism for constructing generalized

joint probability distributions of structure factors. *Acta Crystallogr* A **44:** 517–545.

89. Sakata M, Sato M. (1990) Accurate structure analysis by the maximum-entropy method. *Acta Crystallogr* A **46:** 263–270.
90. Kumazawa S, Kubota Y, Tanaka M, *et al.* (1993) MEED: a program package for electron-density-distribution calculation by the maximum-entropy method. *J Appl Crystallogr* **26:** 453–457.
91. Nagase S, Kobayashi K. (1994) Theoretical study of the lanthanide fullerene CeC_{82}. Comparison with ScC_{82}, YC_{82} and LaC_{82}. *Chem Phys Lett* **228:** 106–110.
92. Schulte J, Boehm MC, Dinse KP. (1996) Electronic structure of endohedral $Y@C_{82}$; an ab initio Hartree-Fock investigation. *Chem Phys Lett.* **259:** 48–54.
93. Rodriguez-Fortea A, Balch AL, Poblet JM. (2011) Endohedral metallofullerenes: a unique host-guest association. *Chem Soc Rev* **40:** 3551–3563.

CHAPTER 2

Synthesis and Isolation of Endohedral Fullerenes – A General Review

*Fupin Liu, Jian Guan, Tao Wei, Song Wang, Shangfeng Yang**

Endohedral fullerenes (ENFs) are a special family of fullerenes with atoms, ions, or clusters entrapped in the interior of the carbon cage. The synthesis of ENFs provides the sample availability as the prerequisite of the follow-up structural and property study, and thus plays a key role in ENF research. Nowadays, the DC arc discharge method is the most widely used approach for the synthesis of ENFs in laboratory, while other methods, including laser ablation, radio frequency (RF) furnace, high-pressure incorporation, ion beam implantation, hot-atom chemistry, glow discharge, "molecular surgery" *et al.*, have been also used for the synthesis of specific ENFs. After the extraction of ENFs from the soot mixture, ENFs generally mix with lots of other fullerene byproducts, thus the isolation of ENFs affording pure samples for structural study is the next yet crucial issue in the ENF research. Up to now, high performance liquid chromatography (HPLC) is the most powerful and frequently used technique for the isolation and purification of ENFs. Besides, great

*Hefei National Laboratory for Physical Science at Microscale, CAS Key Laboratory of Materials for Energy Conversion & Department of Materials Science and Engineering, University of Science and Technology of China (USTC), Hefei 230026, China. Fax: +86(551)63601750; E-mail: sfyang@ustc.edu.cn

efforts have been also devoted to the developments of the alternative non-HPLC solutions for the selective isolation of ENFs, such as chemical or electrochemical methods. This chapter presents a comprehensive and exhaustive review on the synthesis, extraction and isolation of ENFs.

1. INTRODUCTION

Because of the large inner cavity of the fullerene cage, an intuitive inspiration is to stuff atom(s) into the inner cavity. Soon after the discovery of C_{60}, Smalley *et al.* detected the signal of LaC_{60} in the mass spectrum of the mixture synthesized by laser ablation of the graphite disk impregnated with $LaCl_3$.[1] Several years later, Smalley's group succeeded in enriching $La@C_{82}$ by the sublimation method in 1991, and they advocated to use symbol "@" for identifying such specific fullerenes.[2] This study triggered the new research field of endohedral fullerenes (ENFs), which represents a special family of fullerenes with atoms, ions, or clusters entrapped in the interior of the carbon cage.[3]

A breakthrough for the synthesis of fullerenes was accomplished by Krätschmer and Huffman *et al.* in 1990, when they invented the direct current (DC) arc discharge (hereafter named as DC arc discharge for clarity) method for the macroscopic synthesis of fullerenes.[4] Nowadays, the DC arc discharge method is still the most widely used method for the synthesis of fullerenes in laboratory, which is sometimes modified for the synthesis of ENFs. Besides, with the discoveries of some new and specific types of ENFs, several new methods have also been developed for the synthesis of some specific ENFs, including those encapsulating noble gas atoms, group-V atoms, hydrogen molecules, etc. Among all ENFs reported so far,[3] ENFs based on group-IIA, IIIB and IVB metals, including both the conventional endohedral metallofullerenes (EMFs, in which the entrapped species is only metal atom(s))[3a,3e,3n] or endohedral clusterfullerenes (in which the entrapped species is metal cluster composed of metal and nonmetal atoms)[3a-c,3e,f,3i-k], are studied most extensively mainly because of their high yield and stability, fascinating physical and chemical properties, and potential application in many fields such as biomedicine, photonics, *et al.*[3] Thus, we focus on such metal-based ENFs in this review.

Extraction of ENFs from the soot mixture is the next critical issue to be addressed. While the sublimation method was used to enrich ENFs in the early stage, the solvent extraction method is nowadays the most commonly used extraction method due to the considerable solubility of fullerenes (ENFs) in organic solvents. Furthermore, after the extraction of ENFs from the soot mixture, ENFs generally mix with lots of other fullerene byproducts, thus the isolation of ENFs is necessary to obtain the pure samples for structural study. Up to now, high performance liquid chromatography (HPLC) is the most powerful and frequently used technique for the isolation and purification of ENFs. Nevertheless, since the HPLC apparatus is expensive and the isolation process is extremely time-consuming, which are not favorable for large-scale isolation of ENFs, in recent years some alternative non-HPLC methods for the selective isolation of ENFs have been also developed, such as chemical or electrochemical methods.

Several reviews on the general aspects of ENFs have been reported in literatures, including the seminal one by Shinohara in 2000, the monograph edited by Akasaka and Nagase in 2002, the more recent ones by Dunsch and Yang in 2007, Echegoyen *et al.* in 2009, and Akasaka *et al.* in 2012. Given that the synthesis and isolation of ENFs provides the pure sample availability as the prerequisite of the follow-up structural and property study, in this chapter we present a comprehensive and exhaustive review on the synthesis, extraction and isolation of ENFs.

2. SYNTHESIS OF ENFS

Several methods for the synthesis of ENFs were developed directly from the synthesis of empty fullerenes (e.g. laser ablation, DC arc discharge method). However, it should be noted that, so far the commercial synthesis method of empty fullerenes is combustion of hydrocarbons, which holds the highest yield for empty fullerenes such as C_{60}.[5] However, the combustion method has not been used for the synthesis of ENFs up to now probably because the temperature of combustion method is too low (about 1800 K[5]) for the formation of ENFs. Thus, the DC arc discharge method is the most common method for the synthesis of ENFs in laboratory because of the low cost and relatively high yield. Besides, several new methods have also been developed for the synthesis of some specific ENFs, including those encapsulating noble gas atoms, group-V atoms, hydrogen molecules, etc. Table 1 shows an exhaustive list of the synthesis methods of ENFs reported up to now.

2.1. Laser Ablation Method

The first synthesis of EMFs was achieved by Smalley and coworkers using a laser ablation method as shown in Fig. 1. It involves a 2.5-cm diameter and 50-cm long quartz tube mounted in a temperature-controlled tube furnace. This quartz tube is O-ring sealed on the front end to an aluminum block to which is attached a fused silica window and some plumbing for gas addition and pressure monitoring. Another aluminum block attached at the rear of the quartz tube connects via an adjustable valve to a mechanical vacuum pump. The buffer gas is noble gas (e.g. Ar, He), the optimum condition for the buffer gas is at a pressure of several hundred Torr and at a slow rate (0.1-0.2 $cm \cdot s^{-1}$) of flowing over the target rod. The furnace supplies a temperature of 1200 °C surrounding the reaction zone for higher yield of fullerenes.[2,6]

Although the laser ablation method is suitable for the study of the growth mechanism of fullerenes and ENFs, this method was applied only in the early stage of ENF research, and now it has been largely abandoned by researchers because of the high cost and low output efficiency.

2.2 DC Arc Discharge Method

2.2.1 *Traditional DC Arc Discharge Method*

The DC arc discharge method, which was invented by Krätschmer and Huffman in 1990, made a breakthrough in the high-yield synthesis of empty fullerenes such as C_{60}.[4] Soon after such a discovery, this method was successfully applied for the synthesis of EMFs with a considerably high yield, enabling macroscopic amounts of EMFs available.[7] Up to now, this method is still the most widely used method for the synthesis of EMFs in laboratory worldwide. This is mainly because of the high yield and relative low cost. Specifically, for the synthesis of EMFs, the DC arc supplies temperature high enough for the ionization of the reactant metal or metal oxide.

Table 1 A Detailed Classification and Synthesis Methods of ENFs

Categories	Example	Synthesis Method
Conventional EMFs ($M_x@C_{2n}$)	$La@C_{82}$ $La_2@C_{80}$	Laser Ablation DC Arc Discharge[a)]
Metal Nitride Clusterfullerenes (NCFs) ($M_3N@C_{2n}$)	$Sc_3N@C_{80}$	Modified DC Arc Discharge With Addition of N_2, NH_3, Inorganic Nitrogen Containing Compounds, Selective Organic Solid
Metal Carbide Clusterfullerenes (CCFs) ($M_xC_2@C_{2n}$)	$Sc_2C_2@C_{84}$ $Sc_3C_2@C_{80}$ $Sc_4C_2@C_{80}$	DC Arc Discharge, Radio Frequency Furnace
Metal Oxide Clusterfullerenes (OCFs) ($M_4O_{2,3}@C_{2n}$, $M_2O@C_{2n}$)	$Sc_4O_2@C_{80}$ $Sc_4O_3@C_{80}$ $Sc_2O@C_{82}$	Modified DC Arc Discharge With Addition of O_2
Metal Sulfide Clusterfullerenes (SCFs) ($M_2S@C_{2n}$)	$Sc_2S@C_{82}$	Modified DC Arc Discharge With Addition of SO_2, Guanidinium thiocyanate
Metal Hydrocarbide Clusterfullerenes (HCCFs) ($M_3CH@C_{2n}$)	$Sc_3CH@C_{80}$	Modified DC Arc Discharge With Addition of CH_4
Metal Carbonitride Clusterfullerenes (CNCFs) ($M_3CN@C_{2n}$)	$Sc_3CN@C_{80}$	Modified DC Arc Discharge With Addition of N_2
Metal Cyanide Clusterfullerenes (CYCFs) ($MCN@C_{2n}$)	$YCN@C_{82}$	Modified DC Arc Discharge With Addition of N_2
Endohedal Heterofullerenes ($M_x@C_{2n-1}N$, $M_3N@C_{2n-1}N$)	$Y_2@C_{79}N$ $La_3N@C_{79}N$	Modified DC Arc Discharge With Addition of N_2, $Cu(NO_3)_2{\cdot}2.5H_2O$
Endohedral Non-Metallic Fullerenes (ENMFs)	$He@C_{60}$	High-pressure Incorporation Ion Beam Implantation Hot-Atom Chemistry
	$H_2@C_{60}$ $N@C_{60}$	Molecular Surgery High-pressure Incorporation Ion Beam Implantation Glow Discharge Reactor Radio Frequency Furnace

Note: a) DC arc discharge method represents that not only the conventional DC arc discharge method is applicable, the modified DC arc discharge is available even through some times the yield is a little low.

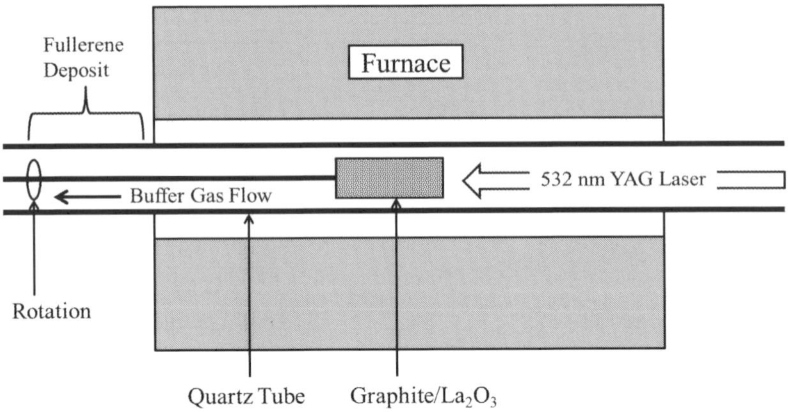

Fig. 1 A cross-sectional view of the apparatus to produce La@C_{82} by laser ablation of a rotating graphite mixed with La_2O_3 compound target in a tube furnace with a flowing argon carrier gas.[6] Reproduced with permission from Cambridge University Press.

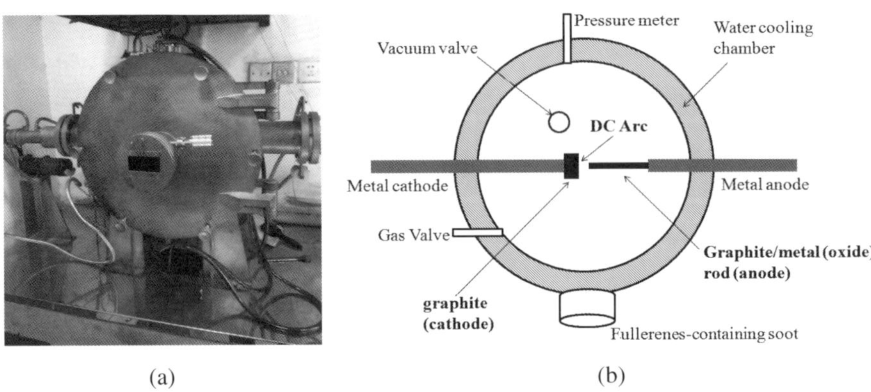

Fig. 2 The photo of the DC arc discharge furnace equipment in our laboratory and (b) the cross-sectional view of the equipment.

Figure 2 (a) illustrates the photo of the DC arc discharge furnace in our laboratory and the cross-sectional view of the equipment is shown in Fig. 2 (b). The reactants, graphite and metal oxide (or metal, alloy, metal chloride *et al.*) are mixed and filled into the core-drilled graphite rod (Φ 8 mm × 150 mm in our laboratory) as an anode, and an undrilled graphite rod (Φ 40 mm × 33 mm in our laboratory) is applied as a cathode. Prior to DC arc discharge, noble gas (e.g. He) should be introduced into the chamber as the cooling gas with an optimized pressure (200–400 mbar in our laboratory). During the DC arc discharge process, the distance between two electrodes is kept at about 5–10 mm to ensure continuous arc generation, and the graphite/metal (oxide) anode is consumed, leading to formations of fullerenes including ENFs.

It should be noted that a high-temperature heating pretreatment (above ca 1000 °C) of graphite/metal (oxide) composite rods are normally required to "activate" the metal (oxide) to form metal carbides in the phase of MC_2, which is crucial to the formation of EMFs.[8] Alternatively, the "*in-situ* activation" technique can be also used.[9] Besides, the "back-burning" technique was also found to be helpful for an efficient synthesis of EMFs.[10] The addition of some transition metals or metal oxides such as Cu,[11] CoO,[12] FeN_x[13] *et al.* as catalysts has been also applied for improving the yield of EMFs.

Using the DC arc discharge method, empty fullerenes,[4] conventional EMFs[7] and metal carbide clusterfullerenes (CCFs)[14] have been synthesized.

2.2.2 *Modified DC Arc Discharge Method with the Addition of Other Compounds*

1) *Addition of Gaseous N_2: the "trimetallic nitride template" (TNT) process*

For the synthesis of ENFs by the DC arc discharge method, it was commonly accepted in the fullerene community that it was necessary to avoid nitrogen as a cooling gas. Unexpectedly, in 1999, Dorn and co-workers introduced trace of N_2 gas into the DC arc discharge furnace by accident, leading to the discovery of $Sc_3N@C_{80}$ as a brand new type of ENF — metal nitride clusterfullerene (NCF). The authors named this method as the "trimetallic nitride template" (TNT) process. By using this method, milligram quantities of product containing 3–5% $Sc_3N@C_{80}$ were synthesized, for which the yield is significantly higher than that of all other ENFs and empty fullerenes except C_{60} and C_{70}.[15] Using this TNT process, a large variety of NCFs have been isolated, including $Er_xSc_{3-x}N@C_{80}$ ($x = 0–3$),[12,15-16] and $A_xSc_{3-x}N@C_{68}$ ($x = 0–2$; A = Tm, Er, Gd, Ho, La).[17] $Sc_3N@C_{78}$,[18] $Lu_3N@C_{80}$,[19] $Lu_{3-x}A_xN@C_{80}$ ($x = 0–2$; A = Gd, Ho),[19] $Y_3N@C_{2n}$ ($2n = 80–88$),[20] $Tb_3N@C_{2n}$ ($2n = 80, 84, 86, 88$),[13,21] $CeSc_2N@C_{80}$,[22] $Gd_3N@C_{2n}$ ($2n = 78–88$).[23] $TiM_2N@C_{80}$ (M = Sc,[24] Y[25]), etc. Besides, empty fullerenes, conventional EMFs, CCFs and endohedral heterofullerenes ($M_2@C_{79}N$, (M = Y,[26] Tb,[26] Gd[27])) have been also synthesized along with NCFs. Moreover, CNCFs ($Sc_3CN@C_{2n}$ (2n = 78,[28] 80[29])) and CYCFs ($YCN@C_{82}$[30]) have been also synthesized along with NCFs.

2) *Addition of Gaseous NH_3: the "reactive gas atmosphere" route*

In 2003, Dunsch and co-workers developed the "reactive gas atmosphere" route by introducing NH_3 as a reactive gas into the DC arc discharge furnace, and successfully produced NCFs as the predominant products in the soot mixture with a relative yield of empty fullerenes and conventional EMFs being less than 5%, thus realizing the high-selectivity synthesis of NCFs for the first time.[31] Later on, this method has been widely used for the synthesis of NCFs since it promises the high selectivity synthesis, which makes the isolation of NCFs more facile. What should be emphasized is that the high selectivity is primarily caused by the dramatic suppression effect for the formation of empty fullerenes and conventional EMFs by the addition of ammonia.[32] Up to now, many NCFs were synthesized by this route, such as large families of clusterfullerenes $M_3N@C_{2n}$ (M = Ho,[33] Tb,[33] Gd,[34] Dy,[35] Tm;[36] $38 \leq n \leq 44$; M = La, Ce, Pr, Nd; $40 \leq n \leq 48$[37]), $Gd_xSc_{3-x}N@C_{80}$ (I, II, x = 0–2),[38] $Sc_3N@C_{70}$,[39] $DySc_2N@C_{76}$,[40] $MSc_2N@C_{68}$ (M = Dy, Lu),[41] $Lu_2ScN@C_{68}$,[41] $Lu_xY_{3-x}N@C_{80}$ (I, x = 0–2),[42] $LuCe_2N@C_{80}$,[43] $Gd_xSc_{3-x}N@C_{2n}$ ($38 \leq n \leq 44$).[44]

3) *Addition of Inorganic Nitrogen-containing Compounds (CaNCN, $Cu(NO_3)_2\cdot 2.5H_2O$, $((NH_4)_xH_{3-x}PO_4$ (x = 0–2), $(NH_4)_2SO_4$, $(NH_4)_2CO_3$, NH_4X (X = F, Cl), NH_4SCN, KSCN, $NaNO_3$, $NaNO_2$)*

For promoting the yield of NCFs and/or the selective synthesis of NCFs, a series of inorganic nitrogen-containing compounds have been added as the nitrogen sources for the synthesis of NCFs. As the first step of modifying the synthesis of NCFs without using gaseous N_2, calciumcyanamide (CaNCN) as the inorganic nitrogen-containing compound that did not contain oxygen was used by Dunsch and co-workers in 2004, revealing that this compound acted as an efficient nitrogen source for the synthesis of $Sc_3N@C_{80}$ with a selectivity ranging from 3 to 42%.[32] Nevertheless, this method has the disadvantage of the low reproducibility of the fullerene yield presumably caused by traces of water and/or hydrocarbons in the raw material.[32] Therefore this inorganic nitrogen source was not widely applied in the synthesis of NCFs afterwards.

In 2007, Stevenson and co-workers applied another inorganic nitrogen-containing compound, copper nitrate ($Cu(NO_3)_2{\cdot}2.5H_2O$) as an additive for the synthesis of $Sc_3N@C_{80}$ with gaseous N_2 used as the nitrogen source, realizing the selective synthesis of $Sc_3N@C_{80}$. This method was named by the authors as "Chemically Adjusting Plasma Temperature, Energy and Reactivity" (CAPTEAR).[45] The CAPTEAR method with $Cu(NO_3)_2{\cdot}2.5H_2O$ applies an exothermic nitrate moiety to suppress the formations of empty fullerenes and conventional EMFs, while the reduced product Cu acts as a catalyst additive to offset the reactive plasma environment and boost the yield of $Sc_3N@C_{80}$. With this method, the high-selectivity (up to 96%) synthesis of $Sc_3N@C_{80}$ was achieved.

Using this CAPTEAR method, the same group also synthesized endohedral heterofullerene, $La_3N@C_{79}N$ with a relative higher yield than that of the dominant product of La-based NCFs, $La_3N@C_{96}$.[46]

More recently, our group successfully applied 12 other nitrogen-containing inorganic solid compounds with variable oxidation states of nitrogen and counter ions, including ammonium salts ($(NH_4)_xH_{3-x}PO_4$ (x = 0–2), $(NH_4)_2SO_4$, $(NH_4)_2CO_3$, NH_4X (X = F, Cl), NH_4SCN), thiocyanate (KSCN), nitrates ($Cu(NO_3)_2$, $NaNO_3$) and nitrite ($NaNO_2$), as new inorganic solid nitrogen sources toward the synthesis of Sc-NCFs. Among them, triammonium phosphate hydrate ($(NH_4)_3PO_4{\cdot}3H_2O$) and ammonium thiocyanate (NH_4SCN) are revealed to behave as better nitrogen sources than others, and the highest yield of Sc-NCFs was achieved when NH_4SCN was used as a nitrogen source.[47] Furthermore, we quantitatively compared the yield of $Sc_3N@C_{80}$ ($I_h + D_{5h}$) per gram Sc_2O_3 by using different nitrogen sources, and investigated the dependence of the production yield of Sc-NCFs on the oxidation state of nitrogen and counter ion (Fig. 3). It was found that the yield of $Sc_3N@C_{80}$ ($I_h + D_{5h}$) per gram Sc_2O_3 by using the N_2-based group of nitrogen sources (thiocyanate, nitrates, and nitrite) was in overall much lower than those by using gaseous N_2 and NH_4SCN, indicating the strong dependence of the yield of Sc-NCFs on the oxidation state of nitrogen, whereas for the NH_3-based group of nitrogen sources (ammonium salts) which exhibits a (-3) oxidation states of nitrogen, their performance as nitrogen sources was sensitively dependent on

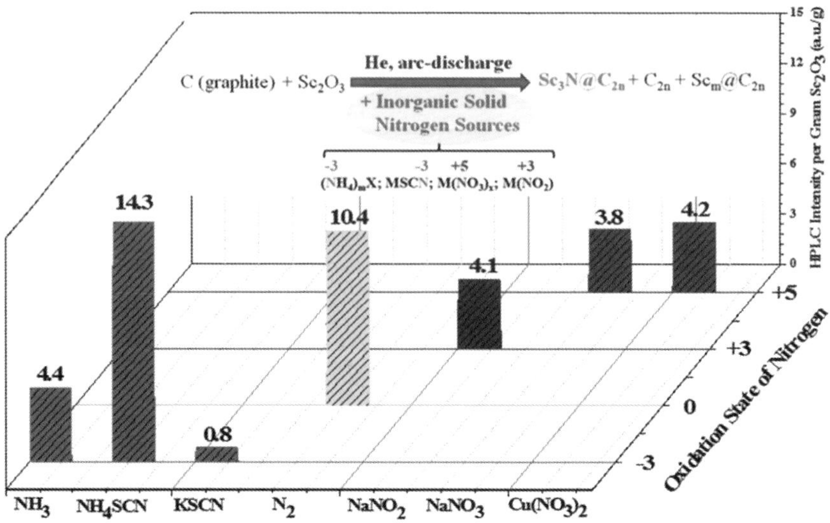

Fig. 3 Comparison of the HPLC intensity of fraction A per gram Sc_2O_3 obtained from different nitrogen sources.[47] Reproduced with permission from American Chemical Society.

the anion.[47] Interestingly, in this work it was proved that copper nitrate ($Cu(NO_3)_2$) can work as an independent nitrogen source.[47]

4) *Addition of Small molecular Organic Solids (Guanidinium salts and urea)*

In 2010, Yang and co-workers developed a new "selective organic solid" (SOS) route for the facile synthesis of NCFs by using two different guanidinium salts (guanidinium thiocyanate and guanidinium hydrochloride) as the new nitrogen sources.[48] Using the SOS routes, a series of NCFs based on Dy, Sc, Y, Gd, Lu and mixed metal Sc/Dy, Sc/Gd, Sc/Lu and Lu/Ce were synthesized. The yield of NCFs synthesized by the SOS route is comparable with that by using the "reactive gas atmosphere" route. Compared with the reported TNT process and "reactive gas atmosphere" route discussed above, the advantage of this SOS route is that no additional preheating treatment is needed, thus simplifying the procedure and being much more facile. What's noteworthy is that the SOS route seems to be a general route for the synthesis of NCFs as demonstrated for different metal systems.[48]

Besides the high-selectivity synthesis of NCFs, an unexpected finding of this work was that a new family of clusterfullerene, metal sulfide clusterfullerenes (SCFs), $M_2S@C_{82}$ (M = Sc, Y, Dy, and Lu) were discovered along with NCFs.[49]

Recently, another small molecular organic solid, urea ($CO(NH_2)_2$), was used successfully as the new nitrogen source by Yang and co-workers for the synthesis of Sc-based NCFs.[50] The yield of $Sc_3N@C_{80}$ (I_h+D_{5h}) per gram Sc_2O_3 by using $CO(NH_2)_2$ as nitrogen source was quantitatively compared to those by using the previously reported nitrogen sources such as N_2, NH_3 and guanidinium thiocyanate, indicating that the yield of $Sc_3N@C_{80}$ (I_h, D_{5h}) per gram Sc_2O_3 by using urea is almost identical to that by using N_2, but much higher than that by NH_3. Compared to the reported organic solid guanidinium thiocyanate, the yield of $Sc_3N@C_{80}$ (I_h, D_{5h}) per gram Sc_2O_3 synthesized by using $CO(NH_2)_2$ is around 1.6 times higher, while $CO(NH_2)_2$ is about 8.5 times cheaper than guanidinium thiocyanate.[50] Interestingly, they also found that melamine ($C_3H_6N_6$) as decomposition product of $CO(NH_2)_2$ could act as an efficient organic solid nitrogen source for the synthesis of Sc-NCFs.[50]

5) *Addition of SO_2*

In 2010, nearly at the same time as the discovery of SCFs $M_2S@C_{82}$ by Dunsch and co-workers, Echegoyen and co-workers introduced a small amount of SO_2 (10% of He atmosphere) as a new sulfide source into the DC arc discharge furnace, and detected a large family of SCFs — $Sc_2S@C_{2n}$ (2n = 80–100).[51] This is obviously different from that synthesized by solid sulfur source, guanidinium thiocyanate, in which only one $M_2S@C_{82}$ (C_{3v}(8)) isomer was observed.[49] The results show the significant influence of the sulfur source on the formation of SCFs and open a new route for the exploration of this novel cluster fullerene family.

In addition to the isolation of two isomers of $Sc_2S@C_{82}$ (C_{3v}:8; *Cs*:6)), more recently the same group successfully isolated two new SCFs including $Sc_2S@C_{72}$[52] and $Sc_2S@C_{70}$.[53]

6) *Addition of CH_4*

In 2007, Dunsch and co-workers introduced a reactive gas, CH_4, to the DC arc discharge furnace, leading to the discovery of a new type of ENF — metal hydrocarbon clusterfullerene (HCCF), $Sc_3CH@C_{80}$.[54] The successful synthesis of $Sc_3CH@C_{80}$ provides a further example that the concept of a "reactive gas atmosphere" route discussed above proves to be a general method for the synthesis of new clusterfullerenes with different cluster structures as well as the number of atoms comprising the cluster. It should be noted that $Sc_3CH@C_{80}$ is the only reported HCCF so far.

7) *Addition of* O_2

In 2008, Stevenson and co-workers introduced a small amount of flowing air as additive during the synthesis of Sc-based endohedral metallofullerenes, and discovered metal oxide cluster fullerene (OCF), $Sc_4O_2@C_{80}$ (I_h), as a new family of clusterfullerenes. Obviously gaseous O_2 as the minor component of air plays the key role as the oxygen source for the formation of OCF.[55] Soon after that, the same group further synthesized and isolated $Sc_4O_3@C_{80}$ (I_h) for which the largest number of atoms (7 atoms) up to now is encaged in the fullerene cage.[56] Interestingly, in the course of optimizing the yield of $Sc_4O_2@C_{80}$ they found that a flow rate of 12 $Torr\cdot min^{-1}$ was optimal. In addition, using the CAPTEAR method discussed above, it was shown that increasing the content of copper nitrate ($Cu(NO_3)_2\cdot 2.5H_2O$) dramatically improved the yield of $Sc_4O_2@C_{80}$ and the optimal conditions for formation of $Sc_4O_2@C_{80}$ were an air flow rate of 0.2 $Torr\cdot min^{-1}$ with graphite rods packed with 80% $Cu(NO_3)_2$ and 20% Sc_2O_3, leading to a 12% yield from the soluble extract.[55]

2.3 Radio Frequency Furnace Method

In search for a substitute for the laser ablation method, Jansen and co-workers developed a new method named the radio frequency (RF) furnace method in 1992 and achieved a high-yield synthesis of empty fullerenes.[57] The model of the production chamber in the radio frequency furnace is shown in Fig. 4. Generally speaking, the carbon and metal source are evaporated by the radio frequency (RF) generator, the evaporated carbon and metal ion are mixed and cooled in the fullerene formation area, and thus fullerene is generated.

Actually, different furnaces based on radio frequency have been designed for the synthesis of fullerenes up to now. In 1992, Yoshie and co-workers developed a hybrid plasma method by combining radio frequency furnace with DC arc discharge, and they got a 7% fullerene yield

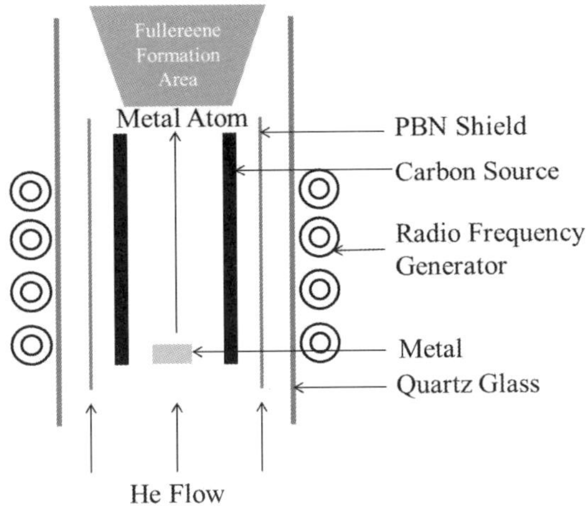

Fig. 4 Typical model of the fullerene production chamber in the radio frequency furnace.[58] Reproduced with permission from the American Chemical Society.

at atmospheric pressure.[59] A radio frequency (RF) inductively coupled thermal plasma (ICTP) furnace was designed in 2001 by Wang and co-workers.[60] After that, Marković Z. and co-workers used this apparatus to produce empty fullerenes and a maximum yield of 4.1% corresponding to a fullerene production rate of 6.4 $g \cdot h^{-1}$ was obtained under optimum conditions.[61]

As a variant of the traditional radio frequency furnace, Kaneko and coworkers designed an apparatus which generated electron cyclotron resonance (ECR) plasma for the high yield synthesis of $N@C_{60}$.[62] Their apparatus applied a microwave generator (2.45 GHz, 800 W) for the ionization of N_2, the generated plasma was divided into two regions, i.e. the ECR and process regions, since electrons in the ECR region were trapped in the mirror magnetic field and accelerated owing to ECR at the bottom of the mirror configuration (875 G), a number of N^* and N^+ are expected to be produced as a result of the effective dissociation and ionization of N_2 gas. The authors claimed that the concentration of $N@C_{60}$ in C_{60} was increased up to 0.03%, which was one order of magnitude larger than that in the conventional radio frequency furnace.

Some catalysts were proved effective for enhancing the yield of fullerene. The synthesis of fullerenes was clearly enhanced by the 4 mol % of Fe added in the starting materials.[63]

Along with the extensive research of ENFs, the radio frequency furnace method was also applied in the synthesis of all kinds of ENFs. In 1995, Jansen and co-workers realized the synthesis of La@C_{82} EMF by this method.[64] Later on, they further succeeded in synthesizing other M@C_{2n} EMFs with smaller cages (2n < 80), and they found that, using this method, a high amount of EMFs with smaller cages such as M@C_{74} and M@C_{76} were obtained in the generated soot.[65]

In 2010, Krokos developed a novel RF furnace for the synthesis of different, and successfully produced not only the empty fullerenes and conventional EMFs but also cluster fullerenes such as CCF-Sc_4C_2@C_{80} and NCF-Sc_3N@C_{2n}.[58]

Moreover, when an additional oven used for the sublimation of C_{60} was added to the bottom of the RF furnace while nitrogen gas was input from the top of the apparatus, metal and carbon source shown in Fig. 4 was withdrawn, N@C_{60} could be also synthesized by the RF furnace as reported by Huang and coworkers in 2002. It is recognized that N@C_{60} and polymerized N@C_{60} could be formed in the RF furnace without isolation and acceleration of nitrogen ions in advance. Compared to the ion beam implantation method typically used for the synthesis of N@C_{60}, they claimed that N@C_{60} could be formed in their chamber at the same concentration level (about 10^{-5}–10^{-4}) but in a much shorter time (about 1/50).[66] Using this method, Miyanaga and co-workers studied the relationship between the yield of N@C_{60} and the sublimation rate of C_{60}, revealing that the increasing amount of nitrogen molecule ions (N_2^+) impinging on the sublimated fullerenes enhanced the synthesis of N@C_{60}.[67]

Based on the above discussions on the considerable success in the synthesis of empty fullerenes, EMFs and N@C_{60}, the radio frequency (RF) furnace method may become a potential substituent of the DC arc discharge method commonly used for the synthesis of empty fullerenes and ENFs currently in lab.

2.4 High-pressure Incorporation

In 1992, Saunders and co-workers demonstrated that fullerenes synthesized by DC arc discharge method contain endohedral noble gas fullerenes (He@C_{60} and Ne@C_{60}) with concentration of about one per 880,000 molecules.[68] They heated C_{60} in several atmospheres of ^{3}He and Ne at 600 °C (the pressure is about three atmospheres) leading to inclusion of these atoms in C_{60} at similar levels. Soon after these discoveries, the same group developed the high-pressure incorporation method for the synthesis of endohedral noble gas fullerenes (He@$C_{60,70}$, Ne@$C_{60,70}$, Ar@$C_{60,70}$, Kr@$C_{60,70}$ and Xe@$C_{60,70}$).[69] They used a 4-inch tube (1/4-inch o.d.) of oxygen-free copper as the inner container of the fullerene sample, in which noble gas was introduced and sealed after a series of careful treatments. Then, the resulting ampoule was put into the treated high pressure vessel with a pressure of about 2700 atm. When the sealed vessel was heated and the pressure was increased, the copper tube was collapsed and compressed the inner samples. Using this high-pressure incorporation method, the yield of X@C_{2n} was enhanced largely. The yield of X@C_{60} was 0.1%, 0.2%, 0.3%, 0.3% and 0.008% with ^{3}He, Ne, Ar, Kr and Xe, respectively, and the yield of X@C_{70} was comparable with that of X@C_{60}, 0.1%, 0.2%, 0.2%, 0.2% and 0.04% with ^{3}He, Ne, Ar, Kr and Xe, respectively.[69]

Aside from the synthesis of X@C_{2n} (X = He, Ne, Ar, Kr, Xe), high-pressure incorporation method was also applied for the synthesis of other non-metal endohedral fullerenes combined with the "molecular surgery" method discussed below, such as H_2@C_{60}.[70]

Recently, Wang and co-workers developed a modified explosive method for the high-yield synthesis of non-metal endohedral fullerenes. Using this method, the yield of He@C_{60} and He_2@C_{60} reached ~6% and 0.4%, respectively.[71] The authors claimed that, this method can be easily extended to the preparation of other non-metal endohedral fullerenes, including N@C_{60}.

2.5 Ion Implantation

The ion implantation method (sometimes called the collisional production method because the procedure needs the two reactants to collide or ion bombardment) for the synthesis of ENFs could be understood as another modified method of high pressure incorporation; this is because the ion implantation method applied the ion incorporation into the fullerene cage by high-energy ion beam. Up to now, there have been several methods to realize the ion implantation. The classical method is illustrated in Fig. 5. The collisional production method was first reported by

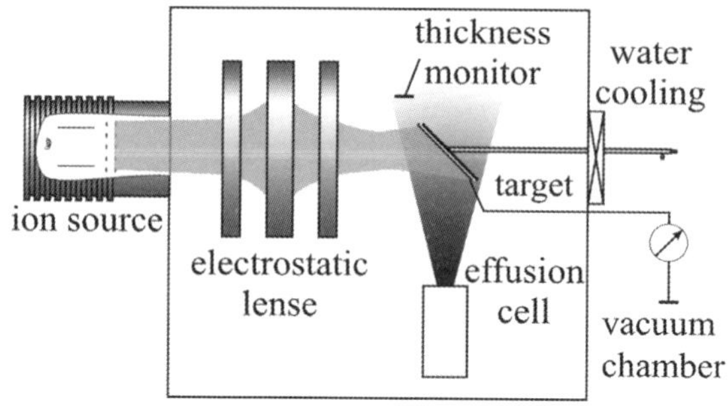

Fig. 5 Schematic view of the experimental set-up for ion beam implantation.[72] Reproduced with kind permission from Springer Science: Applied Physics A: Materials Science & Processing.

Krätschmer and coworkers soon after fullerenes became generally available in 1991; these experiments involved high-energy (5 keV) C_{60}^{+} ions collided with neutral rare gas atoms.[73] This could be understood as the rudiment of the ion implantation method. In 1996, ion implantation was applied by Murphy and coworkers for the synthesis of $N@C_{60}$. In their experiment, the nitrogen ion was generated from a conventional plasma discharge ion source, with the help of an extraction voltage of 800 V, the yielding ions was with several hundreds eV.[74] In the same year, Campbell and coworkers used a similar method for the synthesis of alkali metal endohedral fullerenes, they proved that the values for the energetic thresholds for capture of alkali ions was approximately 6 eV for Li^+ and increased to approximately 40 eV for K^+.[75]

The ion implantation method is a very powerful technique since it has less limitations than other methods. In principle, every element can be implanted into fullerene cages. Currently it has been used specifically for the high-yield synthesis of group-V element-based ENFs (e.g. $N@C_{60}$).[72] Furthermore, by using nitrogen ion beam (about 40 eV), a nitrogen gas molecule could be encapsulated in the C_{60} cage, generating $N_2@C_{60}$.[76]

2.6 Hot-Atom Chemistry

The hot-atom chemistry method applies the nuclear reaction for the synthesis of ENFs, and was mainly used for the synthesis of radioactive ENFs. In 1994, Saunders and coworkers successfully put tritium atoms into C_{60}

using hot-atom chemistry.[77] They irradiated a lithium salt of C_{60} in a low-flux nuclear reactor. ^{6}Li reacts with thermal neutrons to make tritium atoms at high kinetic energy,

$$^{6}Li + {}^{0}n \rightarrow {}^{3}H + {}^{4}He + 4.8\ MeV. \quad (1)$$

The generated ^{3}H atoms move through the sample losing energy by ionizing molecules and stops, then some of them stop inside C_{60} molecule, and the tritium endohedral fullerene is generated. They proved that the tritium was inside by detecting the ^{3}He produced by its decay.

The reaction (1) has two drawbacks: one is that the preparation of lithium salt of C_{60} needs to be done in the absence of air, and the other is that the anion radical makes is susceptible to radiation damage. Later on, the same group developed another modified method for the synthesis of tritium@C_{60}:

$$^{3}He + {}^{1}n \rightarrow {}^{3}H + {}^{1}H + 760\ keV. \quad (2)$$

The nuclear reaction (2) dissipates only 16% of the energy of reaction (1), thus radiation damage should be reduced. The authors proved that reaction (2) was a better reaction for the synthesis of tritium@C_{60}.[78]

Nearly at the same time as Martin's application of hot-atom chemistry, Kikuchi and coworkers successfully used a similar method for the synthesis of radioactive ^{159}Gd and ^{161}Tb ENFs.[79]

After that, this technique was applied for the neutron activation of $^{165}Ho_x$@C_{2n}, generating $^{166}Ho_x$@C_{2n}, which decayed to $^{166}Er_x$@C_{2n}.[80] Ar@C_{60} was also prepared by neutron irradiation of Ar and crystalline C_{60}, and the neutron activated Ar with a high kinetic energy could penetrate the shell of the fullerene molecule in a high-energy bimolecular collision, so the endohedral fullerene Ar@C_{60} was synthesized.[81] ^{7}Be@C_{60},[82] ^{212}Pb@C_{60},[83] *et al.* were also prepared by the hot-atom chemistry method.

Furthermore, the neutron irradiation method was also applied for the synthesis of radioactive fullerenes containing 60–70% of ^{11}C nuclide.[84]

2.7 Glow Discharge Reactor

In 1998, Weidinger and coworkers proposed an alternative set-up using a glow discharge method for the synthesis of group-V ENF N@C_{60}.[72]

They claimed that this was a much simpler and cheaper method and could easily be used in any physical or chemical laboratory. The equipment consists of a quartz tube with two electrodes and a continuous flow of low pressure (ca. 1 mbar) nitrogen gas through the tube. The central part of the tube, in which C_{60} powder is deposited, is surrounded by a furnace. The glow discharge is started after the vapor pressure of C_{60} is high enough to result in efficient sublimation to the water-cooled cathode.

Recently, Ito and coworkers optimized the condition of this method for the synthesis of N@C_{60}. They concluded that the yield was higher at a temperature of about 650 °C and a nitrogen pressure of 1.0 Torr while the discharge voltage is about 850 V. The authors emphasized that the last two parameters correspond to the lowest values in which the electrode can discharge. So, it is probable that the yield of N@C_{60} can be improved further by designing a system having different electrical characteristics, such as smaller distances between the electrodes. Using this method, the yield of N@C_{60} was about 0.0025%. Besides N@C_{60}, N@C_{70} and N_2@C_{70} were also prepared.[85]

Actually, the glow discharge method could be understood as a combination of the ion implantation and DC arc discharge methods. Although this is a simple and cheaper set-up, the application of this method has been scarce up to now.

2.8 Molecular Surgery

Synthesis of ENFs by organic reactions, the so-called "molecular surgery" method, is a method with the rational design for the synthesis of ENFs. This is a totally different pathway to other methods described above because it uses organic reactions that initially provoke bond scission on the fullerene framework, and finally reform the cleaved bonds to afford the intact fullerenes.[86] For to this method, a temporary hole of a fullerene cage is first opened, followed by the introduction of a small atom, molecule or ion in the fullerene cage, and then the holed cage is closed to keep the original fullerene framework. Scheme 1 presents the general procedure of "molecular surgery" method.

The task of cracking open a fullerene cage in a controlled fashion was hard in the early fullerene chemistry research, and a breakthrough was achieved when Wudl and coworkers observed that the enamine-like

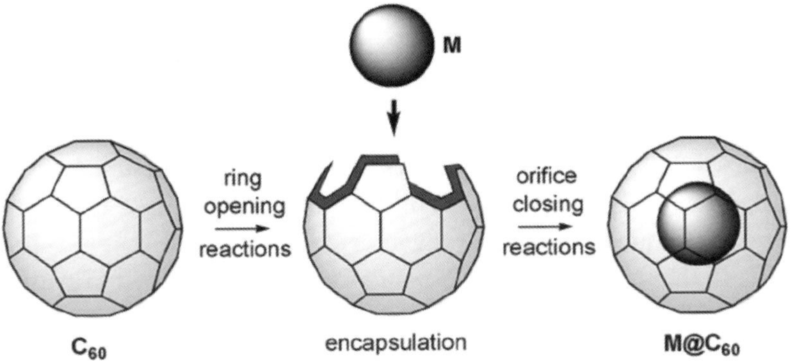

Scheme 1 A General Procedure of the "Molecular Surgery" Method.[86] Reproduced with permission from The Royal Society of Chemistry.

methoxyethoxymethly-azafulleroid reacts in a [2+2] fashion with single oxygen to form, via a dioxetane intermediate, the keto-lactam having an 11-membered ring opening.[87] Up to now, many kinds of methods have been applied for the cleavage of the fullerene cage, and several detailed reviews about the molecular surgery method have been published recently.[86,88]

In 1997, Rubin *et al.* proposed the organic approaches to synthesize EMFs,[89] and realized the encapsulation of He and H_2 into the open-caged C_{60} in 2001. They used high pressure (ca. 475 atm) for the encapsulation of He/H_2 into the open-caged C_{60} with a concentration (1.5% for He, 5% for H_2) higher than that obtained by traditional high-pressure method (0.1% for He@C_{60}), while the endohedral fullerene cage was still open.[70b]

Two years later, Komatsu and coworkers successfully accomplished the 100% encapsulation of H_2 into C_{60} with a larger orifice (13-membered-ring-orifice). The encapsulation was achieved at a high pressure (800 atm) of H_2 at 200 °C, and the ratio of H_2 encapsulation would decrease (90% under 560 atm H_2 and 51% under 180 atm H_2) if the pressure of H_2 was lower. Another breakthrough of this work is that H_2@C_{60} was present when the endohedral open-caged fullerene survived the laser irradiation during the MALDI-TOF MS detection. Thus, the closed cage endohedral fullerene was achieved in the gas phase by self-restoration of open cage endohedral fullerene.[70a] Later on, the same group used a four-step organic reaction that completely closed a 13-membered ring orifice of an open-cage fullerene. This process as the third step of molecular surgery method fulfilled the successful synthesis of $H_2@C_{60}$ with a perfect C_{60} cage. Furthermore, they got the dimer of $H_2@C_{60}$ through mechanochemical solid-state dimerization of $H_2@C_{60}$ by the use of a high-speed vibration milling technique.[90]

Using the same method, recently Murata and coworkers successfully synthesized $H_2O@C_{60}$ ENF and characterized the structure by single-crystal X-ray analysis.[91] More recently, the same group used two stepwise insertion techiques and successfully inserted both He and N atoms into C_{60} and C_{70}. In the first step, molecular surgery method was used to synthesize $He@C_{60}$ or $He@C_{70}$. Then, N atom was inserted by radio-frequency plasma method, affording the first new ENFs encapsulating two non-metal atoms, namely $HeN@C_{60}$ and $HeN@C_{70}$. This work represents a successful combination of the "molecular surgery" and "nitrogen RF plasma" methods.[92]

Except for He, H_2O and H_2, up to now, Ne,[93] Ar,[93] Kr,[93] N_2,[93] CO,[94] NH_3,[95] CH_4[96] *et al.* have been also encapsulated into the fullerene cage via the molecular surgery method.

3. EXTRACTION AND ISOLATION OF ENFs

3.1. Extraction of ENFs

For ENFs synthesized by laser ablation, DC arc discharge or RF furnace methods, the generated carbon soot is typically the mixture of different fullerene structures (mainly empty fullerenes, ENFs with different cage sizes and isomer structures) and amorphous carbon. Thus, the first step of the isolation is the extraction of fullerene-only mixture from the carbon soot. Up to now, several methods have been developed for the efficient extraction of fullerenes, including Soxhlet extraction method,[2] two-step Soxhlet extraction,[97] high temperature/high pressure solvent extraction,[98] ultrasonication extraction,[99] mixed solvent extraction[97] and solid-phase extraction, etc.[100] The extraction of fullerenes generally uses the solvent which dissolves fullerene, and in most cases, the solvent could not dissolve the amorphous carbon, thus the fullerene was extracted from the byproducts in the carbon soot. Solvents play a key role in the extraction of fullerenes, up to now, and in the following we focus on the effect of the solvent on the performance of extraction of ENFs.

At the early stage of fullerene research, the carbon soot generated by the DC arc discharge method was dissolved simply in benzene by Krätschmer and coworkers, and the liquid-dissolved fullerenes were isolated from the soot.[4] Since the first application of the Soxhlet extraction method in 1991 by using boiling toluene as solvent resulting in the isolation of the first EMF La@C_{82},[2] the Soxhlet extraction method has been commonly accepted as a general route for the extraction of fullerenes from the raw carbon soot. Several solvents were used for the Soxhlet extraction of fullerenes, such as xylene, 1,3,5-trimethylbenzene, 1,2,4-trichlorobenzene, α,α,α-trichloro-toluene, 1-methylnaphthalene, o-xylene, supercritical carbon tetrachloride,[98b] pyridine,[101] *N,N*-Dimethylformamide (DMF),[102] a polar solvent mixture that consists of CS_2 and CH_3OH,[97] aniline,[99] and o-dichlorobenzene[10] *et al.* Among them, 1,2,4-trichlorobenzene is particularly promising for extraction of giant fullerenes[98b] and some special "missing" EMFs (e.g. La@$C_{74}(D_{3h})$,[103] La@$C_{82}(C_{3v}(7))$[104]). N-containing solvents such as DMF and pyridine were found to be particularly affinitive to monometallfullerenes, most probably by forming some unclear complexes.[99, 102, 105] However, after evaporation of these solvents, the complexes decomposed and some of the monometallo fullerenes, such as M@C_{60}, remained insoluble in other non-polar solvents, which hinders their further purification and characterization.

The electrochemical reduction method was applied for the extraction of insoluble endohedral fullerenes. Generally speaking, the insoluble endohedral fullerenes are either free radicals or have small HOMO-LUMO gaps; they are kinetically unstable and react readily to form insoluble, polymerized solids. These intermolecular bonds can be broken by electrochemical reduction. After they are reduced to soluble anions, they could be isolated and characterized. As the first success, Diener and coworkers realized the electrochemical reduction extraction of C_{74}, Gd@C_{60} and Gd@C_{74}.[106] On the other hand, $AgPF_6$, $AgSbCl_6$, tris(4-bromo-phenyl)aminium hexachoroantimonate, etc. as mild oxidants and $AlCl_3$ as a stronger oxidant were also applied for the extraction of EMFs insoluble in general organic solvents such as toluene, xylene and CS_2 etc. was also realized.[10]

For the specific ENFs such as non-metal-based ENFs synthesized by other methods discussed above, the as-synthesized ENFs contained less byproducts, making the extraction of ENFs even easier, and the generated product can be dissolved simply in solvent such as carbon disulfide, aniline, pyridine, toluene, o-dichlorobenzene *et al.*, and sometimes, ultrasonic irradiation technology is needed as an assistant for dissolving such ENFs.[107]

3.2 Isolation of ENFs by High Performance Liquid Chromatography (HPLC)

The extracted fullerene product is still a mixture containing mainly empty fullerenes (C_{60}, C_{70}, *et al.*) and different kinds of ENFs, thus the further isolation and purification of ENFs are needed. Up to now, the widely used method for the isolation/purification of ENFs in laboratory is high performance liquid chromatography (HPLC) isolation. This method was firstly used for the isolation of Sc-based EMFs by Shinohara and coworkers.[108]

For HPLC isolation, the column plays a determinative role in the performance of isolation. Several types of commercial HPLC columns, namely Buckyprep, 5PYE, Buckyprep-M, 5-PBB and Buckyclutcher, have been designed specially for the isolation of fullerenes and ENFs, and the differences between these columns are illustrated in Fig. 6; their different isolation performance is based on different selectivity on cage sizes or the molecular polarities.[109]

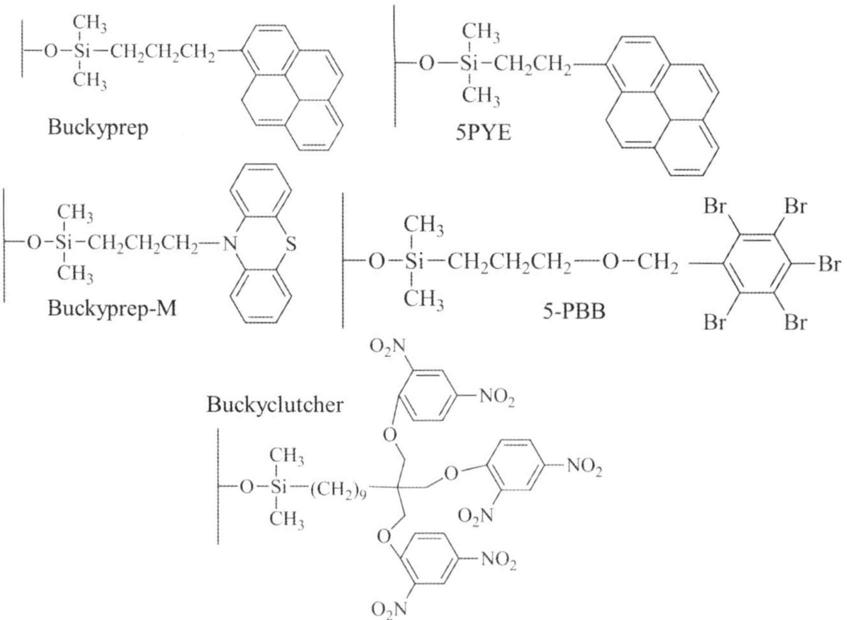

Fig. 6 Some chromatographic stationary phases used in HPLC columns for the isolation of ENFs.

In addition, during the past two decades, linear combinations of these columns, multi-step isolations, or recycling HPLC allow the isolation of different isomers of ENFs.[3a]

3.3. Non-HPLC Isolation of ENFs

Although over the years HPLC has been the most powerful technique to isolate ENFs, this isolation process is usually time-consuming and expensive. The main reason for this is that empty fullerenes (C_{60} and C_{70} etc.) are usually two orders of magnitude more abundant than endohedral fullerenes in the extract of the raw soot synthesized by DC arc discharge, and the ENF abundance in the mixture synthesized by other methods is even lower. On the other hand, different isomers usually have very similar retention times, thus requiring longer isolation and consequently it is necessary to use multi-step and/or recycling HPLC isolations. Thus, alternative isolation techniques which avoid HPLC (non-HPLC isolation techniques) are needed and of high practical importance.

3.3.1 *Sublimation*

In the early stage of fullerene research, the mixture containing ENFs was sublimed from the raw carbon soot generated through laser ablation.[2] This method used the different sublimation temperature between fullerenes and other byproducts, fulfilling the isolation of fullerenes from the raw carbon soot. Although this method played an important role when other isolation techniques were not available, the difference of the sublimation temperatures of different fullerene isomers is generally not large. Therefore this method failed in the isolation of fullerene (ENF) isomers. A mass spectral analysis showed that only negligible amounts of fullerenes remained in the sublimation-treated soot, indicating that sublimation is highly effective in removing fullerenes from the carbon soot mixture.[110]

3.3.2 *Isolation by Electrochemical or Chemical Redox Reaction*

It is possible to exploit the difference on the redox properties of ENFs with different cage sizes or isomers to isolate them from one another and

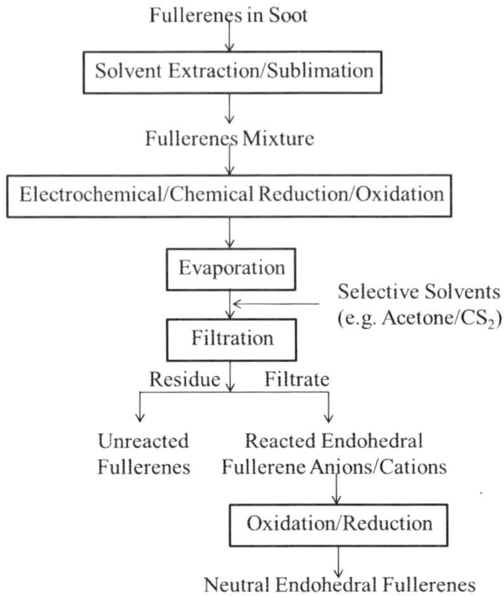

Scheme 2 A General Procedure for Isolation of ENFs by Redox Reaction.

from empty fullerenes. The general procedure for the isolation by redox reaction is shown in Scheme 2.

3.3.2.1 *Isolation by Electrochemical Reduction*

In 1998, Diener and coworkers realized the electrochemical reduction extraction of C_{74}, Gd@C_{60} and Gd@C_{74} etc. for the first time,[106] and they suggested that, with further refinement, the small band-gap and radical species, including La@C_{60}, can be eventually isolated.

In 2004, Akasaka and coworkers reported a convenient method for the isolation of EMFs from solvent-extract of carbon soot, in which selective reduction of the EMFs with low redox potentials were carried out. Reduced EMFs and neutral empty fullerenes may then be isolated based on their different solubility.[111] As a result, they accomplished the isolation of two isomers of La@C_{82} and La_2@C_{80}.

The isolated EMF anions could be electrochemically reoxidized to their neutral states.[106] Alternatively, the reoxidization to their neutral forms could be also made chemically by using the weak acid $CHCl_2COOH$[111] or $FcPF_6$.[106]

3.3.2.2 *Isolation by Electrochemical Oxidation*

Similar with the method of isolation by electrochemical reduction, it's reasonable for the isolation of endohedral fullerenes by electrochemical oxidation based on the different first electrochemical oxidation potentials. In 2005, Echegoyen and coworkers reported the 270 mV difference in the first electrochemical oxidation potentials of two isomers of $Sc_3N@C_{80}$ (I, I_h) and (II, D_{5h}), and designed a strategy for the isolation by the selective chemical oxidation of $Sc_3N@C_{80}$ (II, D_{5h}) based on the 270 mV difference in the first electrochemical oxidation potentials of these two isomers.[112] It is probably due to the unstable property of the oxidized endohedral fullerenes that they abandoned the electrochemical oxidation strategy. Anyhow, electrochemical oxidation for the isolation of endohedral fullerenes is still an untouched and challenging topic.

3.3.2.3 *Isolation by Chemical Oxidation*

In the early 1990s, Lewis acid $AlCl_3$ was applied for the isolation of C_{60} from C_{70} and other higher empty fullerenes by Olah and coworkers.[113] In their method, $AlCl_3$ was added into the CS_2 solution of fullerene extract mixture and stirred at room temperature for 2–6 days. It was found that C_{70} complexed much stronger with $AlCl_3$ and the resulting complexes precipitated out of the CS_2 solution. The remaining CS_2 solution mainly contained C_{60} and the unreacted $AlCl_3$, then the liquid phase was decanted, neutralized with saturated $NaHCO_3$, washed with water, and dried of anhydrous $MgSO_4$. Finally, the purified C_{60} was seized. The authors claimed that they achieved the high concentration of C_{60} (>99.8% confirmed by HPLC). The precipitates contained C_{70} and less C_{60} was treated in a similar way used for the remaining CS_2 solution containing C_{60}, and the concentrated C_{70}/C_{60} mixture was recovered. Because a little water is helpful for the generation of Lewis acid of $AlCl_3$, the stepwise addition of a little water into the fullerene/$AlCl_3$ system was found to expedite the isolation (about 75 hrs). For other Lewis acids, both C_{60} and C_{70} reacted with $AlBr_3$, SbF_5, $TiCl_4$, $SnCl_4$, and $FeCl_3$ were also tried, no reaction was observed in neat SbF_5 at room temperature even after five months.[113]

This method was not applied for the isolation of EMFs until the early 2000s. In 2003, Bolskar and coworkers introduced $AgPF_6$, $AgSbCl_6$, tris(4-bromo-phenyl)aminium hexachoroantimonate etc. as mild oxidants and $AlCl_3$ as stronger oxidant for the enrichment/isolation of EMFs.[10] In 2008, the same group used $AlCl_3$ for the extraction of $Gd@C_{2n}$. They also introduced $NOSbF_6$ as oxidant, which showed similar effect on the oxidation of insoluble fullerenes.[110]

In 2005, Echegoyen and coworkers reported the isolation of two isomers of $Sc_3N@C_{80}$ (I, I_h) and (II, D_{5h}) by the selective oxidation of $Sc_3N@C_{80}$ (II, D_{5h}) based on the 270 mV difference in the first electrochemical oxidation potentials of these two isomers.[112] A suitable chemical oxidant (tris(p-bromophenyl)-aminium hexachloroantimonate (TPBAH)) with an oxidation potential falling between the first two oxidation waves was selected for stabilizing the oxidized $Sc_3N@C_{80}$ (D_{5h}).[112] Very recently, they found a more convenient chemical oxidant acetylferrocinium tetrakis-(pentafluorophenyl)boride ([Fe($COCH_3C_5H_4$)Cp]–[TFAB]) for the complete and selective separation of the $Sc_3N@D_{5h}$-C_{80} and $Sc_3N@C_{78}$ isomers from $Sc_3N@I_h$-C_{80}.[114]

In 2009, Stevenson and coworkers applied a series of Lewis acids for the isolation of NCFs and OCFs.[115] In their conditions, NCFs and OCFs reacted quickly with Lewis acids, but empty fullerenes were largely unreactive. The reactivity order is $Sc_4O_2@C_{80} > Sc_3N@C_{78} > Sc_3N@C_{68} > Sc_3N@C_{80}$ (D_{5h}) > $Sc_3N@C_{80}$ (I_h). Manipulation of the Lewis acids and kinetic differences results in a selective precipitation scheme, which can be used alone for purifying $Sc_3N@C_{80}$ (I_h) or combined with a final HPLC pass for $Sc_4O_2@C_{80}$, $Sc_3N@C_{80}$ (D_{5h}), $Sc_3N@C_{68}$, and $Sc_3N@C_{78}$. One advantage of this method is that the purification process is scalable.

More recently, Shinohara and coworkers used another Lewis acid, $TiCl_4$, for the rapid isolation of EMFs and empty fullerenes.[116] The authors claimed that, irrespective of the kind of metal atom, number of metal atom(s) encapsulated and size of the fullerene cage, virtually all EMFs can be isolated and purified in a single step by this method. Moreover, this method enabled one to separate the EMFs up to 99% purity within 10 min without using any type of HPLC. They also found that the first oxidation potentials of the EMFs primarily determine the isolation/purification efficiency. They believed that any EMFs having a first oxidation potential lower than 0.5–0.6 V versus Fc/Fc^+ can be separated in high purity (>99%) by using $TiCl_4$.

It is noteworthy that, when the Lewis acids were applied for the separation of EMFs, no water was needed to the reactor, probably in order to suppress the activity of the Lewis acid, thus empty fullerenes could not react with the Lewis acid. Thus, using this series of Lewis acids, the isolation of ENFs is perhaps based on the redox reaction between Lewis acids and ENFs. Taking $AlCl_3$ for an example, the reaction products solubilized by $AlCl_3$ treatment would require further characterization to establish their nature, but the reaction likely proceeds with the oxidation of the fullerene by $AlCl_3$ with the *in situ* formation of tetrachloroaluminate ($AlCl_4^-$) as the counter anion.[110,117]

3.3.2.4 *Isolation by Chemical Reduction*

EMFs could be also extracted and isolated from the raw soot by alloy reduction. In 2002, Gu and coworkers developed a method which could selectively reduce $Gd@C_{82}$ and $Gd_2@C_{80}$ with Al-Ni alloy in toluene/tetrahydrofuran (THF).[118] By tuning the ratio of toluene and THF, the isolation performance could be adjusted, and $Gd_2@C_{80}$ with purity higher than 95% was isolated directly from the raw soot without HPLC purification when the volume ratio of toluene and THF was 12:5. Because the method did not need the sublimation of fullerenes in the raw soot, it seems more facile for selective isolation of EMFs.[118a]

Recently, Zn reduction of insoluble empty fullerenes for the recovery of giant, small band-gap fullerenes was reported, and this method could probably be used for the isolation of EMFs.[119]

3.3.3 *Isolation by Chemical Functionalization*

Based on the difference of the chemical reactivity, in 2005 Dorn and coworkers reported Diels–Alder reactions of carbon soot with a reactive resin (cyclopentadiene-functionalized styrene-divinylbenzene resin), affording a single-step isolation of a series of NCFs ($A_3N@C_{80}$, A = Er, Gd, Ho, Lu, Sc, Tb, Tm, Y) from the as-prepared carbon soot (see Fig. 7, left panel).[120] This chemical method takes advantage of the extraordinary kinetic chemical stability of NCFs with respect to the other fullerenes. Taking Sc-NCFs as an example, the isolation results are shown in Fig. 7, right panel. When

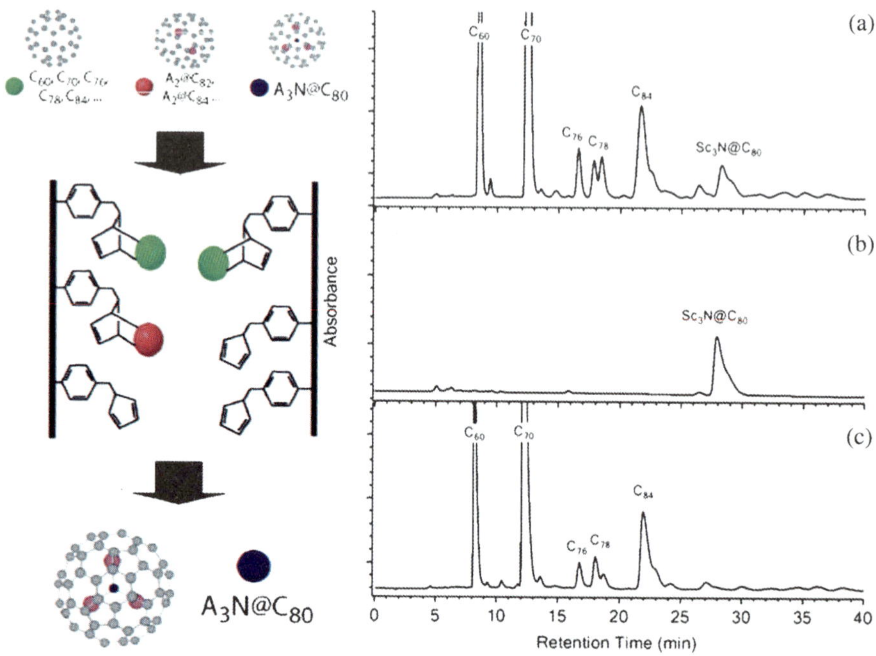

Fig. 7 (Left panel) Separation of NCFs from empty cage fullerenes and classical endohedral metallofullerenes by selective cycloaddition with cyclo-pentadiene-functionalized styrene-divinylbenzene resin. Reproduced with permission from American Chemical Society (Right panel) HPLC (PYE column, toluene, 3 mL/min, detection at 390 nm) traces of (a) the toluene extract of "scandium soot", (b) the room-temperature toluene eluent from application of the extract to a column packed with cyclopentadiene-functionalized resin; the peak at 27.9 min corresponds to $Sc_3N@C_{80}$; there are minor peaks due to $Sc_3N@C_{68}$ (16 min) and $Sc_3N@C_{78}$ (27 min) and retention times of 5–10 min due to unreactive polycyclic aromatic hydrocarbons from the soot; (c) the eluent from the column after eclution (b) upon heating at 85 °C in the presence of maleic anhydride.[120] Reproduced with permission from American Chemical Society.

the toluene-extract (Fig. 7a) passes through the column treated by the reactive resin, empty fullerenes and conventional EMFs react with the resin, whereas $Sc_3N@C_{2n}$ (2n = 68, 78, 80) does not react and passes through the column (Fig. 7b). The species adhered to the resin (empty fullerenes and conventional EMFs) could be recovered simply by heating in the presence of maleic anhydride (Fig. 7c).

In 2006, Stevenson and coworkers proposed another chemical isolation method named as the "stir and filter approach" (SAFA). Without using any chromatography equipment, they used cyclopentadienyl- and amino-functionalized silica to selectively bind contaminant fullerenes (empty fullerenes and non-NCFs) and claimed to obtain purified NCFs under optimum conditions. Later on, the same group further compared the reactivity of the D_{5h} isomer of $Sc_3N@C_{80}$ relative to the I_h isomer of $Sc_3N@C_{80}$ towards a reactive diamino silica used in the "SAFA" method.[121] It was found that the reactivity of C_{70}, C_{60} and $Sc_3N@C_{80}$ (D_{5h}) > $Sc_3N@C_{80}$ (I_h). Accordingly, this significant difference on the reactivity could be used as a rapid, non-oxidative, non-chromatographic method for the rapid removal of the $Sc_3N@C_{80}$ (D_{5h}) isomer from the $Sc_3N@C_{80}$ (I_h).[122]

In 2008, Dorn and coworkers reported the isolation of NCFs by a chemical reaction of congeners with eutectic 9-methylanthracene.[123] This is a solvent-free reaction of crude carbon soot extract (empty fullerenes and Sc- or Lu-NCFs) in molten 9-methylanthracene that affords almost complete conversion of empty fullerenes to 9-methylanthracene adducts, while leaving NCFs unreacted. After the recovered extract is washed with diethyl ether, the washed extract can then be applied to a silica gel column and subsequently flushed with toluene. To illustrate for $Sc_3N@C_{80}$, the sample was purified to ~60% in a time period of <24 hrs. The final one-step HPLC purification provided high-purity NCF samples (>99%) in an expeditious and less costly manner.[123]

3.3.4 *Isolation by Supramolecular Chemistry*

In 1994, Atwood and coworkers firstly used the selective complexation with calixarenes for the purification of C_{60} and C_{70}. Complexation of *p*-Bu^t-calix[8]arene with a mixture of the toluene extract of crude fullerene soot, followed by a series of recrystallizations, afforded >99.5% pure C_{60}.[124]

In 2006, Akasaka and coworkers seized the host–guest complexation of EMFs with azacrown ether, and they successfully realized the selective extraction of ENFs from extracts of carbon soot.[125] Through this method, they extracted $La@C_{82}$ (two isomers) and $La_2@C_{80}$ from the fullerene mixture. Even though this method could not realize the complete isolation of ENFs, it supplied a selective enrichment of ENFs, and this would be make the continuous HPLC isolation simpler.

4. CONCLUSIONS AND PERSPECTIVE

The synthesis of ENFs plays a key role in ENF research. Up to now, there have been several methods developed to synthesize versatile ENFs. The synthesis methods for ENFs could be classified as four types according to the process during the generation of ENFs: 1) "Top-down-top" processes such as laser ablation, DC arc discharge and radio frequency furnace methods. In the first step, the graphite and the encaged species survived the evaporation, and the carbon fragments generated. Such process could be named a "top-down" process. Then the carbon fragments grow to ENFs as a "down-top" process. Thus the overall process could be named as "top-down-top." 2) Incorporation of atom/molecule into the intact fullerene cage. In this process, the intact fullerene (e.g. C_{60}, C_{70}) was applied during the synthesis, and the encaged species was incorporated into the fullerene cage through several methods, including high-pressure incorporation, ion implantation and glow arc discharge. 3) Incorporation of atom/molecule into the open-caged fullerene — the "molecular surgery" method. 4) Transformation of the existing ENFs to other ones by hot-atom chemistry.

Up to now, the yield of ENFs synthesized from all synthesis methods developed is still relatively low, and this largely hinders the advancement of the ENF research. Hence, developing new and high performance methods for the high-yield synthesis of ENFs is a huge challenge. In particular, at present the DC arc discharge method and radio frequency furnace are used widely in laboratory because of the relative high yield for the synthesis of ENFs, but the mechanism of fullerene/ENF formation is still a mystery. Once this mystery is unraveled, developing new methods for the synthesis of ENFs in a controllable manner could be achieved. For instance, it is well known that the commercial synthesis method of empty fullerenes is the combustion of aromatic hydrocarbon, which affords the highest yield of empty fullerenes.[5,126] However, so far this method has not been applied for the synthesis of EMFs yet, and presumably the main reason is that the combustion flame with a low temperature (~1800 K) could not evaporate the metal to be encapsulated in the carbon cage. Therefore, a combination of evaporation of metal (e.g. metal ion sources, radio frequency furnace *et al.*) and combustion might be developed as a modified combustion synthesis of ENFs which would be definitely a breakthrough.

So far the "molecular surgery" method still focuses on the C_{60} cage, and the encaged element mainly contains a non-metal element. How about applying the "molecular surgery" for the synthesis of ENFs? Recently, Wang, Yang and coworkers successfully synthesized a 13-membered ring orifice on the $Sc_3N@C_{80}$ (I_h) cage via chemical functionalization.[127] With such a large orifice, if some non-metal atom/molecule (e.g. H, H_2) could be incorporated into the cage, is it possible to form a new type of metal clusterfullerene upon the chemical reaction between the existing Sc_3N moiety and H_2/H in a fullerene molecule?

With respect to the extraction and isolation of ENFs, obviously the most important issue about extraction of ENFs is the solvent. Although up to now lots of solvents have been successfully applied for the extraction of ENFs, specifically CS_2 due to its high solubility for most fullerenes and ENFs and low boiling point, such solvents as CS_2 do not work for the extraction of some special EMFs such as $La@C_{74}$, $M@C_{60}$ which are insoluble in CS_2 etc. Therefore, other extraction methods based on other or mixed solvents and/or electrochemical/chemical redox extraction should be considered.

Finally, because of the similarity of the cage structure and existence of multiple isomers of ENFs, currently HPLC is still the most powerful and commonly applied method for the isolation/purification of ENFs. However this technique is expensive and the isolation process is extremely time-consuming. Therefore, it is highly desirable to develop other non-HPLC methods for the selective isolation of ENFs. Although up to now some success has been demonstrated, which focused on the difference of the redox property and chemical reactivity between ENFs and empty fullerenes, such methods were used only in limited research labs and their universalities have to be checked further. Thus, much more effort should be devoted to developing alternative isolation strategies, which is a major challenge for ENF research.

ACKNOWLEDGMENTS

We thank the present and former members of our group for their contributions as well as our collaborators on the published works cited in this review. The research work was supported by National Basic Research

Program of China (2010CB923300, 2011CB921401), National Natural Science Foundation of China (90921013, 21132007), and "100 Talents Programme" from the Chinese Academy of Sciences.

REFERENCES

1. Heath JR, O'Brien SC, Zhang Q, *et al.* (1985) Lanthanum Complexes of Spheroidal Carbon Shells. *J Am Chem Soc* **107:** 7779–7780.
2. Chai Y, Guo T, Jin CM, *et al.* (1991) Fullerenes with Metals Inside. *J Phys Chem* **95:** 7564–7568.
3. (a) Popov AA, Yang SF, Dunsch L. (2013) Endohedral Fullerenes. *Chem Rev* **113:** 5989–6113; (b) Zhang J, Stevenson S, Dorn HC. (2013) Trimetallic Nitride Template Endohedral Metallofullerenes: Discovery, Structural Characterization, Reactivity, and Applications. *Acc Chem Res* **46:** 1548–1557; (c) Lu X, Akasaka T, Nagase S. (2013) Carbide Cluster Metallofullerenes: Structure, Properties, and Possible Origin. *Acc Chem Res* **46:** 1627–1635; (d) Yang SF. (2012) Synthesis, Separation, and Molecular Structures of Endohedral Fullerenes. *Curr Org Chem* **16:** 1079–1094; (e) Lu X, Feng L, Akasaka T, *et al.* (2012) Current status and future developments of endohedral metallofullerenes. *Chem Soc Rev* **41:** 7723–7760; (f) Yang SF, Liu FP, Chen CB, *et al.* (2011) Fullerenes encaging metal clusters–clusterfullerenes. *Chem Commun* **47:** 11822–11839; (g) Rodriguez-Fortea A, Balch AL, Poblet JM. (2011) Endohedral metallofullerenes: a unique host-guest association. *Chem Soc Rev* **40**: 3551–3563; (h) Yamada M, Akasaka T, Nagase S. (2010) Endohedral Metal Atoms in Pristine and Functionalized Fullerene Cages. *Acc Chem Res* **43:** 92–102; (i) Chaur MN, Melin F, Ortiz AL, *et al.* (2009) Chemical, electrochemical, and structural properties of endohedral metallofullerenes. *Angew Chem Int Ed* **48:** 7514–7538; (j) Dunsch L, Yang SF. (2007) Metal Nitride Cluster Fullerenes: Their Current State and Future Prospects. *Small* **3:** 1298–1320; (k) Dunsch L, Yang SF. (2007) Endohedral clusterfullerenes — playing with cluster and cage sizes. *Phys Chem Chem Phys* **9:** 3067–3081; (l) Yang SF, Dunsch L. (2008) *Endohedral Fullerenes.* John Wiley & Sons, Ltd., Chichester, United Kingdom; (m) Akasaka T, Nagase S. (2002) *Endofullerenes: a new family of carbon clusters.* Kluwer Academic Publishers, Dordrecht; (n) Shinohara H. (2000) Endohedral metallofullerenes. *Rep Prog Phys* **63:** 843–892.
4. Krätschmer W, Lamb LD, Fostiropoulos K, *et al.* (1990) Solid C_{60}: a New Form of Carbon. *Nature* **347:** 354–358.

5. Howard JB, Mckinnon JT, Makarovsky Y, *et al.* (1991) Fullerenes C_{60} and C_{70} in Flames. *Nature* **352:** 139–141.
6. Haufler R, Chai Y, Chibante L, *et al.* (1990) Carbon arc generation of C_{60}. *MRS Online Proceedings Library* **206:** 627–637.
7. (a) Johnson RD, de Vries MS, Salem J, *et al.* (1992) Electron paramagnetic resonance studies of lanthanum-containing C_{82}. *Nature* **355:** 239–240; (b) Alvarez MM, Gillan EG, Holczer K, *et al.* (1991) Lanthanum carbide (La_2C_{80}): a soluble dimetallofullerene. *J Phys Chem* **95:** 10561–10563.
8. Bandow S, Shinohara H, Saito Y, *et al.* (1993) High-Yield Synthesis of Lanthanofullerenes Via Lanthanum Carbide. *J Phys Chem* **97:** 6101–6103.
9. Liu BB, Xu WG, Liu ZY, *et al.* (1996) High yield synthesis and extraction of La@C_2n. *Solid State Commun* 97**:** 407–410.
10. Bolskar RD, Alford JM. (2003) Chemical oxidation of endohedral metallofullerenes: identification and separation of distinct classes. *Chem Commun* 1292–1293.
11. Stevenson S, Mackey MA, Thompson MC, *et al.* (2007) Effect of copper metal on the yield of $Sc_3N@C_{80}$ metallofullerenes. *Chem Commun* 4263–4265.
12. Olmstead MM, de Bettencourt-Dias A, Duchamp JC, *et al.* (2000) Isolation and crystallographic characterization of $ErSc_2N@C_{80}$: an endohedral fullerene which crystallizes with remarkable internal order. *J Am Chem Soc* **122:** 12220–12226.
13. Beavers CM, Zuo TM, Duchamp JC, *et al.* (2006) $Tb_3N@C_{84}$: An improbable, egg-shaped endohedral fullerene that violates the isolated pentagon rule. *J Am Chem Soc* **128:** 11352–11353.
14. Wang CR, Kai T, Tomiyama T, *et al.* (2001) A scandium carbide endohedral metallofullerene: $(Sc_2C_2)@C_{84}$. *Angew Chem Int Ed* **40:** 397–399.
15. Stevenson S, Rice G, Glass T, *et al.* (1999) Small-bandgap endohedral metallofullerenes in high yield and purity. *Nature* **401:** 55–57.
16. Macfarlane RM, S Bethune D, Stevenson S, *et al.* (2001) Fluorescence spectroscopy and emission lifetimes of Er^{3+} in $Er_xSc_{3-x}N@C_{80}$ (x = 1–3). *Chem Phys Lett* **343:** 229–234.
17. Stevenson S, Fowler PW, Heine T, *et al.* (2000) Materials science: A stable non-classical metallofullerene family. *Nature* **408:** 427–428.
18. Olmstead MH, de Bettencourt-Dias A, Duchamp JC, *et al.* (2001) Isolation and structural characterization of the endohedral fullerene $Sc_3N@C_{78}$. *Angew Chem Int Ed* **40:** 1223–1225.
19. Iezzi EB, Duchamp JC, Fletcher KR, *et al.* (2002) Lutetium-based Trimetallic Nitride Endohedral Metallofullerenes: New Contrast Agents. *Nano Lett* **2:** 1187–1190.

20. (a) Fu WJ, Zhang J, Champion H, *et al.* (2011) Electronic Properties and ^{13}C NMR Structural Study of $Y_3N@C_{88}$. *Inorg Chem* **50:** 4256–4259; (b) Fu W, Xu L, Azurmendi H, *et al.* (2009) [89]Y and ^{13}C NMR cluster and carbon cage studies of an yttrium metallofullerene family, $Y_3N@C_{2n}$ (n = 40–43). *J Am Chem Soc* **131:** 11762–11769.
21. Zuo TM, Beavers CM, Duchamp JC, *et al.* (2007) Isolation and Structural Characterization of a Family of Endohedral Fullerenes Including the Large, Chiral Cage Fullerenes $Tb_3N@C_{88}$ and $Tb_3N@C_{86}$ as well as the I_h and D_{5h} Isomers of $Tb_3N@C_{80}$. *J Am Chem Soc* **129:** 2035–2043.
22. Wang XL, Zuo TM, Olmstead MM, *et al.* (2006) Preparation and structure of $CeSc_2N@C_{80}$: An icosahedral carbon cage enclosing an acentric $CeSc_2N$ unit with buried f electron spin. *J Am Chem Soc* **128:** 8884–8889.
23. (a) Burke BG, Chan J, Williams KA, *et al.* (2010) Investigation of $Gd_3N@C_{2n}$ (40 < n < 44) family by Raman and inelastic electron tunneling spectroscopy. *Phys Rev B* **81:** 115423; (b) Beavers CM, Chaur MN, Olmstead MM, *et al.* (2009) Large metal ions in a relatively small fullerene cage: the structure of $Gd_3N@C_2(22010)$-C_{78} departs from the isolated pentagon rule. *J Am Chem Soc* **131:** 11519–11524; (c) Mercado BQ, Beavers CM, Olmstead MM, *et al.* (2008) Is the isolated pentagon rule merely a suggestion for endohedral fullerenes? The structure of a second egg-shaped endohedral fullerene-$Gd_3N@C_s(39663)$-C_{82}. *J Am Chem Soc* **130:** 7854–7855; (d) Chaur MN, Melin F, Elliott B, *et al.* (2007) $Gd_3N@C_2n$ (n = 40, 42, and 44): remarkably low HOMO-LUMO gap and unusual electrochemical reversibility of $Gd_3N@C_{88}$. *J Am Chem Soc* **129**: 14826–14829; (e) Stevenson S, Phillips JP, Reid JE, *et al.* (2004) Pyramidalization of Gd_3N inside a C_{80} cage. The synthesis and structure of $Gd_3N@C_{80}$. *Chem Commun* 2814–2815.
24. Yang SF, Chen CB, Popov AA, *et al.* (2009) An endohedral titanium(III) in a clusterfullerene: putting a non-group-III metal nitride into the C_{80}-I_h fullerene cage. *Chem Commun* 6391–6393.
25. Chen CB, Liu FP, Li SJ, *et al.* (2012) Titanium/Yttrium Mixed Metal Nitride Clusterfullerene $TiY_2N@C_{80}$: Synthesis, Isolation, and Effect of the Group-III Metal. *Inorg Chem* **51:** 3039–3045.
26. Zuo TM, Xu LS, Beavers CM, *et al.* (2008) $M_2@C_{79}N$ (M = Y, Tb): Isolation and Characterization of Stable Endohedral Metallofullerenes Exhibiting M–M Bonding Interactions inside Aza[80]fullerene Cages. *J Am Chem Soc* **130:** 12992–12997.
27. Fu WJ, Zhang JY, Fuhrer T, *et al.* (2011) $Gd_2@C_{79}N$: Isolation, Characterization, and Monoadduct Formation of a Very Stable Heterofullerene with a Magnetic Spin State of S = 15/2. *J Am Chem Soc* **133:** 9741–9750.

28. Wu JY, Wang TS, Ma YH, *et al.* (2011) Synthesis, Isolation, Characterization, and Theoretical Studies of $Sc_3NC@C_{78}$-C_2. *J Phys Chem C* **115:** 23755–23759.
29. Wang TS, Feng L, Wu JY, *et al.* (2010) Planar Quinary Cluster inside a Fullerene Cage: Synthesis and Structural Characterizations of $Sc_3NC@C_{80}$-I_h. *J Am Chem Soc* **132:** 16362–16364.
30. Yang SF, Chen CB, Liu FP, *et al.* (2013) An Improbable Monometallic Cluster Entrapped in a Popular Fullerene Cage: YCN@C_s(6)-C_{82}. *Sci Rep* **3:** 1487.
31. Dunsch L, Georgi P, Ziegs F, *et al.* (2003) Production of endohedral fullerenes. DE 10301722 A1.
32. Dunsch L, Krause M, Noack J, *et al.* (2004) Endohedral nitride cluster fullerenes — Formation and spectroscopic analysis of $L_{3-x}M_xN@C_{2n}$ (0 <= x <= 3; N = 39, 40). *J Phys Chem Solids* **65:** 309–315.
33. Wolf M, Müller KH, Skourski Y, *et al.* (2005) Magnetic Moments of the Endohedral Cluster Fullerenes $Ho_3N@C_{80}$ and $Tb_3N@C_{80}$: The Role of Ligand Fields. *Angew Chem Int Ed* **44:** 3306–3309.
34. Krause M, Dunsch L. (2005) Gadolinium Nitride Gd_3N in Carbon Cages: The Influence of Cluster Size and Bond Strength. *Angew Chem Int Ed* **44:** 1557–1560.
35. Yang SF, Dunsch L. (2005) A large family of dysprosium-based trimetallic nitride endohedral fullerenes: $Dy_3N@C_{2n}$ (39 <= n <= 44). *J Phys Chem B* **109:** 12320–12328.
36. Krause M, Wong J, Dunsch L. (2005) Expanding the world of endohedral fullerenes — The $Tm_3N@C_{2n}$ (39 <= n <= 43) clusterfullerene family. *Chem Eur J* **11:** 706–711.
37. (a) Chaur MN, Valencia R, Rodriguez-Fortea A, *et al.* (2009) Trimetallic Nitride Endohedral Fullerenes: Experimental and Theoretical Evidence for the $M3N^{6+}@C2n^{6-}$ model. *Angew Chem Int Ed* **48:** 1425–1428; (b) Chaur MN, Melin F, Elliott B, *et al.* (2008) New $M_3N@C_{2n}$ endohedral metallofullerene families (M = Nd, Pr, Ce; n = 40–53): Expanding the preferential templating of the C_{88} cage and approaching the C_{96} cage. *Chem Eur J* **14:** 4594–4599; (c) Chaur MN, Melin F, Ashby J, *et al.* (2008) Lanthanum nitride endohedral fullerenes $La_3N@C_{2n}$ (43 <or = n <or = 55): preferential formation of $La_3N@C_{96}$. *Chem Eur J* **14:** 8213–8219.
38. (a) Yang SF, Popov AA, Kalbac M, *et al.* (2008) The isomers of gadolinium scandium nitride clusterfullerenes $Gd_xSc_{3-x}N@C_{80}$ (x = 1, 2) and their influence on cluster structure. *Chem Eur J* **14:** 2084–2092; (b) Yang SF, Kalbac M, Popov AA, *et al.* (2006) Gadolinium-based mixed metal nitride clusterfullerenes $Gd_xSc_{3-x}N@C_{80}$ (x = 1, 2). *Chemphyschem* **7:** 1990–1995.

39. Yang SF, Popov AA, Dunsch L. (2007) Violating the Isolated Pentagon Rule (IPR): The endohedral Non-IPR C_{70} cage of $Sc_3N@C_{70}$. *Angew Chem Int Ed* **46:** 1256–1259.
40. Yang SF, Popov AA, Dunsch L. (2007) The role of an asymmetric nitride cluster on a fullerene cage: The Non-IPR endohedral $DySc_2N@C_{76}$. *J Phys Chem B* **111:** 13659–13663.
41. Yang SF, Popov AA, Dunsch L. (2008) Large mixed metal nitride clusters encapsulated in a small cage: the confinement of the C-68-based clusterfullerenes. *Chem Commun* 2885–2887.
42. Yang SF, Popov AA, Dunsch L. (2008) Carbon Pyramidalization in Fullerene Cages Induced by the Endohedral Cluster: Non-Scandium Mixed Metal Nitride Clusterfullerenes. *Angew Chem Int Ed* **47:** 8196–8200.
43. Zhang L, Popov AA, Yang SF, *et al.* (2010) An endohedral redox system in a fullerene cage: the Ce based mixed-metal cluster fullerene $Lu_2CeN@C_{80}$. *Phys Chem Chem Phys* **12:** 7840–7847.
44. Svitova AL, Popov AA, Dunsch L. (2013) Gd–Sc-Based Mixed-Metal Nitride Cluster Fullerenes: Mutual Influence of the Cage and Cluster Size and the Role of Scandium in the Electronic Structure. *Inorg Chem* **52:** 3368–3380.
45. Stevenson S, Thompson MC, Coumbe HL, *et al.* (2007) Chemically adjusting plasma temperature, energy, and reactivity (CAPTEAR) method using NO_x and combustion for selective synthesis of $Sc_3N@C_{80}$ metallic nitride fullerenes. *J Am Chem Soc* **129:** 16257–16262.
46. Stevenson S, Ling Y, Coumbe CE, *et al.* (2009) Preferential Encapsulation and Stability of La_3N Cluster in 80 Atom Cages: Experimental Synthesis and Computational Investigation of $La_3N@C_{79}N$. *J Am Chem Soc* **131:** 17780–17782.
47. Liu FP, Guan J, Wei T, *et al.* (2013) A Series of Inorganic Solid Nitrogen Sources for the Synthesis of Metal Nitride Clusterfullerenes: The Dependence of Production Yield on the Oxidation State of Nitrogen and Counter Ion. *Inorg Chem* **52:** 3814–3822.
48. Yang SF, Zhang L, Zhang WF, *et al.* (2010) A Facile Route to Metal Nitride Clusterfullerenes by Using Guanidinium Salts: A Selective Organic Solid as the Nitrogen Source. *Chem Eur J* **16:** 12398–12405.
49. Dunsch L, Yang SF, Zhang L, *et al.* (2010) Metal Sulfide in a C_{82} Fullerene Cage: A New Form of Endohedral Clusterfullerenes. *J Am Chem Soc* **132:** 5413–5421.
50. Jiao MZ, Zhang WF, Xu Y, *et al.* (2012) Urea as a New and Cheap Nitrogen Source for the Synthesis of Metal Nitride Clusterfullerenes: The Role of Decomposed Products on the Selectivity of Fullerenes. *Chem Eur J* **18:** 2666–2673.

51. Chen N, Chaur MN, Moore C, *et al.* (2010) Synthesis of a new endohedral fullerene family, $Sc_2S@C_{2n}$ (n = 40–50) by the introduction of SO_2. *Chem Commun* **46:** 4818–4820.
52. Chen N, Beavers CM, Mulet-Gas M, *et al.* (2012) $Sc_2S@C_s(10528)\text{-}C_{72}$: A Dimetallic Sulfide Endohedral Fullerene with a Non Isolated Pentagon Rule Cage. *J Am Chem Soc* **134:** 7851–7860.
53. Chen N, Mulet-Gas M, Li YY, *et al.* (2013) $Sc_2S@C_2(7892)\text{-}C_{70}$: a metallic sulfide cluster inside a non-IPR C_{70} cage. *Chem Sci* **4:** 180–186.
54. Krause M, Ziegs F, Popov AA, *et al.* (2007) Entrapped bonded hydrogen in a fullerene: The five-atom cluster Sc_3CH in C_{80}. *Chemphyschem* **8:** 537–540.
55. Stevenson S, Mackey MA, Stuart MA, *et al.* (2008) A Distorted Tetrahedral Metal Oxide Cluster inside an Icosahedral Carbon Cage. Synthesis, Isolation, and Structural Characterization of $Sc_4(\mu_3\text{-}O)_2@I_h\text{-}C_{80}$. *J Am Chem Soc* **130:** 11844–11845.
56. Mercado BQ, Olmstead MM, Beavers CM, *et al.* (2010) A seven atom cluster in a carbon cage, the crystallographically determined structure of $Sc_4(\mu_3\text{-}O)_3@I_h\text{-}C_{80}$. *Chem Commun* **46:** 279–281.
57. Peters G, Jansen M. (1992) A New Fullerene Synthesis. *Angew Chem Int Ed* **31:** 223–224.
58. Krokos E. (2010) Plasma Coupled Radio Frequency Furnace: The Synthesis, Separation, and Elucidation of the Elusive Sc_4C_{82} Fullerene. *J Phys Chem C* **114:** 7626–7630.
59. Yoshie K-i, Kasuya S, Eguchi K, *et al.* (1992) Novel method for C_{60} synthesis: A thermal plasma at atmospheric pressure. *Appl Phys Lett* **61:** 2782–2783.
60. Wang C, Imahori T, Tanaka Y, *et al.* (2001) Synthesis of fullerenes from carbon powder by using high power induction thermal plasma. *Thin Solid Films* **390:** 31–36.
61. Todorovic-Marković B, Marković Z, Mohai I, *et al.* (2003) Efficient synthesis of fullerenes in RF thermal plasma reactor. *Chem Phys Lett* **378:** 434–439.
62. Kaneko T, Abe S, Ishida H, *et al.* (2007) An electron cyclotron resonance plasma configuration for increasing the efficiency in the yield of nitrogen endohedral fullerenes. *Physics of Plasmas* **14:** 110705–110703.
63. Cota-Sanchez G, Soucy G, Huczko A, *et al.* (2004) Effect of Iron Catalyst on the Synthesis of Fullerenes and Carbon Nanotubes in Induction Plasma. *J Phys Chem B* **108:** 19210–19217.
64. Jansen M, Peters G, Wagner N. (1995) Zur Bildung von Fullerenen und endohedralen Metallofullerenen: Darstellung im Hochfrequenzofen. *Z Anorg Allg Chem* **621:** 689–693.
65. (a) Reich A, Panthöfer M, Modrow H, *et al.* (2004) The Structure of $Ba@C_{74}$. *J Am Chem Soc* **126:** 14428–14434; (b) Bucher K, Epple L, Mende J, *et al.*

(2006) Synthesis, isolation and characterization of new endohedral fullerenes M@C_{72} (M = Eu, Sr, Yb). *physica status solidi (b)* **243:** 3025–3027; (c) Haufe O, Hecht M, Grupp A, *et al.* (2005) Isolation and Spectroscopic Characterization of New Endohedral Fullerenes in the Size Gap of C_{74} to C_{76}. *Z Anorg Allg Chem* **631:** 126–130.

66. Huang HJ, Ata M, Ramm M. (2002) ^{14}N@C_{60} formation in a nitrogen rf-plasma. *Chem Commun* 2076–2077.
67. Miyanaga S, Kaneko T, Ishida H, *et al.* (2010) Synthesis evaluation of nitrogen atom encapsulated fullerenes by optical emission spectra in nitrogen plasmas. *Thin Solid Films* **518:** 3509–3512.
68. Saunders M, Jiménez-Vázquez HA, Cross RJ, *et al.* (1993) Stable Compounds of Helium and Neon: He@C_{60} and Ne@C_{60}. *Science* **259:** 1428–1430.
69. Saunders M, Jimenez-Vazquez HA, Cross RJ, *et al.* (1994) Incorporation of helium, neon, argon, krypton, and xenon into fullerenes using high pressure. *J Am Chem Soc* **116:** 2193–2194.
70. (a) Murata Y, Murata M, Komatsu K. (2003) 100% Encapsulation of a Hydrogen Molecule into an Open-Cage Fullerene Derivative and Gas-Phase Generation of H_2@C_{60}. *J Am Chem Soc* **125**: 7152–7153; (b) Rubin Y, Jarrosson T, Wang GW, *et al.* (2001) Insertion of Helium and Molecular Hydrogen Through the Orifice of an Open Fullerene This work was supported by grants from the National Science Foundation. *Angew Chem Int Ed* **40:** 1543–1546.
71. Peng RF, Chu SJ, Huang YM, *et al.* (2009) Preparation of He@C_{60} and He_2@C_{60} by an explosive method. *J Mater Chem* **19:** 3602–3605.
72. Weidinger A, Waiblinger M, Pietzak B, *et al.* (1998) Atomic nitrogen in C_{60}:N@C_{60}. *Applied Physics A: Materials Science & Processing* **66:** 287–292.
73. Weiske T, Böhme DK, Hrušák J, *et al.* (1991) Endohedral Cluster Compounds: Inclusion of Helium within $C_{60}^{\cdot\oplus}$ and $C_{70}^{\cdot\oplus}$ through Collision Experiments. *Angew Chem Int Ed* **30:** 884–886.
74. Almeida Murphy T, Pawlik T, Weidinger A, *et al.* (1996) Observation of Atomlike Nitrogen in Nitrogen-Implanted Solid C_{60}. *Phys Rev Lett* **77:** 1075–1078.
75. Tellgmann R, Krawez N, Lin SH, *et al.* (1996) Endohedral fullerene production. *Nature* **382**: 407–408.
76. Suetsuna T, Dragoe N, Harneit W, *et al.* (2002) Separation of N_2@C_{60} and N@C_{60}. *Chem Eur J* **8:** 5079–5083.
77. Jiménez-Vázquez HA, Cross RJ, Saunders M, *et al.* (1994) Hot-atom incorporation of tritium atoms into fullerenes. *Chem Phys Lett* **229:** 111–114.

78. Khong A, Cross RJ, Saunders M. (2000) From $^{3}He@C_{60}$ to $^{3}H@C_{60}$: Hot-Atom Incorporation of Tritium in C_{60}. *J Phys Chem A* **104:** 3940–3943.
79. Kikuchi K, Kobayashi K, Sueki K, *et al.* (1994) Encapsulation of Radioactive ^{159}Gd and ^{161}Tb Atoms in Fullerene Cages. *J Am Chem Soc* **116:** 9775–9776.
80. Cagle DW, Thrash TP, Alford M, *et al.* (1996) Synthesis, Characterization, and Neutron Activation of Holmium Metallofullerenes. *J Am Chem Soc* **118:** 8043–8047.
81. Braun T, Rausch H. (1995) Endohedral incorporation of argon atoms into C_{60} by neutron irradiation. *Chem Phys Lett* **237:** 443–447.
82. Ohtsuki T, Masumoto K, Ohno K, *et al.* (1996) Insertion of Be Atoms in C_{60} Fullerene Cages: $Be@C_{60}$. *Phys Rev Lett* **77:** 3522–3524.
83. Diener MD, Alford JM, Kennel SJ, *et al.* (2007) $^{212}Pb@C_{60}$ and Its Water-Soluble Derivatives: Synthesis, Stability, and Suitability for Radioimmunotherapy. *J Am Chem Soc* **129:** 5131–5138.
84. Ohtsuki T, Masumoto K, Kikuchi K, *et al.* (1996) Production of radioactive fullerene families using accelerators. *Mater Sci Eng, A* **217–218:** 38–41.
85. Ito S, Shimotani H, Takagi H, *et al.* (2008) On the Synthesis Conditions of N and N_2 Endohedral Fullerenes. *Fuller Nanotub Carbon Nanostruct* **16:** 206–213.
86. Vougioukalakis GC, Roubelakis MM, Orfanopoulos M. (2010) Open-cage fullerenes: towards the construction of nanosized molecular containers. *Chem Soc Rev* **39:** 817–844.
87. (a) Hummelen JC, Prato M, Wudl F. (1995) There Is a Hole in My Bucky. *J Am Chem Soc* **117**: 7003–7004; (b) Hummelen JC, Knight B, Pavlovich J, *et al.* (1995) Isolation of the Heterofullerene $C_{59}N$ as Its Dimer $(C_{59}N)_2$. *Science* **269:** 1554–1556.
88. (a) Gan L, Yang D, Zhang Q, *et al.* (2010) Preparation of Open-Cage Fullerenes and Incorporation of Small Molecules Through Their Orifices. *Adv Mater* **22:** 1498–1507; (b) Murata M, Murata Y, Komatsu K. (2008) Surgery of fullerenes. *Chem Commun* 6083–6094.
89. Rubin Y. (1997) Organic Approaches to Endohedral Metallofullerenes: Cracking Open or Zipping Up Carbon Shells? *Chem Eur J* **3:** 1009–1016.
90. Komatsu K, Murata M, Murata Y. (2005) Encapsulation of Molecular Hydrogen in Fullerene C_{60} by Organic Synthesis. *Science* **307:** 238–240.
91. Kurotobi K, Murata Y. (2011) A Single Molecule of Water Encapsulated in Fullerene C_{60}. *Science* **333:** 613–616.
92. Morinaka Y, Sato S, Wakamiya A, *et al.* (2013) X-ray observation of a helium atom and placing a nitrogen atom inside $He@C_{60}$ and $He@C_{70}$. *Nat Commun* **4:** 1554.

93. Stanisky CM, Cross RJ, Saunders M. (2009) Putting Atoms and Molecules into Chemically Opened Fullerenes. *J Am Chem Soc* **131:** 3392–3395.
94. Iwamatsu S-i, Stanisky CM, Cross RJ, *et al.* (2006) Carbon monoxide inside an open-cage fullerene. *Angew Chem Int Ed* **45:** 5337–5340.
95. Whitener KE, Frunzi M, Iwamatsu S, *et al.* (2008) Putting ammonia into a chemically opened fullerene. *J Am Chem Soc* **130:** 13996–13999.
96. Whitener KE, Cross RJ, Saunders M, *et al.* (2009) Methane in an Open-Cage [60]Fullerene. *J Am Chem Soc* **131:** 6338–6339.
97. Fuchs D, Rietschel H, Michel RH, *et al.* (1996) Extraction and Chromatographic Elution Behavior of Endohedral Metallofullerenes: Inferences Regarding Effective Dipole Moments. *J Phys Chem* **100:** 725–729.
98. (a) Capp C, Wood TD, Marshall AG, *et al.* (1994) High-Pressure Toluene Extraction of La@C_n for Even n from 74 to 90. *J Am Chem Soc* **116:** 4987–4988; (b) Creasy WR, Zimmerman JA, Ruoff RS. (1993) Fullerene molecular weight distributions in graphite soot extractions measured by laser desorption Fourier transform mass spectrometry. *J Phys Chem* **97:** 973–979.
99. Kubozono Y, Noto T, Ohta T, *et al.* (1996) Extractions of Ca@C_{60} and Sr@C_{60} with Aniline. *Chem Lett* **25:** 453–454.
100. Huang H, Yang S. (2000) Toward Efficient Synthesis of Endohedral Metallofullerenes by Arc Discharge of Carbon Rods Containing Encapsulated Rare Earth Carbides and Ultrasonic Soxhlet Extraction. *Chem Mater* **12:** 2715–2720.
101. Xiao J, Savina MR, Martin GB, *et al.* (1994) Efficient HPLC Purification of Endohedral Metallofullerenes on a Porphyrin-Silica Stationary Phase. *J Am Chem Soc* **116:** 9341–9342.
102. Ding J, Yang S. (1996) Efficient N,N-Dimethylformamide Extraction of Endohedral Metallofullerenes for HPLC Purification. *Chem Mater* **8:** 2824–2827.
103. Nikawa H, Kikuchi T, Wakahara T, *et al.* (2005) Missing metallofullerene La@C_{74}. *J Am Chem Soc* **127:** 9684–9685.
104. Akasaka T, Lu X, Kuga H, *et al.* (2010) Dichlorophenyl Derivatives of La@C_{3v}(7)-C_{82}: Endohedral Metal Induced Localization of Pyramidalization and Spin on a Triple-Hexagon Junction. *Angew Chem Int Ed* **49:** 9715–9719.
105. Tsuchiya T, Wakahara T, Lian YF, *et al.* (2006) Selective extraction and purification of endohedral metallofullerene from carbon soot. *J Phys Chem B* **110:** 22517–22520.
106. Diener MD, Alford JM. (1998) Isolation and properties of small-bandgap fullerenes. *Nature* **393:** 668–671.

107. Gromov A, Kratschmer W, Krawez N, *et al.* (1997) Extraction and HPLC purification of $Li@C_{60/70}$. *Chem Commun* 2003–2004.
108. Shinohara H, Yamaguchi H, Hayashi N, *et al.* (1993) Isolation and spectroscopic properties of scandium fullerenes ($Sc_2@C_{74}$, $Sc_2@C_{82}$, and $Sc_2@C_{84}$). *J Phys Chem* **97:** 4259–4261.
109. HPLC Column. *https://sesres.com/CosmosilColumns.asp.*
110. Raebiger JW, Bolskar RD. (2008) Improved Production and Separation Processes for Gadolinium Metallofullerenes. *J Phys Chem C* **112:** 6605–6612.
111. Tsuchiya T, Wakahara T, Shirakura S, *et al.* (2004) Reduction of endohedral metallofullerenes: A convenient method for isolation. *Chem Mater* **16:** 4343–4346.
112. Elliott B, Yu L, Echegoyen L. (2005) A simple isomeric separation of D_{5h} and I_h $Sc_3N@C_{80}$ by selective chemical oxidation. *J Am Chem Soc* **127:** 10885–10888.
113. Bucsi I, Aniszfeld R, Shamma T, *et al.* (1994) Convenient separation of high-purity C_{60} from crude fullerene extract by selective complexation with $AlCl_3$. *Proc Nat Acad Sci USA* **91:** 9019–9021.
114. Cerón MR, Li FF, Echegoyen L. (2013) An Efficient Method to Separate $Sc_3N@C_{80}$ I_h and D_{5h} Isomers and $Sc_3N@C_{78}$ by Selective Oxidation with Acetylferrocenium $[Fe(COCH_3C_5H_4)Cp]^+$. *Chem Eur J* **19:** 7410–7415.
115. Stevenson S, Mackey MA, Pickens JE, *et al.* (2009) Selective Complexation and Reactivity of Metallic Nitride and Oxometallic Fullerenes with Lewis Acids and Use as an Effective Purification Method. *Inorg Chem* **48:** 11685–11690.
116. (a) Akiyama K, Hamano T, Nakanishi Y, *et al.* (2012) Non-HPLC Rapid Separation of Metallofullerenes and Empty Cages with $TiCl_4$ Lewis Acid. *J Am Chem Soc* **134:** 9762–9767; (b) Wang Z, Nakanishi Y, Noda S, *et al.* (2012) The Origin and Mechanism of Non-HPLC Purification of Metallofullerenes with TiCl4. *J Phys Chem C* **116:** 25563–25567.
117. Connelly NG, Geiger WE. (1996) Chemical Redox Agents for Organometallic Chemistry. *Chem Rev* **96:** 877–910.
118. (a) Lu X, Li H, Sun B, *et al.* (2005) Selective reduction and extraction of $Gd@C_{82}$ and $Gd_2@C_{80}$ from soot and the chemical reaction of their anions. *Carbon* **43**: 1546–1549; (b) Sun B, Gu Z. (2002) Solvent-dependent Anion Studies on Enrichment of Metallofullerene. *Chem Lett* **31:** 1164–1165.
119. Raebiger JW, Alford JM, Bolskar RD, *et al.* (2011) Chemical redox recovery of giant, small-gap and other fullerenes. *Carbon* **49:** 37–46.
120. Ge Z, Duchamp JC, Cai T, *et al.* (2005) Purification of endohedral trimetallic nitride fullerenes in a single, facile step. *J Am Chem Soc* **127:** 16292–16298.

121. Stevenson S, Harich K, Yu H, *et al.* (2006) Nonchromatographic "stir and filter approach" (SAFA) for isolating $Sc_3N@C_{80}$ metallofullerenes. *J Am Chem Soc* **127:** 8829–8835.
122. Stevenson S, Mackey MA, Coumbe CE, *et al.* (2007) Rapid removal of D_{5h} isomer using the "stir and filter approach" and isolation of large quantities of isomerically pure $Sc_3N@C_{80}$ metallic nitride fullerenes. *J Am Chem Soc* **129:** 6072–6073.
123. Angeli CD, Cai T, Duchamp JC, *et al.* (2008) Purification of Trimetallic Nitride Templated Endohedral Metallofullerenes by a Chemical Reaction of Congeners with Eutectic 9-Methylanthracene. *Chem Mater* **20:** 4993–4997.
124. Atwood JL, Koutsantonis GA, Raston CL. (1994) Purification of C_{60} and C_{70} by selective complexation with calixarenes. *Nature* **368:** 229–231.
125. Tsuchiya T, Sato K, Kurihara H, *et al.* (2006) Host-guest complexation of endohedral metallofullerene with azacrown ether and its application. *J Am Chem Soc* **128:** 6699–6703.
126. McKinnon JT, Bell WL, Barkley RM. (1992) Combustion synthesis of fullerenes. *Combust Flame* **88:** 102–112.
127. Wang GW, Liu TX, Jiao MZ, *et al.* (2011) The Cycloaddition Reaction of I_h-$Sc_3N@C_{80}$ with 2-Amino-4,5-diisopropoxybenzoic Acid and Isoamyl Nitrite to Produce an Open-Cage Metallofullerene. *Angew Chem Int Ed* **50:** 4658–4662.

CHAPTER 3

Crystallographic Study of Endohedral Metallofullerenes

*Yun-Peng Xie, Shasha Zhao, Xing Lu**

1. INTRODUCTION

Endohedral metallofullerenes (EMFs) are a new class of novel hybrid molecules with metal atoms or metallic clusters encapsulated inside the inner cavities of fullerenes.[1] Because of the electron transfer from the metallic species to the fullerene cages, EMFs exhibit unique properties differing distinctly from those of empty fullerenes. The carbon cage of EMFs is always comprised of 12 pentagons and a variable number of hexagons that depends upon the fullerene size, while the encapsulated metallic species are found to cover mostly Groups 3, 4, 5 metals with lanthanide elements the most common.[2]

Structural elucidation is an important but intractable issue in the research of EMFs for precise determination of isomeric structures of the carbon cages. The molecular structures of EMFs are more complicated than those of empty fullerenes due to the existence of the internal metallic species and their uncertain locations. Experimental studies of the structures of EMFs have focused on examination of ^{13}C NMR spectra,[3–6]

*State Key Laboratory of Material Processing and Die & Mould Technology, School of Materials Science and Engineering, Huazhong University of Science and Technology, Wuhan 430074, China. Email: lux@hust.edu.cn synchrotron radiation powder diffraction with Rietveld/maximum entropy method (SRPD/Rietveld/MEM) treatment,[7–9] and single crystal X-ray diffraction. Theoretical calculation[10–13] is often of strong help in the understanding of relative stabilities of various fullerene cage isomers, and provides the general guidelines on the fullerene isomerism, or assists experimental spectroscopic studies. ^{13}C NMR spectroscopy is used to characterize EMF cage structures as common structural tool, and often gives important information about the symmetry of the carbon cage. But the data of many paramagnetic EMFs such as $M@C_{82}$ (M = Sc, Y, La, Ce, Pr, Gd, etc.), are difficult to interpret. Experimental infrared spectroscopy, Raman spectroscopy, absorption spectroscopy, electrochemical spectroscopy, and even high-resolution tunneling electronic microscopy (HRTEM) also provide valuable information to characterize EMFs.

Single crystal X-ray diffraction (XRD) is certainly the most reliable method for structural determination of EMFs. XRD studies can provide information regarding not only the cage structure, but also the positioning and motioning behaviors of the internal atoms. Comparatively, the structural determinations based on NMR spectroscopy and SRPD/Rietveld/MEM frequently appear to be incomplete and ambiguous. Thus, the molecular structures of either newly discovered or previously reported EMFs are mainly determined or updated by single crystal X-ray crystallographic method, and tremendous valuable results or conclusions have been obtained taking advantage of recent single crystal growth, especially in the last decade (Tables 1–3).[14]

In addition to the low availability of single crystals of EMFs of diffraction quality, the carbon cages of fullerenes and EMFs are highly symmetric and the spherical molecules rotate rapidly within the crystal lattice; they are prone to crystallographic disorder and impede direct determination of the carbon cage structures. To overcome this problem, two representative and practical strategies have been developed, and many successful results have been published recently. First, the EMF molecules can form co-crystallized complexes with Ni- or Co-octaethylporphyrines (OEP).[15] Therefore, the rotation of fullerene molecules is largely hindered and consequently the resulted complexes are particularly suitable for XRD measurements. On the other hand, chemical derivatization can also hinder the rotation of EMF molecules in the crystals.[16] Numerous successful examples have been

Table 1 Mono-EMFs Characterized with XRD Crystallography

Mono-EMFs/cocrystals	Space group	Ref.	Mono-EMF derivatives	Space group	Ref.
$Li^{+}@I_h$-C_{60}/$[SbCl_6]^{-}$	Orthorhombic, *Amm2*	43	La@C_2(10612)-C_{72}-$C_6H_3Cl_2$-B	Monoclinic, $P2_1/c$	102
Ba@D_{3h}(1)-C_{74}/Co(OEP)	Monoclinic, *C2*	17	La@D_{3h}-C_{74}-$C_6H_3Cl_2$-IA	Orthorhombic, *Pnma*	18
Yb@C_{2v}(3)-C_{80}/Ni(OEP)	Triclinic, *P*–1	21			
Ca@C_{3v}(134)-C_{94}/Ni(OEP)	Monoclinic, *C2/m*	42	La@D_{3h}-C_{74}-$C_6H_3Cl_2$-IIB	Orthorhombic, *Pnma*	19
Tm@C_{3v}(134)-C_{94}/Ni(OEP)	Tetragonal, $I4_1/a$	42			
La@C_{2v}(9)-C_{82}/Ni(OEP)	Monoclinic, *C2/m*	36	La@C_{2v}(3)-C_{80}-$C_6H_3Cl_2$-B	Monoclinic, $P2_1/c$	20
Gd@C_{2v}(9)-C_{82}/Ni(OEP)	Monoclinic, *C2/m*	35	La@C_{3v}(7)-C_{82}-$C_6H_3Cl_2$-B	Orthorhombic, $P2_12_12_1$	28
Yb@C_{2v}(9)-C_{82}/Ni(OEP)	Triclinic, *P*–1	37			
Sm@C_2(5)-C_{82}/Ni(OEP)	Monoclinic, *C2/m*	38	Y@C_{2v}(9)-C_{82}-$C_{10}H_{14}$-A	Monoclinic, $P2_1/n$	23
Sm@C_s(6)C_{82}/Ni(OEP)	Monoclinic, *C2/m*	38	La@C_{2v}(9)-C_{82}-$C_{10}H_{14}$-A	Monoclinic, $P2_1/n$	22
Sm@C_{3v}(7)-C_{82}/Ni(OEP)	Monoclinic, *C2/m*	38	Gd@C_{2v}(9)-C_{82}-$C_{10}H_{14}$-A	Monoclinic, $P2_1/n$	25
Sm@D_{3d}(19)-C_{84}/Ni(OEP)	Monoclinic, *C2/m*	39	Ce@C_{2v}(9)-C_{82}-$C_{10}H_{14}$-B/TBA	Triclinic, *P*–1	24
Sm@C_2(13)-C_{84}/Ni(OEP)	Monoclinic, *C2/m*	39	Sc@C_{2v}(9)-C_{82}-$C_{10}H_{14}$-A	Monoclinic, $P2_1/n$	26
Sm@C_2(40)-C_{90}/Ni(OEP)	Monoclinic, *C2/m*	40	La@C_{2v}(9)-C_{82}-CBr$(COOC_2H_5)_2$	Monoclinic, $P2_1/c$	33
Sm@C_2(42)-C_{90}/Ni(OEP)	Monoclinic, *C2/m*	40	La@C_{2v}(9)-C_{82}-$[CH(COOC_2H_5)_2]_2$	Monoclinic, $P2_1/c$	34
Sm@C_{2v}(46)-C_{90}/Ni(OEP)	Monoclinic, *I2/a*	40			
Sm@C_2(45)-C_{90}/Ni(OEP)	Monoclinic, *C2/m*	40	La@C_{2v}(9)-C_{82}-$(CHClC_6H_3Cl_2)$-D	Monoclinic, $P2_1/n$	29
Sm@C_1(42)-C_{92}/Ni(OEP)	Monoclinic, *C2/m*	41			
Sm@C_s(24)-C_{92}/Ni(OEP)	Monoclinic, *C2/c*	41	La@C_{2v}(9)-C_{82}-Cp*	Triclinic, *P*–1	30
Sm@C_{3v}(134)-C_{94}/Ni(OEP)	Monoclinic, *C2/m*	41	La@C_{2v}(9)-C_{82}-$C_6H_5CC_3H_6COOCH_3$	Monoclinic, $P2_1/c$	27
			La@C_{2v}(9)-C_{82}-$(C_6H_4)_2NO_2$	Monoclinic, $P2_1/c$	31
			Dy@C_{2v}(9)-C_{82}-CCH$(CO_2Me)_2PPh_3$	Monoclinic, $P2_1/c$	32

Table 2 Di-EMFs Characterized with XRD Crystallography

Di-EMFs/cocrystals	Space group	Ref.	Di-EMF derivatives	Space group	Ref.
$Tb_2@C_{79}N/Ni(OEP)$	Monoclinic, $C2/m$	100	$La_2@C_2(10611)$-C_{72}-$C_{10}H_{14}$-B	Triclinic, P–1	103
$Er_2@C_s(6)$-$C_{82}/Co(OEP)$	Triclinic, P–1	54	$La_2@C_2(10611)$-C_{72}-$C_{10}H_{14}$-C	Triclinic, P–1	103
$Er_2@C_{3v}(8)$-$C_{82}/Co(OEP)$	Triclinic P–1	55	$La_2@C_2(10611)$-C_{72}-$C_{10}H_{14}$-D	Triclinic, P–1	103
$Sm_2@D_2(35)$-$C_{88}/Ni(OEP)$	Monoclinic $C2/m$	56	$La_2@C_2(10611)$-C_{72}-$(C_{10}H_{14})_2$	Triclinic, P–1	104
$Sm_2@C_1(21)$-$C_{90}/Ni(OEP)$	Monoclinic $C2/m$	56	$La_2@D_{3h}(5)$-C_{78}-$C_{10}H_{14}$-A	Hexagonal, P_{63}/m	52
$Sm_2@D_3(85)$-$C_{92}/Ni(OEP)$	Monoclinic $C2/m$	56	$Ce_2@D_{3h}(5)$-C_{78}-$(Mes_2Si)_2CH_2$	Orthorhombic, $P2_12_12_1$	53
$La_2@D_5(450)$-$C_{100}/2Ni(OEP)$	Monoclinic $Pbca$	58			
$Sm_2@D_{3d}(822)$-$C_{104}/2Ni(OEP)$	Monoclinic, $P2_1/n$	57	$La_2@I_h(7)$-C_{80}-$C_{10}H_{14}$	Orthorhombic, $Pbca$	49
			$Ce_2@I_h(7)$-C_{80}-$(Mes_2Si)_2CH_2$	Orthorhombic, $P2_12_12_1$	47
			$La_2@I_h(7)$-C_{80}-$(Ar_2Si)_2CH_2$	Monoclinic, $P2_1/n$	46
			$La_2@I_h(7)$-C_{80}-^{t}Bp$(Dep)_2H$	Monoclinic, $P2_1/n$	48
			$La_2@I_h(7)$-C_{80}-$(CH_2)_2NCPh_3$	Triclinic, P–1	44
			$La_2@I_h(7)$-C_{80}-C_6H_5Cl	Monoclinic, $P2_1/c$	50
			$La_2@I_h(7)$-C_{80}-$C_{10}H_{14}$-CClPh	Monoclinic, $P2_1/c$	50
			$La_2@I_h(7)$-C_{80}-TCNEO	Orthorhombic, $Pnma$	45
			$La_2@I_h(7)$-C_{80}-$C_6H_5CC_3H_6COOCH_3$	Monoclinic, $P2_1/c$	51

Table 3 Cluster EMFs Characterized with XRD Crystallography

Pristine cluster EMFs	Space group	Ref.	Cluster EMF derivatives	Space group	Ref.
$Sc_3N@D_3(6140)\text{-}C_{68}/Ni(OEP)$	Triclinic, $P\text{–}1$	101	$Sc_3N@D_{3h}(5)\text{-}C_{78}\text{-}(CH_2)_2NCPh_3$	Triclinic, $P\text{–}1$	82
$Sc_3N@D_{3h}(5)\text{-}C_{78}/2Co(OEP)$	Monoclinic, $C2/m$	80, 81	$Sc_3N@I_h(7)\text{-}C_{80}\text{-}C_{10}H_{12}O_2$	Monoclinic, $P2_1/n$	16
$Gd_3N@C_2(22010)\text{-}C_{78}/Ni(OEP)$	Monoclinic, $C2/c$	106	$Sc_3N@I_h(7)\text{-}C_{80}\text{-}(Mes_2Si)_2CH_2\text{-}B$	Tetragonal, $I4_1$	69
$Sc_3N@I_h(7)\text{-}C_{80}/Co(OEP)$	Monoclinic, $C2/m$	59	$Sc_3N@I_h(7)\text{-}C_{80}\text{-}(CH_2)_2NCPh_3$	Triclinic, $P\text{–}1$	62
$Sc_3N@D_{5h}(6)\text{-}C_{80}/Ni(OEP)$	Triclinic, $P\text{–}1$	79	$Sc_3N@I_h(7)\text{-}C_{80}\text{-}(CF_3)_{14}\text{-}I$	Monoclinic, $P2_1/n$	64
$ErSc_2N@I_h(7)\text{-}C_{80}/Co(OEP)$	Monoclinic, $C2/m$	83			
$Tb_3N@I_h(7)\text{-}C_{80}/Ni(OEP)$	Monoclinic, $C2/c$	78	$Sc_3N@I_h(7)\text{-}C_{80}\text{-}(CF_3)_{16}\text{-}I$	Triclinic, $P\text{–}1$	64
$Tb_3N@D_{5h}(6)\text{-}C_{80}/Ni(OEP)$	Monoclinic, $C2/m$	78	$Sc_3N@I_h(7)\text{-}C_{80}\text{-}(CF_3)_{10}$	Monoclinic, $P2_1/n$	65
$Tb_3N@D_3(17)\text{-}C_{86}/Ni(OEP)$	Monoclinic, $C2/m$	78			
$Tb_3N@D_2(35)\text{-}C_{88}/Ni(OEP)$	Triclinic, $P\text{–}1$	78	$Sc_3N@I_h(7)\text{-}C_{80}\text{-}(CF_3)_{12}$	Monoclinic, $P2_1/c$	66
$Lu_3N@I_h(7)\text{-}C_{80}/Ni(OEP)$	Triclinic, $P\text{–}1$	60			
$Sc_3N@Ih(7)\text{-}C_{80}/Ni(OEP)$	Triclinic, $P\text{–}1$	60	$Sc_3N@I_h(7)\text{-}C_{80}\text{-}(CF_3)_{14}\text{-}II$	Monoclinic, $P2_1/c$	66
$Gd_3N@I_h(7)\text{-}C_{80}/Ni(OEP)$	Monoclinic, $C2/m$	77			
$CeSc_2N@I_h(7)\text{-}C_{80}/Ni(OEP)$	Monoclinic, $C2/c$	84	$Sc_3N@I_h(7)\text{-}C_{80}\text{-}(CF_3)_{14}\text{-}III$	Triclinic, $P\text{–}1$	66
$Sc_3N@I_h(7)\text{-}C_{80}/Zn_2BisPy\text{-}2$	Monoclinic, $P2(1)/n$	61	$Sc_3N@I_h(7)\text{-}C_{80}\text{-}(CF_3)_{14}\text{-}V$	Monoclinic, $P2_1/c$	66
$Sc_3N@I_h(7)\text{-}C_{80}/Zn_2BisPy\text{-}1$	Monoclinic, $C2/m$	61			
$Sc_3N@I_h(7)\text{-}C_{80}/2p\text{-}xylene$	Monoclinic, $P2_1/c$	61	$Sc_3N@I_h(7)\text{-}C_{80}\text{-}(CF_3)_{14}\text{-}IV$	Triclinic, $P\text{–}1$	66
$GdSc_2N@I_h(7)\text{-}C_{80}/Ni(OEP)$	Monoclinic, $C2/c$	85			
$Gd_2ScN@I_h(7)\text{-}C_{80}/Ni(OEP)$	Monoclinic, $I2/a$	85	$Sc_3N@D_{5h}(6)\text{-}C_{80}\text{-}(CF_3)_{18}$	Triclinic, $P\text{–}1$	66

(Continued)

Table 3 (*Continued*)

Pristine cluster EMFs	Space group	Ref.	Cluster EMF derivatives	Space group	Ref.
$TbSc_2N@I_h(7)$-C_{80}/Ni(OEP)	Monoclinic, *C*2/*m*	85	$Sc_3N@I_h(7)$-C_{80}-C_6H_4-[5,6]	Orthorhombic, *Pnma*	67
$Dy_3N@I_h(7)$-C_{80}/Ni(OEP)	Monoclinic, *C*2/*c*	75			
$Tm_3N@I_h(7)$-C_{80}/Ni(OEP)	Monoclinic, *C*2/*c*	76	$Sc_3N@I_h(7)$-C_{80}-C_6H_4-[6,6]	Orthorhombic, $P2_12_12_1$	67
$Tm_3N@D_{5h}(7)$-C_{80}/Ni(OEP)	Monoclinic, *C*2/*m*	76			
$Gd_3N@C_s(39663)$-C_{82}/Ni(OEP)	Triclinic, *P*–1	107	$Sc_3N@I_h(7)$-C_{80}-$(CH_2C_6H_5)_2$	Orthorhombic, *Pnma*	63
$Tb_3N@C_s(51365)$-C_{84}/Ni(OEP)	Triclinic, *P*–1	108			
$Gd_3N@C_s(51365)$-C_{84}/Ni(OEP)	Triclinic, *P*–1	109	$Sc_3N@I_h(7)$-C_{80}-C_6H_2-$O(OCHCH_3)_2$	Monoclinic, $P2_1/n$	68
$Tm_3N@C_s(51365)$-C_{84}/Ni(OEP)	Triclinic, *P*–1	109	$Sc_3N@I_h(7)$-C_{80}-$NC_6H_4OC_3H_7$	Monoclinic, *C*2/*c*	71
$Sc_2O@C_s(6)$-C_{82}/Ni(OEP)	Monoclinic, *C*2/*m*	97			
$Sc_4O_2@I_h(7)$-C_{80}/Ni(OEP)	Monoclinic, *C*2/*m*	95	$Sc_3N@I_h(7)$-C_{80}-PCBM	Triclinic, *P*–1	70
$Sc_4O_3@I_h(7)$-C_{80}/Ni(OEP)	Monoclinic, *C*2/*m*	96	$Y_3N@I_h(7)$-C_{80}-PCBM	Monoclinic, $P2_1/n$	70
$Gd_2C_2@D_3(85)$-C_{92}/Ni(OEP)	Monoclinic, *C*2/*m*	94			
$Sc_2C_2@C_{2v}(5)$-C_{80}/Co(OEP)	Monoclinic, *C*2	92	$Y_3N@I_h(7)$-C_{80}-$C(CO_2CH_2Ph)_2$	Orthorhombic, $P2_12_12_1$	73
$Sc_2C_2@C_{3v}(8)$-C_{82}/Co(OEP)	Monoclinic, *C*2	92			
$Sc_2C_2@D_{2d}(23)$-C_{84}/2Co(OEP)	Triclinic, *P*–1	92	$Y_3N@I_h(7)$-C_{80}-C_4H_9N	Tetragonal, $P4_12_12$	72
$Sc_3CN@I_h(7)$-C_{80}/Ni(OEP)	Monoclinic, *C*2/*m*	99			

(*Continued*)

Table 3 (*Continued*)

Pristine cluster EMFs	Space group	Ref.	Cluster EMF derivatives	Space group	Ref.
$Sc_2S@C_s(6)\text{-}C_{82}$/Ni(OEP)	Triclinic, *P*–1	98	$Lu_3N@I_h(7)\text{-}C_{80}\text{-}Dep_2Si$	Orthorhombic, *Pccn*	74
$Sc_2S@C_{3v}(8)\text{-}C_{82}$/Ni(OEP)	Triclinic, *P*–1	98			
$Sc_2S@C_s(10528)\text{-}C_{72}$/Ni(OEP)	Triclinic, *P*–1	105			
			$Sc_3C_2@I_h(7)\text{-}C_{80}\text{-}C_{10}H_{14}$	Monoclinic, $P2_1/m$	86
			$Sc_3C_2@I_h(7)\text{-}C_{80}\text{-}C_{10}H_{14}\text{-}n\text{-}Bu_4N^+$	Orthorhombic, *Pnnm*	86
			$Sc_2C_2@C_{2v}(5)\text{-}C_{80}\text{-}C_{10}H_{14}$-A	Triclinic, *P*–1	88
			$Sc_2C_2@C_{2v}(5)\text{-}C_{80}\text{-}C_{10}H_{14}$-B	Triclinic, *P*–1	88
			$Sc_2C_2@C_{2v}(5)\text{-}C_{80}\text{-}C_{10}H_{14}$-C	Monoclinic, $P2_1/n$	88
			$Sc_2C_2@C_{2v}(5)\text{-}C_{80}\text{-}C_{10}H_{14}$-E	Monoclinic, $P2_1/n$	88
			$Sc_2C_2@C_{3v}(8)\text{-}C_{82}\text{-}C_{10}H_{14}$	Monoclinic, $P2_1/n$	89
			$Sc_2C_2@C_{2v}(9)\text{-}C_{82}\text{-}(CH_2)_2NCPh_3$-A	Monoclinic, $P2_1/n$	90
			$Sc_2C_2@C_s(6)\text{-}C_{82}\text{-}(CH_2)_2NCPh_3$	Triclinic, *P*–1	91
			$(\mu\text{-H})_3Re_3(CO)_9\text{-}Sc_2C_2@C_{3v}(8)\text{-}C_{82}$	Orthorhombic, *Pbcm*	93

characterized structurally using single-crystal XRD crystallography in both ways. Notably, exohedral modification inevitably distorts the fullerene cage and thus possibly alters the metal position inside. Therefore, the X-ray structures of EMF-derivatives may differ from the true situation of pristine EMFs, although the change is very slightly in some cases, while cocrystals of EMFs with metal porphyrins can better illustrate the pristine environment with respect to both cage structures and metal positions.

2. X-RAY STRUCTURES OF EMFS

2.1. *X-ray Structures of Mono-EMFs*

Mono-EMFs with only one metal atom are the simplest prototypes and ideal candidates for elucidating metal-cage interactions, in which either two or three electrons are transferred from the internal metal to the cage. Metal positions in EMFs are of special importance because their molecular symmetry and intrinsic properties are strongly influenced by the location and motion of the encapsulated metals. Studies reveal that the metal atom is not located in the center of the fullerene cage, but it tends to coordinate strongly with the cage carbons.

Ba@D_{3h}C$_{74}$·Co^{II}(OEP) is the first example of mono-EMFs that have been elucidated with XRD.[17] Because the structure exhibits a strong disorder, theoretical calculations are conducted to help with the assignment of the metal position. The final results showed that the single metal is situated at one pole of the molecule along one of the three C_3 axes (Fig. 1a).

Recently, Akasaka *et al.* developed a method to extract EMFs from soot using 1,2,4-trichlorobenzene (TCB). For example, some insoluble EMF species (also called missing EMFs), such as La@C$_{74}$ La@C$_{80}$, and La@C$_{802}$, react *in situ* with the dichlorophenyl radicals generated by refluxing TCB, affording the corresponding dichlorophenyl derivatives La@C_{2n}-$C_6H_3Cl_2$ (2n = 74, 80, 82).[18–20] The X-ray structure of La@D_{3h}-C_{74}-$C_6H_3Cl_2$ shows that the La atom resides closely under a [6,6]-bond at one pole of the molecule, while the molecular structure of La@C_{2v}(3)-C_{80}-$C_6H_3Cl_2$ reveals that the encapsulated La atom is located near the addition position and disordered at three positions (Fig. 2b).

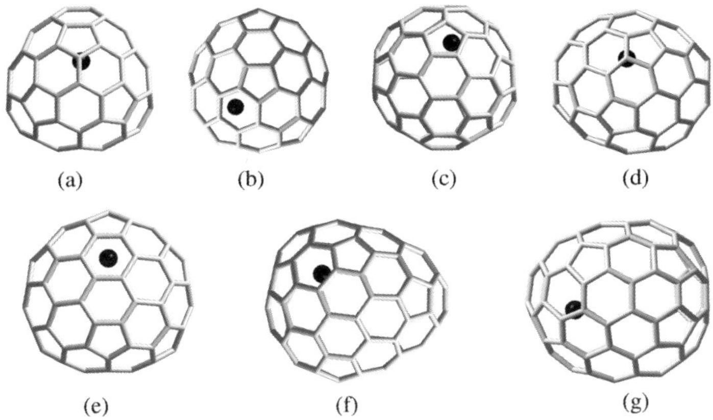

Fig. 1 X-ray structures of pristine mono-EMFs: (a) Ba@D_{3h}-C_{74}; (b) Yb@C_{2v}(3)-C_{80}; (c) Gd@C_{2v}(9)-C_{82}; (d) Sm@C_2(13)-C_{84}; (e) Sm@C_{2v}(46)-C_{90}; (f) Sm@C_1(42)-C_{92}; (g) Sm@C_{3v}(134)-C_{94}.

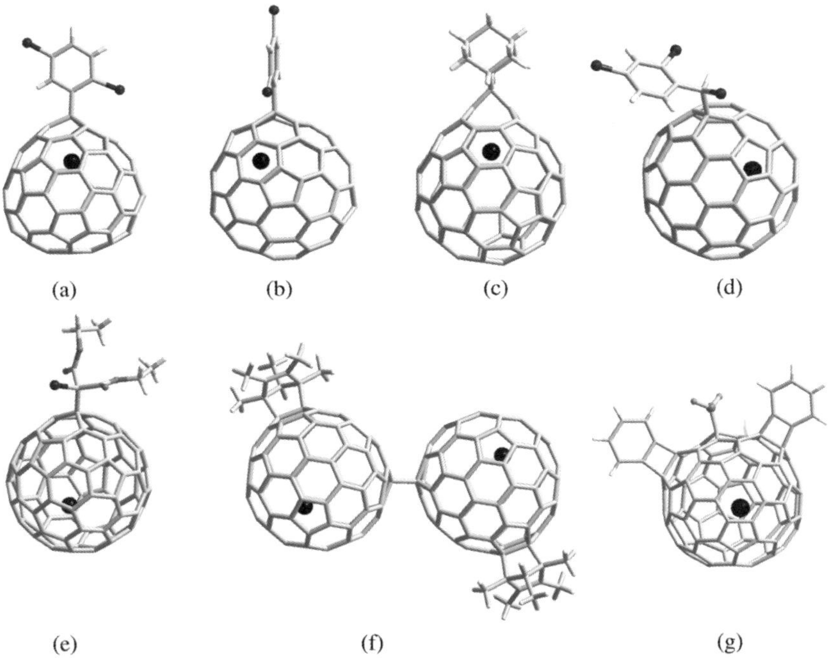

Fig. 2 X-ray structures of selected derivatives of mono-EMFs: (a) La@C_{3v}(7)-C_{82}-$C_6H_3Cl_2$; (b) La@C_{2v}(3)-C_{80}-$C_6H_3Cl_2$; (c) La@C_{2v}(9)-C_{82}-Ad; (d) La@C_{2v}(9)-C_{82}-$CHClC_6H_3Cl_2$; (e) La@C_{2v}(9)-C_{82}-$CBr(COOC_2H_5)_2$; (f) La@C_{2v}(9)-C_{82}-Cp*; (g) La@C_{2v}(9)-C_{82}-$(C_6H_4)_2NO_2$.

Interestingly, Yb@C_{2v}(3)-C_{80} is more soluble than La@C_{2v}(3)-C_{80}, and thus it can be directly extracted. The molecular structure of Yb@C_{2v}(3)-C_{80}·Ni^{II}(OEP) was confirmed by single-crystal XRD.[21] The X-ray result reveals that the Yb atom is located under a hexagonal carbon ring apart from the two-fold axis of the C_{2v}(3)-C_{80} cage, presenting the first example of mono-EMFs with an asymmetrically positioned metal (Fig. 1b). Such an anomalous location of Yb^{2+} is due to its strong ability to pursue a large coordination number and the lack of hexagon along the C_2 axis of C_{2v}(3)-C_{80}. The nearest Yb-C distances vary in a narrow range from 2.476(14) to 2.644(12) Å.

Among mono-EMFs, M@C_{82} is the most abundantly yielded species. Crystallographic results of pristine M@C_{2v}(9)-C_{82} (M = Sc, Y, La, Ce, Gd, etc.) lagged behind the X-ray characterizations of their derivatives, all of which show that the single metal atom always tends to stay under a hexagonal ring along the C_2 axis. In 2004, the first definite structure of La@C_{2v}(9)-C_{82} was confirmed by XRD method performed on its carbene derivative (La@C_{2v}(9)-C_{82}-Ad, Ad = adamantylidene) (Fig. 2c).[22] The C-C bond on the addition site is cleaved by adding the electrophilic carbene to form the [6,6]-open structure. The encapsulated La atom is localized in an off-center position along the C_2 axis of C_{2v}-C_{82} cage and coordinates with two hexagonal rings. Subsequently, the Ad-adducts of Y@C_{2v}(9)-C_{82},[23] Ce@C_{2v}(9)-C_{82},[24] Gd@C_{2v}(9)-C_{82},[25] and Sc@C_{2v}(9)-C_{82},[26] are also obtained by the same reaction, and their structures are confirmed by XRD. Although the encapsulated metals are different, the chemical reactivity and structure of the Ad-adducts are similar to that of La@C_{2v}(9)-C_{82}-Ad. The metal always tends to locate under the hexagonal ring along the C_2 axis. In 2011, the other carbene derivative of La@C_{2v}(9)-C_{82}, La@C_{2v}(9)-C_{82}-$C_6H_5CC_3H_6COOCH_3$,[27] was obtained, and its structure is fully determined by X-ray crystallographic analysis.

By radical reaction, two derivatives of La@C_{82}, La@C_{3v}(7)-C_{82}-$C_6H_3Cl_2$ and La@C_{2v}(9)-C_{82}-$CHClC_6H_3Cl_2$ have been obtained.[28,29] The X-ray results of La@C_{3v}(7)-C_{82}-$C_6H_3Cl_2$ reveal that the dichlorophenyl group is singly bonded to a triple-hexagon junction carbon atom on the C_3 axis (Fig. 2a). The La atom is located closely under C(1), and the La-C(1) distance of 2.72 Å indicates a strong metal-cage interaction. In the molecular structure of La@C_{2v}(9)-C_{82}-$CHClC_6H_3Cl_2$, the trichlorobenzyl group is linked to the fullerene cage by forming one carbon-carbon single bond (Fig. 2d). The La position in the isomer is off the C_2 axis and moves slightly against the hemisphere that undergoes a coupling reaction.

In 2010, the Diels–Alder cycloaddition reaction of La@C_{2v}(9)-C_{82} with 1,2,3,4,5-pentamethylcyclopentadiene (Cp*), afforded a stable monoadduct La@C_{2v}(9)-C_{82}-Cp*, with a four-membered ring attached to the cage surface as revealed by XRD.[30] X-ray crystallographic analysis reveals two orientations in the crystal structure, and 60% of the monoadduct forms a dimer in the solid state (Fig. 2f). The length of the C–C bond connecting two C_{82} cages in the La@C_{2v}(9)-C_{82}-Cp* dimer (1.606 Å) is only about 0.1 Å longer than the typical C-C single bond length.

Other cycloaddition reactions of mono-EMFs have been also reported. For example, the [2+2] cycloaddition and oxidation reaction of La@C_{2v}(9)-C_{82}, occurred to produce an unprecedented closed-cage EMF, La@C_{2v}(9)-C_{82}-$(C_6H_4)_2NO_2$.[31] X-ray crystallographic analysis reveals that the benzyne groups selectively add to the [5,6]-bonds, and the NO_2 group originated from the trace of NO/NO_2 radicals in isoamyl nitrite is singly bonded to a [5,6,6]-ring junction (Fig. 2g). In 2007, single-crystal X-ray structure was reported for the [6,6]-open adduct of Dy@C_{2v}(9)-C_{82} bearing a triphenylphosphonium substituent.[32]

The Bingel–Hirsch reaction is a widely applied reaction in fullerene chemistry. In 2005, the reaction of La@C_{82} with bromomalonate in the presence of 1,8-diazabicyclo[5.4.0]undec-7-ene (DBU) yielded a singly bonded monoadduct, La@C_{2v}(9)-C_{82}-CBr$(COOC_2H_5)_2$.[33] The X-ray results reveal that the La atom is located at a fixed position far from the bromomalonate group (Fig. 2e). When malonate is subjected to the reaction instead of bromomalonate, the bisadduct of La@C_{2v}-C_{82}, La@C_{2v}-C_{82}-[CH(COOC2H5)$_2$]$_2$, has been synthesized, and characterized by XRD.[34] The bis-adduct retains an open-shell electronic structure and crystallizes to form a dimer. The La positions are off the C_2 axis and move slightly toward the hemisphere that undergoes bis-addition.

Gd@C_{2v}(9)-C_{82} and La@C_{2v}(9)-C_{82} are the first two non-derivatives of M@C_{82} whose structures were determined by XRD using the cocrystallization method.[35,36] The single Gd^{3+} (or La^{3+}) cation tends to reside under a hexagonal ring along the C_2 axis (Fig. 1c). Very recently, single crystals of three Yb@C_{82} isomers, namely, Yb@C_2(5)-C_{82}, Yb@C_s(6)-C_{82}, and Yb@

C_{2v}(9)-C_{82}, cocrystallized with Ni^{II}(OEP).[37] The Yb^{2+} cation prefers to reside close to a hexagonal ring in Yb@C_2(5)-C_{82} and Yb@C_{2v}(9)-C_{82}, but a [5,6,6]-junction carbon atom in Yb@C_s(6)-C_{82}. These locations correspond to electrostatic potential minima and give the most stable configurations of these Yb@C_{82} isomers. Thus, electrostatic metal-cage interactions play an important role in both stabilizing the endohedral structure and determining the position of the metal ion inside the carbon cage.

The list of structurally characterized mono-EMFs has been recently extended by three isomers of Sm@C_{82} (C_2(5), C_s(6) and C_{3v}(7)),[38] three isomers of Sm@C_{84} (C_2(11), C_2(13) and D_{3d}(19)),[39] four isomers of Sm@C_{90} (C_2(40), C_2(42), C_2(45), and C_{2v}(46)),[40] two isomers of Sm@C_{92} (C_s(24) and C_1(42))[41] and Sm@C_{3v}(134)-C_{94}[41] cocrystallized with Ni^{II}(OEP) (Figs. 1d–1g). The X-ray results show that it is possible for the single metal atom to rotate inside the fullerene cage and adopt a range of locations upon crystallization. For example, the X-ray crystallographic results of the four isomers of Sm@C_{90} (C_2(40), C_2(42), C_2(45), and C_{2v}(46)) show that the Sm^{2+} atom moves rapidly along a cage belt comprised of continuous hexagonal carbon rings. However, it is not possible to conclude that the dynamic motion of the internal metal is only caused by the large cages, which provide more space for the encapsulated metal ion. Cage shape and symmetry could be another critical factor. For example, the X-ray structure of Ca@C_{3v}(134)-C_{94} reveals that Ca^{2+} is located closely to a [6,6]-junction, which is near the C_3-axis.[42] In sharp contrast, X-ray results of Tm@C_{3v}(134)-C_{94} finds that the Tm^{2+} is located under a hexagonal ring close to the C_3-axis.[42] It is more likely that the coordinating ability of the internal metal and the cage structures are both critical factors for the motion or location of the encapsulated metal.

It is noteworthy that a novel mono-EMF Li@C_{60} was obtained by single-crystal XRD study of its salt [Li@C_{60}^{+}]($SbCl_6^{-}$) in 2010.[43]

2.2. X-ray Structures of Di–EMFs

Fullerenes with two metal atoms inside, di-EMFs, are found for most rare earth metals, but the isolated examples are less than the corresponding mono-EMFs. The most abundantly obtained di-EMFs are M_2@I_h(7)-C_{80} (M = La, Ce), and their derivatives have been characterized by single-crystal XRD.

Two regioisomers of pyrrolidinodi-EMF derivatives of $M_2@I_h(7)\text{-}C_{80}$ (M = La, Ce) were obtained by the thermal 1,3-dipolar cycloaddition between $M_2@I_h(7)\text{-}C_{80}$ and 3-triphenylmethyl-5-oxazolidinone in toluene, and the single crystal structure of the [6,6]-adduct, $M_2@I_h(7)\text{-}C_{80}\text{-}(CH_2)_2NTrt$ (Trt = triphenylmethyl), has been fully determined by X-ray crystallographic analysis.[44] The two metal atoms are localized at the slanted position, which are determined using paramagnetic NMR shift analyses. The metal distance in $La_2@I_h(7)\text{-}C_{80}\text{-}(CH_2)_2NTrt$ (La–La = 3.823 Å) is shorter than that in [Ce2@$I_h(7)\text{-}C_{80}\text{-}(CH_2)_2NTrt$ (Ce–Ce = 3.900 Å). In 2011, Akasaka *et al.* reported the 1,3-dipolar cycloaddition of $La_2@I_h(7)\text{-}C_{80}$ with tetracyanoethylene oxide (TCNEO), which enables the formation of the corresponding adduct having a tetracyanotetrahydrofuran moiety.[45] X-ray crystallographic analysis reveals that the cycloaddition takes place as a [5,6]-addition, and the La atoms have a large-amplitude motion inside the fullerene cage, with the La–La distances of 3.811–3.816 Å (Fig. 4f).

In 2007, the silylation reaction of $La_2@I_h(7)\text{-}C_{80}$ with 1,1,2,2-tetrakisaryl-1,2-disilirane afforded the adduct as a bis-silylated product, $La_2@I_h(7)\text{-}C_{80}\text{-}(Dep_2Si)_2CH_2$ (Dep = 2,6-diethylphenyl)).[46] The two La atoms hop two-dimensionally along the equator of the C_{80} cage with the La–La distance of 3.792 Å. The same reaction of $Ce_2@I_h(7)\text{-}C_{80}$ has been also reported.[47] X-ray crystallographic analysis of $Ce_2@I_h(7)\text{-}C_{80}\text{-}(Mes_2Si)_2CH_2$ shows that the two Ce atoms positions closely resemble those in $La_2@I_h(7)\text{-}C_{80}\text{-}(Mes_2Si)_2CH_2$, with the Ce–Ce distance of 3.829 Å, which are localized at two positions directing the hexagonal ring at the equator (Fig. 4c). In 2010, two diastereomers of the carbosilylated $La_2@I_h(7)\text{-}C_{80}$ by thermal carbosilylation of $La_2@I_h(7)\text{-}C_{80}$ with silacyclopropane have been isolated and characterized.[48] The molecular structure of one diastereomer of the carbosilylated derivatives, $La_2@I_h(7)\text{-}C_{80}\text{-}{}^t Bp(Dep)_2H$ (^{t}Bp = butylphenyl and Dep = diethylphenyl), was determined unambiguously using X-ray crystallographic analysis. The addition of silirane takes place at the [5,6]-bond, yielding formation of silacyclopentane rings on the C_{80} cage (Fig. 4b). The La atoms in the carbosilylated derivatives show dynamic motion, with the La–La distances of 3.795–3.833 Å.

In 2008, Akasaka *et al.* reported that the carbene cycloaddition of the adamantylidene carbene generated by photoirradiation to 2-adamantane-2,3'-[3H]-diazirine occurs at the [6,6]-junction of the $La_2@I_h(7)\text{-}C_{80}$

to afford the formation of $La_2@I_h(7)$-C_{80}-Ad by opening the [6,6]-junction.[49] Crystallographic data reveals that the two La atoms are collinear with the spiro carbon of the Ad moiety. Furthermore, the La–La distance is highly elongated to 4.031 Å due to the bond cleavage reducing the electrostatic repulsion between the positively charged La atoms.

In 2011, the photoreaction of $La_2@I_h(7)$-C_{80} with phenylchlorodiazirine ($PhClCN_2$) afforded a [6,6]-open-cage carbene monoadduct, $La_2@I_h(7)$-C_{80}(CClPh).[50] The two La atoms inside the cage are located at collinear positions relative to the spiro carbon, with the La–La distances of 4.013 Å, as in the case of $La_2@I_h(7)$-C_{80}-Ad. Then, a toluene solution of $La_2@I_h(7)$-C_{80}(CClPh) and adamantane diazirine (AdN_2) was photoirradiated in the same way as the reaction with $PhClCN_2$, This result indicates that all of the reaction products are the bisadduct of $La_2@C_{80}$ connected with phenylchlorocarbene and Ad addends.[50] Single-crystal X-ray crystallographic analysis of $La_2@I_h(7)$-C_{80}-(CClPh)Ad shows that the C–C bond between two six-membered rings was broken by carbene addition to afford the open-cage structure (Fig. 4e). Two La atoms are located underneath each addition site and are collinear with a spiro carbon of carbine addendum. Because of two bond cleavages on the cage, the La–La distance (4.16 Å) is much more elongated. In the same year, the other carbene cycloaddition derivative of $La_2@I_h(7)$-C_{80}, $La_2@I_h(7)$-C_{80}-$C_6H_5CC_3H_6COOCH_3$, was also determined by XRD.[51]

In addition to $M_2@C_{80}$, the single-crystal structure of the derivatives $M_2@C_{78}$ (La and Ce) has been also determined by XRD. The addition reaction of $La_2@D_{3h}(5)$-C_{78} with adamantylidene carbene affords four isomers of the monoadduct $La_2@D_{3h}(5)$-C_{78}-Ad.[52] Single-crystal X-ray crystallographic analysis of one monoadduct reveals that $La_2@D_{3h}(5)$-C_{78}-Ad has an [5,6] open structure with two La atoms on the C_3 axis of the cage with the La–La distance of 4.081 Å (Fig. 4a). In the same year, the bis-silylation reaction of $Ce_2@D_{3h}(5)$-C_{78} proceeded to form a bis-silylated adduct, $Ce_2@D_{3h}(5)$-C_{78}-$(Mes_2Si)_2CH_2$, resulting from the 1,4-addition of the disilirane to the cage.[53] The Ce atoms are located at two positions directed toward the hexagonal ring at the C_3 axis of the ellipsoidal cage with the Ce–Ce distance of 4.036 Å.

For $M_2@C_{82}$, two isomers of $Er_2@C_s(6)$-C_{82} and $Er_2@C_{3v}(8)$-C_{82} were cocrystallized with Co^{II}(OEP), respectively.[54,55] The encapsulated two Er^{3+} ions are presented to be in circular motion with X-ray crystallographic characterizations. In $Er_2@C_s(6)\text{-}C_{82}$, Er^{3+} ions possess multiple positions with occupancies ranging from 0.35 to 0.01. These positions are along a band of 10 contiguous hexagons of the $C_s(6)\text{-}C_{82}$ cages. The major Er^{3+} is close to a [6,6]-junction and a [5,6]-junction with Er–C distance ranging from 2.30 to 2.39 Å. Another isomer, $Er_2@C_{3v}(8)\text{-}C_{82}$ also has similar structural features in regard of the metal motion. It appears that the encapsulated Er^{3+} ions have almost equivalent associations with either [6,6]-junctions or [5,6]-junctions (Fig. 3a).

Recently, some larger di-EMFs such as $Sm_2@D_2(35)\text{-}C_{88}$, $Sm_2@C_1(21)\text{-}C_{90}$, and $Sm_2@D_3(85)\text{-}C_{92}$ were isolated and characterized by single-crystal XRD (Figs. 3b–3d).[56] In these di-EMFs, the Sm–Sm distances are 4.216, 4.146 and 4.619 Å, and the range of Sm–C distances is 2.51–2.66, 2.33–2.72 and 2.50–2.77 Å, respectively. $Sm_2@D_{3d}(822)\text{-}C_{104}$ is the largest isolated and single-crystal structurally characterized di-EMF to date, in which the two primary samarium atoms are located near the threefold axis of the cage and are separated by 5.832 Å (Fig. 3f).[57] Each Sm^{3+} cation is likely under a triple hexagon junction. Recently the second largest di-EMF, $La_2@D_5(450)\text{-}C_{100}$, was characterized by single-crystal XRD.[58] Within the cage, the lanthanum ions are displaced slightly from the fivefold axis of the fullerene cage with the La–La distance of 5.744 Å, and each La^{3+} ion is likely under a [5,6]-junction (Fig. 3e).

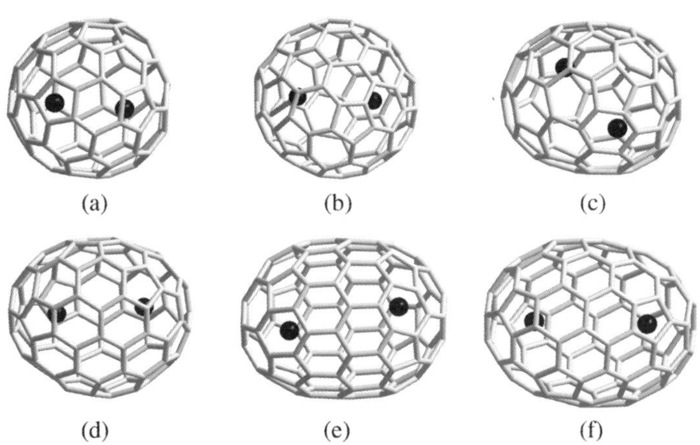

Fig. 3 X-ray structures of pristine di-EMFs: (a) $Er_2@C_{3v}(8)\text{-}C_{82}$; (b) $Sm_2@D_2(35)\text{-}C_{88}$; (c) $Sm_2@C_1(21)\text{-}C_{90}$; (d) $Sm_2@D_3(85)\text{-}C_{92}$; (e) $La_2@D_5(450)\text{-}C_{100}$; (f) $Sm_2@D_{3d}(822)\text{-}C_{104}$.

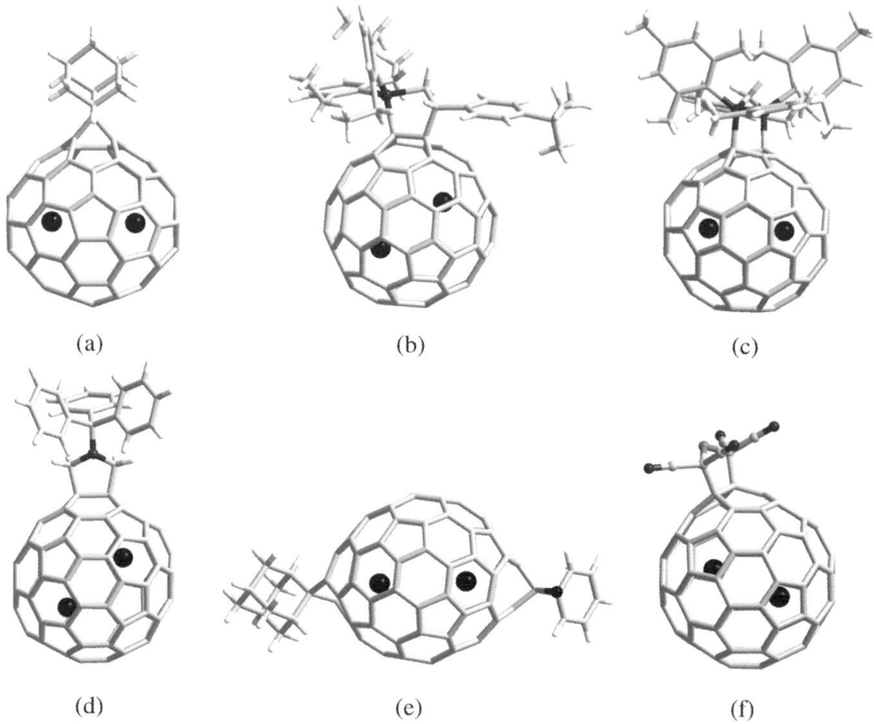

Fig. 4 X-ray structures of selected derivatives of di-EMFs: (a) $La_2@D_{3h}(5)$-C_{78}-Ad; (b) $La_2@I_h(7)$-C_{80}-^{t}Bp(Dep)$_2$H; (c) $Ce_2@I_h(7)C_{80}$-$(Mes_2Si)_2CH_2$; (d) $La_2@I_h(7)$-C_{80}-$(CH_2)_2$NTrt; (e) $La_2@I_h(7)$-C_{80}-(CClPh)Ad; (f) $La_2@I_h(7)$-C_{80}-TCNEO.

2.3. *X-ray Structures of Cluster EMFs*

Metal clusters, such as a cluster of metal carbide ($M_2C_2/M_3C_2/M_4C_2$), metal nitride (M_3N), metal oxide ($M_2O/M_4O_2/M_4O_3$), metal cyanide (M_3NC), and metal sulfide (M_2S), have been successfully encapsulated in fullerenes, and most of them have been structurally characterized with single-crystal XRD.

2.3.1. *X-ray Structures of Metal Nitride Cluster EMFs*

The molecular structure of metal nitride cluster EMFs, $Sc_3N@C_{80}$, was first determined by single-crystal XRD in 1999.[59] In $Sc_3N@I_h(7)$-C_{80}, three Sc atoms are found to form an equilateral triangle with a nitrogen atom in the center of the planar cluster Sc_3N, with the Sc–N distances of 1.966–2.011Å, and face three pentagons within the C_{80} cage. Subsequently, single-crystal XRD study was also performed for the o-xylene solvate of $Sc_3N@I_h(7)$-C_{80}.[60] For the major sites, the Sc–C distances range from 2.145–2.205 Å, and the Sc–N distances span the narrow range 1.981–2.0526 Å. In 2001, the X-ray structures of $Sc_3N@I_h(7)$-C_{80} encapsulated by cyclic Zn-bisporphyrins was isolated.[61]

Because of its high availability for the further studies of chemical reactivity, various types of chemical transformations such as Diels–Alder reactions, Prato reactions, Bingel reactions, [2+2] cycloadditions with benzynes, [2+1] cycloadditions with carbenes, disilylations, and free-radical reactions have been utilized to modify $Sc_3N@I_h(7)$-C_{80}. Thus numerous single-crystal XRD studies of $Sc_3N@I_h(7)$-C_{80} derivatives have been reported.

The first single crystal structure of $Sc_3N@I_h(7)$-C_{80} derivative, $Sc_3N@I_h(7)$-C_{80}-$C_{10}H_{12}O_2$, was obtained by the reaction between $Sc_3N@I_h(7)$-C_{80} and 6,7-dimethoxyisochroman-3-one in 1,2,4-trichlorobenzene solution.[16] The reaction occurred *via* a [4 + 2] Diels–Alder cycloaddition to the [5,6]-ring junction (Fig. 6b). The Sc_3N unit is positioned well away from the site of external modification.

In 2006, Dorn *et al.* reported that the 1,3-dipolar cycloaddition reaction of the $Sc_3N@I_h(7)$-C_{80} with N-triphenylmethyl-5-oxazolidinone yields [5,6]- and [6,6]-monoadducts, $Sc_3N@I_h(7)$-C_{80}-$(CH_2)_2NCPh_3$ characterized by X-ray crystallography.[62] The kinetic product [6,6]-adduct can be converted to the thermodynamic product 5,6-adduct at elevated temperatures (Fig. 6a).

In 2008, the photochemically radical addition reaction of $Sc_3N@I_h(7)$-C_{80} with benzyl radicals produced a dibenzyl adduct $Sc_3N@I_h(7)$-C_{80}-$(CH_2C_6H_5)_2$ in high yield.[63] The X-ray crystallographic analysis of $Sc_3N@C_{80}(CH_2C_6H_5)_2$ confirmed unambiguously its 1,4-addition pattern on the hexagon ring (Fig. 6f).

By trifluoromethylation reactions, a series of multi-CF_3 derivatives of $Sc_3N@I_h(7)$-C_{80}, $Sc_3N@I_h(7)$-$C_{80}(CF_3)_n$ (n = 10, 12, 14, 16),[64–66] were successfully synthesized and isolated. In 2009, Strauss, Boltalina *et al.* reported that the trifluoromethylation reactions of the $Sc_3N@I_h(7)$-C_{80} with $Ag(CF_3CO_2)$ yields $Sc_3N@I_h(7)$-$C_{80}(CF_3)_n$ (n = 14, 16), and their molecule structures were determined by XRD (Fig. 6g).[64] The X-ray crystallographic analysis of these EMFs reveals their unprecedented addition patterns. In 2011, the same group reported the structure of $Sc_3N@I_h(7)\text{-}C_{80}(CF_3)_{10}$ and $Sc_3N@I_h(7)\text{-}C_{80}(CF_3)_{12}$, which show a significant distortion of the Sc_3N cluster shape from the lateral triangle.[65] In the same year, Yang and Troyanov *et al.* reported the synthesis and isolation of $Sc_3N@I_h(7)\text{-}C_{80}(CF_3)_{16}$ and characterized their structures by XRD.[66]

In 2011, the first [2+2] cycloaddition reaction of $Sc_3N@I_h(7)\text{-}C_{80}$ with benzyne successfully formed both the [5,6]- and [6,6]-monoadducts, $Sc_3N@I_h(7)\text{-}C_{80}\text{-}C_6H_4$, with a four-membered ring attached to the cage surface on [5,6]- and [6,6]-ring fusions as revealed by XRD, respectively.[67] The two adducts contain planar cyclobutenyl units, in which the cage C–C bond is elongated to 1.65–1.66 Å (Fig. 6c). In the same year, under an aerobic atmosphere, the [2+2] cycloaddition and oxidation reaction of $Sc_3N@I_h(7)\text{-}C_{80}$ with 4,5-diisopropoxybenzyne occurred to produce an unprecedented open-cage EMF, $Sc_3N@I_h(7)\text{-}C_{80}\text{-}C_6H_2\text{-}O(OCHCH_3)_2$.[68] Additions of both a benzyne unit and an oxygen atom have occurred at [5,6]-ring junctions in a single pentagon. One scandium atom is located near the orifice that has formed, lies close to four of the carbon atoms.

In 2006, the photochemical silylation reaction of $Sc_3N@I_h(7)\text{-}C_{80}$ with 1,1,2,2-tetramesityl-1,2-disilirane afforded the adduct as a bis-silylated product, $Sc_3N@I_h(7)\text{-}C_{80}\text{-}(Mes_2Si)_2CH_2$ with both 1,2- and 1,4-cycloadducts isolated and characterized by XRD.[69] The 1,2-addition was shown to proceed to the [5,6]-bond, but this adduct was thermodynamically less stable and isomerized to the 1,4- adduct upon heating (Fig. 6d).

In 2009, the carbene cycloaddition reaction (diazoalkane addition) of $Sc_3N@I_h(7)\text{-}C_{80}$ with methyl 4-benzoylbutyrate *p*-tosylhydrazone, $Sc_3N@I_h(7)\text{-}C_{80}$-PCBM, was synthesized and characterized by XRD.[70]

Very recently, the azide addition reaction of $Sc_3N@I_h(7)\text{-}C_{80}$ with 4-isopropoxyphenyl azide in the dark, afforded formation of two product, [5,6]-open and [6,6]-open azafulleroids. The structure of the [5,6]-open adduct was elucidated by XRD.[71] The thermal reaction of $Sc_3N@I_h(7)\text{-}C_{80}$ provides both isomeric [5,6]- and [6,6]-open adduct, while photoirradiation generates only the [6,6]-open adduct.

For single-crystal X-ray structure of $Y_3N@I_h(7)\text{-}C_{80}$, only their derivatives, $Y_3N@I_h(7)\text{-}C_{80}C_4H_9N$, $Y_3N@I_h(7)\text{-}C_{80}$-PCBM and $Y_3N@I_h(7)\text{-}C_{80}\text{-}C(CO_2CH_2Ph)_2$, have been studied so far. In 2006, the single-crystal structure of [5,6]-pyrrolidino-$Y_3N@C_{80}$ was reported by Echegoyen *et al.*[72] The crystallographic study of $Y_3N@I_h(7)$-$C_{80}C_4H_9N$ shows that the nitrogen atom was slightly displaced (by 0.13 Å) out of the plane of the three Y atoms. The synthesis of the 5,6-N-ethyl pyrrolidine monoadduct of $Y_3N@C_{80}$ involved the thermal rearrangement of the [6,6]-isomer. The Y–N distances of 2.022–2.067 Å are generally longer than the Sc–N distances of 2.020–2.032 in the functionalized adduct, $Sc_3N@I_h(7)$-C_{80}-$C_{10}H_{12}O_2$.

One year later, Echegoyen *et al.* reported the synthesis and single-crystal structure of malonate monomethanofullerene, $Y_3N@I_h(7)C_{80}$-$C(CO_2CH_2Ph)_2$.[73] The crystallographic study of the Bingel–Hirsch adduct showed that the [6,6]-bond at the site of addition opens up and one of the yttrium atoms resides close to the addition site (Fig. 6e). In 2009, the diazoalkane addition of $Yc_3N@I_h(7)$-C_{80} with methyl 4-benzoylbutyrate p-tosylhydrazone, $Yc_3N@I_h(7)$-C_{80}-PCBM, was synthesized and characterized by XRD.[70]

It is noted that $Sc_3N@C_{80}$ is obviously less reactive than $Y_3N@C_{80}$ in both Prato and Bingel–Hirsch reactions due to the smaller cluster size of Sc_3N.

For larger lanthanides metal nitride cluster EMFs, $M_3N@I_h(7)$-C_{80} (M = Gd, Tb, Dy, Tm, and Lu), single-crystal XRD has been reported.

In 2002, the o-xylene solvate of $Lu_3N@I_h(7)$-C_{80}·Ni^{II}(OEP) was characterized by XRD.[60] The Lu_3N cluster is nearly planar corresponding Sc_3N cluster, and the Lu–N (1.980–2.081 Å) and Lu–C (2.112–2.220 Å) distances are similar to the Sc–N (1.981–2.052 Å) and Sc–C (2.145–2.205 Å) distances, respectively. More recently, the silylation reaction of $Lu_3N@I_h(7)$-C_{80} with a thermally generated silylenen in the dark afforded a sole product, [6,6]-silacyclopropane.[74] Then it is transformed slowly into isomeric product [5,6]-silafulleroid under ambient light, and the structure of the latter has been determined by XRD. The X-ray crystallographic analysis of $Lu_3N@I_h$-C_{80}-($SiDep_2$) shows that the Lu_3N cluster is planar, with the Lu–N bond lengths 2.03–2.09 Å.

Subsequently, the structure of $M_3N@I_h(7)$-C_{80}·Ni^{II}(OEP) (M = Gd, Tb, Dy and Tm) has been determined by X-ray crystallography. In $Dy_3N@I_h(7)$-C_{80}·Ni^{II}(OEP),[75] the Dy_3N cluster can be regarded as nearly planar with Dy–N distances that span the range of 2.00–2.07 Å (Fig. 5a). Three Dy atoms are likely under a respective hexagon, with the Dy–C distances that lie in the range of 2.37–2.46 Å. In $Tm_3N@I_h(7)$-C_{80}·Ni^{II}(OEP),[76] the Tm_3N cluster has a planar structure with Tm–N distances that range from

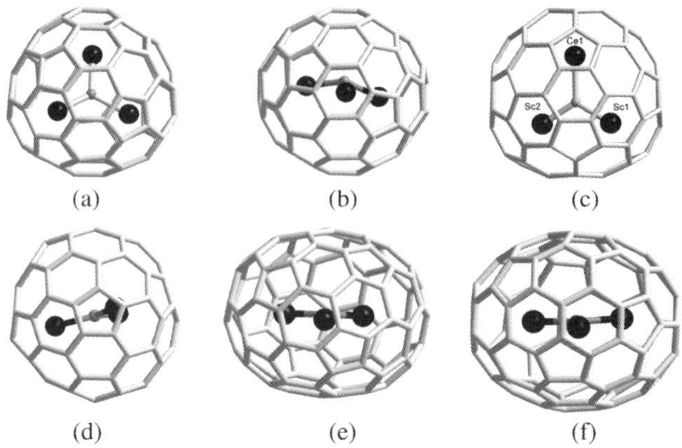

Fig. 5 X-ray structures of pristine metal nitride cluster EMFs: (a) $Dy_3N@I_h(7)\text{-}C_{80}$; (b) $Gd_3N@I_h(7)\text{-}C_{80}$; (c) $CeSc_2N@I_h(7)\text{-}C_{80}$; (d) $Sc_3N@D_{3h}(5)\text{-}C_{78}$; (e) $Tb_3N@D_3(17)\text{-}C_{86}$; (f) $Tb_3N@D_3(17)\text{-}C_{88}$.

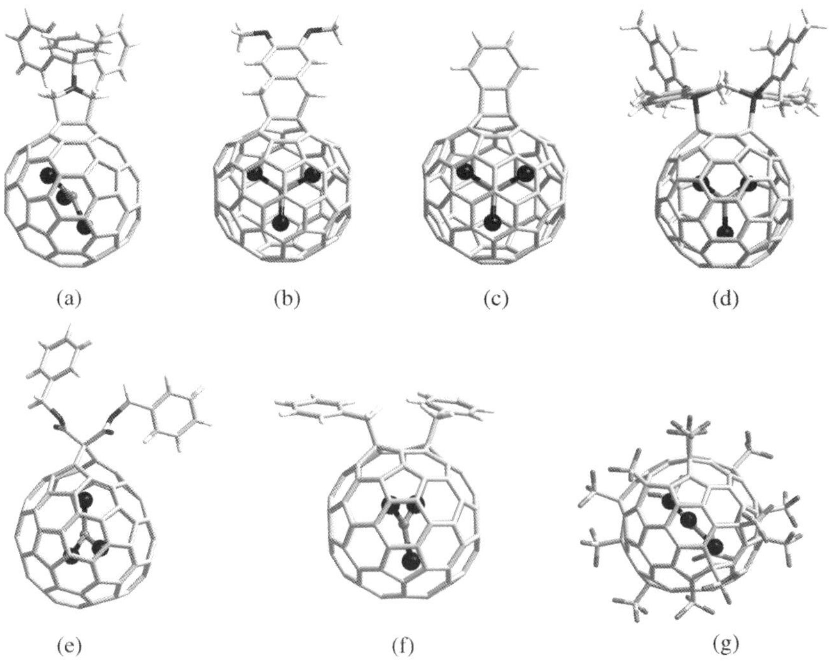

Fig. 6 X-ray structures of selected derivatives of metal nitride cluster EMFs: (a) $Sc_3N@I_h(7)\text{-}C_{80}\text{-}N(CH_2)_2Trt$; (b) $Sc_3N@I_h(7)\text{-}C_{80}\text{-}C_{10}H_{12}O_2$; (c) $Sc_3N@I_h(7)\text{-}C_{80}\text{-}C_6H_4$; (d) $Sc_3N@I_h(7)\text{-}C_{80}\text{-}(Mes_2Si)_2CH_2$; (e) $Y_3N@I_h(7)C_{80}\text{-}C(CO_2CH_2Ph)_2$; (f) $Sc_3N@I_h(7)\text{-}C_{80}\text{-}(CH_2C_6H_5)_2$; (g) $Sc_3N@I_h(7)\text{-}C_{80}(CF_3)_{16}$.

2.020–2.058 Å. Three Tm atoms are likely under a respective hexagon, with the Tm–C distances that range from 2.349–2.539 Å in the major orientation. The X-ray crystal structure of $Gd_3N@I_h(7)$-C_{80}·Ni^{II}(OEP) shows that the Gd_3N cluster is pyramidal, with Gd–N distances that range from 2.038–2.117 (Fig. 5b).[77] All Gd atoms are under six-member rings with Gd–C distances that range from 2.344–2.600 Å. In $Tb_3N@I_h(7)$-C_{80}·Ni^{II}(OEP), the Tb_3N unit also is pyramidalized, with Tb–N distances of 2.038–2.089 Å.[78]

The structure of the less abundant isomer of $Sc_3N@C_{80}$ with D_{5h} was first determined through XRD on $Sc_3N@D_{5h}(6)$-C_{80}·Ni^{II}(OEP) in 2006.[79] The Sc_3N portion of $Sc_3N@D_{5h}(6)$-C_{80} is strictly planar, but the plane of these four atoms is tipped out of the noncrystallographic, horizontal mirror plane of the fullerene by 30°. One scandium ion is under a [6,6]-junction, and the other two are under [5,6]-junctions. More recently, the first multi-CF_3 derivative of $Sc_3N@D_{5h}(6)$-C_{80} with 18 CF_3 groups was characterized by XRD. The $Tm_3N@D_{5h}(6)$-C_{80}·Ni^{II}(OEP) and $Tb_3N@D_{5h}(6)$-C_{80}·Ni^{II}(OEP) have been determined by XRD.[66] In $Tm_3N@D_{5h}(6)$-C_{80}·Ni^{II}(OEP),[76] the Tm_3N cluster has a planar structure with Tm–N distances that range from 2.025–2.062 Å. Three Tm atoms are likely under a respective [6,6]-junction, with Tm–C distances that range from 2.198–2.345 Å. The structure of $Tb_3N@D_{5h}(6)$-C_{80}·Ni^{II}(OEP) is similar to that of the D_{5h} isomers $Sc_3N@C_{80}$ and $Tm_3N@C_{80}$.[78] The Tb_3N cluster is pyramidal, and the Tb–N and Tb–C distances are similar to those in $Tb_3N@I_h(7)$-C_{80}·Ni^{II}(OEP).

The geometries of M_3N units within a C_{80} cage depend upon the size of the metal ions. Smaller ions, such as Sc, Lu, Tm, Y and Dy, form metal nitride cluster with planar structures, while larger metal ions, such as Gd and Tb, exhibit pyramidal structures.

In addition to $Sc_3N@C_{80}$, the $Sc_3N@D_{3h}(5)$-C_{78} cocrystallized with Co^{II}(OEP), was isolated and characterized by XRD in 2001.[80,81] In $Sc_3N@D_{3h}(5)$-C_{78}·Co^{II}(OEP), each scandium atom lies over a [6,6]-ring junction within a pyracylene patch, with the shortest Sc–C distances falling in a narrow range of 2.024–2.107 Å (Fig. 5d). In 2007, the N-tritylpyrrolidino monoadduct of $Sc_3N@D_{3h}(5)$-C_{78}, $Sc_3N@D_{3h}(5)$-C_{78}-$(CH_2)_2NCPh_3$ was successfully synthesized *via* 1,3-dipolar cycloaddition of tritylazomethine ylides, and characterized by XRD.[82] The Sc_3N unit remains in the horizontal plane of the D_{3h} cage and does not interact directly with the site of addition.

Two largest metal nitride cluster EMFs, $Tb_3N@D_3(17)$-C_{86}·Ni^{II}(OEP) and $Tb_3N@D_2(35)$-C_{88}·Ni^{II}(OEP),[78] have been crystallographically characterized, in which the Tb_3N clusters are planar (Fig. 5e and 5f). The Tb–N (2.148–2.159 Å) distances in $Tb_3N@C_{86}$ are shorter than the Tb–N (2.171–2.209 Å) distances found in $Tb_3N@C_{88}$. But all the Tb–N distances are longer than those in $Tb_3N@I_h(7)$-C_{80} and $Tb_3N@D_{5h}(6)$-C_{80}, and the lengthening of the Tb–N distances allows the Tb_3N units to adopt planar structures in the larger cages. In contrast to I_h and D_{5h} isomers of $Tb_3N@C_{80}$, $Tb_3N@C_{86}$ and $Tb_3N@C_{88}$ both have a flattened aspect that provides added space for the planar M_3N unit.

In contrast to the homonuclear M_3N cluster, there is a paucity of mixed metal cluster M_xL_{3-x} reported by X-ray crystallographic structures. In 2000, the structure of a metal nitride cluster EMFs, $ErSc_2N@I_h(7)$-C_{80}·Co^{II}(OEP), was first determined by single-crystal XRD.[83] In the planar $ErSc_2N$ unit, the Er–N distance is 2.089 Å, and the Sc–N distance is 1.968 Å. Each metal is likely under [5,6,6]-junctions, and it makes close contact with individual carbon atoms of the cage with shortest Sc–C distances, in the range of 2.03–2.12 Å, and shortest Er–C distances of 2.20 and 2.22 Å.

Another series of mixed metal species of TNT-EMFs, $CeSc_2N$, $GdSc_2N$, $TbSc_2N$, and Gd_2ScN nitride clusters, have been also characterized by XRD.[84,85]

The crystallographic characterization of $CeSc_2N@I_h(7)$-C_{80}·Ni^{II}(OEP) indicates that the $CeSc_2N$ unit is planar, with the Ce–N distance of 2.184(2) Å and Sc–N distances of 1.942–1.933 Å, respectively (Fig. 5c).[84] Due to the large size of the Ce ion, the N atom is pushed 0.36 Å away from the central position toward the two scandium ions, which makes the compressed Sc–N bond lengths even shorter than those in $ErSc_2N@I_h(7)$-C_{80}. The Ce atom is under a hexagon and the two Sc atoms are near [5,6]-junctions, with Sc–C distances of 2.227–2.383 Å, and Er–C distances of 2.579–2.610 Å.

In 2008, structural characterizations of three nitride cluster EMFs, $GdSc_2N@I_h(7)$-C_{80}, $Gd_2ScN@I_h(7)$-C_{80}, and $TbSc_2@I_h(7)$-C_{80}, were obtained by single-crystal XRD.[85] All three have planar MM'_2N units. The central nitride ion is positioned further from the larger Gd^{3+} or Tb^{3+} ions and closer to the smaller Sc^{3+} ions. In $GdSc_2N@I_h(7)-C_{80}·Ni^{II}(OEP)$, the Gd atom is under a hexagon, and the two Sc atoms are near [5,6]-junctions The Gd–C distances in the nearest six-membered ring range from 2.459 to 2.509 Å. The Sc–C distances are 2.268–2.381 Å. In $Gd_2ScN@I_h(7)-C_{80}·Ni^{II}(OEP)$, the two Gd atoms are under hexagons, and Sc is near a [5,6]-junction. The Gd–C distances in the nearest six-membered ring that range from 2.394 to 2.641 Å. The shortest Sc–C distances are 2.246–2.269 Å. In $TbSc_2@I_h(7)-C_{80}·Ni^{II}(OEP)$, the Tb atom is under a hexagon, one Sc atom is near a [5,6]-junction and the other is near a [6,6]-junction. The Tb–C distances in the nearest six-membered ring range from 2.416 to 2.507 Å. The Sc–C distances are 2.138–2.312 Å.

2.3.2. *X-ray Structures of Metal Carbide Cluster EMFs*

Recently, $Sc_3@C_{3v}(7)-C_{82}$ was revealed to be a carbide cluster EMF, namely, $Sc_3C_2@I_h(7)-C_{80}$ characterized with both NMR and single crystal XRD methods.[86] In the Sc_3C_2 cluster, scandium atoms form a triangle, while the C_2 unit can adopt different configurations either in-plane or perpendicular to the Sc_3 plane (Fig. 7f).

Moreover, the previously assigned $Sc_2@C_{82}$ isomers have been recently determined to also be carbide EMFs, $Sc_2C_2@C_{2v}(5)-C_{80}$. In 2011, $Sc_2C_2@C_{80}$ with $C_{2v}(5)$ cage structure was proved by single-crystal XRD of the adamantylidene adduct.[87] Notably, NMR results disclose that the dynamic motion of the internal Sc_2C_2 cluster depends strongly on temperature. At 335 K or lower, the cluster is fixed in the cage, and two nonequivalent Sc atoms interact strongly with the cage carbons, engendering C_s symmetry for the whole molecule. Above this temperature, the cluster rotates rapidly inside the cage, featuring two equivalent Sc atoms and weaker metal-cage interactions, and developing C_{2v} symmetry of the molecule. Subsequently, in 2012 five isolable $Sc_2C_2@C_{2v}(5)-C_{80}$-Ad isomers were determined by XRD.[88] The addition mostly takes place at a cage region on top of the arch of the cluster, and after functionalization, the two Sc atoms are more likely to move, although the C_2-unit is almost fixed relative to the cage.

On the other hand, Sc_2C_{84} isomers, are also identified as metal carbide cluster EMFs, $Sc_2C_2@C_{3v}(8)-C_{82}$-Ad, $Sc_2C_2@C_{2v}(9)-C_{82}-N(CH_2)_2$Trt (Fig. 7e) and $Sc_2C_2@C_s(6)-C_{82}-N(CH_2)_2$Trt.[89–91] In $Sc_2C_2@C_s(6)-C_{82}$, the internal

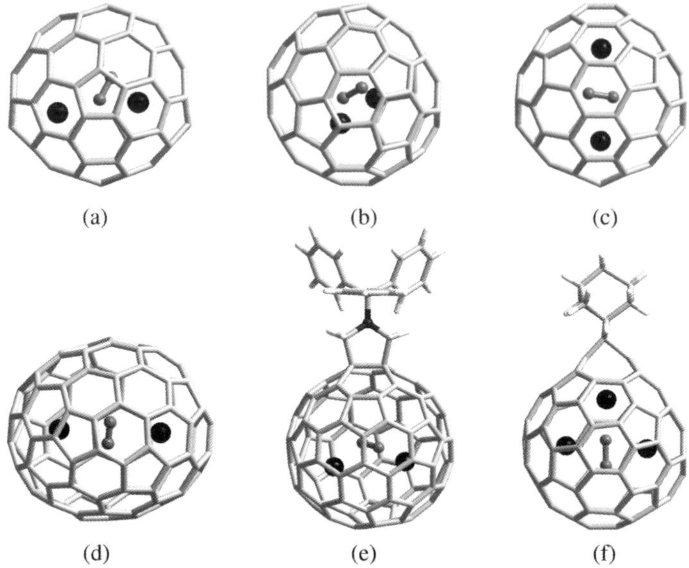

Fig. 7 X-ray structures of selected metal carbide cluster EMFs: (a) $Sc_2C_2@C_{2v}(5)$-C_{80}; (b) $Sc_2C_2@C_{3v}(8)$-C_{82}; (c) $Sc_2C_2@D_{2d}(23)$-C_{84}; (d) $Gd_2C_2@D_3(85)$-C_{92}; (e) $Sc_2C_2@C_{2v}(9)$-C_{82}-$N(CH_2)_2$Trt; (f) $Sc_3C_2@I_h(7)$-C_{80}-Ad.

Sc_2C_2 cluster is far from the addition position, with Sc–C distances of the cluster ranging from 2.293 to 2.349 Å. In contrast, the closest contacts between the Sc atoms and the cage carbons are even shorter (2.179 and 2.199 Å), indicating that the bent cluster prefers a flat conformation as much as possible, thus the metals are balanced between the internal C_2 unit and the cage.

In 2011, X-ray analyses of the cocrystals of a series of metal carbide cluster EMFs: $Sc_2C_2@C_{2v}(5)$-C_{80}, $Sc_2C_2@C_{3v}(8)$-C_{82}, and $Sc_2C_2@D_{2d}(23)$-C_{84} with cobalt(II) octaethylporphyrin, further reveal that a clear correlation between the cluster strain and cage size is observed in metal carbide cluster EMFs (Figs. 7a–7c): Sc–Sc distances and dihedral angles of the bent cluster increase along with cage expansion, indicating that the bending strain within the cluster makes it pursue a planar structure to the greatest degree possible.[92] Moreover, the cluster motion is strongly dependent on the cage size; it is nearly fixed in C_{80} but oscillates inside C_{82} and rotates in C_{84}. The preferential formation of cluster EMFs for scandium might be associated with its small ionic radius and the strong coordination ability of Sc^{3+} as well. More recently, the single-crystal structure of $(\mu\text{-H})_3Re_3(CO)_9$-$Sc_2C_2@C_{3v}(8)$-$C_{82}$ was reported, in which a Re_3 cluster was added regiospecifically to one hexagon of the $Sc_2C_2@C_{3v}(8)$-C_{82} cage.[93]

The largest carbon cage among structurally characterized metal carbide cluster EMF is found in $Gd_2C_2@D_3(85)$-C_{92}, whose structure was elucidated by XRD.[94] The Gd_2C_2 unit has a butterfly shape, and the Gd–C distances for the two the major sites fall in the range 2.484–2.527 Å (Fig. 7d). These gadolinium sites are located near the centers of hexagons on the inside of the fullerene cage.

2.3.3. *X-ray Structures of Metal Oxide Cluster EMFs*

Metal oxide cluster EMFs were discovered by Stevenson *et al.*, and X-ray structures were also reported recently. For instance, $Sc_4(\mu_3\text{-O})_2@I_h(7)$-$C_{80}$ is the first metal oxide cluster EMF whose structure is unambiguously determined by XRD,[95] indicating that the I_h-C_{80} cage encapsulates a distorted tetrahedral $Sc_4(\mu_3\text{-O})_2$ cluster disordered over three orientations (Fig. 8b). For the major site, the Sc–O distances are within the range of 1.834–2.090 Å. The scandium to carbon contacts are similar to those seen in related EMFs, with Sc–C distances of 2.016–2.348 Å

More recently, the structures of both $Sc_4(\mu_3\text{-O})_3@I_h(7)$-$C_{80}$ and $Sc_2(\mu_2\text{-O})@C_s(6)$-$C_{82}$ were also determined by XRD by Balch *et al.*[96,97] For $Sc_4(\mu_3\text{-O})_3@I_h(7)$-$C_{80}$, it is revealed that four Sc atoms form a distorted

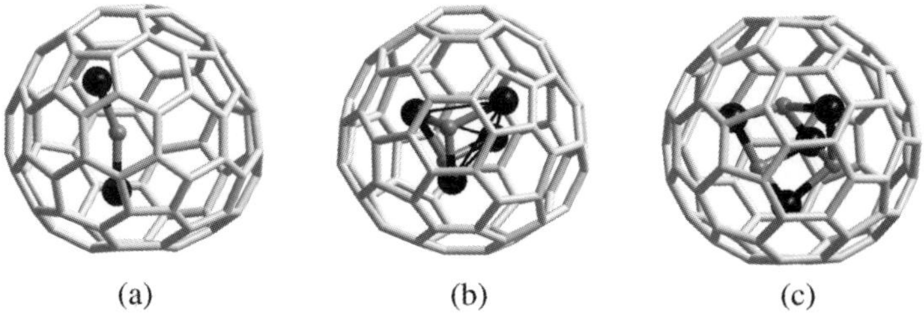

Fig. 8 X-ray structures of metal oxide cluster EMFs: (a) $Sc_2(\mu_2\text{-O})@C_s(6)$-$C_{82}$; (b) $Sc_4(\mu_3\text{-O})_2@I_h(7)$-$C_{80}$; (c) $Sc_4(\mu_3\text{-O})_3@I_h(7)$-$C_{80}$.

tetrahedron (Fig. 8c). The Sc–Sc and Sc–O distances in $Sc_4(\mu_3\text{-}O)_3$@I_h-C_{80} are similar to those in $Sc_4(\mu_3\text{-}O)_2$@I_h(7)-C_{80}. On the other hand, X-ray crystallographic study clearly revealed the C_s(6) isomeric structure of the C_{82} cage of $Sc_2(\mu_2\text{-}O)$@C_s(6)-C_{82} (Fig. 8a). The close-up study reveals that the C_s(6)-C_{82} cage possessed a unique band of 10 contiguous hexagons, and the Sc ions are located along that set of 10 contiguous hexagons. The Sc_2O unit inside the fullerene cage is disordered. For the major site, the Sc–O distances are 1.867 and 1.934 Å, and the Sc–O–Sc angle is 156.6°.

2.3.4. *X-ray Structures of Metal Sulfide Cluster EMFs*

Very recently, metal sulfide cluster EMFs have been discovered and shown to have different structures than the most abundant oxide cluster EMFs. Echegoyen and Balch *et al.* determined successfully the structures of the two isomers of Sc_2S@C_{82}: $Sc_2(m_2\text{-}S)$@C_s(6)-C_{82} and $Sc_2(m_2\text{-}S)$@C_{3v}(8)-C_{82}.[98] The crystallographic data for $Sc_2(m_2\text{-}S)$@C_s(6)-C_{82}·Ni^{II}(OEP) shows two remarkable features with the presence of two slightly different cage sites and a fully ordered molecule $Sc_2(m_2\text{-}S)$@C_s(6)-C_{82} in one of these sites (Fig. 9b). Interestingly, it is found that the Sc–S–Sc angles in $Sc_2(m_2\text{-}S)$@C_s(6)-C_{82} and $Sc_2(m_2\text{-}S)$@C_{3v}(8)-C_{82} differ: in $Sc_2(m_2\text{-}S)$@C_s(6)-C_{82} the Sc–S–Sc angle is 113.84(3)°, whereas in the major site in $Sc_2(m_2\text{-}S)$@C_{3v}(8)-C_{82} that angle is narrower, 97.34(13)°. The Sc–S–Sc angle in $Sc_2(m_2\text{-}S)$@C_s(6)-C_{82} appears to be narrower than the Sc–O–Sc angle (156.6°) in $Sc_2(m_2\text{-}O)$@C_s(6)-C_{82}, and the Sc–S distances are longer than the Sc–O distances in these two EMFs.

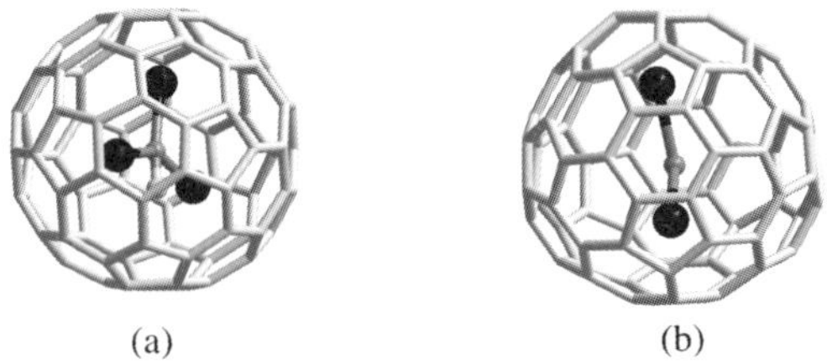

(a) (b)

Fig. 9 X-ray structures of metal cyano and sulfide cluster EMFs: (a) Sc_3CN@I_h(7)-C_{80}; (b) $Sc_2(\mu_2\text{-}S)$@C_s(6)-C_{82}.

2.3.5. *X-ray Structures of Metal Cyano Cluster EMFs*

The metal cyano cluster EMF $Sc_3CN@C_{80}$ is isolated and its structure is definitely confirmed as $Sc_3CN@I_h(7)$-C_{80} by XRD.[99] The Sc_3NC cluster is planar with the nitrogen atom in the center and the C atom residing on one side of the triangle formed by the Sc atoms, with Sc–N bond lengths of 2.072–2.182 Å. (Fig. 9a) of Sc atoms are more likely located under respective hexagons, with the Sc–C distances of 2.048–2.875 Å.

2.4. *X-ray Structures of Hetero EMFs*

In 2008, dimetallo-heterofullerenes $M_2@C_{79}N$ (M = Y, Tb) were isolated and structurally characterized for the first time by Dorn and co-workers.[100] Single-crystal X-ray diffraction has shown that two Tb ions are encapsulated in the pseudo-$I_h(7)$ carbon cage; because of the severe cage disorder in the crystal, the exact position of the nitrogen atom could not be elucidated by XRD, but DFT study showed that it likely to be located on the pentagon-hexagon-hexagon junction.

2.5. *X-ray Structures of Non-IPR EMFs.*

For a given fullerene cage, the arrangement of the pentagonal and hexagonal carbon rings of the cage framework is controlled by the well-known isolated pentagon rule (IPR), which states that each pentagon must be surrounded by hexagons to give chemically stable fullerene isomers.

Recently, several non-IPR EMFs have been isolated and characterized. For example, in 2003 the non-IPR cage structure of $Sc_3N@D_3(6140)$-C_{68} was confirmed by XRD.[101] The three fused pentagon pairs are symmetric to the D_3 axis, and the three Sc^{3+} ions are adjacent to the three [5,5]-junctions, respectively (Fig. 10a). The Sc_3N unit is planar, and Sc–N bond lengths are 1.961, 1.974 and 2.022 Å, respectively. In $Sc_3N@C_{68}$, the Sc–C bond lengths fall in a narrower range of 2.225–2.450 Å, and the shortest metal-to-carbon separations involve the carbon atoms at the [5,5]-junctions.

The X-ray structure of the dichlorophenyl adduct of mono-EMFs La@$C_2(10612)$-C_{72} definitely shows a non-IPR C_2-C_{72} cage, in which the La^{3+} ion is localized closely to the fused [5,5]-junction, indicating their specific interactions (Fig. 11a).[102] The distances between La and two carbons of the [5,5]-junction are 2.615 Å and 2.606 Å, respectively.

In 2008, di-EMFS $La_2@C_{72}$ was definitely confirmed to have a non-IPR structure as $La_2@D_2(10611)$-C_{72} by crystallographic analysis performed on its mono- and bis-adamantylidene adducts (Figs. 11b and 11c).[103,104] There are two pairs of fused pentagons on the two poles of $D_2(10611)$-C_{72} cage. The two encapsulated La atoms are highly localized and close to the two [5,5]-junctions, respectively. The shortest La–C distances (2.495 Å and 2.501 Å) also involve the carbons on the fused [5,5]-junctions, indicating the strong associations between La^{3+} ions and their adjacent pentagon pairs.

More recently, a non-IPR cluster EMF $Sc_2S@C_s(10528)$-C_{72} was characterized by XRD.[105] The fullerene cage appears almost triangular as viewed orthogonally to the Sc_2S unit, while it shows a typical roundness as viewed parallel to the plane of the Sc_2S unit (Fig. 10b). Each Sc atom position is near [5,5]-junctions, with Sc–C bond lengths of 2.206–2.314 Å.

In 2009, the non-IPR cage structure of $Gd_3N@C_2(22010)$-C_{78} was confirmed by XRD.[106] Two of Gd atoms of the planar Gd_3N unit are located within the two [5,5]-junctions, with the Gd–N distances of 2.037–2.133 Å (Fig. 10c). The third Gd atom resides at the center of a hexagonal face of

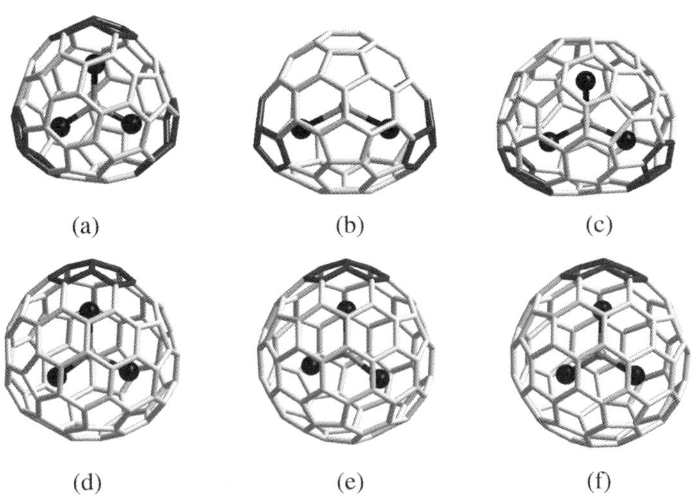

Fig. 10 X-ray structures of pristine non-IPR EMFs: (a) $Sc_3N@D_3(6140)$-C_{68}; (b) $Sc_2S@C_s(10528)$-C_{72}; (c) $Gd_3N@C_2(22010)$-C_{78}; (d) $Gd_3N@C_s(39663)$-C_{82}; (e) $Tb_3N@C_s(51365)$-C_{84}; (f) $Tm_3N@C_s(51365)$-C_{84}.

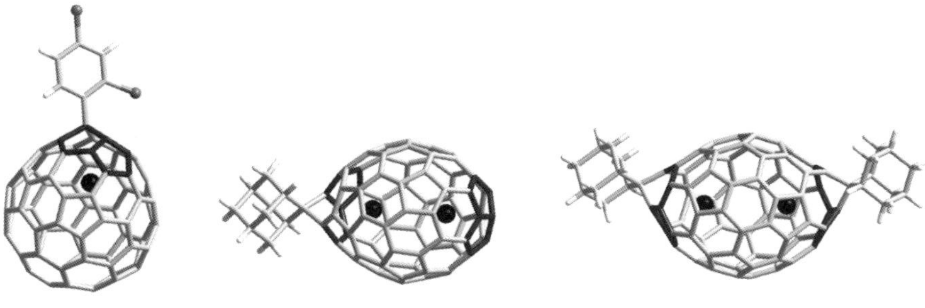

(a) (b) (c)

Fig. 11 X-ray structures of derivatives of non-IPR EMFs EMFs: (a) La@C_2(10612)-C_{72}; (b) La_2@D_2(10611)-C_{72}-Ad; (c) La_2@D_2(10611)-C_{72}-Ad_2.

the fullerene. The distances between Gd and four carbons of the two [5,5]-junctions lie in the range of 2.365 Å–2.434 Å.

In addition, several species of non-IPR EMFs, such as Gd_3N@C_s(39663)-C_{82} and M_3N@C_s(51365)-C_{84} (M = Gd, Tb, Tm) have also been confirmed by XRD[107–109] which have only one fused pentagon pair on the pole of their fullerene cage (Figs. 10d–10f). Likewise, in each molecule, there is one metal atom of M_3N approaching to the only [5,5]-bond. Thus, in all these non-IPR metal nitride cluster EMFs, the shortest M–C separations always involve the carbon atoms at [5,5]-junctions, which highlights the significant interactions between the metal ions and the fused pentagon pairs.

3. CONCLUSIONS AND OUTLOOK

EMFs, as a new class of hybrid materials with unusual structures, fascinating properties, and promising applications, have been obtained and characterized using numerous careful and systematic investigations. In particular, the X-ray crystallographic method clearly elucidated new concepts of molecular architecture and molecular jailing. Taking advantage of this, some valuable results relating to the metal-cage interactions and metal location/motion have been obtained. We believe that the diversity of molecular structures of both the fullerene cage and the internal metallic species makes these molecules particularly promising candidates because of their combined functionality.

ACKNOWLEDGMENTS

This work is supported by The National Thousand Talents Program of China, and The National Natural Science Foundation of China (21171061 and 21201067).

REFERENCES

1. Akasaka T, Nagase S. (2002) *Endofullerenes: A New Family of Carbon Clusters*, Kluwer, Dordrecht.
2. Lu X, Akasaka T, Nagase S. (2009) In: Huang CH (ed.), *Rare Earth Coordination Chemistry: Fundamentals and Applications*, Wiley-Blackwell, Singapore.
3. Akasaka T, Nagase S, Kobayashi K, *et al.* (1997) *^{13}C and ^{139}La Nmr Studies of $La_2@C_{80}$: First Evidence for Circular Motion of Metal Atoms in Endohedral Dimetallofullerenes, Angew Chem Int Ed Engl* **36:** 1643–1645.
4. Kodama T, Ozawa N, Miyake Y, *et al.* (2002) *Structural Study of Three Isomers of $Tm@C_{82}$ by ^{13}C NMR Spectroscopy, J Am Chem Soc* **124:** 1452–1455.
5. Fu WJ, Wang XL, Azuremendi H, *et al.* (2011) *^{14}N and ^{45}Sc NMR Study of Trimetallic Nitride Cluster $(M_3N)^{6+}$ Dynamics inside a Icosahedral C_{80} Cage, Chem Commun* **47:** 3858–3860.
6. Fu WJ, Xu LS, Azurmendi H, *et al.* (2009) *^{89}Y and ^{13}C NMR Cluster and Carbon Cage Studies of an Yttrium Metallofullerene Family, $Y_3N@C_{2n}$ ($n = 40$–43), J Am Chem Soc* **131:** 11762–11769.
7. Takata M, Umeda B, Nishibori E, *et al.* (1995) *Confirmation by X-Ray Diffraction of the Endohedral Nature of the Metallofullerene $Y@C_{82}$, Nature* **377:** 46–49.
8. Nishibori E, Ishihara M, Takata M, *et al.* (2006) *Bent $(Metal)_2C_2$ Clusters Encapsulated in $(Sc_2C_2)@C_{82}(III)$ and $(Y_2C_2)@C_{82}(III)$ Metallofullerenes, Chem Phys Lett* **433:** 120–124.
9. Nishibori E, Takata M, Sakata M, *et al.* (1998) *Determination of the Cage Structure of $Sc@C_{82}$ by Synchrotron Powder Diffraction, Chem Phys Lett* **298:** 79–84.
10. Nagase S, Kobayashi K, Kato T, *et al.* (1993) *A Theoretical Approach to C_{82} and LaC_{82}, Chem Phys Lett* **201:** 475–480.
11. Nagase S, Kobayashi K. (1994) *Theoretical Study of the Dimetallofullerene $Sc_2@C_{84}$, Chem Phys Lett* **231:** 319–324.
12. Nagase S, Kobayashi K. (1994) *Theoretical Study of the Lanthanide Fullerene CeC_{82} Comparison with ScC_{82}, YC_{82} and LaC_{82}, Chem Phys Lett* **228:** 106–110.

13. Nagase S, Kobayashi K, Akasaka T. (1999) *Unconventional Cage Structures of Endohedral Metallofullerenes, J Mol Struc: THEOCHEM* **462:** 97–104.
14. Iiduka Y, Wakahara T, Nakahodo T, *et al.* (2005) *Structural Determination of Metallofuilerene* Sc_3C_{82} *Revisited: A Surprising Finding, J Am Chem Soc* **127:** 12500–12501.
15. Stevenson S, Fowler PW, Heine T, *et al.* (2000) *Materials Science — a Stable Non-Classical Metallofullerene Family, Nature* **408:** 427–428.
16. Lee HM, Olmstead MM, Iezzi E, *et al.* (2002) *Crystallographic Characterization and Structural Analysis of the First Organic Functionalization Product of the Endohedral Fullerene* $Sc_3N@C_{80}$, *J Am Chem Soc* **124:** 3494–3495.
17. Reich A, Panthofer M, Modrow H, *et al.* (2004) *The Structure of* $Ba@C_{74}$, *J Am Chem Soc* **126:** 14428–14434.
18. Nikawa H, Kikuchi T, Wakahara T, *et al.* (2005) *Missing Metallofullerene* $La@C_{74}$, *J Am Chem Soc* **127:** 9684–9685.
19. Lu X, Nikawa H, Kikuchi K, *et al.* (2011) *Radical Derivatives of Insoluble* $La@C_{74}$: *X-Ray Structures, Metal Positions, and Isomerization, Angew Chem Int Ed* **50:** 6356–6359.
20. Nikawa H, Yamada T, Cao BP, *et al.* (2009) *Missing Metallofullerene with* C_{80} *Cage, J Am Chem Soc* **131:** 10950–10954.
21. Lu X, Lian Y, Beavers CM, *et al.* (2011) *Crystallographic X-Ray Analyses of* $Yb@C_{2v}(3)$-C_{80} *Reveal a Feasible Rule That Governs the Location of a Rare Earth Metal inside a Medium-Sized Fullerene, J Am Chem Soc* **133:** 10772–10775.
22. Maeda Y, Matsunaga Y, Wakahara T, *et al.* (2004) *Isolation and Characterization of a Carbene Derivative of* $La@C_{82}$, *J Am Chem Soc* **126:** 6858–6859.
23. Lu X, Nikawa H, Feng L, *et al.* (2009) *Location of the Yttrium Atom in* $Y@C_{82}$ *and Its Influence on the Reactivity of Cage Carbons, J Am Chem Soc* **131:** 12066–12067.
24. Takano Y, Aoyagi M, Yamada M, *et al.* (2009) *Anisotropic Magnetic Behavior of Anionic* $Ce@C_{82}$ *Carbene Adducts, J Am Chem Soc* **131:** 9340–9346.
25. Akasaka T, Kono T, Takematsu Y, *et al.* (2008) *Does* $Gd@C_{82}$ *have an Anomalous Endohedral Structure? Synthesis and Single Crystal X-Ray Structure of the Carbene Adduct, J Am Chem Soc* **130:** 12840–12841.
26. Hachiya M, Nikawa H, Mizorogi N, *et al.* (2012) *Exceptional Chemical Properties of* $Sc@C_{2v}(9)$-C_{82} *Probed with Adamantylidene Carbene, J Am Chem Soc* **134:** 15550–15555.
27. Feng L, Slanina Z, Sato S, *et al.* (2011) *Covalently Linked Porphyrin-*$La@C_{82}$ *Hybrids: Structural Elucidation and Investigation of Intramolecular Interactions, Angew Chem Int Ed* **50:** 5909–5912.

28. Akasaka T, Lu X, Kuga H, *et al.* (2010) *Dichlorophenyl Derivatives of La@C_{3v}(7)-C_{82}: Endohedral Metal Induced Localization of Pyramidalization and Spin on a Triple-Hexagon Junction*, *Angew Chem Int Ed* **49:** 9715–9719.
29. Takano Y, Yomogida A, Nikawa H, *et al.* (2008) *Radical Coupling Reaction of Paramagnetic Endohedral Metallofullerene La@C_{82}*, *J Am Chem Soc* **130:** 16224–16230.
30. Maeda Y, Sato S, Inada K, *et al.* (2010) *Regioselective Exohedral Functionalization of La@C_{82} and its 1,2,3,4,5-Pentamethylcyclopentadiene and Adamantylidene Adducts*, *Chem Eur J* **16:** 2193–2197.
31. Lu X, Nikawa H, Tsuchiya T, *et al.* (2010) *Nitrated Benzyne Derivatives of La@C_{82}: Addition of NO_2 and Its Positional Directing Effect on the Subsequent Addition of Benzynes*, *Angew Chem Int Ed* **49:** 594–597.
32. Li XF, Fan LZ, Liu DF, *et al.* (2007) *Synthesis of a Dy@C_{82} Derivative Bearing a Single Phosphorus Substituent Via a Zwitterion Approach*, *J Am Chem Soc* **129:** 10636–10637.
33. Feng L, Nakahodo T, Wakahara T, *et al.* (2005) *A Singly Bonded Derivative of Endohedral Metallofullerene: La@$C_{82}CBr(COOC_2H_5)_2$*, *J Am Chem Soc* **127:** 17136–17137.
34. Feng L, Tsuchiya T, Wakahara T, *et al.* (2006) *Synthesis and Characterization of a Bisadduct of La@C_{82}*, *J Am Chem Soc* **128:** 5990–5991.
35. Suzuki M, Lu X, Sato S, *et al.* (2012) *Where Does the Metal Cation Stay in Gd@C_{2v}(9)-C_{82}? A Single-Crystal X-Ray Diffraction Study*, *Inorg Chem* **51:** 5270–5273.
36. Sato S, Nikawa H, Seki S, *et al.* (2012) *A Co-Crystal Composed of the Paramagnetic Endohedral Metallofullerene La@C_{82} and a Nickel Porphyrin with High Electron Mobility*, *Angew Chem Int Ed* **51:** 1589–1591.
37. Suzuki M, Slanina Z, Mizorogi N, *et al.* (2012) *Single-Crystal X-Ray Diffraction Study of Three Yb@C_{82} Isomers Cocrystallized with Ni^{II}(Octaethylporphyrin)*, *J Am Chem Soc* **134:** 18772–18778.
38. Yang H, Jin HX, Wang XQ, *et al.* (2012) *X-Ray Crystallographic Characterization of New Soluble Endohedral Fullerenes Utilizing the Popular C_{82} Bucky Cage. Isolation and Structural Characterization of Sm@C_{3v}(7)-C_{82}, Sm@C_s(6)-C_{82}, and Sm@C_2(5)-C_{82}*, *J Am Chem Soc* **134:** 14127–14136.
39. Yang H, Yu M, Jin H, *et al.* (2012) *Isolation of Three Isomers of Sm@C_{84} and X-Ray Crystallographic Characterization of Sm@D_{3d}(19)-C_{84} and Sm@C_2(13)-C_{84}*, *J Am Chem Soc* **134:** 5331–5338.
40. Yang H, Jin HX, Zhen HY, *et al.* (2011) *Isolation and Crystallographic Identification of Four Isomers of Sm@C_{90}*, *J Am Chem Soc* **133:** 6299–6306.

41. Jin HX, Yang H, Yu ML, *et al.* (2012) *Single Samarium Atoms in Large Fullerene Cages. Characterization of Two Isomers of Sm@C_{92} and Four Isomers of Sm@C_{94} with the X-Ray Crystallographic Identification of Sm@C_1(42)-C_{92}, Sm@C_s(24)-C_{92}, and Sm@C_{3v}(134)-C_{94}*, J Am Chem Soc **134:** 10933–10941.
42. Che YL, Yang H, Wang ZM, *et al.* (2009) *Isolation and Structural Characterization of two very Large, and Largely Empty, Endohedral Fullerenes: Tm@C_{3v}-C_{94} and Ca@C_{3v}-C_{94}*, Inorg Chem **48:** 6004–6010.
43. Aoyagi S, Nishibori E, Sawa H, *et al.* (2010) *A Layered Ionic Crystal of Polar Li@C_{60} Superatoms*, Nat Chem **2:** 678–683.
44. Yamada M, Okamura M, Sato S, *et al.* (2009) *Two Regioisomers of Endohedral Pyrrolidinodimetallofullerenes M_2@I_h-C_{80}(CH_2)$_2$NTrt (M = La, Ce; Trt = Trityl): Control of Metal Atom Positions by Addition Positions*, Chem Eur J **15:** 10533–10542.
45. Yamada M, Minowa M, Sato S, *et al.* (2011) *Regioselective Cycloaddition of La_2@I_h-C_{80} with Tetracyanoethylene Oxide: Formation of an Endohedral Dimetallofullerene Adduct Featuring Enhanced Electron-Accepting Character*, J Am Chem Soc **133:** 3796–3799.
46. Wakahara T, Yamada M, Takahashi S, *et al.* (2007) *Two-Dimensional Hopping Motion of Encapsulated La Atoms in Silylated La_2@C_{80}*, Chem Commun 2680–2682.
47. Yamada M, Nakahodo T, Wakahara T, *et al.* (2005) *Positional Control of Encapsulated Atoms inside a Fullerene Cage by Exohedral Addition*, J Am Chem Soc **127:** 14570–14571.
48. Yamada M, Minowa M, Sato S, *et al.* (2010) *Thermal Carbosilylation of Endohedral Dimetallofullerene La_2@ I_h-C_{80} with Silirane*, J Am Chem Soc **132:** 17953–17960.
49. Yamada M, Someya C, Wakahara T, *et al.* (2008) *Metal Atoms Collinear with the Spiro Carbon of 6,6-Open Adducts, M_2@C_{80}(Ad) (M = La and Ce, Ad = Adamantylidene)*, J Am Chem Soc **130:** 1171–1176.
50. Ishitsuka MO, Sano S, Enoki H, *et al.* (2011) *Regioselective Bis-Functionalization of Endohedral Dimetallofullerene, La_2@C_{80}: Extremal La-La Distance*, J Am Chem Soc **133:** 7128–7134.
51. Feng L, Radhakrishnan SG, Mizorogi N, *et al.* (2011) *Synthesis and Charge-Transfer Chemistry of La_2@$I_h$$C_{80}$/$Sc_3$N@$I_h$$C_{80}$ Zinc-Porphyrin Conjugates: Impact of Endohedral Cluster*, J Am Chem Soc **133:** 7608–7618.
52. Cao B, Nikawa H, Nakahodo T, *et al.* (2008) *Addition of Adamantylidene to La_2@C_{78}: Isolation and Single-Crystal X-Ray Structural Determination of the Monoadducts*, J Am Chem Soc **130:** 983–989.

53. Yamada M, Wakahara T, Tsuchiya T, *et al.* (2008) *Location of the Metal Atoms in $Ce_2@C_{78}$ and Its Bis-Silylated Derivative, Chem Commun* 558–560.
54. Olmstead MM, De Bettencourt-Dias A, Stevenson S, *et al.* (2002) *Crystallographic Characterization of the Structure of the Endohedral Fullerene {$Er_2@C_{82}$ Isomer I} with C_s Cage Symmetry and Multiple Sites for Erbium Along a Band of Ten Contiguous Hexagons, J Am Chem Soc* **124:** 4172–4173.
55. Olmstead MM, Lee HM, Stevenson S, *et al.* (2002) *Crystallographic Characterization of Isomer 2 of $Er_2@C_{82}$ and Comparison with Isomer 1 of $Er_2@C_{82}$, Chem Commun* 2688–2689.
56. Yang H, Jin HX, Hong B, *et al.* (2011) *Large Endohedral Fullerenes Containing Two Metal Ions, $Sm_2@D_2(35)$-C_{88}, $Sm2@C_1(21)$-C_{90}, and $Sm_2@D_3(85)$-C_{92}, and Their Relationship to Endohedral Fullerenes Containing Two Gadolinium Ions, J Am Chem Soc* **133:** 16911–16919.
57. Mercado BQ, Jiang A, Yang H, *et al.* (2009) *Isolation and Structural Characterization of the Molecular Nanocapsule $Sm_2@D_{3d}(822)$-C_{104}, Angew Chem Int Ed* **48:** 9114–9116.
58. Beavers CM, Jin H, Yang H, *et al.* (2011) *Very Large, Soluble Endohedral Fullerenes in the Series La_2C_{90} to La_2C_{138}: Isolation and Crystallographic Characterization of $La_2@D_5(450)$-C_{100}, J Am Chem Soc* **133:** 15338–15341.
59. Stevenson S, Rice G, Glass T, *et al.* (1999) *Small-Bandgap Endohedral Metallofullerenes in High Yield and Purity, Nature* **401:** 55–57.
60. Stevenson S, Lee HM, Olmstead MH, *et al.* (2002) *Preparation and Crystallographic Characterization of a New Endohedral, $Lu_3N@C_{80}$·5(O-Xylene), and Comparison with $Sc_3N@C_{80}$ ·5(O-Xylene), Chem Eur J* **8:** 4528–4535.
61. Hernandez-Eguia LP, Escudero-Adan EC, Pinzon JR, *et al.* (2011) *Complexation of $Sc_3N@C_{80}$ Endohedral Fullerene with Cyclic Zn-Bisporphyrins: Solid State and Solution Studies, J Org Chem* **76:** 3258–3265.
62. Cai T, Slebodnick C, Xu L, *et al.* (2006) *A Pirouette on a Metallofullerene Sphere: Interconversion of Isomers of N-Tritylpyrrolidino I_h -$Sc_3N@C_{80}$, J Am Chem Soc* **128:** 6486–6492.
63. Shu C, Slebodnick C, Xu L, *et al.* (2008) *Highly Regioselective Derivatization of Trimetallic Nitride Templated Endohedral Metallofullerenes via a Facile Photochemical Reaction, J Am Chem Soc* **130:** 17755–17760.
64. Shustova NB, Chen YS, Mackey MA, *et al.* (2009) *$Sc_3N@(C_{80}$-$I_h(7))(CF_3)_{14}$ and $Sc_3N@(C_{80}$-$I_h(7))(CF_3)_{16}$. Endohedral Metallofullerene Derivatives with Exohedral Addends on Four and Eight Triple-Hexagon Junctions. Does the Sc_3N Cluster Control the Addition Pattern or Vice Versa?, J Am Chem Soc* **131:** 17630–17637.

65 Shustova NB, Peryshkov DV, Kuvychko IV, *et al.* (2011) *Poly(Perfluoroalkylation) of Metallic Nitride Fullerenes Reveals Addition-Pattern Guidelines: Synthesis and Characterization of a Family of $Sc_3N@C_{80}(CF_3)_n$ (n = 2–16) and Their Radical Anions*, *J Am Chem Soc* **133:** 2672–2690.
66. Yang SF, Chen CB, Jiao MZ, *et al.* (2011) *Synthesis, Isolation, and Addition Patterns of Trifluoromethylated D_{5h} and I_h Isomers of $Sc_3N@C_{80}$: $Sc_3N@D_{5h}$-$C_{80}(CF_3)_{18}$ and $Sc_3N@I_h$-$C_{80}(CF_3)_{14}$*, *Inorg Chem* **50:** 3766–3771.
67. Li F-F, Pinzon JR, Mercado BQ, *et al.* (2011) *[2 + 2] Cycloaddition Reaction to $Sc_3N@I_h$-C_{80}. The Formation of Very Stable [5,6]- and [6,6]-Adducts*, *J Am Chem Soc* **133:** 1563–1571.
68. Wang G-W, Liu T-X, Jiao M, *et al.* (2011) *The Cycloaddition Reaction of I_h-$Sc_3N@C_{80}$ with 2-Amino-4,5-Diisopropoxybenzoic Acid and Isoamyl Nitrite to Produce an Open-Cage Metallofullerene*, *Angew Chem Int Ed* **50:** 4658–4662.
69. Wakahara T, Iiduka Y, Ikenaga O, *et al.* (2006) *Characterization of the Bis-Silylated Endofullerene $Sc_3N@C_{80}$*, *J Am Chem Soc* **128:** 9919–9925.
70. Shu CY, Xu W, Slebodnick C, *et al.* (2009) *Syntheses and Structures of Phenyl-C_{81}-Butyric Acid Methyl Esters (Pcbms) from $M_3N@C_{80}$*, *Org Lett* **11:** 1753–1756.
71. Liu TX, Wei T, Zhu SE, *et al.* (2012) *Azide Addition to an Endohedral Metallofullerene: Formation of Azafulleroids of $Sc_3N@I_h$-C_{80}*, *J Am Chem Soc* **134:** 11956–11959.
72. Echegoyen L, Chancellor CJ, Cardona CM, *et al.* (2006) *X-Ray Crystallographic and EPR Spectroscopic Characterization of a Pyrrolidine Adduct of $Y_3N@C_{80}$*, *Chem Commun* 2653–2655.
73. Lukoyanova O, Cardona CM, Rivera J, *et al.* (2007) *"Open Rather Than Closed" Malonate Methano-Fullerene Derivatives. The Formation of Methanofulleroid Adducts of $Y_3N@C_{80}$*, *J Am Chem Soc* **129:** 10423–10430.
74. Sato K, Kako M, Suzuki M, *et al.* (2012) *Synthesis of Silylene-Bridged Endohedral Metallofullerene $Lu_3N@I_h$-C_{80}*, *J Am Chem Soc* **134:** 16033–16039.
75. Yang SF, Troyanov SI, Popov AA, *et al.* (2006) *Deviation from the Planarity — a Large Dy_3N Cluster Encapsulated in an I_h-C_{80} Cage: An X-Ray Crystallographic and Vibrational Spectroscopic Study*, *J Am Chem Soc* **128:** 16733–16739.
76. Zuo TM, Olmstead MM, Beavers CM, *et al.* (2008) *Preparation and Structural Characterization of the I_h and the D_{5h} Isomers of the Endohedral Fullerenes $Tm_3N@C_{80}$: Icosahedral C_{80} Cage Encapsulation of a Trimetallic Nitride Magnetic Cluster with Three Uncoupled Tm^{3+} Ions*, *Inorg Chem* **47:** 5234–5244.
77. Stevenson S, Phillips JP, Reid JE, *et al.* (2004) *Pyramidalization of Gd_3N inside a C_{80} Cage. The Synthesis and Structure of $Gd_3N@C_{80}$*, *Chem Commun* 2814–2815.

78. Zuo TM, Beavers CM, Duchamp JC, *et al.* (2007) *Isolation and Structural Characterization of a Family of Endohedral Fullerenes Including the Large, Chiral Cage Fullerenes* $Tb_3N@C_{88}$ *and* $Tb_3N@C_{86}$ *as Well as the* I_h *and* D_{5h} *Isomers of* $Tb_3N@C_{80}$, *J Am Chem Soc* **129:** 2035–2043.
79. Cai T, Xu LS, Anderson MR, *et al.* (2006) *Structure and Enhanced Reactivity Rates of the* D_{5h} $Sc_3N@C_{80}$ *and* $Lu_3N@C_{80}$ *Metallofullerene Isomers: The Importance of the Pyracylene Motif, J Am Chem Soc* **128:** 8581–8589.
80. Olmstead MH, De Bettencourt-Dias A, Duchamp JC, *et al.* (2001) *Isolation and Structural Characterization of the Endohedral Fullerene* $Sc_3N@C_{78}$, *Angew Chem Int Ed* **40:** 1223–1225.
81. Campanera JM, Bo C, Olmstead MH, *et al.* (2001) *Bonding within the Endohedral Fullerenes* $Sc_3N@C_{78}$ *and* $Sc_3N@C_{80}$ *as Determined by Density Functional Calculations and Reexamination of the Crystal Structure of* $\{Sc_3N@C_{78}\cdot Co(OEP)\}\cdot 1.5(C_6C_6)\cdot 0.3(CHCl_3)$, *J Phys Chem A* **106:** 12356–12364.
82. Cai T, Xu L, Gibson HW, *et al.* (2007) $Sc_3N@C_{78}$*: Encapsulated Cluster Regiocontrol of Adduct Docking on an Ellipsoidal Metallofullerene Sphere, J Am Chem Soc* **129:** 10795–10800.
83. Olmstead MM, De Bettencourt-Dias A, Duchamp JC, *et al.* (2000) *Isolation and Crystallographic Characterization of* $ErSc_2N@C_{80}$*: An Endohedral Fullerene Which Crystallizes with Remarkable Internal Order, J Am Chem Soc* **122:** 12220–12226.
84. Wang XL, Zuo TM, Olmstead MM, *et al.* (2006) *Preparation and Structure of* $CeSc_2N@C_{80}$*: An Icosahedral Carbon Cage Enclosing an Acentric* $CeSc_2N$ *Unit with Buried F Electron Spin, J Am Chem Soc* **128:** 8884–8889.
85. Stevenson S, Chancellor CJ, Lee HM, *et al.* (2008) *Internal and External Factors in the Structural Organization in Cocrystals of the Mixed-Metal Endohedrals* ($GdSc_2N@I_h$-C_{80}, $Gd_2ScN@I_h$-C_{80}, *and* $TbSc_2N@I_h$-C_{80}) *and Nickel(II) Octaethylporphyrin, Inorg Chem* **47:** 1420–1427.
86. Iiduka Y, Ikenaga O, Sakuraba A, *et al.* (2005) *Chemical Reactivity of* $Sc_3N@C_{80}$ *and* $La_2@C_{80}$, *J Am Chem Soc* **127:** 9956–9957.
87. Kurihara H, Lu X, Iiduka Y, *et al.* (2011) $Sc_2C_2@C_{80}$ *Rather Than* $Sc_2@C_{82}$*: Templated Formation of Unexpected* $C_{2v}(5)$-C_{80} *and Temperature-Dependent Dynamic Motion of Internal* Sc_2C_2 *Cluster, J Am Chem Soc* **133:** 2382–2385.
88. Kurihara H, Lu X, Iiduka Y, *et al.* (2012) *Chemical Understanding of Carbide Cluster Metallofullerenes: A Case Study on* $Sc_2C_2@C_{2v}(5)$-C_{80} *with Complete X-Ray Crystallographic Characterizations, J Am Chem Soc* **134:** 3139–3144.
89. Iiduka Y, Wakahara T, Nakajima K, *et al.* (2007) *Experimental and Theoretical Studies of the Scandium Carbide Endohedral Metallofullerene* $Sc_2C_2@C_{82}$ *and Its Carbene Derivative, Angew Chem Int Ed* **46:** 5562–5564.

90. Lu X, Nakajima K, Iiduka Y, *et al.* (2012) *The Long-Believed $Sc_2@C_{2v}(17)$-C_{84} Is Actually $Sc_2C_2@C_{2v}(9)$-C_{82}: Unambiguous Structure Assignment and Chemical Functionalization, Angew Chem Int Ed* **51:** 5889–5892.
91. Lu X, Nakajima K, Iiduka Y, *et al.* (2011) *Structural Elucidation and Regioselective Functionalization of an Unexplored Carbide Cluster Metallofullerene $Sc_2C_2@C_s(6)$-C_{82}, J Am Chem Soc* **133:** 19553–19558.
92. Kurihara H, Lu X, Iiduka Y, *et al.* (2012) *X-Ray Structures of $Sc_2C_2@C_{2n}$ (n = 40–42): In-Depth Understanding of the Core–Shell Interplay in Carbide Cluster Metallofullerenes, Inorg Chem* **51:** 746.
93. Chen CH, Yeh WY, Liu YH, *et al.* (2012) *[$(\mu$-$H)_3Re_3(CO)_9$(•^{2},•^{2},•^{2}-$Sc_2C_2@C_{3v}(8)$-C_{82})]: Face-Capping Cluster Complex of an Endohedral Fullerene, Angew Chem Int Ed* **51:** 13046–13049.
94. Yang H, Lu C, Liu Z, *et al.* (2008) *Detection of a Family of Gadolinium-Containing Endohedral Fullerenes and the Isolation and Crystallographic Characterization of One Member as a Metal-Carbide Encapsulated inside a Large Fullerene Cage, J Am Chem Soc* **130:** 17296–17300.
95. Stevenson S, Mackey MA, Stuart MA, *et al.* (2008) *A Distorted Tetrahedral Metal Oxide Cluster inside an Lcosahedral Carbon Cage. Synthesis, Isolation, and Structural Characterization of $Sc_4(\mu_3$-$O)_2@I_h$-C_{80}, J Am Chem Soc* **130:** 11844–11845.
96. Mercado BQ, Olmstead MM, Beavers CM, *et al.* (2010) *A Seven Atom Cluster in a Carbon Cage, the Crystallographically Determined Structure of $Sc_4(\mu_3$-$O)_3@I_h$-C_{80}, Chem Commun* **46:** 279–281.
97. Mercado BQ, Stuart MA, Mackey MA, *et al.* (2010) *$Sc_2(\mu_2$-$O)$ Trapped in a Fullerene Cage: The Isolation and Structural Characterization of $Sc_2(\mu_2$-$O)@C_S(6)$-C_{82} and the Relevance of the Thermal and Entropic Effects in Fullerene Isomer Selection, J Am Chem Soc* **132:** 12098–12105.
98. Mercado BQ, Chen N, Rodriguez-Fortea A, *et al.* (2011) *The Shape of the $Sc_2(\mu_2$-$S)$ Unit Trapped in C_{82}: Crystallographic, Computational, and Electrochemical Studies of the Isomers, $Sc_2(\mu_2$-$S)@Cs(6)$-C_{82} and $Sc_2(\mu_2$-$S)@C_{3v}(8)$-C_{82}, J Am Chem Soc* **133:** 6752–6760.
99. Wang TS, Feng L, Wu JY, *et al.* (2010) *Planar Quinary Cluster inside a Fullerene Cage: Synthesis and Structural Characterizations of $Sc_3NC@C_{80}$-I_h, J Am Chem Soc* **132:** 16362–16364.
100. Zuo TM, Xu LS, Beavers CM, *et al.* (2008) *$M_2@C_{79}N$ (M = Y, Tb): Isolation and Characterization of Stable Endohedral Metallofullerenes Exhibiting M-M Bonding Interactions inside Aza[80]Fullerene Cages, J Am Chem Soc* **130:** 12992–12997.
101. Olmstead MM, Lee HM, Duchamp JC, *et al.* (2003) *$Sc_3N@C_{68}$: Folded Pentalene Coordination in an Endohedral Fullerene That Does Not Obey the Isolated Pentagon Rule, Angew Chem Int Ed* **42:** 900–903.

102. Wakahara T, Nikawa H, Kikuchi T, *et al.* (2006) *La@C_{72} Having a Non-Ipr Carbon Cage*, *J Am Chem Soc* **128:** 14228–14229.
103. Lu X, Nikawa H, Nakahodo T, *et al.* (2008) *Chemical Understanding of a Non-Ipr Metallofullerene: Stabilization of Encaged Metals on Fused-Pentagon Bonds in La_2@C_{72}*, *J Am Chem Soc* **130:** 9129–9136.
104. Lu X, Nikawa H, Tsuchiya T, *et al.* (2008) *Bis-Carbene Adducts of Non-Ipr La_2@C_{72}: Localization of High Reactivity around Fused Pentagons and Electrochemical Properties*, *Angew Chem Int Ed* **47:** 8642–8645.
105. Chen N, Beavers CM, Mulet-Gas M, *et al.* (2012) *Sc_2S@C_s(10528)-C_{72}: A Dimetallic Sulfide Endohedral Fullerene with a Non Isolated Pentagon Rule Cage*, *J Am Chem Soc* **134:** 7851–7860.
106. Beavers CM, Chaur MN, Olmstead MM, *et al.* (2009) *Large Metal Ions in a Relatively Small Fullerene Cage: The Structure of Gd_3N@C_2(22010)-C_{78} Departs from the Isolated Pentagon Rule*, *J Am Chem Soc* **131:** 11519–11524.
107. Mercado BQ, Beavers CM, Olmstead MM, *et al.* (2008) *Is the Isolated Pentagon Rule Merely a Suggestion for Endohedral Fullerenes? The Structure of a Second Egg-Shaped Endohedral Fullerene-Gd_3N@C_s(39663)-C_{82}*, *J Am Chem Soc* **130:** 7854–7855.
108. Beavers CM, Zuo TM, Duchamp JC, *et al.* (2006) *Tb_3N@C_{84}: An Improbable, Egg-Shaped Endohedral Fullerene That Violates the Isolated Pentagon Rule*, *J Am Chem Soc* **128:** 11352–11353.
109. Zuo T, Walker K, Olmstead MM, *et al.* (2008) *New Egg-Shaped Fullerenes: Non-Isolated Pentagon Structures of Tm_3N@C_s(51365)-C_{84} and Gd_3N@C_s(51365)-C_{84}*, *Chem Commun* 1067–1069.

CHAPTER 4

Metal Nitride Clusterfullerenes – New Advances and Challenges

Tao Wei,[a] *Song Wang,*[a] *Fupin Liu,*[a] *Jian Guan,*[a] *Alexey A. Popov,*[*,b] *Lothar Dunsch*[*,b] *and Shangfeng Yang*[*,a]

Metal nitride clusterfullerenes (NCFs) represent a special type of endohedral fullerene in which metal nitride clusters are encaged in the robust fullerene cages. The discovery of the first NCF, $Sc_3N@C_{80}$, in 1999 initiated the new era of endohedral fullerene research and led to the successive findings of several other types of clusterfullerenes with encaged carbide, oxide, sulfide, hydrocarbide, and carbonitride/cyanide clusters. Among all types of clusterfullerenes reported to date, NCFs show the largest diversity in molecular structures in terms of the inner nitride cluster and the outer carbon cage, with non-group-III metal encaged and the size of carbon cage extended to C_{96} nowadays. In this chapter we review the current state of NCFs focusing on the new advances after 2007, including the synthesis methods, isolation and structures of NCFs. Finally the unique properties and applications of NCFs are briefly discussed as well.

*Corresponding authors.

[a]Hefei National Laboratory for Physical Sciences at Microscale, CAS Key Laboratory of Materials for Energy Conversion & Department of Materials Science and Engineering, University of Science and Technology of China (USTC), Hefei 230026, China. Fax: +86(551)63601750; E-mail: sfyang@ustc.edu.cn

[b]Group of Electrochemistry and Conducting Polymers, Leibniz-Institute for Solid State and Materials Research (IFW) Dresden, D-01171 Dresden, Germany. Fax: +49(351)4659811; E-mail: A.Popov@ifw-dresden.de, l.dunsch@ifw-dresden.de

1. INTRODUCTION

Endohedral fullerenes represent a special family of fullerenes with novel nanostructures, featuring the encapsulation of atoms, ions, or clusters in the fullerene cage.[1] After the first detection of LaC_{2n}^+ ions in the mass spectrum of a sample prepared by vaporization of graphite rods impregnated with $LaCl_3$ by Smalley *et al.*[2] in 1991, the same group accomplished the isolation of the first endohedral metallofullerene La@C_{82}.[3] From then on, endohedral fullerenes have been attracting great interest from scientists all over the world, and a large number of endohedral fullerenes have been synthesized with their endohedral nature proved by versatile experimental studies. One of the remarkable features of differentiating endohedral fullerenes from empty fullerenes is the unique electronic property with the electron transfer from the encaged species to the carbon cage. Such an electron transfer as the key feature of the electronic property of endohedral fullerene has been the focus of the characterization of endohedral fullerenes, and endows endohedral fullerenes some unique and fascinating electronic properties, such as peculiar redox and photoelectrochemical behaviors, paramagnetism, and enhanced nonlinear optical response etc., which are inaccessible for the empty fullerenes.

The world of endohedral fullerenes was significantly extended by the discovery of metal nitride clusterfullerenes (NCFs) in 1999, which encapsulate a four-atomic metal nitride cluster in the interior of the carbon cage resulting in their mutual stabilization. As the first member of the NCFs, Sc_3N@C_{80}, which is also the most abundant NCF up to now, was synthesized by Dorn *et al.* via introducing a small portion of nitrogen gas into the Krätschmer–Huffman generator during vaporization of graphite rods containing metal oxides.[4] The discovery of NCFs is so significant that the idea of stabilizing both metal and nonmetal atoms in the form of metal cluster within a carbon cage led to the successive findings of several other types of clusterfullerenes, including metal carbide, hydrocarbide, oxide, sulfide, and carbonitride/cyanide clusterfullerenes.

A comprehensive review on NCFs was presented by Dunsch and Yang in 2007, accounting for all aspects of NCFs including the synthesis separation, molecular structures, electronic and chemical properties of NCFs as well as their potential applications.[1b] Since 2007, significant advances in the research field of NCFs have been achieved, including specifically both the inner and outer aspects. In addition to the success in encapsulating the non-group-III metal in the form of metal nitride into the carbon cage, the size of the NCF carbon cage has been extended to C_{96}. Herein we review the new advances of NCFs since 2007, focusing on the synthesis and isolation of novel NCFs based on new metals and cages. The unique molecular structures of NCFs are discussed so as to understand the stability of the NCF molecules. Then we discuss briefly the electronic, physical and chemical properties of NCFs as well as their potential applications.

2. SYNTHESIS, ISOLATION AND STRUCTURE OF NCFs

2.1 NCFs Based on Homogenous Metals

2.1.1 *NCFs Encaging Small-Size Metals (Sc, Y, Lu-Gd)*

The first NCF, $Sc_3N@C_{80}$, was discovered by chance, representing a milestone in the development of endohedral fullerene research.[4] In 1999, Dorn *et al.* discovered $Sc_3N@C_{80}$ by introducing a small portion of nitrogen gas into the Krätschmer–Huffman generator during vaporization of graphite rods containing scandium oxide. Up to now, $Sc_3N@C_{80}$ is the most abundant member of the family of endohedral fullerenes and is stabilized by the electron transfer from the inner Sc_3N nitride cluster to the outer icosahedral carbon cage, which are both not stable as the single moiety. Another noteworthy characteristic of $Sc_3N@C_{80}$ NCF is that its carbon cage has an isomeric form (I_h or D_{5h}), which has not been isolated in the empty fullerene. The diversities of the isomers of carbon cage have also enlarged the family of endohedral fullerenes.

It is known that the ionic radius of the encaged metal strongly influences the yield of the corresponding endohedral fullerenes.[1a–c] For scandium, which has the smallest atomic radius among rare earth metals (see Table 1), the yield of the corresponding $Sc_3N@C_{80}$ NCF is the highest

Table 1 Ionic Radius (r(M^{3+}), Å) of Rare Earth Metals[16]

M	Sc	Y	La	Ce	Pr	Nd	Gd	Tb	Dy	Ho	Er	Tm	Lu
r	0.75	0.90	1.05	1.03	1.01	0.98	0.94	0.92	0.91	0.90	0.88	0.87	0.85

among all NCFs reported to date. By using Dorn's method, named as trimetallic nitride template (TNT) process later, a series of new NCFs base on the different small-size metals have been synthesized, including $Sc_3N@C_{2n}$ (n = 34, 39),[5,6] $Gd_3N@C_{78}$,[7a,7b] $Y_3N@C_{80}$,[8] $Lu_3N@C_{80}$,[9] $Er_3N@C_{80}$,[4] $Ho_3N@C_{80}$,[10] $Tb_3N@C_{80}$,[11a,11b] $Tb_3N@C_{84}$ etc.[12] For Gd, which has a larger ionic radius than that of Sc, the yield of Gd-based NCFs is much lower and this hindered its isolation. In 2005, using the "reactive gas atmosphere" route, Dunsch *et al.* successfully synthesized a family of Gd-based NCFs, $Gd_3N@C_{2n}$ ($40 \le n \le 44$) and isolated the most abundant member, $Gd_3N@C_{80}$ (I).[7] Later on, Echegoyen *et al.* used the TNT method and successfully isolated $Gd_3N@C_{2n}$ (n = 42, 44) in 2007.[13] In 2009, Dorn's group synthesized and isolated Y-based NCFs with the cage size beyond C_{80} and extended to C_{86}. In the synthesis procedure, graphite rods containing a mixture of Y_2O_3, graphite powder, and Cu (as the catalyst, to be discussed in detail in the next section) with a ratio of 1.1:1.0:2.1 were vaporized in the Krätschmer–Huffman generator with a dynamic flow of the He and N_2 (the flow rate ratio=3:100).[8] Recently, based on the TNT process, Wang *et al.* synthesized and isolated $Y_3N@C_{78}$ with the C_2-C_{78} cage isomer, which is different from that of the reported $Sc_3N@C_{78}$ (D_{3h}). Note that for $Y_3N@C_{78}$, the Y_3N cluster is encaged in a non-IPR C_2-C_{78} cage with two pairs of fused pentagons, as determined by FTIR spectra in combination with DFT calculation.[14] This type of endohedral fullerene with the special cage structure is particularly interesting, as discussed further below.

2.1.2 *NCFs Encaging Large-Size Metals (Nd-La)*

As it is the smallest rare earth metal (see Table 1), scandium-based NCFs preferentially take the C_{80} cage at a yield higher than those based on other cages. Over the years, the encaged metals of the family of NCFs have been rapidly extended to rare earth metals with relatively larger sizes, including Nd, Pr, Ce and La. According to the analysis on the yield of the series of

NCFs based on the small-size metals discussed above and its relationship with the ionic radius of the encaged metals, it is obvious that, with the increase of the size of the encapsulated metal (from Sc to Gd), the yield of NCFs generally decreases, and C_{80}-based NCFs are dominantly abundant. Upon increasing the size (ionic radius) from Sc to Y, NCFs based on larger carbon cages such as C_{82}, C_{84}, C_{86} and C_{88} formed as well despite that the C_{80}-based NCF ($Y_3N@C_{80}$) is still the predominant NCF product.[8] A similar tendency was also observed for dysprosium (Dy).[15,16] In the case of gadolinium (Gd), the ionic radius increases further compared to Dy and Y, as a result the relative abundance of Gd-NCFs based on larger cages (C_{82}-C_{88}) becomes higher.[13] An abrupt change happens when the size of metal increases further to neodymium (Nd): the cage size distribution of $Nd_3N@C_{2n}$ is extended to C_{80}-C_{98} according to the mass spectroscopic results, which are dramatically different from $Gd_3N@C_{2n}$, and C_{88} cage turns into the most abundant cage for Nd-NCFs instead of C_{80} cage (see Fig. 1).[17] The reason is that the Nd_3N cluster is more suitable for accommodating in the larger cage because of its larger size. These results revealed that the small-size metals with the upper threshold of Gd tend to take preferentially the C_{80} cage, whereas the large-size metals with the lower threshold of Nd prefer the larger cages such as C_{88}.

On the basis of the high-selectivity production of NCFs through the "reactive gas atmosphere" method, in 2008 Echegoyen *et al.* synthesized new NCFs based on even larger metals such as Pr[18] and Ce[18] by using a very high ratio (1:1) of metal oxide to graphite powder. Note that the largest cages of $Pr_3N@C_{2n}$ and $Ce_3N@C_{2n}$ NCFs detected in the mass spectra are even extended to C_{104} and C_{106}, respectively, which are much larger than that of Nd (C_{98}). Moreover, in the case of Pr, the relative abundance of the C_{80}-NCF ($Pr_3N@C_{80}$) was lower than that of $Nd_3N@C_{80}$, while only a small amount of C_{86}-NCF was found and C_{80}-NCF ($Ce_3N@C_{80}$) completely disappeared in the case of $Ce_3N@C_{2n}$. Nevertheless, the C_{88} cage is still the preferential template for Pr_3N and Ce_3N nitrides, and a gradual increase in the abundance of C_{96} cage with the increase of size of metal was revealed by the mass spectroscopic and HPLC chromatogram analysis. Specifically $Ce_3N@C_{96}$ is the second most abundant NCF for the $Ce_3N@C_{2n}$ family. All these features suggest that with the further increase of the size of metal (cluster), the relative abundance of $M_3N@C_{96}$ may surpass that of C_{88} cage

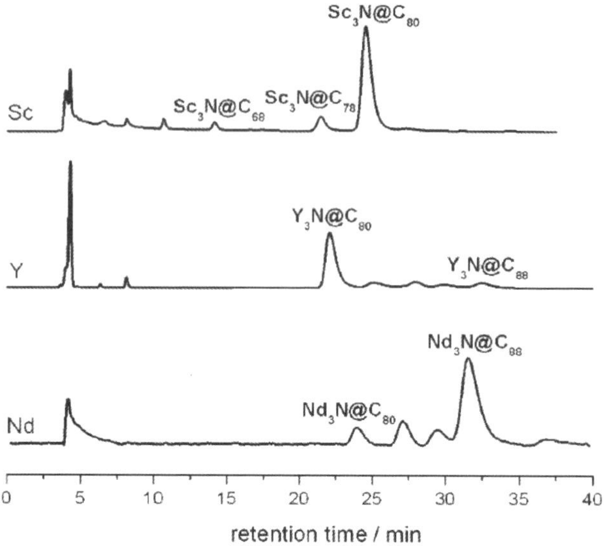

Fig. 1 HPLC traces of the $Sc_3N@C_{2n}$, $Y_3N@C_{2n}$, and $Nd_3N@C_{2n}$ families obtained by arc burning graphite rods containing 10% metal oxide in a mixture of ammonia (20 mbar) and helium (200 mbar); column: 10 × 250 mm Buckyprep-M; eluent: toluene; flow rate: 4 mL min^{-1}; detection at 372 nm. Reproduced with permission from Ref. 17. Copyright 2007 Wiley-VCH Verlag GmbH & Co. KGaA.

and hence the C_{96} cage becomes the preferential template. This presumption was experimentally confirmed by the later synthesis of $La_3N@C_{2n}$ by the same group, revealing an even larger cage size distribution from C_{86} to C_{110} as shown in the mass spectra. For NCF based on La_3N as the largest metal nitride cluster encapsulated in the carbon cage so far, $La_3N@C_{96}$ shows the highest yield followed by the species with C_{88} cage.[19] Thus, when the metal cluster size increases, a new template effect appears, and the strong dependence of the cage size distribution of NCF on the size of the encaged metal (cluster) has been demonstrated.

2.2 Mixed Metal Nitride Clusterfullerenes (MMNCFs)

2.2.1 *Scandium/Lanthanide Metal-Based MMNCFs*

In the course of the discovery of $Sc_3N@C_{80}$ in 1999, Dorn *et al.* also used a mixture of Er_2O_3, Sc_2O_3 and graphite as raw material for NCF synthesis.

Interestingly, they found that, in addition to the formation of homogenous metal-based NCFs $Sc_3N@C_{80}$ and $Er_3N@C_{80}$, Sc/Er mixed metal NCFs (MMNCFs) in the form of $Er_xSc_{3-x}N@C_{80}$ (x = 1, 2) were also detected in the mass spectrum of the mixture product. A qualitative comparison of the relative yield of $Er_xSc_{3-x}N@C_{80}$ (x = 1, 2) with those of $Sc_3@C_{82}$ and $Sc_3N@C_{80}$ was also made, indicating that the relative abundance of $Sc_xEr_{3-x}N@C_{80}$ (x = 1, 2) was one or two orders of magnitude higher than that of $Sc_3@C_{82}$, and the relative yield of $ErSc_2N@C_{80}$ was even higher than that of $Sc_3N@C_{80}$.[4] Based on the TNT process, a series of Sc-based MMNCFs have been synthesized, including $Er_xSc_{3-x}N@C_{80}$ (x = 1, 2), $CeSc_2N@C_{80}$,[20] $TbSc_2N@C_{80}$,[21] and the non-IPR structure of C_{68}-based MMNCFs, $A_xSc_{3-x}N@C_{68}$ (x = 1,2; A=Tm, Er, Gd, Ho, La).[22] The concept of mixing two different metals within the NCF cage has been also extended to encage up to three different metals in the form of MMNCFs. So far the only report on the MMNCFs encaging three different metals, ScYErN@C_{80}, was presented by Wang *et al.* by mixing Sc_2O_3, Y_2O_3, Er_2O_3 and graphite as the raw materials.[23] In this process two other kinds of MMNCFs, $Sc_xY_{3-x}N@C_{80}$ (x = 1, 2), $Y_xEr_{3-x}N@C_{80}$ (x = 1, 2), were also synthesized and detected in the mass spectrum.[24]

With the advantage of high-selectivity synthesis, the "reactive gas atmosphere" method was widely used to synthesize MMNCFs as well. Yang and Dunsch *et al.* synthesized and isolated a series of MMNCFs based on mixing of different lanthanide metals with Sc, including $M_xSc_{3-x}N@C_{80}$ (I, II, M = Lu, Gd, Dy, Nd, x = 1, 2),[25–27] and several MMNCFs with such non-IPR cages as $DySc_2N@C_{76}$,[28] $DySc_2N@C_{68}$, $Lu_xSc_{3-x}N@C_{68}$ (x = 1, 2).[29] Note that in 2006, Yang and Dunsch reported the first quantitative analysis of the relative yield of $Gd_xSc_{3-x}N@C_{80}$ (I_h, x = 1, 2) to that of $Sc_3N@C_{80}$ (I_h), which was determined to be 0.56:2.2:1 ($Gd_2ScN@C_{80}$: $GdSc_2N@C_{80}$: $Sc_3N@C_{80}$), confirming quantitatively that the yield of $GdSc_2N@C_{80}$ is much higher than that of $Sc_3N@C_{80}$.[25,26] In a follow-up study, Yang and Dunsch succeeded in isolating the second isomer, D_{5h}, of $Gd_xSc_{3-x}N@C_{80}$ (x = 1, 2), for which the relative yield was also compared quantitatively with that of $Sc_3N@C_{80}$ (D_{5h}) ($Gd_2ScN@C_{80}$ (D_{5h}): $GdSc_2N@C_{80}$ (D_{5h}): $Sc_3N@C_{80}$ (D_{5h}) = 0.36:0.8:1). Furthermore, the cage structures of the two isomers of $Gd_xSc_{3-x}N@C_{80}$ (I, II, x = 1, 2) were determined by FTIR and Raman vibrational spectroscopic studies combined with the DFT calculations, which revealed also the dependence of the cluster–cage interactions and geometric structures of the encaged $Gd_xSc_{3-x}N$ clusters on the composition (x value) of cluster. Interestingly, it was proposed that, while a pyramidal Gd_3N cluster configuration was experimentally confirmed for $Gd_3N@C_{80}$ (I, I_h) and predicted for $Gd_3N@C_{80}$ (II, D_{5h}), both the $GdSc_2N$ and Gd_2ScN clusters adopted the planar structures instead either in I_h and D_{5h} cage isomers of $Gd_xSc_{3-x}N@C_{80}$ MMNCFs (Fig. 2).[26]

Recently, Dunsch *et al.* successfully synthesized and isolated two isomers of $Ho_xSc_{3-x}N@C_{80}$ (x = 1, 2) MMNCFs, for which the isomeric structures were studied by UV/Vis/NIR, FTIR and Raman spectroscopy. They further carried out the first comprehensive ^{13}C and ^{45}Sc NMR spectroscopic study of $Ho_xSc_{3-x}N@C_{80}$ (I, II; x = 1, 2), not only confirming the I_h-symmetric cage structure for $Ho_xSc_{3-x}N@C_{80}$ (I, x = 1, 2) but also revealing the remarkably broad NMR peaks and strong paramagnetic chemical shifts of $Ho_xSc_{3-x}N@C_{80}$ (I, II; x = 1, 2) due to the strong influence of the

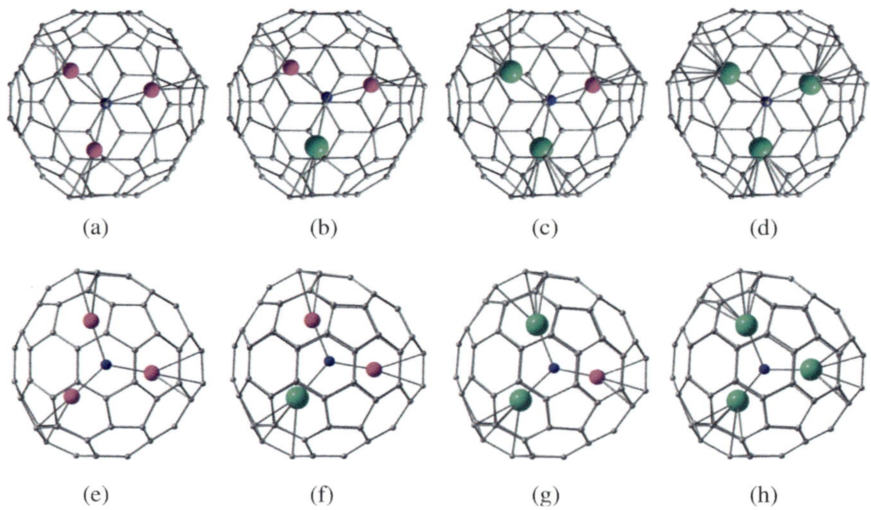

Fig. 2 DFT optimized structures of $Sc_3N@C_{80}$ (I_h, a), $GdSc_2N@C_{80}$ (I_h, b), $Gd_2ScN@C_{80}$ (I_h, c), $Gd_3N@C_{80}$ (I_h, d), $Sc_3N@C_{80}$ (D_{5h}, e), $GdSc_2N@C_{80}$ (D_{5h}, f), $Gd_2ScN@C_{80}$ (D_{5h}, g), and $Gd_3N@C_{80}$ (D_{5h}, h). The structures in a–d are viewed along the C_3 axis of the I_h-C_{80} cage, and for those in e–h the C_5 axis of D_{5h}-C_{80} cage is lying in the plane of the paper. The Sc, Gd, and N atoms are drawn in violet, green, and blue, respectively. Reproduced with permission from Ref. 26. Copyright 2008 Wiley-VCH Verlag GmbH & Co. KGaA.

$4f^{10}$ electrons located on the Ho atom. Besides, a variable-temperature ^{45}Sc NMR spectroscopic study pointed out that the Ho-induced paramagnetic shift was dominated by the pseudo-contact term.[30] More recently, Popov and Dunsch *et al.* isolated successfully a large family of $Gd_xSc_{3-x}N@C_{2n}$ (2n = 78–88) MMNCFs and studied the mutual influence of the cluster and cage size on the preferable fullerene structure systematically. It was found that Gd_3N-NCFs were not the preferable products in cages up to C_{88}, and there was a gradual decrease of the number of Sc atoms for the larger cages. As a result, both $GdSc_2N$ and Gd_2ScN clusters were obtained for C_{84} and C_{86} cages, whereas MMNCF with C_{88} cage encaging $GdSc_2N$ cluster was not detected at all.[31]

By mixing Sc with the largest lanthanide metal, La, Stevenson *et al.* recently prepared and isolated successfully $LaSc_2N@C_{80}$ MMNCF and elucidated its molecular structure by single-crystal X-ray diffraction, revealing that, despite of the relatively large size, the encaged $LaSc_2N$ unit is planar.[32]

2.2.2 *Non-Scandium MMNCFs*

According to the extensive study of the Sc-based MMNCFs, it appears that the Sc atom can direct the formation of MMNCFs by promoting the encapsulation of other metals within carbon cage due to the smallest ionic radius of Sc. As a result, for Sc-based MMNCFs ($M_xSc_{3-x}N@C_{2n}$) the yield of the homogenous Sc-NCFs dominates that of the M-based homogenous NCFs for which the formation is dramatically suppressed. In 2008, Yang *et al.* successfully synthesized and isolated the first MMNCFs without the involvement of Sc atom (non-Sc MMNCFs) — $Lu_xY_{3-x}N@C_{80}$ (x = 1, 2). A quantitative analysis of the relative yield of $LuY_2N@C_{80}$ and $Lu_2YN@C_{80}$ in comparison with those of the corresponding homogenous NCFs was made, which was 2.3: 4.6: 3.6 : 1 ($Y_3N@C_{80}$: $LuY_2N@C_{80}$: $Lu_2YN@C_{80}$: $Lu_3N@C_{80}$). Thus the enhancement of the yield upon the formation of MMNCFs was confirmed for non-Sc MMNCFs and a complicated competition effect when two metals with comparable ionic radii were mixed to form the MMNCFs was revealed. According to the ^{13}C NMR spectroscopic study of $LuY_2N@C_{80}$ and $Lu_2YN@C_{80}$, the I_h-C_{80} isomeric cage structure was determined. Furthermore, they also carried out a systematic study of the ^{13}C NMR chemical shifts of $Lu_xY_{3-x}N@C_{80}$ (x = 0–3) in combination with those of $Lu_xSc_{3-x}N@C_{80}$ (x = 0–3), and found that the two characteristic NMR lines experienced appreciable downshifts with the shrinking of the cluster size (see Fig. 3), which was interpreted by the correlation of the shifts of the ^{13}C NMR lines with the pyramidalization of the carbon atoms as analyzed in details by DFT theoretical computations.[33]

The next member of the non-Sc MMNCFs is $Lu_2CeN@C_{80}$, isolated by Dunsch *et al.* The electronic configuration of $Lu_2CeN@C_{80}$ was proposed to be $(Lu_2CeN)^{6+}@(C_{80})^{6-}$ with trivalent Ce, which however changed to a four-valent ion of Ce^{4+} upon electrochemical oxidation of $Lu_2CeN@C_{80}$, demonstrating an interesting endohedral redox behavior as discussed in detail in Section 3.2. Moreover, localization of the f-electron at Ce ion results in paramagnetic temperature-dependent shifts of ^{13}C NMR signals compared to the analogous values in diamagnetic $M_3N@C_{80}$.[34] Very

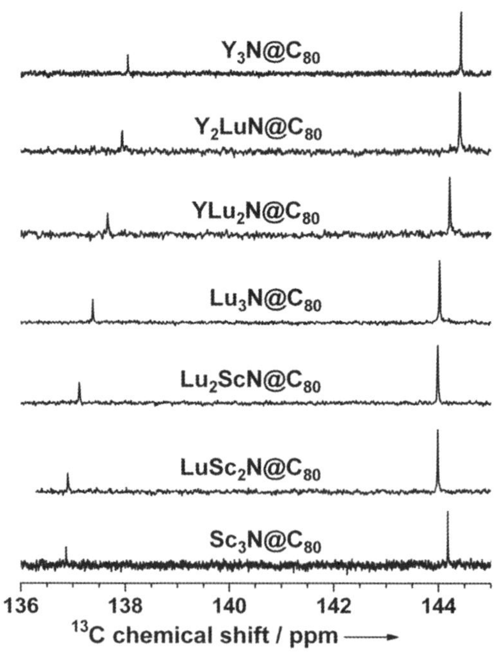

Fig. 3 The 150 MHz ^{13}C NMR spectra of $Lu_xY_{3-x}N@C_{80}$ and $Lu_xSc_{3-x}N@C_{80}$ (x = 0–3) in CS_2/d_6-Acetone at room temperature. Reproduced with permission from Ref. 33. Copyright 2008 Wiley-VCH Verlag GmbH & Co. KGaA.

recently, $CeY_2N@C_{80}$ was also synthesized using the "selective organic solid" (SOS) route and likewise exhibited peculiar redox behaviour.[35]

For non-Sc MMNCFs $A_xB_{3-x}N@C_{2n}$, because of the identity of the cage symmetry upon varying the composition of the $A_xB_{3-x}N$ nitride cluster especially when the sizes of A and B are comparable, their isolation becomes even more difficult and tedious than the Sc-based MMNCFs and homogenous NCFs.[1,33] This feature hampers the study of non-Sc MMNCFs for which the members reported so far are limited to $Lu_xY_{3-x}N@C_{80}$ (x = 1, 2), and $CeM_2N@C_{80}$ (M = Y, Lu).[33–35]

2.2.3 *MMNCFs Encapsulating Non-Group-III Metal Nitride*

Up till now, a wealth of NCFs including homogenous NCFs and MMNCFs have been synthesized and isolated by using versatilely modified Krätschmer–Huffman methods. However, the encaged metals for these reported NCFs is mainly limited to group-III metals. This fact significantly confines the availability of metals for encapsulation within NCFs and consequently limits diversities of NCFs compared to the conventional endohedral metallofullerenes. In 2009, for the first time, Yang *et al.* successfully put titanium (Ti) as a non-group-III metal into NCF carbon cage by synthesizing and isolating a new MMNCF based on the non-group-III metal – $TiSc_2N@C_{80}$ (Fig. 4).[36] Note that in this procedure the formation of Ti_3N and Ti_2ScN-NCFs were not detected, suggesting that, unlike other lanthanide metals, the Ti atom is more difficult to encapsulate into the carbon cage in the form of a nitride cluster, and this can however be conquered by the direction effect of Sc in some degree as discussed above. This work demonstrated also the possibility of encaging Ti into the C_{80}-I_h fullerene cage in a form other than metal carbide and dimetal, and greatly enriched the type of NCFs family. Interestingly, an endohedral redox behavior similar to that for $Lu_2CeN@C_{80}$ was also found for $TiSc_2N@C_{80}$,[37] and the details are to be discussed in Section 3.2. More recently, the same group reported the first Ti-containing non-Sc MMNCFs – $Y_2TiN@C_{80}$, not only extending the non-Sc MMNCFs to other groups but also suggesting that encapsulation of Ti into NCFs may be a general route for the synthesis of novel MMNCFs.[38]

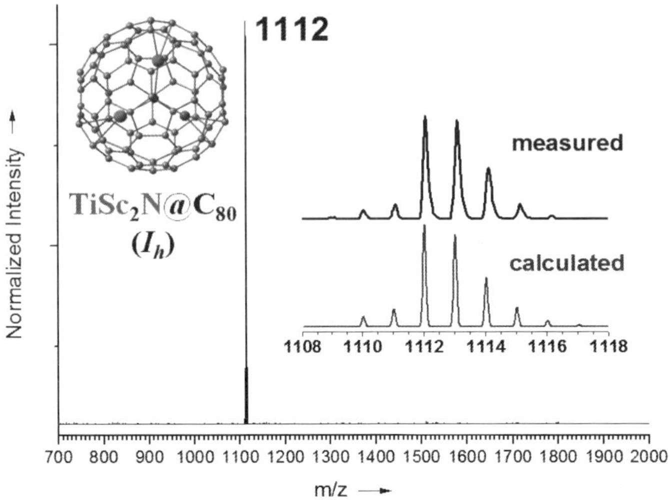

Fig. 4 Positive ion laser desorption time-of-flight (LD-TOF) mass spectrum and the schematic structure of $TiSc_2N@C_{80}$. The inset shows the measured and calculated isotope distributions of $TiSc_2N@C_{80}$. Reproduced with permission from Ref. 36. Copyright 2009 Royal Society of Chemistry.

2.2.4 *Isolation of MMNCFs*

Despite the fact that versatile separation techniques have been developed for NCFs, as discussed in detail in the review of Dunsch and Yang in 2007, high performance liquid chromatography (HPLC) has been demonstrated to be one of the most powerful and frequently used techniques for separation of NCFs, which allows a separation of fullerenes according to their molecular weight, size, shape or other properties.[1b–d] Through a judicious choice of the HPLC columns based on the polarity and number of phenyl rings etc., separation of the various isomers of NCFs is also possible. Specifically, for MMNCFs, their isolation becomes even more difficult and time consuming due to the identity of the cage symmetry upon varying the composition of the nitride cluster.[1,25,33] Thus, multi-step and/or recycling HPLC separation using different modified columns is required in most cases. As the first success, by using recycling HPLC, Yang *et al.* isolated two isomers of $Gd_xSc_{3-x}N@C_{80}$ ($x = 1, 2$)[25,26] with high purity (≥99%). In addition, a series of MMNCFs, including $DySc_2N@C_{76}$,[28] $MSc_2N@C_{68}$

(M = Dy, Lu),[29] $Lu_2ScN@C_{68}$,[29] $Lu_xY_{3-x}N@C_{80}$ (I, x = 0–2),[33] $M_xSc_{3-x}N@C_{80}$ (M = Dy, Lu, Nd),[27] $Ce_xSc_{3-x}N@C_{80}$,[20] $Tb_xSc_{3-x}N@C_{80}$,[21] $M_2TiN@C_{80}$,[36,38] $Ho_xSc_{3-x}N@C_{80}$[30] and $CeM_2N@C_{80}$ (M = Y, Lu)[34,35] have also been isolated. Indeed, these successes indicate that the separation of cage isomers of NCFs which relies on multi-step HPLC becomes is even simplified by using recycling HPLC.

For the isolation of the more recently synthesized MMNCF, $LaSc_2N@C_{80}$, Stevenson *et al.* applied the "stir and filter approach" (SAFA method), which was developed by themselves earlier and can be used to remove the empty cage fullerenes from a carbon soot efficiently,[39] and claimed to accomplish the selective purification of $LaSc_2N@C_{80}$ by two steps with the second step fulfilled by a single pass on the HPLC.[32]

2.3 New Nitrogen Sources for the Synthesis of NCFs

2.3.1 *"Selective Organic Solid" (SOS) Route*

As the crucial component atom of NCFs, nitrogen (N) was originated from the nitrogen sources used in the raw materials or gas atmospheres. All of the NCFs reported so far were synthesized primarily by two modified Krätschmer–Huffman DC arc discharging methods: the "trimetallic nitride template" (TNT) process named by Dorn *et al.* in 1999[4] and the "reactive gas atmosphere" method developed by Dunsch *et al.* in 2003,[40] with N_2 and NH_3 being used as the nitrogen source, respectively. Considering the practical need to simplify the method, solid nitrogen sources appear advantageous in terms of the simplicity of both the preparation of the raw mixture materials and the design of fullerene generator. Stimulated by the first attempt of using calciumcyanamide (CaNCN) as a inorganic solid nitrogen source by Dunsch *et al* in 2004,[41] which led to a strong increase of the selectivity of the NCFs (3 ~ 42%) but suffered from the low reproducibility of the fullerene yield, in 2010, Yang *et al.* succeeded in using a new organic nitrogen source, namely, guanidinium salts (guanidinium thiocyana and guanidinium hydrochloride) to accomplish the selective synthesis of a variety of NCFs based on different metals, including Dy, Sc, Y, Gd, Lu and mixed metals Sc/Dy, Sc/Gd, Sc/Lu and Lu/Ce. On the basis of the high selectivity of NCFs, which is comparable to that for the "reactive gas atmosphere" route, this approach was named the "selective organic solid" (SOS) method.[42] A possible reaction mechanism of guanidinium salts was proposed, featuring the *in situ* generation of NH_3 and $N = C(NH_2)$ through thermal decomposition, which contributes concurrently to the formation of NCFs. Obviously, the SOS route seems to be a general route for the synthesis of NCFs as demonstrated for different metal systems, and the advantage of the SOS route compared to the reported TNT process and the "reactive gas atmosphere" route is that no additional heating pretreatment is needed, thus simplifying the procedure and being much more facile.

2.3.2 *Urea ($CO(NH_2)_2$) as a Cheap Organic Nitrogen Source*

In 2012, Yang *et al.* developed a new organic nitrogen source ($CO(NH_2)_2$) to synthesize Sc-NCFs.[43] A quantitative comparison of the yield of $Sc_3N@C_{80}$ (I_h+D_{5h}) per gram Sc_2O_3 obtained by using $CO(NH_2)_2$ as new nitrogen source with those obtained by using other reported gaseous nitrogen sources revealed that the yield of $Sc_3N@C_{80}$ (I_h+D_{5h}) per gram Sc_2O_3 by using $CO(NH_2)_2$ was almost identical to that obtained by using N_2, but much higher than that obtained by using NH_3. Moreover, compared with the reported solid nitrogen source of guanidinium thiocyanate, the yield of $Sc_3N@C_{80}$ (I_h+D_{5h}) per gram Sc_2O_3 synthesized with $CO(NH_2)_2$ is around 1.6 times higher, while $CO(NH_2)_2$ is ca. 8.5 times cheaper than guanidinium thiocyanate. In this study, they also proposed the possible decomposition process of $CO(NH_2)_2$ during discharging, and concluded

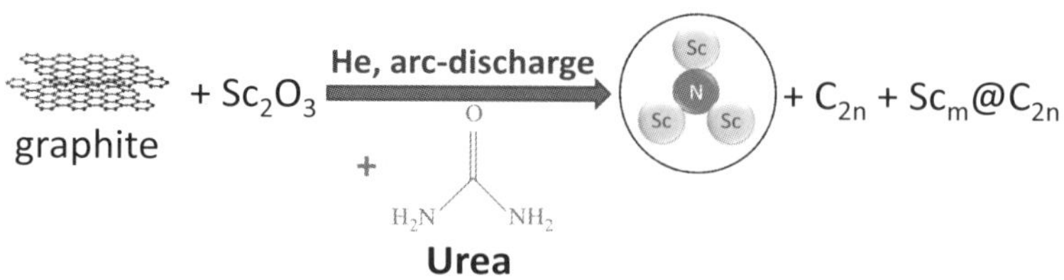

Scheme 1 A schematic diagram of the synthesis of Sc-based NCFs by a modified Krätschmer–Huffman DC arc discharging method using $CO(NH_2)_2$ as the new nitrogen source. C_{2n} and $Sc_m@C_{2n}$ represent the empty fullerenes (C_{60}, C_{70}, C_{76}, C_{78}, C_{84}, etc.) and Sc-based metallofullerenes ($Sc@C_{82}$, $Sc_2@C_{80}$, $Sc_3C_2@C_{80}$, etc.). Reproduced with permission from Ref. 43. Copyright 2012 Wiley-VCH Verlag GmbH & Co. KGaA.

that $CO(NH_2)_2$ decomposed and generated NH_3 *in situ*, which was responsible for the formation of Sc-NCFs. In addition, when two rods were used for arc discharge, NH_3 generated *in situ* from decomposition of $CO(NH_2)_2$ was mainly responsible for the formation of NCFs in the discharging process of first rod, whereas in the second rod $CO(NH_2)_2$ decomposed into melamine ($C_3H_6N_6$) during discharging of first rod.[43] It was shown recently that melamine itself can be also used as a source of nitrogen in the synthesis of NCFs.[43]

2.3.3 *Inorganic Solid Nitrogen Sources*

Inorganic compounds typically have a higher thermal stability and are more readily available than organic compounds; an intuitive question is whether nitrogen-containing inorganic compounds could act as nitrogen sources for NCF synthesis or not. In the first stage of modifying the synthesis of NCFs, calciumcyanamide (CaNCN) was added to the graphite mixture as a inorganic solid nitrogen source by Dunsch *et al.* in 2004, resulting in a strong increase of the selectivity of the Sc-NCFs (3 ~ 42 %). However, CaNCN has the disadvantage in the low reproducibility of the fullerene yield, which might be due to traces of water and/or hydrocarbons in the raw material.[41] To modify the "inorganic solid nitrogen" route, Yang *et al.* more recently successfully developed a series of nitrogen-containing inorganic solid compounds as new nitrogen sources for the synthesis of Sc-based NCFs, including 12 compounds falling into four categories: ammonium salts ($(NH_4)_xH_{3-x}PO_4$ (x = 0–2), $(NH_4)_2SO_4$, $(NH_4)_2CO_3$, NH_4X (X = F, Cl), NH_4SCN), thiocyanates (KSCN), nitrates ($Cu(NO_3)_2$, $NaNO_3$) and nitrite ($NaNO_2$).[45] On the basis of a quantitative comparison on the yield of Sc-NCFs, ammonium thiocyanate (NH_4SCN) was revealed to behave as the best nitrogen source. On the other hand, among these 12 new inorganic nitrogen sources, $Cu(NO_3)_2{\cdot}2.5H_2O$ was previously used by Stevenson *et al.* as an additive for the synthesis of Sc-NCFs via the TNT process with gaseous N_2 as the nitrogen source, resulting in the selective synthesis of Sc-NCFs.[44] Nevertheless, in that report the major nitrogen source was still gaseous N_2, and $Cu(NO_3)_2$ was believed to behave as an additive.[44] The more recent study by Yang *et al.* demonstrated clearly that $Cu(NO_3)_2$ can work as an independent nitrogen source.[45]

In order to understand the different performances of these 12 new inorganic nitrogen sources toward the synthesis of Sc-NCFs, the authors discussed their thermal decomposition products and revealed the dependence of the production yield of Sc-NCFs on the oxidation state of nitrogen and counter ion. Interestingly, the yield of $Sc_3N@C_{80}$ ($I_h + D_{5h}$) per gram Sc_2O_3 by using the N_2-based group of nitrogen sources (thiocyanate, nitrates, and nitrite) was in overall much lower than those using gaseous N_2 and NH_4SCN, indicating the strong dependence of the yield of Sc-NCFs on the oxidation state of nitrogen, which was attributed to the *in situ* redox reaction taking place for the N_2-based group of nitrogen sources during discharging. For NH_3-based group of nitrogen sources (ammonium salts) which exhibits (−3) oxidation states of nitrogen, their performance as nitrogen sources was found to be sensitively dependent on the anion.[45]

2.4 New Carbon Cages

2.4.1 *Non-IPR NCFs*

It is well known that all empty fullerenes and many endohedral fullerenes follow the isolated pentagon rule (IPR), which can determine the stability of all carbon fullerenes. The presence of adjacent pentagon pairs in the fullerene cage (which is thus non-IPR, i.e. violating the IPR) increases the strain and destabilizes the π-system of the carbon cage, which results in the instability of the entire fullerene molecule.[46] Owing to the electron transfer from the encaged species to the carbon cage, the formation of endohedral fullerenes provides an ideal way to stabilize the adjacent pentagons and consequently violate the IPR. In 2000, Shinohara's group reported the first non-IPR endohedral fullerene – $Sc_2@C_{66}$.[47] Almost at the same time, Dorn's group reported the first non-IPR NCF – $Sc_3N@C_{68}$, for which the cage structure was determined by ^{13}C NMR to be D_3:6140 and confirmed by X-ray crystallography later.[22] From then on, several non-IPR NCFs were isolated, including $Sc_3N@C_{70}$ (C_{2v}:7854),[48] $Tm_3N@C_{78}$ and $Dy_3N@C_{78}$ (II) (both have C_{78}:C_2 (22010) cages),[49] $Tb_3N@C_{84}$ (C_s:51365),[12] and $DySc_2N@C_{76}$ (C_s:17490-I),[28] but the reported numbers of non-IPR NCFs are still very limited by 2007. Among them, $DySc_2N@C_{76}$ (C_s:17490-I), isolated by Yang *et al.*, is of particular importance because it provides the first example of stabilization of the non-IPR C_{76} cage by encapsulation of an asymmetric $DySc_2N$ mixed cluster which has a more suitable geometry for the inner space of the C_{76} cage compared to that of a homogeneous cluster like Sc_3N or Dy_3N. Thus the role of the cluster structure on the stability of the non-IPR fullerene cage was revealed.[28] In 2008, Echegoyen *et al.* successfully synthesized and isolated a new non-IPR NCF, $Gd_3N@C_{82}$-C_s:39663, whose structure was elucidated by X-ray crystallography study revealing the existence of one pair of adjacent pentagons.[50] Later on, Echegoyen and Dorn *et al.* isolated and characterized the most abundant isomers of $Tm_3N@C_{84}$ and $Gd_3N@C_{84}$, both involving a non-IPR C_{84} cage with C_s:51365 symmetry which were similar to the previously reported $Tb_3N@C_{84}$.[51] Note that the C_{84} (C_s:51365) cage revealed in these non-IPR

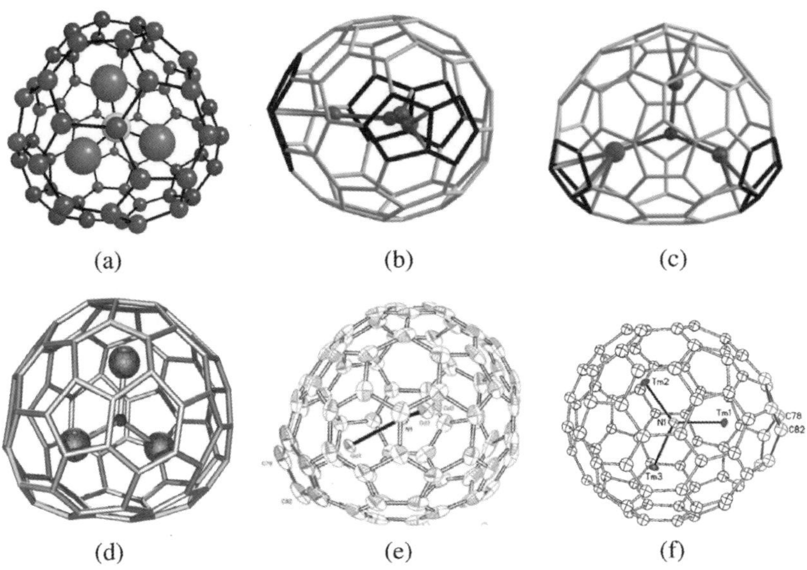

Fig. 5 An exhaustive list of the non-IPR NCFs reported up till now. (a) $Sc_3N@C_{68}$ (D_3:6140). Reproduced with permission from Ref. 22. Copyright 2000 Nature Publishing Group. (b) $Sc_3N@C_{70}$ (C_{2v}:7854). Reproduced with permission from Ref. 48. Copyright 2007 Wiley-VCH Verlag GmbH & Co. KGaA., (c) $DySc_2N@C_{76}$ (C_s: 17490-I). Reproduced with permission from Ref. 28. Copyright 2007 American Chemical Society. (d) $M_3N@C_{78}$ (C_2: 22010) (M = Y[14], Dy[49], Tm[49], Gd[7b]). Reproduced with permission from Ref. 7b. Copyright 2009 American Chemical Society. (e) $M_3N@C_{82}$ (C_s:39663) (M = Gd[50], Y[8]). Reproduced with permission from Ref. 50. Copyright 2008 American Chemical Society. (f) $M_3N@C_{84}$ (C_s:51365) (M = Tm[12], Tb[51], Gd[51], Y[8], Reproduced with permission from Ref. 51. Copyright 2008 Royal Society of Chemistry.

NCFs appears to be the largest cage so far that can adopt the non-IPR structure. Meanwhile, Dorn's group also reported two new Y-based non-IPR NCFs, namely $Y_3N@C_{84}$-C_s:51365, $Y_3N@C_{82}$-C_s:39663, for which the cage structures were determined by ^{13}C NMR spectroscopy.[8] More recently, Echegoyen's group further isolated $Gd_3N@C_{78}$ (C_2:22010) with two pairs of adjacent pentagons. The structure has been determined by X-ray diffraction study with two gadolinium atoms located within the pentalene folds formed by the adjacent pentagons and the third gadolinium atom residing at the center of a hexagonal face of the fullerene. The structure of $Gd_3N@C_{78}$ (C_2:22010) is unique as a rare example of encapsulating large metal ions in a relatively small fullerene cage.[7b] An analogous non-IPR C_{78} cage-based NCF was $Y_3N@C_{78}$ (C_2:22010) isolated by Wang *et al.* in 2011, for which the cage structure was determined by UV-vis and vibrational spectrosopies in combination with DFT calculations.[14]

2.4.2 *Metal Nitride Azafullerene*

Endohedral azafullerenes as a special type of endohedral fullerene with the carbon atom(s) of the carbon cage substituted by nitrogen atom(s) have been predicted theoretically, but the experimental detection was not reported until 1999 when Akasaka *et al.* reported the gas-phase formation of $[La@C_{81}N]^+$ and $[La_2@C_{79}N]^+$ by the fast atom bombardment mass fragmentation of the $La@C_{82}$ adducts of $La@C_{82}(NCH_2Ph)$ and $La_2@C_{80}(NCH_2Ph)$.[52] However, the synthesis of endohedral azafullerenes in bulk quantities was not fulfilled until the discovery and successful isolation of $M_2@C_{79}N$ (M = Y, Tb) by Dorn *et al.* in 2008, using the identical condition (using N_2 as the nitrogen source) to that used for the synthesis of NCFs.[53] One year later, using the "Chemically Adjusting Plasma Temperature, Energy, and Reactivity" (CAPTEAR) method, i.e., adding $Cu(NO_3)_2 \cdot 2.5H_2O$ as an exothermic nitrate moiety to suppress empty-cage fullerene formation, Stevenson *et al.* detected the formation of a metal nitride azafullerene $La_3N@C_{79}N$. Surprisingly, the corresponding $La_3N@C_{80}$ NCF was not detected, suggesting that the La_3N cluster selects a smaller 80-atom cage only if one of the carbon cage atoms is substituted with nitrogen atom. Besides, the yield of $La_3N@C_{79}N$ even surpasses that of $La_3N@C_{96}$.[54] Thus, the authors concluded that, for the large La_3N cluster, the reduction of the preferential cage size from 96 to 80 atoms was due to electronic effects in lieu of ionic radius.[54]

More recently, Dorn's group has synthesized and isolated another kind of azafullerene – $Gd_2@C_{79}N$ – with huge magnetism, which will show great potential for application in the biomedicine field.[55]

3. PROPERTIES OF NCFS

3.1 Electronic Absorption Properties

The electronic absorptions of NCFs arise predominantly from the carbon cage excitations, which are mainly dependent on the structure and charge state of carbon cage. An important criterion for the electronic stability of NCFs is the energy difference between the highest occupied and the lowest unoccupied electronic state, i.e. the HOMO-LUMO gap, which roughly corresponds to the optical band-gap. Experimentally, since the absorption bands are usually quite broad and sometimes several broad bands may overlap, an accurate determination of the optical band-gap is difficult, thus the optical band-gap is often estimated from the onset of electronic absorption.[4,15–16,25–30,33,36,42,48] For instance, for $Sc_3N@C_{80}$, the absorption spectral onset is located at ca. 820 nm, and the optical band-gap is calculated to be 1.51 eV (1240/820 (nm)). Using the value of 1.0 eV to distinguish large and small band-gap fullerenes, $Sc_3N@C_{80}$ is categorized as a large band-gap fullerene. Note that all the C_{80}-based NCFs reported till now are assigned as large band-gap fullerenes. For MMNCFs, an intriguing question is whether the change of the composition of the encaged mixed metal nitride would sensitively affect the absorption property. In 2009, Yang *et al.* compared the UV-vis-NIR spectra of a series of $M_xSc_{3-x}N@C_{80}$ (I) (M=Sc, Lu, Er, Dy, Gd, Nd; x=1, 2), indicating that the overall absorption features are very similar, while a detectable difference was found in the spectral region of 600–800 nm.[27] For $M_2ScN@C_{80}$ (Fig. 6a), the two maximum absorption peaks at 691/666 in $Lu_2ScN@C_{80}$ are gradually red-shifted to 698/669, 707/674, 703/672, 711/677 nm in $Er_2ScN@C_{80}$, $Dy_2ScN@C_{80}$ (I), $Gd_2ScN@C_{80}$ (I) and $Nd_2ScN@C_{80}$, respectively. The absorption spectra of $MSc_2N@C_{80}$ (I) (Fig. 8b) show almost the same phenomenon with the absorption maximum at 706 nm in $LuSc_2N@C_{80}$

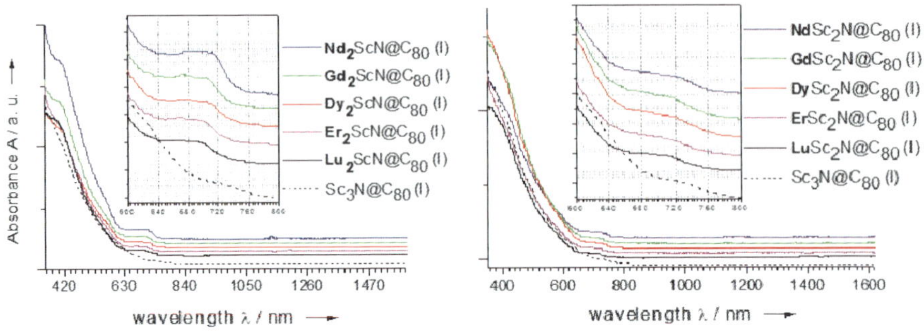

Fig. 6 UV-vis-NIR spectra of $M_2ScN@C_{80}$ (I) (a) and $MSc_2N@C_{80}$ (I) (b) dissolved in toluene (M=Lu, Er, Dy, Gd, Nd) in comparison with that of $Sc_3N@C_{80}$ (I). The insets show the enlarged spectral region. Reproduced with permission from Ref. 27. Copyright 2009 American Chemical Society.

red-shifted to 711 nm in $ErSc_2N@C_{80}$, 715 in $DySc_2N@C_{80}$, 712 nm in $GdSc_2N@C_{80}$, and 718 nm in $NdSc_2N@C_{80}$. However, the absorption peak at 685 nm in $LuSc_2NC_{80}$ is blue-shifted to 682 and 680 nm in $ErSc_2N@C_{80}$, $DySc_2N@C_{80}$ and even disappears in $GdSc_2N@C_{80}$ and $NdSc_2N@C_{80}$. All these results suggest that the electronic spectra of $M_xSc_{3-x}N@C_{80}$ are mainly dominated by the C_{80} carbon cage, while the endohedral nitride cluster imposes a distinct but not strong perturbation of the cage excitations.

3.2 Electrochemical Properties

Electrochemical characterization is the most commonly used method to investigate the electronic properties of NCFs, and can provide a direct insight into their electronic structures and the interaction between the encaged clusters and outer carbon cages. Since the first study on the redox property of NCFs, $Sc_3N@C_{80}$ (I_h), by Dunsch *et al.* in 2004,[56] extensive electrochemical studies of different NCFs have been reported, and the detailed discussion on the electrochemical properties of NCFs reported before 2009 has been summarized in the general review by Dunsch and Yang,[1b,c] as well as a more dedicated review by Echegoyen *et al.*[57] In the following discussion we focus on new advances after 2009, especially the endohedral electrochemistry of NCFs.

The unique electronic feature of endohedral fullerenes, i.e. electron transfer from the encaged species to the carbon cage, brings about the unusual electrochemical property of endohedral fullerenes not accessible for empty fullerenes. Since the encaged species is entrapped within the interior of carbon cage and isolated from the outer environment by the carbon cage, direct contact of endohedral species with the electrode and electrolyte are impossible. However recent electrochemical studies of endohedral fullerenes indicate that the endohedral species can be still electrochemically active, thus exhibiting an interesting endohedral redox behavior specifically for NCFs.[58] For the pristine NCFs, the first report on the endohedral electrochemistry was reported by Popov, Yang, and Dunsch *et al.* for Ti-based MMNCF $TiSc_2N@C_{80}$. According to the cyclic voltammogram of $TiSc_2N@C_{80}$ measured in o-dichlorobenzene (o-DCB), one reversible oxidation and three reversible reduction steps at +0.16, −0.94, −1.58, −2.21 V vs Fc/Fc^+ was observed (Fig. 7). Further comparative study with the electrochemical property of the analogous $Sc_3N@C_{80}$ pointed out that the overall redox process of $TiSc_2N@C_{80}$ was completely different from that of $Sc_3N@C_{80}$ in terms of the potential and reversibility of redox peaks. Based on DFT computational study, this phenomenon was

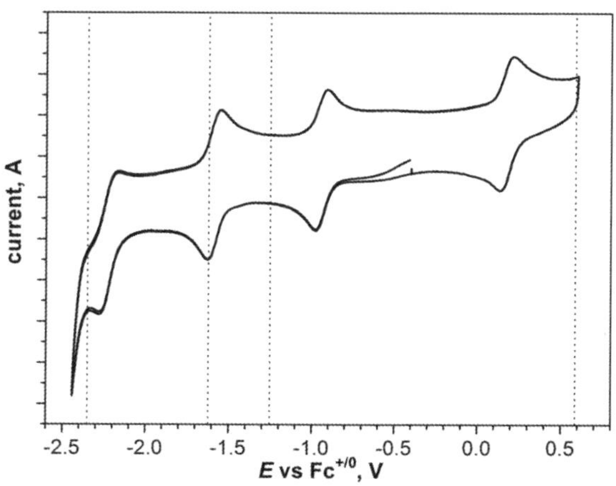

Fig. 7 Cyclic voltammetry of $TiSc_2N@C_{80}$ measured in *o*-DCB solution (room temperature, $TBABF_4$ as supporting electrolyte) at a scan rate of 20 mV s^{-1}. Dotted vertical bars denote reversible redox potentials of $Sc_3N@C_{80}$. Reproduced with permission from Ref. 37. Copyright 2010 American Chemical Society.

interpreted by considering the endohedral redox of the Ti atom: for the pristine $TiSc_2N@C_{80}$ NCF, Ti is in the Ti^{III} state, whereas in the first oxidation and reduction step the valence state of the endohedral Ti changes to Ti^{II} in $TiM_2N@C_{80}^-$ and Ti^{IV} in $TiM_2N@C_{80}^+$.[36,37] Thus, Ti in $TiSc_2N@C_{80}$ represents the first example of an endohedral fullerene with an endohedral redox activity both in the reduction and oxidation. As a result, the redox potentials of $TiSc_2N@C_{80}$ are quite different from those of other $M_3N@C_{80}$, and a considerably smaller electrochemical gap (1.10 V) is obtained for $TiSc_2N@C_{80}$.[36,37] More recently, a similar phenomenon was reported for another Ti-based MMNCF $TiY_2N@C_{80}$, which exhibits one reversible oxidation step with $E_{1/2}$ at 0.00V and a reversible first reduction step with $E_{1/2}$ at −1.11V.[38] Interestingly, it was found that both the first oxidation and the first reduction steps of $TiY_2N@C_{80}$ were obviously more negative than those of $TiSc_2N@C_{80}$, while the electrochemical energy gap of $TiY_2N@C_{80}$ (1.11 V) is almost the same as that of $TiSc_2N@C_{80}$ (1.10 V). These results indicate clearly the strong influence of the encaged group-III metal on the electronic property of $TiM_2N@C_{80}$ (M = Y, Sc).

A similar situation takes place in $Lu_2CeN@C_{80}$ MMNCF reported by Dunsch *et al.*[34] While the valence state of Ce within $CeLu_2N@C_{80}$ was determined to be Ce^{III} with one localized 4f-electron, the electrochemical study showed that the oxidation potential of $CeLu_2N@C_{80}$ was 0.6 V less positive than the oxidation potentials of all other $M_3N@C_{80}$ NCFs. Such an intriguing oxidation behavior was analyzed by computational study and interpreted by the removal of the $4f^1$ electron from Ce under formation of the $Ce^{IV}Lu_2N@C_{80}$ state which was more energetically favourable than the oxidation of the C_{80} carbon cage as occurring in all other $M_3N@C_{80}$ NCFs.[34] Accordingly, the authors concluded that $CeLu_2N@C_{80}$ was so far the only known endohedral fullerene in which an endohedral Ce^{III} exhibited a redox activity and can be reversibly oxidized to a Ce^{IV} state.[34] Very recently this study was extended to the analogous $CeY_2N@C_{80}$ and $CeSc_2N@C_{80}$.[35] Endohedral oxidation of Ce was observed thus in all $CeM_2N@C_{80}$ NCFs (M = Sc, Y, Lu) and confirmed by NMR studies of the pristine NCFs and their cationic forms (which showed the absence of paramagnetic shift). Furthermore, oxidation potential of the endohedral Ce^{III} ion was found to be strongly dependent on the second cluster metal. With the help of DFT calculations, this dependence was explained by the release of the strain of the inner cluster when Ce^{III} is oxidized (Ce^{IV} has much smaller ionic radius). The stronger the strain in the pristine NCF (ie. the larger the cluster), the easier it is to oxidize.[35]

3.3 Magnetic Properties

One of the most intriguing features of endohedral fullerenes is the large magnetic moments and a variety of unique magnetic properties resulting from the encaged lanthanide ions which have the unfilled 4f shells. In particular, with the encapsulation of three magnetic metal ions with triangular configuration, NCFs show much more interesting magnetic properties. Following the first SQUID study of the magnetic property of NCFs ($M_3N@C_{80}$ (M = Ho, Tb)) by Dunsch and Wolf *et al.* in 2005,[59] in recent years the studies of the magnetic property of NCFs by SQUID have been reported on several different NCFs.[59–65] Tiwari *et al.* measured the magnetic moments of $Er_xSc_{3-x}N@C_{80}$ (x = 0–3) MMNCFs in 2008 and found a paramagnetic behavior for all NCFs with negligible Weiss temperatures. Interestingly, the effective magnetic moment per Er atom was gradually decreasing with the increase of the number of Er atoms (x value) in $Er_xSc_{3-x}N@C_{80}$, from 9.8±0.5 μ_B in $ErSc_2N@C_{80}$ to 6.7 ± 0.2 μ_B in $Er_3N@C_{80}$. This behavior was interpreted by the quenching of the angular momentum due to the crystal-field effect from the Er^{3+} ions.[60] In the same year, Smirnova *et al.* reported the SQUID study of $Er_3N@C_{80}$ NCF as well, revealing that the total effective magnetic moment of $Er_3N@C_{80}$ determined either from temperature-dependent susceptibility (at 2–300 K) or isothermal magnetization measurements (at 2 K) was only 10.2 μ_B (i.e. only 3.4 μ_B per Er),[61] which was even smaller than the value determined by Tiwari *et al.*[60] and corresponded to the complete quenching of the orbital momentum ($J \approx S = 3/2$ per Er^{3+} ion).[60] Another independent study on the magnetic property of $Tm_3N@C_{80}$ was reported by Dorn *et al.* in the same year, revealing that the total magnetic moment of $Tm_3N@C_{80}$ at $T > 150$ K was equal to three times of the magnetic moment of an isolated Tm^{3+} ion without any degree of quenching.[62] Obviously this result is quite different to the case of $Er_3N@C_{80}$ discussed above.

Two SQUID studies on the magnetic peroperties of $Gd_3N@C_{80}$ were reported recently. According to the the on-chip SQUID measurements of

$Gd_3N@C_{80}$ at the temperatures down to 0.1 K by Chen *et al.* in 2011, the magnetization curves exhibited no hysteresis in the whole temperature range confirming a completely paramagnetic behavior. In the low temperature region (T < 1.2 K), the magnetic behavior of $Gd_3N@C_{80}$ was well described by three Gd^{3+} ions in their ground $^8S_{7/2}$ state ($\mu_{eff} = 7.9\ \mu_B$) with weak antiferromagnetic exchange interaction (J_{ij} = 7.6 mK), whereas at T > 1.2 K the data were better described by considering that one of the Gd ions is in the excited 2+ state with $^7F_6$ configuration ($L = S = 3$, $J = 6$, $\mu_{eff} = 9.7\ \mu_B$).[63] More recently, Jánossy studied the magnetic properties of $Gd_3N@C_{80}$ by using both SQUID and high-frequency ESR techniques in 2012, indicating that at low temperature the effective magnetic moment corresponds to three coupled $S = 7/2$ spins with total $\mu_{eff} = 21.9\ \mu_B$, whereas the spins are uncorrelated and μ_{eff} of the molecule is approaching the values of 13.7 μ_B at higher temperature.[64]

Very recently, Greber *et al.* reported the first single-molecular magnetism study of NCFs. Both XMCD and SQUID studies of $DySc_2N@C_{80}(I_h)$ MMNCF revealed a characteristic butterfly-shape hysteresis of magnetization curves at temperatures below 6 K, pointing to the long relaxation time of magnetization and showing that $DySc_2N@C_{80}$ behaves as a mononuclear single-molecular magnet.[65]

As another powerful technique to investigate the magnetic properties of NCFs, electron spin resonance (ESR) has been also frequently used even at the early stage of endohedral fullerene research.[36–38] As the only non-group-III metal-based paramagnetic NCFs isolated up to now, $TiM_2N@C_{80}$ (M = Sc, Y) show peculiar ESR features. The room-temperature ESR spectrum of $TiSc_2N@C_{80}$ exhibits a very broad signal with a g-factor of 1.9454, which shifted considerably compared to from the value of the free electron, suggesting the localization of the spin density on Ti atom.[36,37] However, the line width decreased and the signal intensity gradually increased with the lowering of the measurement temperature till the freezing temperature of toluene. The strong ESR signal in the low temperature could be understood by taking into account the valence state of Ti^{3+} with one unpaired electron located on the 3d orbital ($3d^1$) leading to paramagnetic state of $TiSc_2N@C_{80}$. Upon substituting Sc with Y, the ESR spectrum of $TiY_2N@C_{80}$ was found to be almost the same with that of

$TiSc_2N@C_{80}$, exhibiting a broad signal with g-factor of 1.9579 at room temperature which became more intense and narrower with the gradual decrease of the measurement temperature as well.[38] Interestingly, the line width of the ESR spectrum of $TiY_2N@C_{80}$ was dramatically smaller than that of $TiSc_2N@C_{80}$, suggesting the different electronic structures of these two Ti-based MMNCFs resulted from the different electronic configurations between Y^{3+} and Sc^{3+} as well as the different geometric structures of TiM_2N.[38]

3.4 Chemical Properties

Exohedral chemical functionalization of fullerenes plays a significant role in fullerene research not only for the commercialized and potential applications but also because of the advantage of characterizing fullerenes for which the pristine structures could not be directly investigated by NMR or X-ray crystallography. Since the first exohedral functionalization of endohedral fullerenes undertaken by Akasaka *et al.* in 1995 by a photochemical reaction of $La@C_{82}$-A (i.e. $C_{2v}(9)$ isomer) with disilirane leading to the formation of a 1:1 adduct,[66] several different types of exohedral functionalization reactions of NCFs have been developed, including photochemical reaction, Diels–Alder reactions, Prato reaction, Bingel–Hirsch reaction, free-radical addition reaction and some other addition reactions.[57] The detailed discussion on the chemical properties of NCFs has been comprehensively summarized in several reviews by Echegoyen *et al.* in 2009,[57] Lu and Akasaka *et al.* in 2011 and 2012.[67,1g] Here we discuss only two recent supplementary yet important reactions of NCF – trifluoromethylation reaction and azide addition reaction.

By using CF_3I reacting with $Sc_3N@C_{80}$-D_{5h} and $Sc_3N@C_{80}$-I_h, recently Yang and Troyanov *et al.* successfully synthesized and isolated the first multi-CF_3 derivatives of $Sc_3N@C_{80}$-D_{5h}, $Sc_3N@C_{80}$-$D_{5h}(CF_3)_{18}$ and three new isomers of $Sc_3N@C_{80}$-$I_h(CF_3)_{14}$. All the structures have been unambiguously determined by X-ray crystallographic study and in combination with DFT calculation, indicating that the exohedral additions of CF_3 moiety have a strong influence on the behavior of the Sc_3N cluster inside the C_{80} cage.[68]

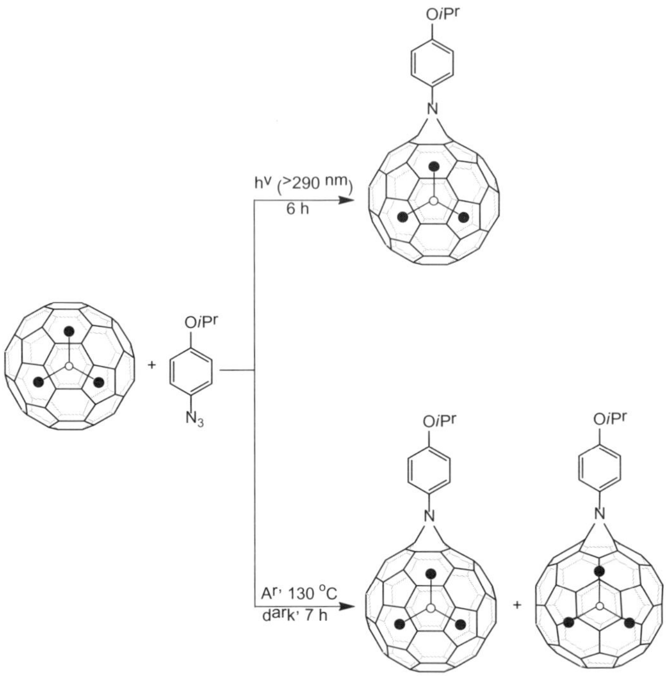

Scheme 2 Photoaddition and thermal Reaction of $Sc_3N@I_h$-C_{80} with 4-isopropoxyphenyl azide under different conditions. Reproduced with permisssion from Ref. 69. Copyright 2012 American Chemical Society.

In 2012, Wang, Yang and Balch *et al.* reported the first reaction of an organic azide (4-isopropoxyphenyl azide) with $Sc_3N@I_h$-C_{80} NCF. Both the thermal and photochemical reaction of $Sc_3N@I_h$-C_{80} with 4-isopropoxyphenyl azide were studied, revealing that the thermal reaction afforded both [6,6]- and [5,6]-azafulleriods while the corresponding photochemical reaction generates only the [6,6]-azafulleroid isomer (Scheme 2). While the structure of the [5,6]-azafulleriod product was determined unambiguously by X-ray crystallography, the structure of [6,6]-azafulleroid isomer was proposed by the computational studies, which indicated further that the [6,6]-azafulleriod isomer was less stable than the [5,6]-azafulleriod isomer and the *in-situ* generated nitrene preferentially reacts with the [6,6] ring junction of $Sc_3N@I_h$-C_{80} to give a kinetic product.[69]

4. POTENTIAL APPLICATIONS OF NCFS

For endohedral fullerenes, the intriguing electronic property arising from the electron transfer from the encaged species to the carbon cage brings about many fascinating and potential applications of endohedral fullerenes in such fields as biomedicine, molecular electronics and photonics, etc.[1,70,71] In particular, the capability of encapsulating three magnetic metals in either homogenous or mixed form enables NCFs superior performance in some areas, thus making them even more promising. The detailed exploration of the potential applications of endohedral fullerenes in general has been given in the previous reviews[1]; here we discuss only the two most promising applications.

4.1 Biomedical Applications as Magnetic Resonance Imaging (MRI) Contrast Agents

The extensive study on the potential applications of NCFs as MRI contrast agents has been reported since 2006 mainly because of their higher yield combined with the higher relaxivity compared to conventional endohedral fullerenes. As the first *in vitro* and *in vivo* imaging study of $Gd_3N@C_{80}$ NCF, Fatouros and Dorn *et al.* first functionalized $Gd_3N@C_{80}$ with poly(ethylene glycol) (PEG) units and then hydroxylated the derivative ($Gd_3N@C_{80}$[DiPEG5000$(OH)_x$]), which exhibited a remarkedly improved water solubility and biodistribution compared to the commercial MRI contrast agent gadodiamide.[72] Later on, the same groups synthesized a series of water-soluble PEG functionalized and hydroxylated $Gd_3N@C_{80}$ by varying the molecular weights of PEG (350–5000 Da) and measured their relaxivities at three magnetic field strengths (0.35, 2.4 and 9.4 T). According to their results, the highest relaxivity r_1, 237/232 mM^{-3} s^{-1} at a clinical-range magnetic field of 2.4 T, was obtained for the 350/750 Da PEG derivatives, which was ca. 60 times higher than that of commercially used Gd^{3+} ion chelated contrast agents such as Gd-DTPA.[73]

In addition to the PEG functionalization of $Gd_3N@C_{80}$ NCF, several other groups developed different protocols to prepare different types of $Gd_3N@C_{80}$-based water-soluble derivatives as MRI contrast agents. In 2008, MacFarland *et al.* synthesized a series of novel derivatives of $Gd_3N@C_{80}$

termed hydrochalarones, $Gd_3N@C_{80}$-Hydrochalarone ($Gd_3N@C_{80}-R_x$, where $R = [N(OH)(CH_2CH_2O)_nCH_3]_x$, n = 1, 3, 6 and x = 10–22), and found that Hydrochalarone-6 (n=6) as the optimal member of this series exhibiting the high relaxivity r_1 of 205 mM^{-1} s^{-1}.[74] Dorn *et al.* reported the synthesis of $Gd_3N@C_{80}(OH)_{\sim 26}(CH_2CH_2COOM)_{\sim 16}$ (M = Na or H) in 2009 and determined their longitudinal and transverse relaxivities r_1 and r_2 to be 207 and 282 mM^{-1} s^{-1} (per C_{80} cage) at 2.4 T, respectively, which are 50 times larger than those of such Gd^{3+} poly(aminocarboxylate) complexes as commercial Omniscan and Magnevist.[75] Besides, Gd-based MMNCFs with higher yield than $Gd_3N@C_{80}$ were also considered to overcome the problem of low yield of $Gd_3N@C_{80}$. In 2007, Wang *et al* synthesized two water-soluble hydroxylated derivatives of Gd/Sc mixed metal NCFs [$Sc_xGd_{3-x}N@C_{80}O_m(OH)_n$ (x = 1, 2; $m \approx 12$; $n \approx 26$)], and determined their relaxivities r1 as 20.7 and 17.6 mM^{-1} s^{-1} for $Sc_2GdN@C_{80}O_{12}(OH)_{26}$ and $ScGd_2N@C_{80}O_{12}(OH)_{26}$, respectively.[76] In 2010, Braun *et al.* synthesized a modular system based on $GdSc_2N@C_{80}$, abbreviated as "Gd-cluster@-BioShuttle," which showed very high proton relaxivity and high signal enhancement at very low Gd concentrations compared to the commonly used contrast agents such as Gd-DTPA, resulting in an incredible 500-fold gain of sensitivity.[77] Recently a Bioshuttle system based on $GdSc_2N@C_{80}$ was designed to target *c-myc* mRNA-expressing cells, and facile intracellular transport and high selectivity of this system were confirmed by confocal fluorescence microscopy.[78]

4.2 Applications as New Acceptors in Organic Photovoltaics (OPV)

An organic photovoltaic (OPV) cell is a photovoltaic cell using organic semiconductors (conductive organic polymers or small organic molecules) as photoactive molecules for light absorption and charge transport, and holds several advantages compared to the traditional inorganic solar cells, including the low cost of the organic semiconductor materials, the ease of fabrication of solar cells and the compatibility with mass printing techniques. In specific, polymer solar cells (PSCs) based on the composites of an electron-donating conjugated polymer and an electron-accepting fullerene are quite promising for the realization of a low-cost, printable, portable and flexible renewable energy source. Currently, a C_{60} derivative, [6,6]-phenyl-C61-butyric acid methyl ester (PCBM), has been most commonly used as the acceptor material in the state-of-art PSCs.[79] Compared to the empty fullerenes like C_{60}, the molecular orbitals of NCFs, which are an important factor determining the open circuit voltage of PSC, can be more readily adjusted by encapsulating different types of metal atoms and varying the cage size. In 2009, Ross *et al.* synthesized the first methano derivatives of $Lu_3N@C_{80}$ which were applied as novel acceptor materials in PSCs. Their results revealed that an optimized high open circuit voltage of 890 mV can be obtained for 1-(3-hexoxycarbonyl)propyl-1-phenyl-[6,6]-$Lu_3N@C_{81}$ ($Lu_3N@C_{80}$–PCBH), which was 260 mV above the reference devices made with the commonly used C_{60}-PCBM acceptor and the highest value for P3HT/fullerene PSC devices. As a result a power conversion efficiency (PCE) of >4% was obtained by using $Lu_3N@C_{80}$–PCBH as the new acceptor. The higher overall PCE of the P3HT/$Lu_3N@C_{80}$-PCBH devices compared to P3HT/C_{60}-PCBM one is attributed to a better alligned LUMO level that captures more of the energy associated with each absorbed photon.[80] A more detailed optimization work was published by the same authors a little later, and the influence of the blending ratio of $Lu_3N@C_{80}$–PCBH with P3HT on the active layer morphology and the performance of PSC devices were studied, revealing that an optimal blending ratio of P3HT and $Lu_3N@C_{80}$-PCBH is 50 wt% fullerene loading, which differs from a molecular equivalent of an optimized P3HT/C_{60}–PCBM active layer.[81] More promisingly, the authors proposed that, using the improved LUMO level offset of $Lu_3N@C_{80}$-PCBH acceptor materials and combining it with the new low-bandgap donor polymers which have been extensively studied recently, PCE greater than 10% may now be feasible for PSCs.[81]

More recently, Feng *et al.* successfully designed and synthesized a covalently linked $Lu_3N@C_{80}$-PDI (perylenebisimide) electron donor–acceptor conjugate and found a photoinduced electron transfer from the ground state of $Lu_3N@C_{80}$ to the excited state of PDI. Furthermore, they applied $Lu_3N@C_{80}$-PCBM as an electron donor in a bilayer heterojunction solar cell device with PDI derivative as electron acceptor, and obtained a moderate efficiency of 0.054%, suggesting that $Lu_3N@C_{80}$ also possesses an electron-donating property.[82]

5. CONCLUSION

Since the discovery of the first NCF $Sc_3N@C_{80}$ in 1999, a large variety of NCFs have been synthesized and isolated by different methods discussed above, featuring the encapsulation of variable-size group-III and non-group-III metals, a wide range of cage size distribution from C_{68} to C_{96} and versatile isomeric cage structures including non-IPR cages as reviewed here. Their molecular structures as well as their electronic, physical and chemical properties have been extensively studied. Moreover, some potential applications of NCFs in the fields of biomedicine and photonics have also been investigated, demonstrating a bright future for this field.

Although the field of NCFs has been achieving rapid advance in the past decade, there still remain some intriguing questions concerning all aspects of NCFs, including their synthesis, isolation, functionalization and application. For instance, so far the members of the NCF family are still more limited compared to the endohedral fullerenes in general. Thus, exploring novel structures of NCFs with new metals encapsulated and/or new cages is highly desired. From the point of view of applications, since the amount of NCF material synthesized by the current methods is still too low to meet the need of practical applications, how can the yield of NCFs be enhanced? With the relatively limited sample amount available, can the application of NCFs be broadened to other fields? Therefore, both synthesis and applications are the two major challenges in the future research of NCFs.

ACKNOWLEDGEMENTS

We thank the present and former members of our groups in USTC and IFW-Dresden for their contributions as well as our collaborators on the published works cited in this review. The research work was supported by National Basic Research Program of China (2010CB923300, 2011CB921401), National Natural Science Foundation of China (90921013, 21132007, 21311120060), "100 Talents Programme" from Chinese Academy of Sciences [to SFY], and Deutsche Forschungsgemeinschaft (grant PO 1602/11– to AAP).

REFERENCES

1. (a) Shinohara H. (2000) Endohedral metallofullerenes. *Rep Prog Phys* **63:** 843–892; (b) Dunsch L, Yang SF. (2007) Metal nitride cluster fullerenes: Their current state and future prospects. *Small* **3:** 1298–1320; (c) Dunsch L, Yang SF. (2007) Endohedral clusterfullerenes — playing with cluster and cage sizes. *Phys Chem Chem Phys* **9:** 3067–3081; (d) Yang SF, Dunsch L. (2008) In: Lukehart CM, Scott RA (eds), *Nanomaterials: Inorganic and Bioinorganic Perspectives*, John Wiley & Sons, Ltd., Chichester, United Kingdom, p. 189; (e) Yang SF, Liu FP, Chen CB, *et al.* (2011) Fullerenes Encaging Metal Clusters — Clusterfullerenes. *Chem Commum* **47:** 11822–11839; (f) Yang SF. (2012) Synthesis, Separation, and Molecular Structures of Endohedral Fullerenes. *Curr Org Chem* **16:** 1079–1094; (g) Lu X, Feng L, Akasaka T, Nagase S. (2012) Current status and future developments of endohedral metallofullerenes. *Chem Soc Rev* **41:** 7723–7760; (h) Popov AA, Yang SF, Dunsch L. (2013) Endohedral Fullerenes. *Chem Rev* **113:** 5989–6113.
2. Heath JR, Brien SC, Zhang Q, *et al.* (1985) Lanthanum complexes of spheroidal carbon shells. *J Am Chem Soc* **107:** 7779–7780.
3. Chai Y, Guo T, Jin CM, *et al.* (1991) Fullerenes with metals inside. *J Phys Chem* **95:** 7564–7586.
4. Stevenson S, Rice G, Glass T, *et al.* (1999) Small-bandgap endohedral metallofullerenes in high yield and purity *Nature* **401:** 55–57.
5. Stevenson S, Fowler PW, Heine T, *et al.* (2000) A stable non-classical metallofullerene family *Nature* **408:** 427–428.
6. Olmstead MH, de Bettencourt-Dias A, Duchamp JC, *et al.* (2001) Isolation and structural characterization of the endohedral fullerene $Sc_3N@C\text{-}_{78}$. *Angew Chem Int Ed* **40:** 1223–1225.
7. (a) Krause M, Dunsch L. (2005) Gadolinium Nitride Gd_3N in Carbon Cages: The Influence of Cluster Size and Bond Strength. *Angew Chem Int Ed* **47:** 1557–1560; (b) Beavers CM, Chaur MN, Olmstead MM, *et al.* (2009) Large Metal Ions in a Relatively Small Fullerene Cage: The Structure of $Gd_3N@C_2(22010)\text{-}C_{78}$ Departs from the Isolated Pentagon Rule. *J Am Chem Soc* **131:** 11519–11524.
8. Fu W, Xu L, Azurmendi H, *et al.* (2009) Y-89 and C-13 NMR Cluster and Carbon Cage Studies of an Yttrium Metallofullerene Family, $Y_3N@C_{2n}$ (n = 40–43). *J Am Chem Soc* **131:** 11762–11769.
9. Iezzi EB, Duchamp JC, Fletcher KR, *et al.* (2002) Lutetium-based Trimetallic Nitride Endohedral Metallofullerenes: New Contrast Agents. *Nano Lett* **2:** 1187–1190.

10. Dunsch L, Georgi P, Krause M, Wang CR. (2003) New clusters in endohedral fullerenes: the metalnitrides . *Synth Met* **135:** 761–762.
11. (a) Feng L, Xu JX, Shi ZJ, *et al.* (2002) Isolation and characterization of the endohedral metallofullerene $Tb_3N@C_{80}$. *Chem J Chin Univ* **23:** 996–998; (b) Zuo T, Beavers CM, Duchamp JC, *et al.* (2007) Isolation and Structural Characterization of a Family of Endohedral Fullerenes Including the Large, Chiral Cage Fullerenes $Tb_3N@C_{88}$ and $Tb_3N@C_{86}$ as well as the Ih and D_{5h} Isomers of $Tb_3N@C_{80}$. *J Am Chem Soc* **129:** 2035–2043.
12. Beavers CM, Zuo TM, Duchamp JC, *et al.* (2006) $Tb_3N@C_{84}$: An Improbable, Egg-Shaped Endohedral Fullerene that Violates the Isolated Pentagon Rule. *J Am Chem Soc* **128:** 11352–11353.
13. Chaur MN, Melin F, Elliott B, *et al.* (2007) $Gd_3N@C_{2n}$ (n = 40, 42, and 44): Remarkably Low HOMO–LUMO Gap and Unusual Electrochemical Reversibility of $Gd_3N@C_{88}$. *J Am Chem Soc* **129:** 14826–14829.
14. Ma YH, Wang TS, Wu JY, *et al.* (2011) Size effect of endohedral cluster on fullerene cage: Preparation and structural studies of $Y_3N@C_{78}$-C_2. *Nanoscale* **3:** 4955–4957.
15. Yang SF, Dunsch L. (2005) A Large Family of Dysprosium-based Trimetallic Nitride Endohedral Fullerenes: $Dy_3N@C_{2n}$ ($39 \leq n \leq 44$). *J Phys Chem B* **109:** 12320–12328.
16. Yang SF, Troyanov SI, Popov AA, *et al.* (2006) Deviation from the Planarity — a Large Dy_3N Cluster Encapsulated in an I_h-C_{80} Cage: An X-ray Crystallographic and Vibrational Spectroscopic Study. *J Am Chem Soc* **128:** 16733–16739.
17. Melin F, Chaur MN, Engmann S, *et al.* (2007) The Large $Nd_3N@C_{2n}$ ($40 \leq n \leq 49$) Cluster Fullerene Family: Preferential Templating of a C_{88} Cage by a Trimetallic Nitride Cluster. *Angew Chem Int Ed* **46:** 9032–9035.
18. Chaur MN, Melin F, Elliott B, *et al.* (2008) New $M_3N@C_{2n}$ Endohedral Metallofullerene Families (M = Nd, Pr, Ce; n = 40–53): Expanding the Preferential Templating of the C_{88} Cage and Approaching the C_{96} Cage. *Chem Eur J* **14:** 4594–4599.
19. Chaur MN, Melin F, Ashby J, *et al.* (2008) Lanthanum Nitride Endohedral Fullerenes $La_3N@C_{2n}$ ($43 \leq n \leq 55$): Preferential Formation of $La_3N@C_{96}$. *Chem Eur J* **14:** 8213–8219.
20. Wang XL, Zuo TM, Olmstead MM, *et al.* (2006) Preparation and structure of $CeSc_2N@C_{80}$: An icosahedral carbon cage enclosing an acentric $CeSc_2N$ unit with buried f electron spin. *J Am Chem Soc* **128:** 8884–8889.
21. Stevenson S, Chancellor CJ, Lee HM, *et al.* (2008) Internal and External Factors in the Structural Organization in Cocrystals of the Mixed-Metal

Endohedrals ($GdSc_2N@I_h$-C_{80}, $Gd_2ScN@I_h$-C_{80}, and $TbSc_2N@I_h$-C_{80}) and Nickel(II) Octaethylporphyrin. *Inorg Chem* **47:** 1420–1427.
22. Olmstead M, Dorn H, Balch A L., *et al* (2003) $Sc_3N@C_{68}$: Folded Pentalene Coordination in an Endohedral Fullerene that Does Not Obey the Isolated Pentagon Rule. *Angew Chem Int Ed* **115:** 928–931.
23. Chen N, Zhang EY, Wang CR. (2006) C80 Encaging Four Different Atoms: The Synthesis, Isolation, and Characterizations of $ScYErN@C_{80}$. *J Phys Chem B* **110:** 13322–13325.
24. Chen N, Fan LZ, Tan K, *et al.* (2007) Comparative Spectroscopic and Reactivity Studies of $Sc_{3-x}Y_xN@C_{80}$ (x = 0–3). *J Phys Chem C* **111:** 11823–11828.
25. Yang SF, Kalbac M, Popov A, Dunsch L. (2006) Gadolinium-Based Mixed-Metal Nitride Clusterfullerenes $Gd_xSc_{3-x}N@C_{80}$ (x = 1, 2). *Chem Phys Chem* **7:** 1990–1995.
26. Yang SF, Popov A, Kalbac M, Dunsch L. (2008) The Isomers of Gadolinium-Scandium Nitride Clusterfullerenes $Gd_xSc_{3-x}N@C_{80}$ (x = 1, 2) and Their Influence on the Cluster Structure. *Chem Eur J* **14:** 2084–2092.
27. Yang SF, Popov AA, Chen CH, Dunsch L. (2009) Mixed Metal Nitride Clusterfullerenes in Cage Isomers: $LuxSc_{3-x}N@C_{80}$ (x = 1, 2) As Compared with $M_xSc_{3-x}N@C_{80}$ (M = Er, Dy, Gd, Nd), *J Phys Chem C* **113:** 7616–7623.
28. Yang SF, Popov AA, Dunsch L. (2007) The Role of an Asymmetric Nitride Cluster on Fullerene Cage: the Non-IPR Endohedral $DySc_2N@C_{76}$. *J Phys Chem B* **111:** 13659–13663.
29. Yang SF, Popov AA, Dunsch L. (2008) Large mixed metal nitride clusters encapsulated in a small cage: The confinement of the C_{68}-based clusterfullerenes. *Chem Commun* 2885–2887
30. Zhang Y, Popov AA, Schiemenz S, Dunsch L. (2012) Synthesis, Isolation, and Spectroscopic Characterization of Holmium-Based Mixed-Metal Nitride Clusterfullerenes: $Ho_xSc_{3-x}N@C_{80}$ (x = 1, 2) *Chem Eur J* **18:** 9691–9698.
31. Svitova AL, Popov AA, Dunsch L. (2013) Gd–Sc-Based Mixed-Metal Nitride Cluster Fullerenes: Mutual Influence of the Cage and Cluster Size and the Role of Scandium in the Electronic Structure. *Inorg Chem* **52:** 3368–3380.
32. Stevenson S, Rose CB, Maslenikova JS, *et al.* (2012) Selective Synthesis, Isolation, and Crystallographic Characterization of $LaSc_2N@I_h$-C_{80}. *Inorg Chem* **51:** 13096–13102.
33. Yang SF, Popov AA, Dunsch L. (2008) Carbon Pyramidalization in Fullerene Cages Induced by the Endohedral Cluster: Non-Scandium Mixed Metal Nitride Clusterfullerenes. *Angew Chem Int Ed* **47:** 8196–8200.

34. Zhang L, Popov AA, Yang SF, *et al.* (2010) An endohedral redox sytem in a fullerene cage: The Ce based mixed cluster fullerene $Lu_2CeN@C_{80}$. *Phys Chem Chem Phys* **12:** 7840–7847.
35. Zhang Y, Schiemenz S, Popov AA, Dunsch L. (2013) Strain-Driven Endohedral Redox Couple CeIV/CeIII in Nitride Clusterfullerenes $CeM_2N@C_{80}$ (M = Sc, Y, Lu). *J Phys Chem Lett* **4:** 2404–2409.
36. Yang SF, Chen CB, Popov AA, *et al.* (2009) An Endohedal Titanium (III) in a Clusterfullerene: Putting a Non-Group-III Metal Nitride into the C_{80}-I_h Fullerene Cage. *Chem Commun* 6391–6393.
37. Popov AA, Chen CB, Yang SF, *et al.* (2010) The Spin-Flow Vibrational Spectroscopy of Molecules with Flexible Spin Density: Electrochemistry, ESR, Cluster and Spin Dynamics, and Bonding in $TiSc_2N@C_{80}$. *ACS Nano* **4:** 4857–4871.
38. Chen CB, Liu FP, Li SJ, *et al.* (2012) Titanium/Yttrium Mixed Metal Nitride Clusterfullerene $TiY_2N@C_{80}$: Synthesis, Isolation, and Effect of the Group-III Metal. *Inorg Chem* **51:** 3039–3045.
39. Stevenson S, Harich K, Yu H, *et al.* (2006) Nonchromatographic "Stir and Filter Approach" (SAFA) for Isolating $Sc_3N@C_{80}$ Metallofullerenes. *J Am Chem Soc* **128:** 8829–8835.
40. Dunsch L, Georgi P, Ziegs F, Zoeller H. (2003) *German Patent DE 10301722* A1.
41. Dunsch L, Krause M, Noack J, Georgi P. (2004) Endohedral nitride cluster fullerenes: Formation and spectroscopic analysis of $L_{3-x}M_xN@C_{2n}$ ($0 \leq x \leq 3$; N = 39, 40). *J Phys Chem Solids* **65:** 309–315.
42. Yang SF, Zhang L, Zhang W, Dunsch L. (2010) A Facile Route to Metal Nitride Clusterfullerenes Using Guanidinium Salts: A Selective Organic Solid as the Nitrogen Source. *Chem Eur J* **16:** 12398–12405.
43. Jiao MZ, Zhang WF, Xu Y, *et al.* (2012) Urea as a New and Cheap Nitrogen Source for the Synthesis of Metal Nitride Clusterfullerenes: the Role of Decomposed Product on Selectivity of Fullerenes. *Chem Eur J* **18:** 2666–2673.
44. Stevenson S, Thompson MC, Coumbe HL, *et al.* (2007) Chemically adjusting plasma temperature, energy, and reactivity (CAPTEAR) method using NOx and combustion for selective synthesis of $SC_3N@C_{80}$ metallic nitride fullerenes. *J Am Chem Soc* **129:** 16257–16262.
45. Liu FP, Guan J, Wei T, *et al.* (2013) A Series of Inorganic Solid Nitrogen Sources for the Synthesis of Metal Nitride Clusterfullerenes: The Dependence of Production Yield on the Oxidation State of Nitrogen and Counter Ion. *Inorg Chem* **52:** 3814–3822.
46. Tan YZ, Xie SY, Huang RB, Zheng LS. (2009) The stabilization of fused-pentagon fullerene molecules *Nature Chem* **1:** 450–460.

47. Wang CR, Kai T, Tomiyama T, *et al.* (2000) Materials science: C_{66} fullerene encaging a scandium dimer. *Nature* **408:** 426–427.
48. Yang SF, Popov AA, Dunsch L. (2007) Violating the Isolated Pentagon Rule (IPR): The Endohedral Non-IPR C_{70} Cage of $Sc_3N@C_{70}$. *Angew Chem Int Ed* **46:** 1256–1259.
49. Popov A, Yang SF, Krause M, *et al.* (2007) C_{78} Cage Isomerism Defined by Trimetallic Nitride Cluster Size: A Computational and Vibrational Spectroscopic Study. *J Phys Chem B* **111:** 3363–3369.
50. Mercado BQ, Beavers CM, Olmstead MM, *et al.* (2008) Is the Isolated Pentagon Rule Merely a Suggestion for Endohedral Fullerenes? The Structure of a Second Egg-Shaped Endohedral Fullerene — $Gd_3N@C_s(39663)\text{-}C_{82}$. *J Am Chem Soc* **130:** 7854–7855.
51. Zuo TM, Walker K, Olmstead MM, *et al.* (2008) New egg-shaped fullerenes: non-isolated pentagon structures of $Tm_3N@C_s(51365)\text{-}C_{84}$ and $Gd_3N@C_s(51365)\text{-}C_{84}$. *Chem Commun* 1067–1069.
52. Akasaka T, Okubo S, Wakahare T, *et al.* (1999) Endohedrally Metal-Doped Heterofullerenes: $La@C_{81}N$ and $La_2@C_{79}N$. *Chem Lett* **9:** 945–946.
53. Zuo TM, Xu L, Beavers CM, *et al.* (2008) $M_2@C_{79}N$ (M = Y, Tb): Isolation and Characterization of Stable Endohedral Metallofullerenes Exhibiting M–M Bonding Interactions inside Azafullerene Cages. *J Am Chem Soc* **130:** 12992–12997.
54. Stevenson S, Ling Y, Coumbe CE, *et al.* (2009) Preferential Encapsulation and Stability of La_3N Cluster in 80 Atom Cages: Experimental Synthesis and Computational Investigation of $La_3N@C_{79}N$. *J Am Chem Soc* **131:** 17780–17782.
55. Fu WJ, Zhang JY, Fuhrer T, *et al.* (2011) $Gd_2@C_{79}N$: Isolation, Characterization, and Monoadduct Formation of a Very Stable Heterofullerene with a Magnetic Spin State of S = 15/2. *J Am Chem Soc* **133:** 9741–9750.
56. Krause M, Dunsch L. (2004) Isolation and Characterisation of Two $Sc_3N@C_{80}$ Isomers. *Chem Phys Chem* **5:** 1445–1449.
57. Chaur MN, Melin F, Ortiz AL, Echegoyen L. (2009) Chemical, Electrochemical, and Structural Properties of Endohedral Metallofullerenes. *Angew Chem Int Ed* **48:** 7514–7538.
58. Popov AA, Dunsch L. (2011) Electrochemistry In Cavea: Endohedral Redox Reactions of Encaged Species in Fullerenes. *J Phys Chem Lett* **7:** 786–794.
59. Wolf M, Muller KH, Skourski Y, *et al.* (2005) Magnetic moments of the endohedral cluster fullerenes $Ho_3N@C_{80}$ and $Tb_3N@C_{80}$: The role of ligand fields. *Angew Chem Int Ed* **21:** 3306–3309.
60. Tiwari A, Dantelle G, Porfyrakis K, *et al.* (2008) Magnetic properties of $ErSc_2N@C_{80}$, $Er_2ScN@C_{80}$ and $Er_3N@C_{80}$ fullerenes. *Chem Phys Lett* **466:** 155–158.

61. Smirnova TI, Smironov AI, Chadwick TG, Walker KL. (2008) Characterization of magnetic and electronic properties of trimetallic nitride endohedral fullerenes by SQUID magnetometry and electron paramagnetic resonance. *Chem Phys Lett* **453:** 233–237.
62. Zuo T, Olmstead MM, Beavers CM, *et al.* (2008) Preparation and Structural Characterization of the I_h and the D_{5h} Isomers of the Endohedral Fullerenes $Tm_3N@C_{80}$: Icosahedral C_{80} Cage Encapsulation of a Trimetallic Nitride Magnetic Cluster with Three Uncoupled Tm^{3+} Ions. *Inorg Chem* **47:** 5234–5244.
63. Chen L, Carpenter EE, Hellberg CS, *et al.* (2011) Spin transition in $Gd_3N@C_{80}$, dected by low-temperature on-chio SQUID technique. *J Appl Phys* **109:** 101–103.
64. Nafradi B, Antal A, Paszror A, *et al.* (2012) Molecular and Spin Dynamics in the Paramagnetic Endohedral Fullerene $Gd_3N@C_{80}$. *J Phys Chem Lett* **3:** 3291–3296.
65. Weaterström R, Dreiser J, Piamonteze C, *et al.* (2012) An Endohedral Single-Molecule Magnet with Long Relaxation Times: $DySc_2N@C_{80}$. *J Am Chem Soc* **134:** 9840–9843.
66. Akasaka T, Kato T, Kobayashi K, *et al.* (1995) exohedral adducts of La-C_{82}. *Nature* **374:** 600–601.
67. Lu X, Akasaka T, Nagase S. (2011) Chemistry of endohedral metallofullerenes: the role of metals. *Chem Commun* **47:** 5942–5957.
68. Yang SF, Chen CB, Jiao MZ, *et al.* (2011) Trifluoromethylated D_{5h} and Ih Isomers of $Sc_3N@C_{80}$: Synthesis, Isolation, and Addition Patterns of $Sc_3N@D_{5h}$-$C_{80}(CF_3)_{18}$ and $Sc_3N@I_h$-$C_{80}(CF_3)_{14}$. *Inorg Chem* **50:** 3766–3771.
69. Liu TX, Wei T, Zhu SE, *et al.* (2012) Azide Addition to An Endohedral Metallofullerene: Formation of Azafulleroids of $Sc_3N@I_h$-C_{80}. *J Am Chem Soc* **134:** 11956–11959.
70. Wilson LJ, Cagle DW, Thrash TP, *et al.* (1999) Metallofullerene drug design. *Chem Rev* **190:** 199–207.
71. Wilson LJ. (1999) Medical applications of fullerenes and metallofullerenes. *Electrochem Soc Interface* **8:** 24–28.
72. Fatouros PP, Corwin FD, Chen ZJ, *et al.* (2006) In Vitro and in Vivo Imaging Studies of a New Endohedral Metallofullerene Nanoparticle. *Radiology* **240:** 756–764.
73. Zhang JF, Fatouros PP, Shu CY, *et al.* (2010) High Relaxivity Trimetallic Nitride (Gd_3N) Metallofullerene MRI Contrast Agents with Optimized Functionality. *Bioconjugate Chem* **21:** 610–615.
74. Macharland DK, Walker KL, Lenk PP, *et al.* (2008) Hydrochalarones: A Novel Endohedral Metallofullerene Platform for Enhancing Magnetic Resonance Imaging Contrast . *J Med Chem* **51:** 3681–3683.

75. Shu C, Corwin FD, Zhang J, *et al.* (2009) Facile Preparation of a New Gadofullerene-Based Magnetic Resonance Imaging Contrast Agent with High 1H Relaxivity. *Bioconjugate Chem* **20:** 1186–1193.
76. Zhang EY, Shu CY, Feng L, Wang CR. (2007) Preparation and Characterization of Two New Water-Soluble Endohedral Metallofullerenes as Magnetic Resonance Imaging Contrast Agents. *J Phys Chem B* **111:** 14223–14226.
77. Braun K, Dunsch L, Pipkorn R, *et al.* (2010) Gain of a 500-fold sensitivity on an intravital MR Contrast Agent based on an endohedral Gadolinium-Cluster-Fullerene-Conjugate: A new chance in cancer diagnostics *J Med Sci* **7:** 136–146.
78. Svitova A, Braun K, Popov AA, Dunsch L. (2012) A Platform for Specific Delivery of Lanthanide–Scandium Mixed-Metal Cluster Fullerenes into Target Cells. *Chemistry Open* **1:** 207–210.
79. (a) Hummelen JC, Knight BW. (1995) Preparation and Characterization of Fulleroid and Methanofullerene Derivatives. *J Org Chem* **60:** 532–538; (b) Gunes S, Neugebauer H, Sariciftci NS. (2007) Conjugated Polymer-Based Organic Solar Cells. *Chem Rev* **107:** 1324–1338.
80. Russel BR, Cardona MC, Guldi DM, *et al.* (2009) Endohedral fullerenes for organic photovoltaic devices. *Nature Mater* **8:** 208–212.
81. Ross RB, Cardona CM, Swain FB, *et al.* (2009) Tuning Conversion Efficiency in Metallo Endohedral Fullerene-Based Organic Photovoltaic Devices. *Adv Funct Mater* **19:** 2332–2337.
82. Feng L, Rudolf M, Wolfrum S, *et al.* (2012) A Paradigmatic Change: Linking Fullerenes to Electron Acceptors. *J Am Chem Soc* **134:** 12190–12197.

CHAPTER 5

Metal Carbide Clusterfullerenes

*Taishan Wang, Chunru Wang**

Information about the family of metal carbide clusterfullerenes is summarized in this chapter. The discovery of $Sc_2C_2@C_{84}$ proclaimed the emergence of metal carbide clusterfullerenes. It is also the outstanding representative of di-metal carbide clusterfullerenes, $M_2C_2@C_{2n}$, which has been found exhibiting various geometries and properties, such as the variable M_2C_2 cluster and flapping motion of two metal atoms. The multi-metal carbide clusters construct much more complex clusterfullerenes, including $Sc_3C_2@I_h$-C_{80} and $Lu_3C_2@C_{88}$. A landmark of structure in metal carbide clusterfullerenes should be ascribed to the $Sc_4C_2@I_h$-C_{80}, where at the center a C_2 unit is nested inside a tetrahedron of four scandium atoms and the whole molecule exhibits a Russian-doll-like nesting structure. Furthermore, the dynamics in metal carbide clusterfullerenes are of great importance as these motions will bring more and more alluring properties in addition to the C_2 quantum gyroscope. Based on these studies, it can be expected that there is a broader development space for the metal carbide clusterfullerenes, such as new structures constructed by C_2 or even $C_n(n>2)$, novel dynamic modes for the metal carbide cluster, and interesting paramagnetic and electronic properties.

*Beijing *National Laboratory for Molecular Sciences and Laboratory of Molecular Nanostructure and Nanotechnology, Institute of Chemistry, Beijing 100190, China* E-mail: crwang@iccas.ac.cn

1. INTRODUCTION

The unique structural characteristic of fullerenes is the confined space enclosed with dozens or even hundreds of carbon atoms. Such nanoscaled fullerene capsules have been discovered to be able to accommodate atoms, clusters, and even molecules, and thus, a new species, named endohedral fullerenes, has emerged. The first experimental evidence of endohedral fullerenes owes to the detection of La@C_{60}[1] in mass spectrum in 1985, which undoubtedly opened the door to the world of endohedral fullerenes. Another landmark is the preparation of La@C_{82}[2,3] with macroscopic quantity, whose yield was greatly improved via the famous arc discharging method. Since then, a large family called endohedral metallofullerenes with different kinds of fullerene cages and endohedral metal-based moieties has been started and expanded. Up to date, many types of metal-containing clusters, e.g. metal atom(s),[1–3] metal nitride,[4] metal carbide,[5] metal carbonitride,[6] metal oxide[7] and metal sulfide,[8] have been encapsulated inside various fullerene cages.

Among these metal clusterfullerenes, the metal carbide clusterfullerenes have attracted wide interest due to their unique structures, endohedral dynamics, and chemical/physical properties. The discovery of Sc_2C_2@C_{84}[5] proclaimed the emergence of metal carbide clusterfullerenes. Thereafter, many newly synthesized metallofullerene molecules were characterized to be metal carbide clusterfullerenes. Moreover, the previously reported metal clusterfullerenes, such as Sc_2@C_{84},[3] Sc_3@C_{82},[3] etc. were re-identified as metal carbide clusterfullerenes.[9–11] Such special geometries lead to their nested electronic structures, which are composed of negatively charged shells and metal cations next to shell, as well as negatively charged C_2 cores, such as C_2^{2-}, C_2^{3-}, and C_2^{6-}. Another big characteristic of metal carbide clusterfullerenes is their endohedral dynamics, which contain two types of motion, the flapping of metal atoms and the rotating of the C_2 unit. The next sections will present the discovery, structures, and endohedral motions of metal carbide clusterfullerenes in detail.

2. DISCOVERY OF METAL CARBIDE CLUSTERFULLERENES

After La@C_{82}, lots of metallofullerenes with varied fullerene cages and endohedral moieties have been synthesized and isolated, such as mono-metal endofullerene $M@C_{2n}$ (M = Sc, Y, lanthanide elements, Ca, Sr, Ba, etc.; 2n = 60, 74, 80, 82, etc.) and di-metal endofullerene $M_2@C_{2n}$ (M = Sc, Y, lanthanides, etc.; 2n = 66, 74, 76, 78, 80, 82 etc.).[3] In 1999, the first trimetallic nitride clusterfullerene, $Sc_3N@I_h$-C_{80},[4] was discovered, which revealed that a more complex endohedral cluster with varying composition and multiple atoms can be encapsulated into a fullerene cage.

In 2001, a metallofullerene having a molecular formula of Sc_2C_{86} was synthesized and isolated.[5] The ^{13}C NMR spectrum of this new molecule consists of 10 lines of equal intensity, one line with half intensity. However, for the hollow C_{86} cage isomers, the most symmetrical isomer of D_3-C_{86} has 15 ^{13}C NMR signals. Therefore, this molecule of Sc_2C_{86} with only 12 ^{13}C NMR lines was not assigned to metal clusterfullerene $Sc_2@C_{86}$ but to a new kind of clusterfullerene, $Sc_2C_2@D_{2d}$-C_{84}. The endohedral metal carbide was further confirmed by synchrotron X-ray diffraction. The discovery of $Sc_2C_2@D_{2d}$-C_{84} not only represents a new model of metal compound clusterfullerene, but also promotes reconsideration of the previously synthesized metal endofullerenes because of the possibility of isomers.

Figure 1 shows the theoretically optimized static structure of $Sc_2C_2@D_{2d}$-C_{84}, in which a planar Sc_2C_2 cluster is encapsulated within a large C_{84}

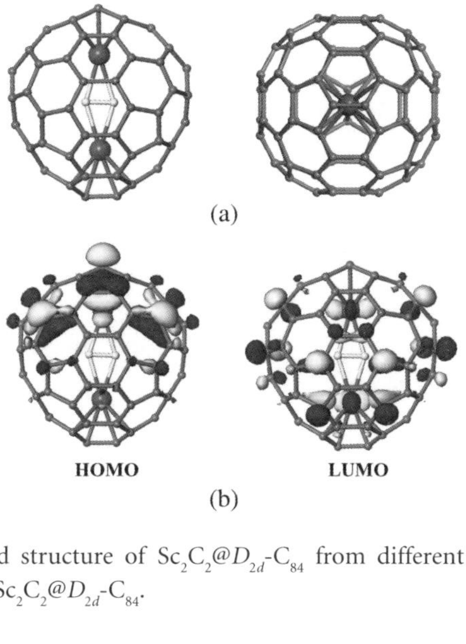

Fig. 1 (a) Optimized structure of $Sc_2C_2@D_{2d}$-C_{84} from different sides. (b) Calculated molecular orbitals of $Sc_2C_2@D_{2d}$-C_{84}.

cage. The theoretical calculations showed that the electronic structure of $Sc_2C_2@D_{2d}$-C_{84} can be described as $(Sc_2)^{6+}(C_2)^{2-}@(C_{84})^{4-}$, in which each scandium contributes three electrons and the C_2 unit is negatively divalent. The HOMOs of $Sc_2C_2@D_{2d}$-C_{84} mainly localize on the cage, and the LUMOs are distributed both on the cage and on the scandium atoms.

3. DI-METAL CARBIDE CLUSTERFULLERENES

$Sc_2C_2@C_{84}$ opens new avenues to understand the composition of endohedral moiety, and many already reported metallofullerenes such as Sc_2C_{84}, Y_2C_{84}, etc. have to be re-examined. The following studies have disclosed that many di-metal-containing endofullerenes were determined as typical $M_2C_2@C_{2n}$ type of metal carbide clusterfullerenes. Moreover, these di-metal carbide clusterfullerenes have been found exhibiting various geometries and properties, such as the bending M_2C_2 cluster, non-IPR cage, and flapping motion like butterfly.

Yttrium carbide clusterfullerenes, $Y_2C_2@C_{82}$ with three isomers, were identified by ^{13}C NMR measurements in detail, and their fullerene cage symmetries were determined to be $C_s(6)$, $C_{2v}(9)$, and $C_{3v}(8)$, respectively.[12] Furthermore, three isomers of Sc_2C_{84}, i.e. $Sc_2C_2@C_s(6)$-C_{82}, $Sc_2C_2@C_{2v}(9)$-C_{82}, and $Sc_2C_2@C_{3v}(8)$-C_{82}, were also characterized one after another to be scandium carbide clusterfullerenes by ^{13}C NMR spectroscopy and single crystal X-ray analysis.[11,13,14] Thus, the metal carbide M_2C_2 unit can be considered as an endohedral template for constructing many other clusterfullerenes with different metal elements and fullerene cages.

Figure 2 shows the theoretically optimized static structures of $Sc_2C_2@C_s$-C_{82}, $Sc_2C_2@C_{2v}$-C_{82}, and $Sc_2C_2@C_{3v}$-C_{82}. Different from the planar Sc_2C_2 cluster in $Sc_2C_2@D_{2d}$-C_{84}, the Sc_2C_2 moieties in these three C_{82} cages all display bent geometry due to the scaling-down of the carbon cage. The electronic structures of three $Sc_2C_2@C_{82}$ molecules can be described as $(Sc_2)^{6+}(C_2)^{2-}@(C_{82})^{4-}$, like that of $Sc_2C_2@D_{2d}$-C_{84}, and the C_2 unit is still negatively divalent.

In addition, the discovery of $Sc_2C_2@C_{2v}(5)$-C_{80} disclosed that the metal carbide can template an unusual fullerene cage.[15] The C_{80} cage is often found in di-metal endofullerenes and trimetallic nitride clusterfullerenes, such as $La_2@I_h$-C_{80}, $Sc_3N@I_h$-C_{80}, etc., in which the C_{80} cage is negatively

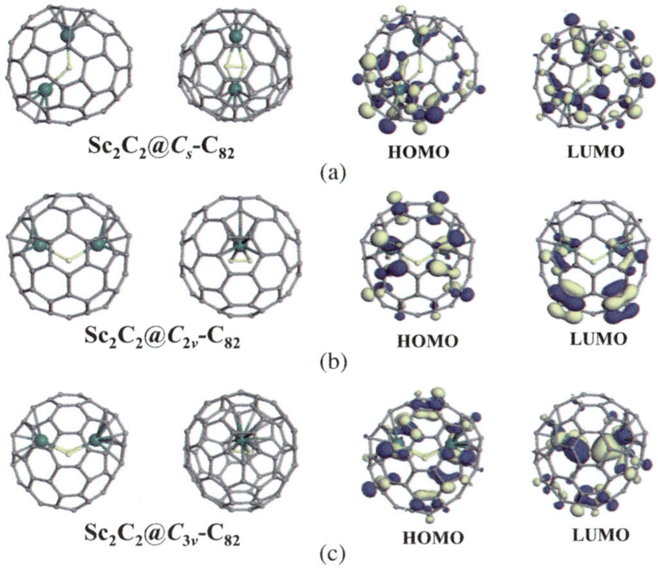

Fig. 2 Optimized structures and calculated molecular orbitals of a) $Sc_2C_2@C_s$-C_{82}, b) $Sc_2C_2@C_{2v}$-C_{82}, and c) $Sc_2C_2@C_{3v}$-C_{82}.

hexavalent with I_h-symmetry. However, the $Sc_2C_2@C_{80}$ has a $C_{2v}(5)$-symmetric C_{80} cage with negative tetravalence. The typical $(M_2C_2)^{4+}$ cluster was believed to selectively template the $C_{2v}(5)$-C_{80} cage.

The M_2C_2 template was also found to be encapsulated in a big fullerene cage. The crystallographic characterization of Gd_2C_{94} reveals that it possesses the gadolinium carbide endohedral structure, i.e. $Gd_2C_2@D_3$-C_{92}, and moreover, Dy_2C_{94} and Er_2C_{94} may well be isostructural with $Gd_2C_2@D_3$-C_{92} revealed by absorption spectra.[16] Unlike the favoring of C_{84} and C_{82} for Sc_2C_2 and Y_2C_2 clusters, the lanthanide carbide M_2C_2 template has a tendency to choose a bigger carbon cage to be stabilized due to the larger lanthanide cation size. Such metal-size-induced cage selectivity plays an important role in the formation of clusterfullerenes, and this effect also does so in the trimetallic nitride clusterfullerenes. That is because along with the increasing of the endohedral cluster, the outer cage suffers much higher internal pressure, and consequently, the fullerene cage grows to adapt the larger cluster.

Interestingly, when $Y_2C_2@C_{3v}$-C_{82}, $Y_2C_2@D_3$-C_{92} and $Y_2C_2@D_5$-C_{100} were systematically compared, it can be found that the endohedral Y_2C_2 clusters change from constrained butterfly structure to linear stretched geometry with the increasing size of the carbon cages.[17] During this process, the endohedral C_2 unit undergoes kinds of transformation along with the variable fullerene cages and metal carbide clusters.

Apart from Sc, Y, and lanthanide elements, the titanium in group IVA was found to be encased in the form of metal carbide within a fullerene cage. The structure of titanium carbide clusterfullerene $Ti_2C_2@C_{78}$ was assigned by theoretical calculation.[18] Interestingly, the calculations predicted two configurations for the endohedral Ti_2C_2 cluster, one being an end-on cluster having a linear Ti-C-C-Ti array, and the other a side-on cluster having a butterfly-like structure. Moreover, these two configurations of the Ti_2C_2 cluster are visualized in the single-molecule images of $Ti_2C_2@C_{78}$ by high-resolution transmission electron microscopy.[19] Further theoretical analysis indicated that the weak covalent bonds between Ti atoms and C atoms make the C_2 unit rotate freely over a wide range.

The unconventional fullerenes violating the isolated pentagon rule (IPR) have attracted wide interest due to their distinctive structures and properties. One of the efficient strategies is through encaging a metal

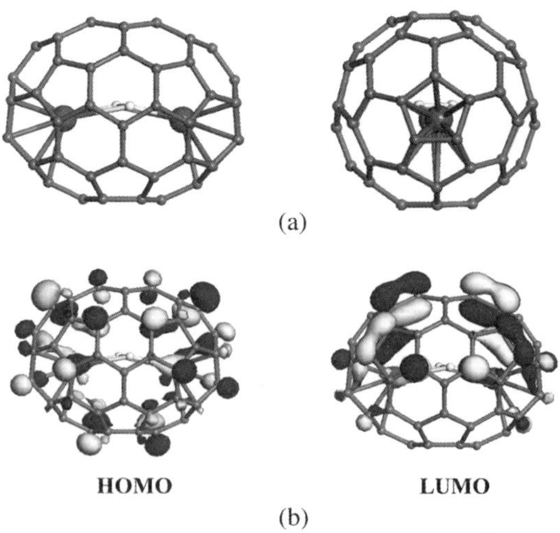

Fig. 3 (a) Optimized structure of $Sc_2C_2@C_{2v}$-C_{68} from different sides. (b) Calculated molecular orbitals of $Sc_2C_2@C_{2v}$-C_{68}.

cluster inside a fullerene cage with pentalene motif as a basic unit, which can be stabilized by strong interaction between fused pentagons and metal ion. $Sc_2C_2@C_{2v}$-C_{68}, having two pairs of pentalene motifs, is the first metal carbide clusterfullerene with a non-IPR cage.[20] Theoretical calculations showed that $Sc_2C_2@C_{68}$ has the valence state of $(Sc^{3+})_2(C_2)^{2-}@(C_{68})^{4-}$.

For the endohedral metal carbide cluster, it is interesting to investigate the chemical shifts of the C_2 unit by means of NMR spectroscopy. Table 1 lists the ^{13}C NMR chemical shifts of endohedral C_2 units for several metal carbide clusterfullerenes. It can be seen that the C_2 units are all significantly deshielded, and the different kinds of metal elements can affect the chemical shifts, for example, 253.2 ppm in $Sc_2C_2@C_{3v}$-C_{82}[21] and 257.0 ppm in $Y_2C_2@C_{3v}$-C_{82},[17] respectively. Moreover, along with increasing the fullerene cage, the C_2 NMR signals become less deshielded and shift to the high field that may be caused by the leaving away metal nuclei in stretched metal carbide cluster.

4. MULTI-METAL CARBIDE CLUSTERFULLERENES

The C_2 unit can be regarded as a backbone in the framework of the metal carbide cluster; besides the butterfly-like quaternary cluster discussed above, more complex endohedral structures can be expected. An amazing tri-metal carbide clusterfullerene, $Sc_3C_2@I_h$-C_{80}, was undoubtedly confirmed by ^{13}C NMR and single crystal X-ray analysis.[10] However, this tri-metal metallofullerene, Sc_3C_{82}, was once considered to consist of three scandium atoms enclosed in a C_{82} cage. This interesting Sc_3C_{82} species has

Table 1 The ^{13}C NMR Chemical Shifts of Endohedral C_2 Units for Several Metal Carbide Clusterfullerenes

	^{13}C NMR chemical shifts (ppm)
$Sc_2C_2@C_{3v}$-C_{82}[21]	253.2
$Sc_2C_2@D_{2d}$-C_{84}[21]	249.2
$Y_2C_2@C_s$-C_{82}[17]	256.2
$Y_2C_2@C_{3v}$-C_{82}[17]	257.0
$Y_2C_2@D_{2d}$-C_{84}[17]	247.8
$Y_2C_2@D_3$-C_{92}[17]	227.3

one unpaired electron, which makes this molecule paramagnetic and NMR silent. Upon one-electron reduction, the anion of Sc_3C_{82} becomes a closed-shell species and can exhibits two ^{13}C NMR signals originated from 20 triphenylenic carbon sites and 60 corannulenic carbon sites, indicating an I_h-symmetric C_{80} cage. Theoretical analysis revealed that the encased Sc_3C_2 cluster is elusory, fluctuating between the planar structure and trifoliate structure. The structure of Sc_3C_{82} is also determined by the MEM/Rietveld method by using synchrotron radiation powder data, which revealed that at the center of the scandium triangle the C_2 unit exhibits a spherical charge distribution.[22] These results show the Sc_3C_2 cluster prefers a trigonal bipyramidal structure. Density functional theory calculations have shown that the trimetal-carbide endofullerene $Sc_3C_2@C_{80}$ has the valence state $(Sc^{3+})_3(C_2)^{3-}@(C_{80})^{6-}$.

This open-shell $Sc_3C_2@I_h$-C_{80} brings about unique paramagnetic properties. Splitting of 6.256 G (3 nuclei) and a *g*-value of 2.0006 for pristine $Sc_3C_2@C_{80}$ were observed in *O*-dichlorobenzene (*o*-DCB).[23] Moreover, the ESR spectrum of $Sc_3C_2@C_{80}$ exhibits high-symmetrical 22 lines due to the

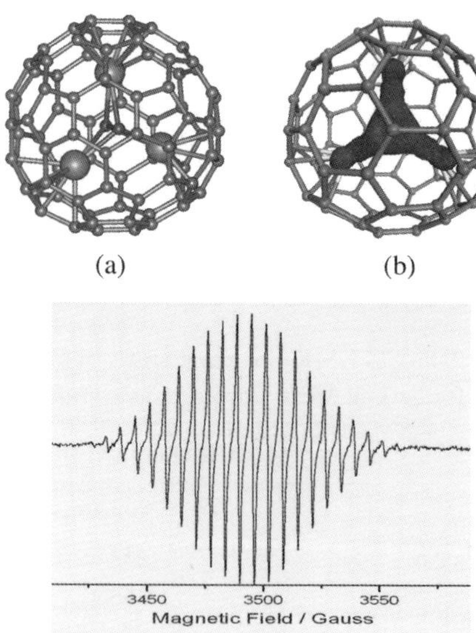

Fig. 4 (a) Optimized structure of $Sc_3C_2@I_h$-C_{80}; (b) Calculated unpaired spin distributions of $Sc_3C_2@I_h$-C_{80}; (c) ESR spectrum of $Sc_3C_2@I_h$-C_{80}.

hyperfine interaction of radical electron with three equivalent scandium nuclei ($I = 7/2$). Theoretical calculations disclosed that the unpaired electron is mainly delocalized on the endohedral C_2 unit, and it is consistent with the observed g factor close to that of a free electron. These three equivalent scandium nuclei also confirmed that the Sc_3C_2 cluster prefers a trigonal bipyramidal structure with higher symmetry.

The chemical modification is an effective method to control the molecular property, and $Sc_3C_2@C_{80}$ has also been derivatized to study its structural changes as well as property modulation. $Sc_3C_2@C_{80}$ fulleropyrrolidine was synthesized by Prato reaction and the 5,6-ring junction on I_h-C_{80} cage was determined to be the reaction site for the $Sc_3C_2@C_{80}$ fulleropyrrolidine mono-adduct.[23] Most importantly, the pyrrolidine addend changes the spin density distributions and alters the paramagnetic properties of $Sc_3C_2@C_{80}$ fulleropyrrolidine as a result. Moreover, the bis-adducts of $Sc_3C_2@C_{80}$ fulleropyrrolidines and ESR studies showed that these bis-addition fulleropyrrolidines have varied paramagnetism resulting from diverse reaction sites for the second pyrrolidine addend.[24] For $Sc_3C_2@C_{80}$ bis-adducts fulleropyrrolidines, each of the

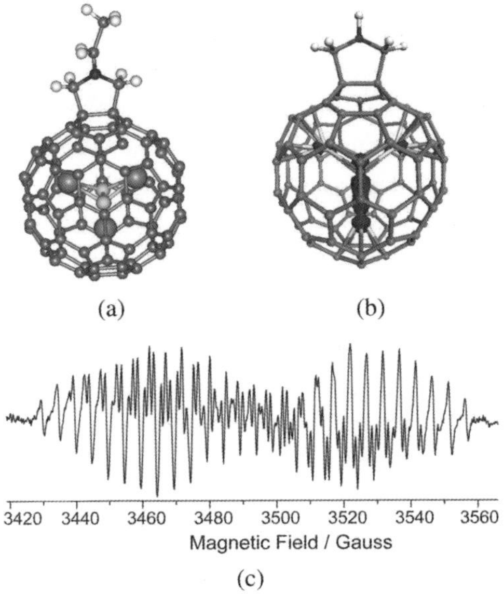

Fig. 5 (a) Calculated structure of $Sc_3C_2@C_{80}$ fulleropyrrolidine; (b) Calculated spin density distributions of $Sc_3C_2@C_{80}$ fulleropyrrolidine; (c) ESR spectrum of $Sc_3C_2@C_{80}$ fulleropyrrolidine.

three Sc nuclei has a distinctive coupling constant, which results in a more complex ESR spectrum. Such controllable molecular paramagnetism is of great significance to construct novel molecular devices.

In $Sc_3C_2@C_{80}$-Ad, a cycloadduct derivative with adamantylidene carbene(Ad), the endohedral Sc_3C_2 cluster distinctly takes on a planar quinary-structure with rigid intra-molecular motion, indicating the considerable influence of the exohedral modification on encaging cluster.[10] In addition, a bisfulleroid derivative of $Sc_3C_2@C_{80}$ was synthesized when reacted with tetrazine.[25] Moreover, both the $Sc_3C_2@C_{80}$-Ad and $Sc_3C_2@C_{80}$ bisfulleroid derivative show a changed ESR spectrum. It is implied that the primary homogenous unpaired spin on three scandium nuclei goes through a change to be uneven after the outer chemical modification.

To enhance the potential application of metal carbide clusterfullerenes on the construction of electronic devices, the solid-state structures and charge-carrier mobility of $Sc_3C_2@C_{80}$ were experimentally investigated.[26] Compared with $La_2@C_{80}$ and $Sc_3N@C_{80}$, the thin film of $Sc_3C_2@C_{80}$ exhibits a highest electron mobility ($\mu = 0.13$ $cm^2V^{-1}s^{-1}$) under normal temperature and atmospheric pressure determined using flash photolysis time-resolved microwave conductivity measurements. The electron mobility of $Sc_3C_2@C_{80}$ is two orders of magnitude higher than that of $La_2@C_{80}$ or $Sc_3N@C_{80}$. Theoretical calculations revealed that the open-shell electronic structure of $Sc_3C_2@C_{80}$ plays an important role in modifying the conduction character.

For $Lu_3C_2@D_2$-C_{88}, a tri-lutetium carbide cluster encased in a large fullerene C_{88} cage, this molecule has a planar tri-metal carbide Lu_3C_2

Table 2 The Hyperfine Coupling Constants (*a*) and g Factors for $Sc_3C_2@C_{80}$ and Its Several Derivatives

	a **(Gauss)**
$Sc_3C_2@C_{80}$[23]	6.256(three nuclei)
$Sc_3C_2@C_{80}$-Ad[10]	7.39(two nuclei); 1.99(one nucleus)
$Sc_3C_2@C_{80}$ fulleropyrrolidine(mono)[23]	8.602(one nucleus); 4.822(two nuclei)
$Sc_3C_2@C_{80}$ fulleropyrrolidine (bis)[24]	8.207; 4.822; 4.734
$Sc_3C_2@C_{80}$ bisfulleroid[25]	6.73(one nucleus); 4.00(two nuclei)

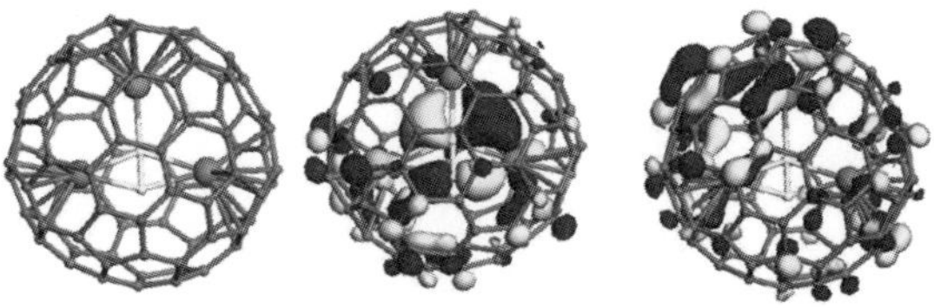

Fig. 6 Optimized structure of $Lu_3C_2@D_2$-C_{88} and molecular orbitals of $Lu_3C_2@D_2$-C_{88}.

endohedral cluster, which is different from that of $Sc_3C_2@C_{80}$.[27] This deformed tri-metal carbide may be induced by the large C_{88} cage. Theoretical calculations revealed that the $Lu_3C_2@C_{88}$ also has an open-shell electronic structure as $(Lu^{3+})_3(C_2)^{3-}@(C_{88})^{6-}$ similar to $Sc_3C_2@C_{80}$, and the unpaired electron is localized on the internal $[Lu_3C_2]^{6+}$ moiety.

An important discovery in metal carbide clusterfullerenes should be ascribed to the report of $Sc_4C_2@I_h$-C_{80} exhibiting a Russian-doll-like nesting structure.[28] For this molecule, at the center a C_2 unit is nested inside a tetrahedron of four scandium atoms, and then this metal-carbide cluster is trapped inside an I_h-C_{80} cage. ^{13}C NMR of $Sc_4C_2@I_h$-C_{80} gives two signals with a ratio of 1: 3, and this pattern reveals a high symmetrical C_{80} cage caused by free rotation of the endohedral Sc_4C_2 cluster. Revealed by quantum chemical calculations, $Sc_4C_2@I_h$-C_{80} possesses a nested electronic structure of $C_2^{6-}@Sc_4^{12+}@C_{80}^{6-}$. The central hexavalent C_2 anion is a typical case of unconventional species stabilized by forming a metallofullerene.

5. SPECIAL MOTION IN METAL CARBIDE CLUSTERS

For many metallofullerenes, especially for IPR-satisfied clusterfullerenes, the motional behavior of endohedral metal or clusters has attracted much interest as the endohedral dynamics have significant effects on carbon cage and molecular properties, such as for $Sc_4C_2@C_{80}$, the free rotation of encaged cluster makes the electronic structure of I_h-C_{80} spherical and finally to exhibit two ^{13}C NMR signals.[28] For the same reason, the $Sc_3C_2@C_{80}$ shows a high symmetrical ESR spectrum.[23] In brief, in metal carbide clusterfullerenes, there are three important types of motion: 1) the rotation of the whole endohedral cluster; 2) the flapping motion of two metal atoms; 3) the rotation of the C_2 moiety.

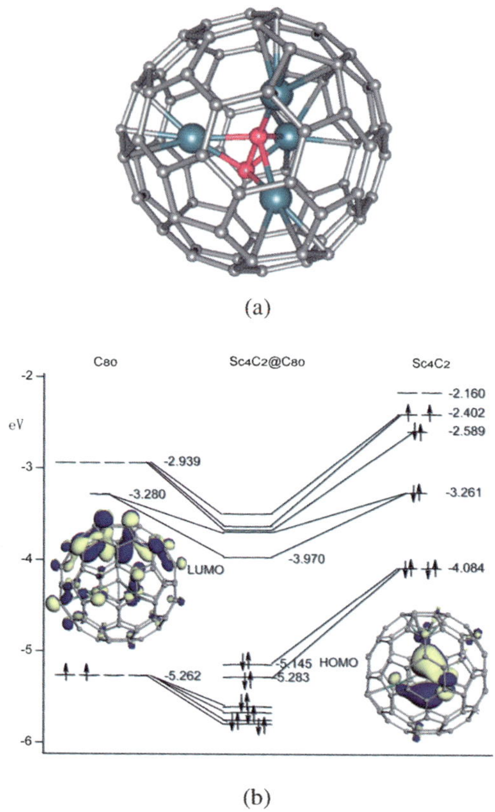

Fig. 7 (a) Optimized structure of $Sc_4C_2@I_h$-C_{80}. (b) Molecular orbitals of $Sc_4C_2@I_h$-C_{80}.

The free rotation of the whole endohedral cluster is often seen in I_h-C_{80}-based clusterfullerenes, such as the above mentioned $Sc_3C_2@C_{80}$ and $Sc_4C_2@I_h$-C_{80}, both of which have a spherical fullerene cage. In addition, in $Sc_2C_2@C_{84}$, which has a large carbon cage, the endohedral Sc_2C_2 cluster can also freely rotate.[29] However, in other metal carbide clusterfullerenes, their fullerene cages have smaller size or ellipsoid-like shapes, thus their endohedral clusters will not rotate freely due to higher energy barrier, such as $Sc_2C_2@C_s$-C_{82}, $Sc_2C_2@C_{3v}(8)$-C_{82}, $Sc_2C_2@C_{2v}(5)$-C_{80}, etc.[29] However, at increasing temperatures, the metal carbide in these clusterfullerenes can break the energy barrier and begin to rotate. ^{45}Sc NMR spectroscopy is an important method to detect the endohedral motion for scandium carbide

Table 3 The ^{45}Sc NMR Chemical Shifts for Sc_2C_2@Cs-C_{82} and Sc_2C_2@C_{2v}(5)-C_{80} under Changed Temperatures

Clusterfullerenes	Chemical shifts (ppm)[a]		
Sc_2C_2@C_s-C_{82}[11]	200, 240 (293K)	230 (373K)	220 (413K)
Sc_2C_2@C_{2v}(5)-C_{80}[15]	130, 170 (293K)	140, 160 (353K)	150 (413K)

[a] The values are estimated according to the NMR spectra in literatures.

clusterfullerenes. Table 3 lists the ^{45}Sc NMR chemical shifts for Sc_2C_2@C_s-C_{82} and Sc_2C_2@C_{2v}(5)-C_{80} under variable temperatures. It can be seen that at elevated temperatures the Sc_2C_2 cluster can get rid of the control from the cage and rotate freely to give two equivalent Sc nuclei.

Crystallographic studies can also map the metal positions within fullerene cages to some extent even though the measured temperature is often around 100 K. In crystallographic studies of Sc_2C_2@C_{84}, several metal sites as well as kinds of C_2 pairs disclose a rotational Sc_2C_2 cluster. In contrast, for Sc_2C_2@C_{3v}(8)-C_{82} and Sc_2C_2@C_{2v}(5)-C_{80}, the relatively fixed orientations of Sc_2C_2 cluster tell of a non-free motion in these two C_{82} cages.[29]

Besides the free rotation, the metal carbide clusters have another unique motion which is the flap of two metal atoms around the C_2 unit revealed by crystallographic studies. Especially in Gd_2C_2@D_3-C_{92}, the Gd_2C_2 cluster is in a rapid rocking motion inside the D_3-C_{92} cage, which is elongated along the three-fold axis.[16] Therefore, in crystallographic results, the motional endohedral Gd_2C_2 cluster has a butterfly shape and it is hard to determine the dihedral angle between the two Gd_2C potions.

For the large metal carbide clusterfullerenes, i.e. Sc_2C_2@C_{84}, Y_2C_2@D_3-C_{92}, etc., the stretched M_2C_2 cluster provides the C_2 unit an opportunity to rotate, and such planar C_2 rotor can be considered as a quantum gyroscope with a cardanic suspension. The quantized rotational states of the planar C_2 rotor were firstly observed by Raman spectroscopy in solid Sc_2C_2@C_{84}, where the rotational transitions induce a periodic line pattern in the low energy Raman spectrum.[30] The low frequency spectral region from 10 to 90 cm^{-1} is described to the motion of the C_2 rotor. The rotational transitions of the diatomic C_2 unit of Y_2C_2@D_3-C_{92} were also observed in the low-energy Raman spectrum and these low-energy quantized rotational states of a free non-rigid C_2 rotor were identified from 24 to 107 cm^{-1}.[31]

6. CONCLUSIONS AND OUTLOOK

In summary, the metal carbide clusterfullerenes have become a large family with different kinds of cages, endohedral clusters, structures, and properties. The discovery of $Sc_2C_2@C_{84}$ proclaimed the emergence of metal carbide clusterfullerenes, and it is also the outstanding representative of typical $M_2C_2@C_{2n}$ type metal carbide clusterfullerenes. These di-metal carbide clusterfullerenes have been found exhibiting various geometries and properties, such as the differing electronic structures, variable M_2C_2 cluster, flapping motion of two metal atoms, etc. The multi-metal carbide clusters construct much more complex clusterfullerenes. For the two paramagnetic tri-metal carbide clusterfullerenes, $Sc_3C_2@I_h$-C_{80} has a Sc_3C_2 cluster with trigonal bipyramidal structure, and for $Lu_3C_2@D_2$-C_{88} it has planar Lu_3C_2 cluster. A landmark of structure in metal carbide clusterfullerenes should be ascribed to the $Sc_4C_2@I_h$-C_{80}, where at the center a C_2 unit is nested inside a tetrahedron of four scandium atoms and the whole molecule exhibits a Russian-doll-like nesting structure. Furthermore, the dynamics in metal carbide clusterfullerenes are of great importance as these motions will bring more and more alluring properties in addition to the C_2 quantum gyroscope. Based on these studies, it can be expected that there is a broader development space for the metal carbide clusterfullerenes, such as new structures constructed by C_2 or even $C_n(n>2)$, a novel dynamic modes for the metal carbide cluster, and interesting paramagnetic and electronic properties.

ACKNOWLEDGMENTS

This work was supported by the National Basic Research Program (2012CB932900), National Natural Science Foundation of China (21121063, 21203205, and 21241004), NSAF (11179006, 11076027), and the China Postdoctoral Science Foundation (201104153).

REFERENCES

1. Kroto HW, Heath JR, O'Brien SC, *et al.* (1985) C_{60}: Buckminsterfullerene. *Nature* **318:** 162–163.
2. Johnson RD, Devries MS, Salem J, *et al.* (1992) Electron-Paramagnetic Resonance Studies of Lanthanum-Containing C_{82}. *Nature* **355:** 239–240.

3. Shinohara H. (2000) Endohedral metallofullerenes. *Rep Prog Phys* **63:** 843–892.
4. Stevenson S, Rice G, Glass T, *et al.* (1999) Small-bandgap endohedral metallofullerenes in high yield and purity. *Nature* **401:** 55–57.
5. Wang CR, Kai T, Tomiyama T, *et al.* (2001) A Scandium Carbide Endohedral Metallofullerene: $(Sc_2C_2)@C_{84}$. *Angew Chem Int Ed* **40:** 397–399.
6. Wang T-S, Feng L, Wu J-Y, *et al.* (2010). Planar Quinary Cluster inside a Fullerene Cage: Synthesis and Structural Characterizations of $Sc_3NC@C_{80}$-I_h. *J Am Chem Soc* **132:** 16362–16364.
7. Stevenson S, Mackey MA, Stuart MA, *et al.* (2008) A distorted tetrahedral metal oxide cluster inside an icosahedral carbon cage. Synthesis, isolation, and structural characterization of $Sc_4(\mu_3\text{-}O)_2@I_h\text{-}C_{80}$. *J Am Chem Soc* **130:** 11844–11845.
8. Dunsch L, Yang S, Zhang L, *et al.* (2010) Metal Sulfide in a C_{82} Fullerene Cage: A New Form of Endohedral Clusterfullerenes. *J Am Chem Soc* **132:** 5413–5421.
9. Iiduka Y, Wakahara T, Nakajima K, *et al.* (2007) Experimental and theoretical studies of the scandium carbide endohedral metallofullerene $Sc_2C_2@C_{82}$ and its carbene derivative. *Angew Chem Int Ed* **46:** 5562–5564.
10. Iiduka Y, Wakahara T, Nakahodo T, *et al.* (2005) Structural determination of metallofullerene Sc_3C_{82} revisited: a surprising finding. *J Am Chem Soc* **127:** 12500–12501.
11. Lu X, Nakajima K, Iiduka Y, *et al.* (2011) Structural Elucidation and Regioselective Functionalization of An Unexplored Carbide Cluster Metallofullerene $Sc_2C_2@C_s(6)\text{-}C_{82}$. *J Am Chem Soc* **133:** 19553–19558.
12. Inoue T, Tomiyama T, Sugai T, *et al.* (2004) Trapping a C_2 radical in endohedral metallofullerenes: Synthesis and structures of $Y_2C_2@C_{82}$ (Isomers I, II, and III). *J Phys Chem B* **108:** 7573–7579.
13. Lu X, Nakajima K, Iiduka Y, *et al.* (2012) The Long-Believed $Sc_2@C_{2v}(17)\text{-}C_{84}$ is Actually $Sc_2C_2@C_{2v}(9)\text{-}C_{82}$: Unambiguous Structure Assignment and Chemical Functionalization. *Angew Chem Int Ed* **51:** 5889–5892.
14. Iiduka, Y, Wakahara, T. *et al.* (2006). ^{13}C NMR spectroscopic study of scandium dimetallofullerene, $Sc_2@C_{84}$ vs. $Sc_2C_2@C_{82}$. *Chem Commun* **19:** 2057–2059.
15. Kurihara H, Lu X, Iiduka Y, *et al.* (2011) $Sc_2C_2@C_{80}$ Rather than $Sc_2@C_{82}$: Templated Formation of Unexpected $C_{2v}(5)\text{-}C_{80}$ and Temperature-Dependent Dynamic Motion of Internal Sc_2C_2 Cluster. *J Am Chem Soc* **133:** 2382–2385.
16. Yang H, Lu C, Liu Z, *et al.* (2008) Detection of a Family of Gadolinium-Containing Endohedral Fullerenes and the Isolation and Crystallographic Characterization of One Member as a Metal-Carbide Encapsulated inside a Large Fullerene Cage. *J Am Chem Soc* **130:** 17296–17300.
17. Zhang JY, Fuhrer T, Fu WJ, *et al.* (2012) Nanoscale fullerene compression of an yttrium carbide cluster. *J Am Chem Soc* **134:** 8487–8493.

18. Tan K, Lu X. (2005) Ti_2C_{80} is more likely a titanium carbide endohedral metallofullerene $Ti_2C_2@C_{78}$. *Chem Commun* 4444–4446.
19. Sato Y, Yumura T, Suenaga K, *et al.* (2006) Direct imaging of intracage structure in titanium-carbide endohedral metallofullerene. *Phys Rev B* **73:** 173401.
20. Shi Z-Q, Wu X, Wang C-R, *et al.* (2006) Isolation and characterization of $Sc_2C_2@C_{68}$: A metal-carbide endofullerene with a non-IPR carbon cage. *Angew Chem Int Ed* **45:** 2107–2111.
21. Valencia R, Rodríguez-Fortea A, Poblet JM. (2008) Molecular and electronic structures of endohedral fullerenes, $Sc_2C_2@C_{3v}$-C_{82} and $Sc_2@C_{3v}$-C_{82}: Benchmark for SCC-DFTB and proposal of new inner cluster structures. *J Phys Chem A* **112:** 4550.
22. Nishibori E, Terauchi I, Sakata M, *et al.* (2006) High-resolution analysis of $Sc_3C_2@C_{80}$ metallofullerene by third generation synchrotron radiation X-ray powder diffraction. *J Phys Chem B* **110:** 19215–19219.
23. Wang T, Wu J, Xu W, *et al.* (2010) Spin divergence induced by exohedral modification: ESR study of $Sc_3C_2@C_{80}$ fulleropyrrolidine. *Angew Chem Int Ed* **49:** 1786–1789.
24. Wang TS, Wu JY, Feng YQ, *et al.* (2012) Preparation and ESR study of $Sc_3C_2@C_{80}$ bis-addition fulleropyrrolidines. *Dalt Trans* **41:** 2567–2570.
25. Kurihara H, Iiduka Y, Rubin Y, *et al.* (2012) Unexpected Formation of a $Sc_3C_2@C_{80}$ Bisfulleroid Derivative. *J Am Chem Soc* **134:** 4092–4095.
26. Sato S, Seki S, Luo GF, *et al.* (2012) Tunable Charge-Transport Properties of I_h-C_{80} Endohedral Metallofullerenes: Investigation of $La_2@C_{80}$, $Sc_3N@C_{80}$, and $Sc_3C_2@C_{80}$. *J Am Chem Soc* **134:** 11681–11686.
27. Xu W, Wang T-S, Wu J-Y, *et al.* (2010) Entrapped Planar Trimetallic Carbide in a Fullerene Cage: Synthesis, Isolation, and Spectroscopic Studies of $Lu_3C_2@C_{88}$. *J Phys Chem C* **115:** 402–405.
28. Wang T-S, Chen N, Xiang J-F, *et al.* (2010) Russian-Doll-Type Metal Carbide Endofullerene: Synthesis, Isolation, and Characterization of $Sc_4C_2@C_{80}$. *J Am Chem Soc* **132:** 16362–16647.
29. Kurihara H, Lu X, Iiduka Y, *et al.* (2012) X-ray Structures of $Sc_2C_2@C_{2n}$ (n = 40–42): In-Depth Understanding of the Core-Shell Interplay in Carbide Cluster Metallofullerenes. *Inorg Chem* **51:** 746–750.
30. Krause M, Hulman M, Kuzmany H, *et al.* (2004) Fullerene quantum gyroscope. *Phys Rev Lett* **93:** 137403.
31. Burke BG, Chan J, Williams KA, *et al.* (2011) Vibrational spectrum of the endohedral $Y_2C_2@C_{92}$ fullerene by Raman spectroscopy: Evidence for tunneling of the diatomic C_2 molecule. *Phys Rev B* **83:** 115457.

CHAPTER 6

The Discovery of Non-IPR Fullerenes

*Wei Xu, Chunying Shu, Chunru Wang**

The world of fullerenes has been significantly enlarged over the last decade due to the continuous synthesis of new structural endohedral fullerenes and fullerene derivatives. In particular, a family of unconventional fullerenes violating the "isolated–pentagon–rule" (IPR) – it was originally believed that fullerenes with adjacent pentagons were extremely unstable than those with isolated pentagons – was recently synthesized and characterized either in endohedral fullerene form or in fullerene derivative form. Compared with IPR fullerenes, the unconventional fullerenes exhibit many advantages such as: (1) thousands more fullerene isomers that largely increase the functional fullerene materials; (2) many unique structural and electronic properties due to the high strains around the adjacent pentagon sites and peculiar charge distribution in the cage. Besides, they also provide a valuable clue on how C_{60} and other stable fullerenes grow up from some small unconventional fullerenes step–by–step via the "fullerene road" mechanism. To date, there have been a lot of reviews focusing on fullerenes, endohedral fullerenes, and their derivatives, but rarely on the unconventional fullerenes. Herein we present a detailed account of the recent advances in unconventional fullerenes along with their elaborate characterization. The unconventional fullerenes may be attainable by several strategies,

*Beijing National Laboratory for Molecular Sciences, Institute of Chemistry, Chinese Academy of Sciences, Beijing, China 100190, China. E–mail: crwang@iccas.ac.cn

i.e. (1) encapsulating one or several metal ions; (2) encapsulating metal carbide; (3) encapsulating metal nitride; (4) exohedrally modifying the fullerenes on the fused-pentagon neighboring sites. Correspondingly, the preparation techniques of unconventional fullerenes include either the traditional arc discharge process or Stone–Wales transforming of IPR fullerenes. As for the structural characterization of unconventional fullerenes, NMR, IR, and Raman spectroscopic techniques as well as DFT calculations have proved to be substantial tools if X-ray crystallography is not available. The unconventional fullerenes are expected to form a new branch in fullerene research and make a great contribution to the structural chemistry and material sciences.

1. INTRODUCTION

Fullerenes C_{2n} are spherical carbon cage molecules constructed from exactly 12 pentagons and (n/2-10) hexagons;[1,2] and the pentagons should be surrounded by hexagons (Figs. 1a and 1b) as one of the tenets of carbon nanoscience, namely, the so-called isolated pentagon rule (IPR).[3] Those with neighboring pentagons are believed to be energetically unstable and usually are not isolable due to the strong steric strain and less local-aromaticity (see Figs. 1c, 1d and 1f). Since so far all the isolated empty fullerenes have been shown to strictly obey the IPR, we name them conventional fullerenes.[3] Endohedral fullerenes are fullerenes with additional atoms, ions, or clusters enclosed within their spheres, which not only enrich the fullerene family dramatically but also exhibit exotic properties over empty fullerenes. Several reviews have appeared to summarize recent progress mainly on endohedral fullerenes.[4–7] Notably, due to the electron transfer from encaged clusters to carbon cages, adjacent pentagons (APs) appeared in the endohedral fullerenes, so those violating the IPR are named *unconventional fullerenes* (Figure 1c, 1d and 1f).[8,9]

The unconventional fullerenes greatly increase the number of available fullerene isomers. For example, C_{60} and C_{70} only have one isomer satisfying the IPR, but the total numbers of unconventional isomers are 1811 and 8148, respectively. For simplification, the nomenclature of unconventional fullerenes herein was specified by Fowler's spiral algorithm, such as $Sc_2@C_{66}$ (4348) and $^{\#1,809}C_{60}$. As every fullerene isomer has its unique structure and electronic properties, the development of unconventional fullerenes

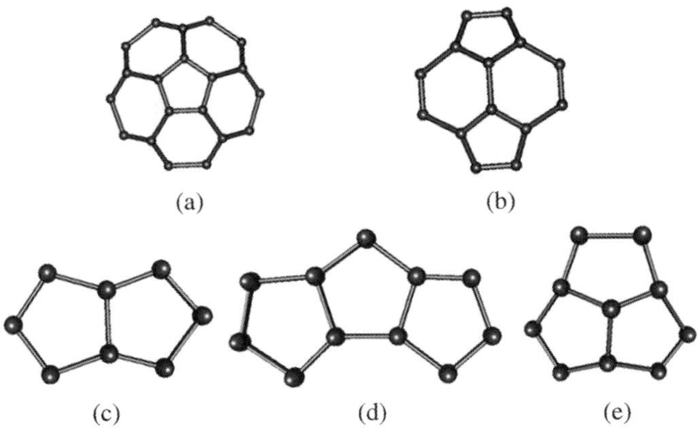

Fig. 1 Two types of isolated pentagons in conventional fullerenes: (a) and (b); and basic types of adjacent-pentagon (AP) configurations in unconventional fullerenes with (c) one pair of APs, (d) two sequentially pairs of APs, (e) triple directly APs.

will largely extend the fullerene material reservoir and provide various potential applications in material science.[10,11]

In 1997, Kobayashi *et al.*[10] calculated the structure and relative stability of Ca@C_{72} based on a previous report on the Ca–based endohedral metallofullerenes (EMFs)[12] and found that two Ca@C_{72} structures containing APs are more stable than the IPR structure. In 2000, Wang and Stevenson *et al.* independently reported two stable unconventional fullerenes, Sc_2@C_{66}[13] and Sc_3N@C_{68},[14] providing a solid proof that unconventional fullerenes can be stabilized by encapsulating metal atoms or cluster.

Since the yields of unconventional EMFs from the arc discharge method are relatively low, it is more favorable to get unconventional fullerenes via exohedral modification. This has been confirmed by the first successful isolation of stable non–IPR fullerene derivative $C_{50}Cl_{10}$,[15] and recently obtained unconventional $C_{76}Cl_{24}$.[16]

The unconventional fullerenes are highly attractive not only because of the high curvatures around the APs,[17] but also because they provide valuable clues on how the fullerenes grow up considering that conventional fullerenes must derive from small unconventional fullerenes according to the "fullerene road" mechanism.[8,18,19]

In this review we present a detailed account of the recent advances in unconventional fullerenes, including non-IPR endohedral fullerenes with metal, metal carbide, metal nitride doping and exohedral modification, and expect that readers can find some clues to produce and characterize more unconditional fullerenes for their potential applications.

2. UNCONVENTIONAL FULLERENES STABILIZED BY ENDOHEDRAL SPECIES

2.1 Encapsulated Metals

EMFs have attracted extensive attention due to their special structures and novel electronic properties.[19,20] Since the electron transfer from metal atoms to the fullerene cages, even the "missing" fullerenes, C_{66}, C_{72} and C_{74} are extracted by metal doping, such as $Sc_2@C_{66}$,[13] $La_2@C_{72}$,[21,22] $Ca@C_{72}$,[23] $Ce_2@C_{72}$,[24] $Sc_2@C_{74}$[25] and $Ca@C_{74}$,[23] etc. This in turn established the mechanism to synthesize unconventional fullerenes through endo-encapsulation.

The production and extraction of unconventional fullerenes are essentially the same as that of conventional EMFs. Briefly, soot containing metallofullerenes was produced by the arc discharge method, and the EMFs were isolated using multistage HPLC.[16]

a) $Sc_2@C_{66}$

The first IPR-violating EMF was $Sc_2@C_{66}$. The high-resolution ^{13}C NMR spectrum of $Sc_2@C_{66}$ showed a total of 19 distinct lines. For the C_{66} fullerene, there are 4,478 possible non-IPR isomers including $2 \times D_3$, $1 \times C_{3v}$, $18 \times C_{2v}$, $112 \times C_s$, $211 \times C_2$ and $4{,}134 \times C_1$.[2] All these isomers contain APs to form closed cage structures, but only eight isomers with C_{2v} symmetry can satisfy the observed ^{13}C NMR pattern.

The geometrical structure of the EMF $Sc_2@C_{66}$ (4348, Fig. 2) was revealed by the maximum-entropy method (MEM)/Rietveld analysis of the synchrotron X-ray powder diffraction data, which contains two pairs of two-fold APs tightly bonding with two scandium atoms. An optimized geometry of $Sc_2@C_{66}$ calculations confirmed this stable structure and revealed that the Sc_2 dimer donates two electrons to the C_{66} cage, resulting in a formal electronic state of $(Sc_2)^{2+}@C_{66}^{2-}$.[13] The charge transfer between

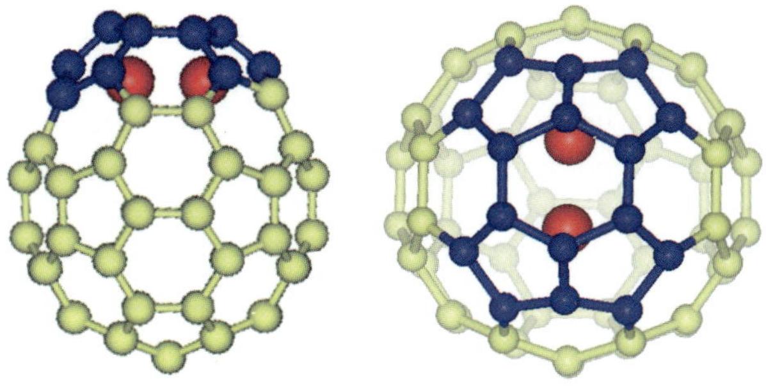

Fig. 2 Side (a) and top (b) views of the $Sc_2@C_{66}$-C_{2v} (4384).

the dimer and APs considerably decreased strain energy caused by a pair of APs and therefore stabilized the structure.

In 2002, Kobayashi *et al.* made *ab initio* calculations on the eight possible isomers of C_{66} and suggested another possibility of the stable structure for $Sc_2@C_{66}$.[26] They argued that the density functional calculations of $Sc_2@C_{66}$-C_{2v} (4348, Fig. 2) did not correspond to an energy minimum but collapsed without barrier to a more stable C_1 structure. Instead a much more stable but unconventional C_{2v} structure with two pairs of threefold APs (4059, Fig. 2) was suggested as a most probable candidate. To make clear the controversial structures on $Sc_2@C_{66}$, more experiments, even X-ray crystallographic analysis, are needed.

b) $M@C_{72}$ and $M_2@C_{72}$

C_{72} has one IPR isomer but it has never been isolated and characterized." Recently, with metal encapsulating or exohedral modification, some EMFs based on C_{72} become isolable, such as $Ca@C_{72}$,[23] $La@C_{72}(C_6H_3Cl_2)$,[27] $La_2@C_{72}$,[22] and $Ce_2@C_{72}$.[24]

In fact, $Ca@C_{72}$ was the first fullerene assigned theoretically to have a non-IPR structure,[23] but this has not yet been experimentally determined. Alternatively, the synthesis of $La@C_{72}(C_6H_3Cl_2)$ allowed characterization of its structure.[27] $La@C_{72}$ is indeed a non-IPR carbon cage with only one pair of APs bonding with the encapsulated La, and the dichlorophenyl group is singly bonded to a carbon at the joint of two hexagons and one pentagon ([5,6,6] junction) on the APs.[27]

Compared with $M@C_{72}$, $M_2@C_{72}$ usually shows higher stability and solubility. The first structural analysis of the C_{72} fullerene encapsulating two La atoms, $La_2@C_{72}$, was fulfilled by ^{13}C NMR.[22] It showed 18 lines in a diamagnetic chemical shift range, which does not correspond to that of the only IPR structure of C_{72}–D_{6d}. According to the number of NMR lines and one NMR resonance locating at very low field (158.1 ppm), it was suggested the existence of APs in the fullerene cage.

To unambiguously assign the unconventional structure of $La_2@C_{72}$, Lu functionalized $La_2@C_{72}$ to yield $La_2@C_{72}Ad$ (Ad = adamantylidene).[28] X–ray structural analysis disclosed that $La_2@C_{72}$ is non-IPR isomer 10611 with two pairs of APs, and the addition occurs at a junction between a five- and a six-membered ring ([5,6] junction) of $La_2@C_{72}$. The calculated charge densities and π -orbital axis vector analysis POAV ($\theta = \pi - 90$) values of the carbons of $La_2@C_{72}$ suggested that the AP sites are very reactive toward carbene.

In addition, Lu also reported a series of bis–carbene adducts of the non-IPR $La_2@C_{72}$.[29] The structural characterizations demonstrated that the two Ad groups are covalently bonded to the [5,6]–junction adjacent to the two fused-pentagon bonds and the addition bond is broken.[29] Interestingly, electrochemical data show that the Ad group has a mild reductive effect on $La_2@C_{72}$ and the electrochemical properties of $La_2@C_{72}$ can be tuned by stepwise addition of the Ad groups.[29] These Ad derivatives may find fundamental applications in photovoltaics and donor–acceptor systems due to the improved solubility and elevated lowest unoccupied molecular orbital (LUMO) levels.

Similarly, the EMF $Ce_2@C_{72}$ with a non-IPR fullerene cage was experimentally and theoretically characterized by Akasaka *et al.*[24] The ^{13}C NMR spectrum of $Ce_2@C_{72}$ showed 18 lines, revealing that $Ce_2@C_{72}$ has a D_2 symmetry, as $La_2@C_{72}$ does.[22] The four ^{13}C NMR signals in the higher magnetic field region, 25.64, 86.88, 92.33 and 128.31 ppm, associated with two fused pentagons, **a**, **b** (or **c**), **c** (or **b**), and **d**, respectively (Fig. 3), are clearly assigned by means of the paramagnetic ^{13}C NMR shift analysis and theoretical calculation. Therefore the two encapsulated Ce atoms are tightly located near the two fused pentagons.

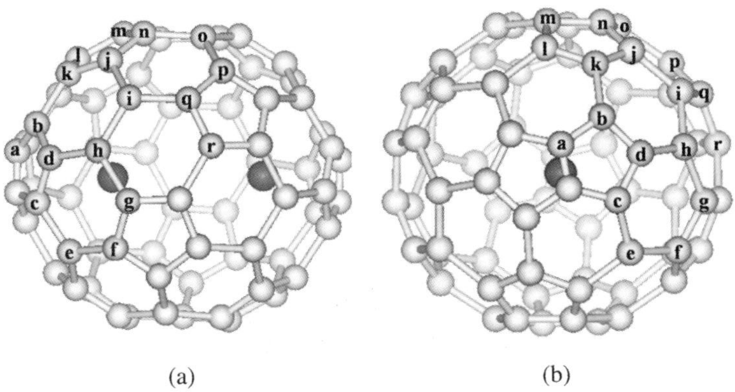

Fig. 3 (a) Front and (b) side views of the optimized structure of $Ce_2@C_{72}$ –D_2 (10611).

2.2 Encapsulate Metal Nitride

In 1999, Dorn *et al.* reported a trimetallic nitride template (TNT) endohedral fullerene $Sc_3N@C_{80}$.[30] Since then the family of $M_3N@C_{2n}$ molecules has been rapidly expanded.[4] The technique to produce $M_3N@C_{2n}$ is essentially similar to that for the production of other EMFs, except for intro ducing a small portion of nitrogen gas into the Krätschmer–Huffman generator.

a) $A_xSc_{3-x}N@C_{68}$ (x = 0-2, A Represents Rare Earth Metal)

$Sc_3N@C_{68}$, as well as $A_xSc_{3-x}N@C_{68}$, was first reported in 2000.[30] ^{13}C NMR spectral study revealed 12 NMR signals, suggesting a high symmetry of this molecule. The size of C_{68} locates between the smallest IPR-fullerene C_{60} and the second smallest IPR-fullerene C_{70}, so it must be an unconventional fullerene. The non-IPR C_{68} fullerene cage has 6332 isomers, but the ^{13}C NMR spectrum of $Sc_3N@C_{68}$ (12 lines) restricts the candidates to only 11 isomers that have either D_3 or S_6 symmetry. On the basis of density functional calculations, two isomers (6140[14] and 6275[31]) with D_3 symmetry but with different shapes have emerged as the possible cage structures (Fig. 5).

X–ray crystallography data showed that the Sc_3N cluster is encapsulated in a D_3 non-IPR C_{68} (6140) cage.[32] There are three equivalent sites where pairs of pentagons share a side ([5,5] junction or pentalene units), and three scandium metals were bonded tightly to these sites (Fig. 4). In addition, due to the strong interaction between encapsulated Sc and pentalene on the cage, the pyramidalization of the carbon atoms at the [5,5]

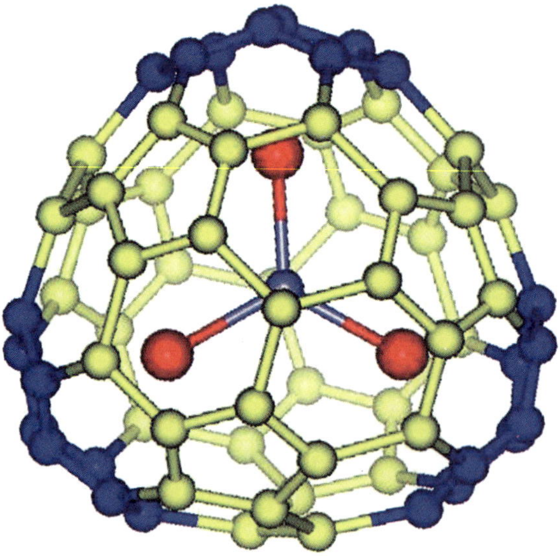

Fig. 4 Molecular Structure of $Sc_3N@C_{68}$-D_3 (6140).

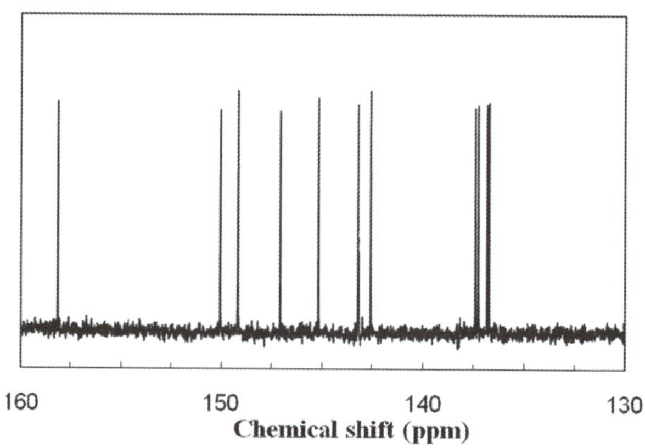

Fig. 5 High-resolution ^{13}C NMR spectrum of $Sc_3N@C_{68}$.

junction is significantly higher than it is for the other carbon atoms of the cage.[32]

Theoretical simulation of the ^{13}C NMR pattern of $Sc_3N@C_{68}$ (6140) presented by Reveles *et al.*[33] agreed reasonably with the experimental data, except for the low-intensity line which was calculated at 143.4 ppm, but was experimentally located at 136.8 ppm. For clarity we revisited the $Sc_3N@C_{68}$ structure by performing a high-resolution ^{13}C NMR spectrometry and found that the relevant line in previous ^{13}C NMR spectrum was indeed wrong-assigned. Our measured low-intensity line at 143.2 ppm (No. 7, Fig. 5) is very close to the theoretical simulation.

The chemical reactivity of non-IPR $Sc_3N@C_{68}$ was studied by using a Bingel reaction.[34] The proposed addition site of $Sc_3N@C_{68}$ monoadduct was regioselective at the C22–C12 bond (Fig. 6), which is close to the unique [5,5] junction according to the ^{13}C NMR simulation. Due to the UV–VIS spectrum of $Sc_3N@C_{68}$ monoadduct being almost identical to that of pristine $Sc_3N@C_{68}$, it was then suggested that the addition bond is broken, namely an "open" structure. This study provides first insight into understanding the chemical reactivity of unconventional fullerenes based on TNT family and also valuable data for the structural characterization.

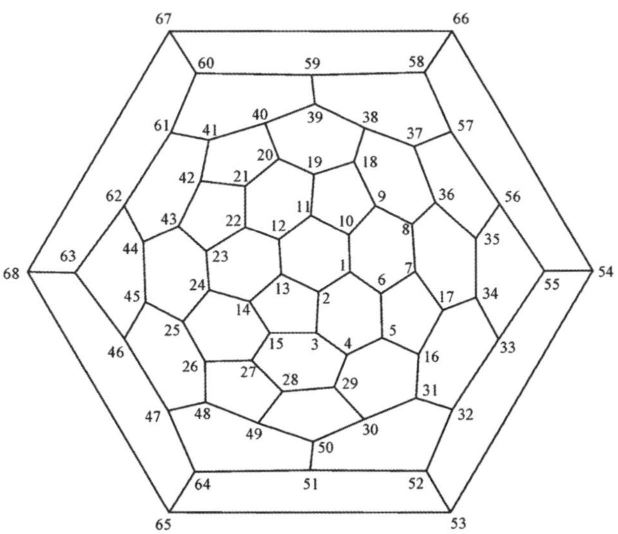

Fig. 6 Schlegel diagram of the C_{68}–*D3 non-IPR isomer* (6140).

b) $Sc_3N@C_{70}$

C_{70} as a second abundant fullerene is very important.[35] The metal atom or cluster inside C_{70} was little known until the recent synthesis of $Sc_3N@C_{70}$.[36] The spectroscopic studies of $Sc_3N@C_{70}$ as well as the elucidation of its structure by a combination of DFT computations suggested that this fullerene cage does not satisfy the IPR. Since fused-pentagons without stabilization by encapsulating metal atoms must induce extra bending energies, DFT optimizations were performed on the 116 hexaanions of C_{70} isomers with three or fewer pairs of APs. The results revealed that C_{2v}:7854 is the only isomer among the 116 isomers that has a HOMO–LUMO gap above 1 eV, and it is about 43 kJ/mol lower in energy than the IPR C_{70}-D_{5h}. This was consistent with the optical band gap of 1.3 eV from the UV/Vis/NIR spectrum of $Sc_3N@C_{70}$.

c) $DySc_2N@C_{76}$

The first non-IPR C_{76} cage based on a mixed metal nitride cluster, $DySc_2N@C_{76}$, was synthesized and isolated recently.[37] Taking account into the six-electron transfer from the entrapped species to the fullerene cage, DFT calculations on the hexaanionic states of all IPR isomers and some non-IPR isomers revealed that only two IPR-obeying isomers C_{76}^{6-} (D_2: 19150 and T_d: 19151) have reasonable HOMO–LUMO gaps (0.75 eV and 0.14 eV, respectively), but much higher energies than the non-IPR isomers. The C_{76}^{6-} (D_2: 19150) is even 101.0 kJ/mol less stable than the non-IPR C_{76}^{6-}-C_s isomer:17490.

Experimentally, $DySc_2N@C_{76}$ is a stable fullerene with a small band gap of 0.96 eV based on the onset of its UV–Vis–NIR spectrum at ~1300 nm. According to the FTIR spectroscopic study in combination with extensive DFT calculations, the cage structure of $DySc_2N@C_{76}$ was then assigned as the non-IPR C_s:17490, in which there are two pairs of APs.

d) $M_3N@C_{78}$ (M = Tm, Dy, Gd)

Up to now, several $M_3N@C_{78}$ (M = Gd,[38] and Tm,[39] Dy[40]) EMFs have been isolated. As established by X-ray crystallographic study, $Gd_3N@C_{78}$ was assigned as $Gd_3N@C_{78}$ (22010)[38] with an unconventional carbon cage. Two of the gadolinium atoms on the planar Gd_3N unit are located within the pentalene folds, and the other gadolinium atom resides at the center of a hexagonal face of the fullerene (Fig. 7).

Unfortunately, neither X-ray nor ^{13}C NMR data is available yet for $Tm_3N@C_{78}$ and $Dy_3N@C_{78}$ because of the paucity of samples. Alternatively, IR and Raman spectroscopic studies were performed and revealed that the structures of $Tm_3N@C_{78}$ and $Dy_3N@C_{78}$ (II) are much different from that of $Sc_3N@C_{78}$-D_{3h} (IPR: isomer 24,109), but similar to the unconventional $Gd_3N@C_{78}$.[38]

The DFT computations found that the non-IPR isomer of C_{78}-C_2 (22,010) has the lowest energy and a suitable cavity to host the planar lanthanide-based M_3N clusters, so the experimentally isolated $Tm_3N@C_{78}$ and $Dy_3N@C_{78}$ (II) may be assigned as the non-IPR C_{78} (C_2: 22,010).[41] This study indicates that vibrational spectroscopy supplemented with reliable theoretical modeling would provide a powerful tool in the structural assignments of endohedral fullerenes.

e) $Gd_3N@C_{82}$

For a C_{82} there are nine isomers that conform to the IPR and 39709 isomeric structures that do not conform to the IPR. Poblet *et al.* predicted

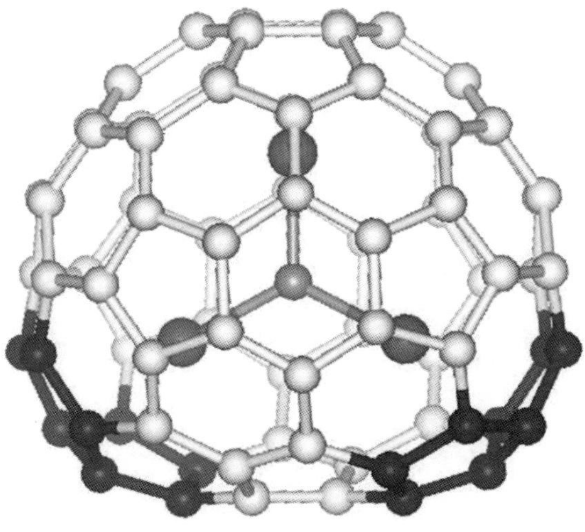

Fig. 7 Molecular structure of $Gd_3N@C_{78}$–C_2(22010).

that $M_3N@C_{82}$ may violate IPR because no hexaanions of IPR C_{82} isomers were calculated to have appropriate large band gaps.[42] Later, detailed computational studies by Popov and Dunsch suggested that the non-IPR structures $M_3N@C_{2v}$:39705-C_{82} and $M_3N@C_s$: 39663-C_{82} are the most probable isomers for the C_{82} cages.[43]

The first C_{82}-based $M_3N@C_{82}$ fullerene, $Gd_3N@C_{82}$, was prepared and characterized by Echegoyen *et al.*[44] C_{82} adopts an egg-shaped carbon cage $Gd_3N@C_s$:39663-C_{82} (Fig. 8), which was indeed the non-IPR C_{82} isomer as predicted by Popov *et al.* The $Gd_3N@C_s$-C_{82} is nestled within the eight ethyl groups of the Ni^{II}(OEP) molecule and the Gd_3N unit is planar with Gd1 situating within the fold of the pentalene unit.

f) $M_3N@C_{84}$ (M = Tb, Tm, Gd)

There are 24 isomeric C_{84} that conform to the IPR. However, the recently isolated $Tb_3N@C_{84}$ does not adopt any these IPR cages but takes an egg-shaped non-IPR C_{84} (C_s:51365)[45] (Fig. 9). There is only one pair of fused pentagons in C_{84}-C_s. The 64 carbon atoms closest to the fused pentagon pair are well behaved, while the other 20 carbon atoms at the opposite end are observed in two sets of positions. The Tb_3N unit is planar, in which Tb1 situated within the fold of the pentalene unit is ordered while the

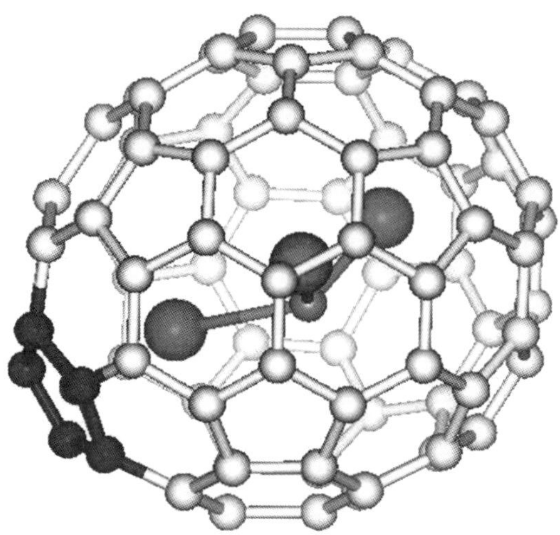

Fig. 8 The Molecular structure of $Gd_3N@C_{82}$–*Cs*(*39633*).

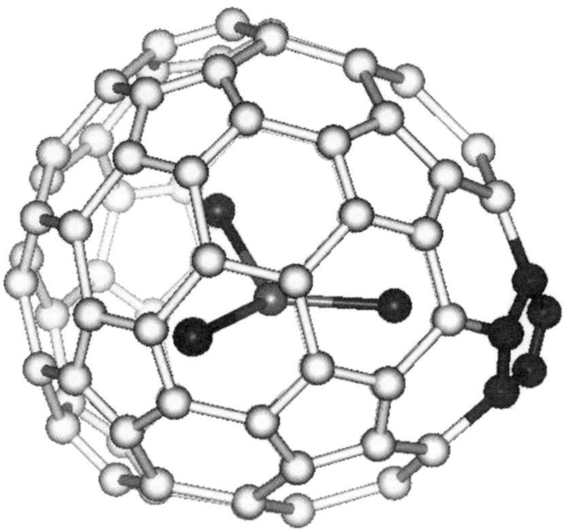

Fig. 9 The molecular structure of $Tb_3N@C_{84}$–*Cs*(*51365*)

other two Tb atoms are disordered. This kind of non-IPR structure was expanded by subsequently isolated $Tm_3N@C_s$:51,365-C_{84} and $Gd_3N@C_s$:51,365-C_{84},[46] which is apparently different from those non-IPR $A_3N@C_n$ (n = 68, 70) with three pairs of APs and thus ordered metal atoms. The different cage sizes should contribute to this phenomenon.

2.3 Metal-Carbide Endohedral Fullerene

Besides $M_3N@C_{2n}$, metal-carbide endohedral fullerenes form another large family of EMFs. So far many metal-carbide endohedral fullerenes with IPR-satisfying fullerene cages have been synthesized and characterized exclusively, e.g. $Sc_2C_2@C_{84}$.[47,48] However, the recent isolation and structural characterizations of $Sc_2C_2@C_{68}$[49] revealed that unconventional fullerenes are also possible through encaging metal carbide.

The synthesis of $Sc_2C_2@C_{68}$ was essentially analogous to those of other scandium-containing EMFs such as $Sc_2@C_{66}$ mentioned above. The structure of Sc_2C_{70} was determined based on spectrometry studies in combination with theoretical calculations. The results suggested that $Sc_2C_2@C_{68}$ would preferentially adopt the non-IPR C_{68}-C_{2v} (6073) cage and has the valence state $(Sc^{3+})_2(C_2)^{2-}@C_{68}^{4-}$.

Interestingly, the isomeric structure and valence state of the non-IPR C_{68} cage proposed for $Sc_2C_2@C_{68}$ were different from that of $Sc_3N@C_{68}$. The former has the C_{2v}-symmetric C_{68} (6073) cage with two pairs of APs accommodating two Sc^{3+} ions, and the latter has a valence state $(Sc^{3+})_3N^{3-}@C_{68}^{6-}$ and D_3-symmetric C_{68} (6140) cage with three pairs of APs. In both cases each pentalene unit coordinates to a Sc^{3+} ion. Therefore, it was suggested that endohedral clusters such as Sc_2C_2 and Sc_3N function as templates for the synthesis of different non-IPR C_{68} cages.

2.4 Metal-Sulfide Endohedral Fullerene

Another type of interesting cluster EMFs is metal sulfide cluster EMFs, which are synthesized by the introduction of sulfur into the discharge reactor. Using CH_5N_3·HSCN, a solid sulfur source, Dunsch and co-workers reported several sulfids cluster EMFs.[50] More recently, Echegoyen *et al.* reported that an extensive family of novel sulfids cluster EMFs with cages ranging from C_{70} to C_{100} can be obtained in macroscopic quantities by introducing SO_2 instead of a solid compound as the sulfur source.[51] Among the family of $Sc_2S@C_{2n}$ (n = 35-50), $Sc_2S@C_{72}$[52] and $Sc_2S@C_{70}$[53] (Fig. 10) are unconventional fullerenes. The single crystal diffraction study shows that the Sc_2S cluster is encaged in a non-IPR Cs(10528)-C_{72} cage. Computational studies show that the Sc_2S cluster transfers four electrons to the C_{72} cage and the Sc_2S unit coordinates the two scandium ions to the two APs on Cs(10528)-C_{72}, and the special geometry of the Sc_2S cluster helps to stabilize a new carbon cage. Besides $Sc_2S@Cs(10528)–C_{72}$, $Sc_2S@$

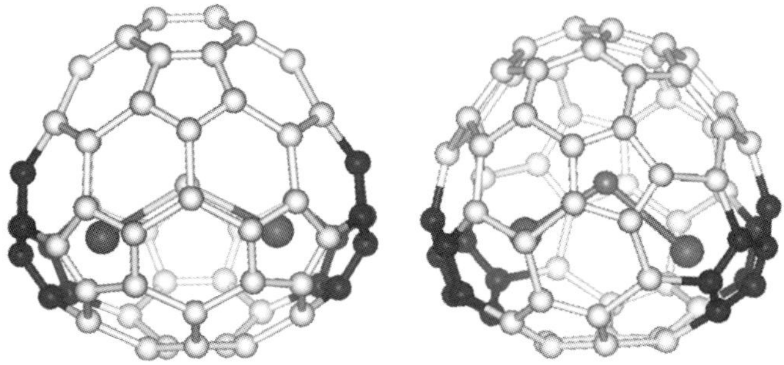

Fig. 10 The structures of $Sc_2S@Cs(10528)–C_{72}$ (left)[52] and $Sc_2S@C_2(7892)–C_{70}$ (right).

$C_2(7892)–C_{70}$ provides an additional example of stabilization of a new non-IPR cage resulting from the combined effect of a four electron transfer and the unique geometrical fit between the Sc_2S cluster and the cage.

3. UNCONVENTIONAL FULLERENES STABILIZED BY EXOHEDRAL MODIFICATION

For a long time, scientists have been exploring a way to stabilize small fullerenes both theoretically and experimentally through exohedrally modifying the small fullerene cages. The first solid form of a small fullerene derivative is $C_{50}Cl_{10}$.[15] Afterwards, several other unconventional fullerene derivatives such as $C_{60}Cl_8$, $C_{60}Cl_{12}$,[54] $C_{64}H_4$[55] and $C_{76}Cl_{24}$[16] were synthesized, revealing an efficient technique to obtain unconventional fullerenes via exohedral modification. Correspondingly, there are three techniques to induce the isomerism of the carbon cage. The first one is a modified arc discharging method, such as the chlorine-containing fullerene derivatives[54] and $C_{64}H_4$[55] formed by introducing a small amount of carbon tetrachloride or chlorine gas, and methane in the helium atmosphere, respectively. The second technique is Stone–Wales transforming of IPR fullerenes, such as $C_{76}Cl_{24}$[16]. And the third is low energy ion beam deposition of vibronically excited C_{60}^+ onto graphite followed by selective sublimation of C_{60}-I_h from the deposited isomer mixture.[56]

3.1 $C_{50}Cl_{10}$ and $C_{64}X_4$ (X = H, Cl)

$C_{50}Cl_{10}$ is the first stable unconventional fullerene derivative isolated and characterized unambiguously.[12] The molecular structure was extensive characterized by MS, ^{13}C NMR etc.[15,57] $C_{50}Cl_{10}$ has five pairs of APs with 10 chlorine atoms attached on the [5,5] junctions on fullerene cage. Since all the carbon atoms in the [5,5] junctions of fullerene cages adopt a sp^3 hybridization that have smaller favorite bonding angles (109°), the strain energy in fused-pentagon sites is largely released, leading to an extreme stable structure. Quite recently, the crystallographic studies of the $C_{50}Cl_{10}$ confirm the previously reported IPR-violating chlorofullerene structure, D_{5h} (Fig. 11).[58]

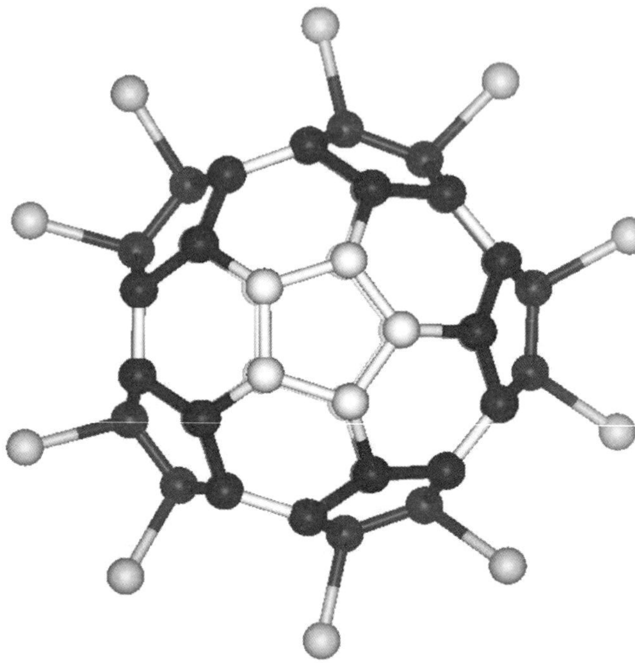

Fig. 11 The molecular structure of $C_{50}CI_{10}$.

The unconventional fullerene derivative $C_{64}H_4$ was synthesized and structurally characterized by our group in 2006.[55] The structural model (Fig. 12) is constructed based on NMR studies and optimized by *ab initio* calculations. The results disclosed that $C_{64}H_4$ holds a C_{3v} symmetry that bears triply fused-pentagons. The H substituents were added exclusively to the carbons of [5,5] junctions, revealing that the reactivity of abutted-pentagon carbons is much higher than that of carbons at other sites as a result of the high surface curvatures.

One of the $C_{64}H_4$ analogues, i.e. $C_{64}Cl_4$, was recently produced and isolated by Xie *et al.*[58] The X-ray crystallographic data and theoretical calculations revealed that $C_{64}Cl_4$ has a similar structure to $C_{64}H_4$, strongly supporting the structural assignments of $C_{64}X_4$.

The fact that $C_{50}Cl_{10}$ and $C_{64}H_4$ are stabilized by the chemical modification on the pentagon–pentagon fusion vertex gives a hint that the modification of the electronic structures near the fused pentagon pairs may be a promising way to stabilize non-IPR fullerenes, further enriching the storage of stable fullerenes for practical applications.

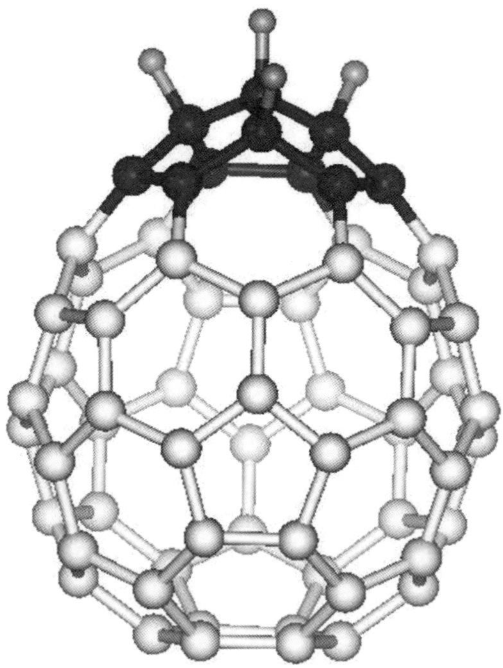

Fig. 12 A view of the molecular structure of $C_{64}H_4$.

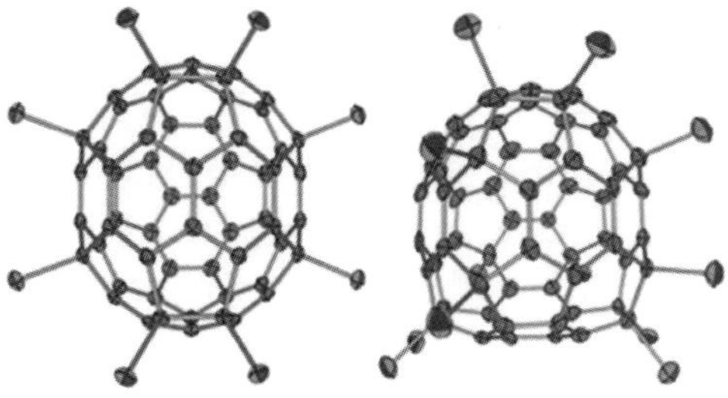

$^{\#1,809}C_{60}Cl_8$ $^{\#1,804}C_{60}Cl_{12}$

Fig. 13 Thermal ellipsoid plot drawings of the crystal structures of $^{\#1,809}C_{60}Cl_8$ and $^{\#1,804}C_{60}Cl_{12}$.[54]

3.2 $C_{60}Cl_8$ and $C_{60}Cl_{12}$

The IPR isomer C_{60} is undoubtedly the most stable empty fullerene known so far. However, after calculations on a large number of C_{60} derivatives with both IPR and non-IPR cages, Alcam *et al.* suggested that it is always possible to generate non-IPR structures with a uniform distribution of APs and pyrene motifs formed by four adjacent hexagonal rings, and sometimes the unconventional C_{60} derivatives are much more stable than the IPR ones.[59]

Most recently, the hypotheses were experimentally proved true by Tan *et al.*[54] Two I_h-symmetry-breaking C_{60} isomers, $^{\#1,809}C_{60}Cl_8$ and $^{\#1,804}C_{60}Cl_{12}$, were synthesized and isolated. Various spectrometries and X-ray crystallography were performed to characterize the relevant structures, and revealed that both the two isomeric C_{60} cages present APs, two in $^{\#1,809}C_{60}$ and three in $^{\#1,804}C_{60}$ (Fig. 16). Therefore, the findings of $C_{60}Cl_8$ and $C_{60}Cl_{12}$ represent that there are at least two C_{60} isomers violating IPR, which may be formed via Stone-Wales pathway (Fig. 16b).[3]

3.3 Other Unconventional Exohedral Fullerene Derivatives

Recently, a fullerene derivative $C_{76}Cl_{24}$ was synthesized.[16] Surprisingly, single-crystal analysis revealed that $C_{76}Cl_{24}$ has a non-IPR structure (Fig. 14), even though the reactant is IPR C_{76}-D_2. It is hypothesized that the D_2-C_{76} isomer takes a series of Stone–Wales rearrangements in the chemical reaction process, leading to the unconventional fullerene structure $^{\#18917}C_{76}Cl_{24}$. In fact, it has long been a dream for chemists to reconstruct the fullerenes by Stone–Wales arrangements, but previously only deletion or addition of a C_2 unit on the C_{60} fullerene cage succeeded.[60,61] Notice that there are usually thousands more non-IPR isomers of fullerenes than IPR isomers, so once this strategy is generally extended to other fullerenes, the available fullerene materials would increase exponentially. In this sense, the synthesis of $^{\#18917}C_{76}Cl_{24}$ opens a new era on the preparation of unconventional fullerenes.

In situ chlorination using CCl_4 or C_{12} as an additional reagent in conventional fullerene synthesis has been found to be an effective method for stabilizing non-IPR fullerene cages.[3] Several unconventional fullerenes have been obtained using this approach, including $^{\#540}C_{54}Cl_8$, $^{\#864}C_{56}Cl_{12}$, $^{\#4,169}C_{66}Cl_6$, $^{\#4,169}C_{66}Cl_{10}$,[62] $C_{78}Cl_8$[63], $^{\#11188}C_{72}Cl_4$[64,65], $C_{84}Cl_{32}$,[66] (Fig. 15) $^{\#916}C_{56}Cl_{12}$,[67] $^{\#913}C_{56}Cl_{12}$,[68] and $^{\#6094}C_{68}Cl_8$ [69]. Among the numerous fullerene species, $^{\#540}C_{54}$, $^{\#864}C_{56}$ and $^{\#4,169}C_{66}$ are the first experimentally established fullerene derivatives featuring the triple sequentially fused pentagons (TSFP). And the example of chlorinated non-IPR fullerenes, $C_{84}Cl_{32}$ like $^{\#18917}C_{76}Cl_{24}$, has been realized by elegant Stone–Wales rearrangement of

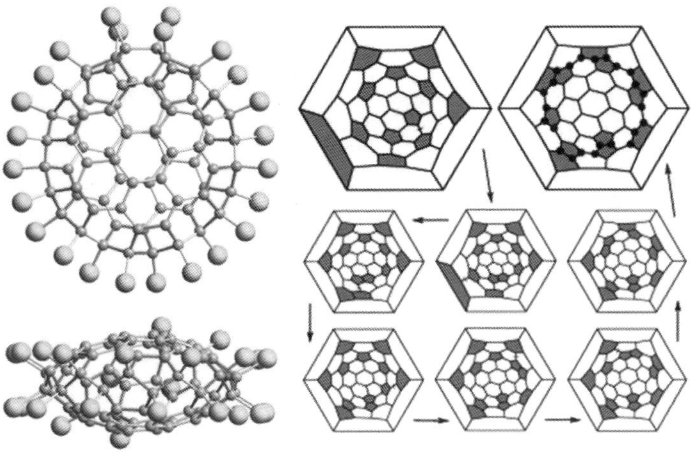

Fig. 14 Molecular XRD structure of $^{\#18917}C_{76}Cl_{24}$ presented in two projections (left) and Stone–Wales pathway (right).

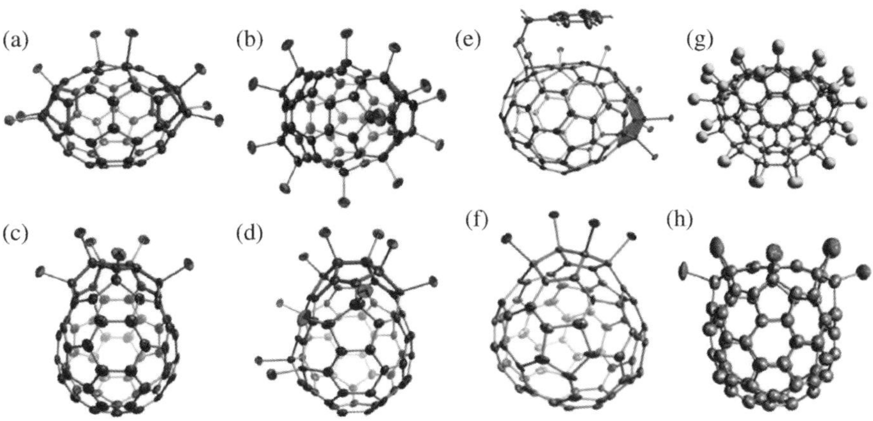

Fig. 15 The structures of $^{\#540}C_{54}Cl_8$ (a), $^{\#864}C_{56}Cl_{12}$ (b), $^{\#4,169}C_{66}C_{16}$ (c), $^{\#4,169}C_{66}Cl_{10}$ (d),[62] $^{\#23863}C_{78}(OOCH_2C_6H_5)Cl_7$ molecule, which shares the same fullerene core with $^{\#23863}C_{78}Cl_8$(e),[63] $^{\#11188}C_{72}Cl$(f),[64,65] and $C_{84}Cl_{32}$(g).[66]

higher IPR fullerenes into non-IPR cages under strong chlorination conditions. Among the three non-IPR isomers of $C_{56}Cl_{12}$ fullerenes, the $^{\#916}C_{56}$ cage has been predicted to be the most thermodynamically stable over a wide temperature range, and $^{\#864}C_{56}Cl_{12}$ and $^{\#913}C_{56}Cl_{12}$ isomers can be easily derived from the most-stable $^{\#916}C_{56}$ by Stone–Wales transformations

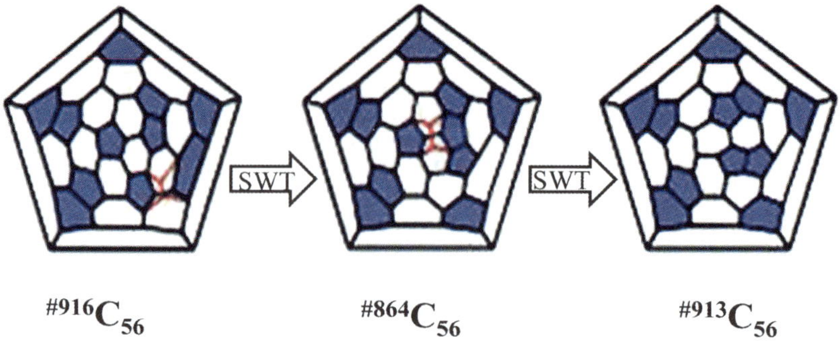

Fig. 16 The Stone–Wales transformation (SWT) of the $^{\#916}C_{56}$ to the less-stable $^{\#864}C_{56}$ and $^{\#913}C_{56}$ fullerenes. The rearrangement of carbon connectivity for the respective step is shown with red lines.

(Fig. 16). In addition to these, $^{\#11188}C_{72}Cl_4$ is a peculiar one since it represents the only known example where a non-IPR cage is expected to be more stable than its IPR isomer; the $^{\#6094}C_{68}Cl_8$ case proved the first experimental evidence for the existence of highly antiaromatic carbon cages. This is an outstanding finding in fullerene science that has been simultaneously and independently reported by the research groups of Xie and Jansen. On the other hand, the ease of access to new chlorination open a new avenue for the study of the chemical reactivity, optical and electronic properties, and practical applications of the most unusual non-IPR fullerenes.

4. CONCLUSION AND PROSPECTS

More and more unconventional fullerenes can be obtained by endo-encapsulation or exo-modification as well as their combination. Besides the released strain energy induced by the bonding of the endocluster with pentlene unit or exo-modification of fused pentagons, peculiar charge distribution in the cage controlled by the APs and pyrene motifs should contribute to the stabilization of unconventional fullerenes. This should be a rule rather than the exception for the existence of unconventional fullerenes, and may direct the fabrication of unconventional fullerenes and make them more common than IPR ones.

The experimental available techniques such as X-ray crystallography supplemented with spectroscopic characterization and reliable theoretical modeling have proved to be powerful tools in the structural assignments and characteristic analysis of unconventional fullerenes. The data listed here undoubtedly would provide basic clues for the identification of oncoming unconventional fullerenes. Although the physical and chemical investigations of unconventional fullerenes are still in their infancy, the tenets of fabricating unconventional fullerenes along with the scientific curiosity would promise them a perspective future, not only in preparing and understanding these new nanoscale materials but also in finding special properties and practical applications.

ACKNOWLEDGMENTS

The authors thank all of the co-workers who have contributed to this research as the literature cited herein. C.R.W. thanks fund support from 973 Program (2006CB300402) and NSFC (20821003, 20702053), the Ministry of Science and Technology (2008ZX05013-004) and the Chinese Academy of Sciences.

REFERENCES

1. Kroto HW, Heath JR, O'Brien SC, *et al.* (1985) C_{60}: buckminsterfullerene. *Nature* **318:** 162–163.
2. Fowler PW, Manolopoulos DE (eds). (1995) *An Atlas of Fullerenes,* Clarendon Press, Oxford.
3. Kroto HW. (1987) The stability of the fullerenes C_n, with n = 24, 28, 32, 36, 50, 60 and 70. *Nature* **329:** 529–531.
4. Dunsch L, Yang S. (2007) Metal nitride cluster fullerenes: Their current state and future prospects. *Small* **3:** 1298–1320
5. a) Popov AA. (2009) Metal-cage bonding, molecular structures and vibrational spectra of endohedral fullerenes: bridging experiment and theory. *J Comp Theor Nanosci* **6:** 292–317; b) Chaur MN, Melin F, Ortiz AL, Echegoyen L. (2009) Chemical, electrochemical, and structural properties of endohedral metallofullerenes. *Angew Chem Int Ed* **48:** 7514–7538.
6. a) Yamada M, Akasaka T, Nagase S. (2010) Endohedral metal atoms in pristine and functionalized fullerene cages. *Acc Chem Res* **43:** 92–102;

b) Lu X, Akasaka T, Nagase S. (2011) Chemistry of endohedral metallofullerenes: the role of metals. *Chem Commun* **47:** 5942–5957; c) Rodríguez-Fortea A, BalchAL, Poblet JM. (2011) Endohedral metallofullerenes: a unique host–guest association. *Chem Soc Rev* **40:** 3551–3563.

7. Lu X, Feng L, Akasaka T, Nagase S. (2012) Current status and future developments of endohedral metallofullerenes. *Chem Soc Rev* **41:** 7723–7760.
8. a) Heath JR. (1998) Fullerenes — C_{60}'s smallest cousin. *Nature* **393:** 730–731; b) Fowler PW, Heine T. (2001) Stabilisation of pentagon adjacencies in thelower fullerenes by functionalisation. *J Chem Soc Perkin Trans* 487–490.
9. Tan YZ, Xie SY, Huang RB, Zheng LS. (2009) The stabilization of fused pentagon fullerene molecules. *Nat Chem* **1:** 450–460.
10. Kobayashi K, Nagase S, Yoshida M, Osawa E. (1997) Endohedral metallofullerenes. Are the isolated pentagon rule and fullerene structures always satisfied? *J Am Chem Soc* **119:** 12693–12694.
11. Nagase S, Kobayashi K, Akasaka T. (1999) Unconventional cage structures of endohedral metallofullerenes. *J Mol Struct Theochem* **461–462:** 97–104.
12. Xu ZD, Nakane T, Shinohara H. (1996) Production and isolation of Ca@C_{82}(I–IV) and Ca@C_{84} (I, II) metallofullerenes. *J Am Chem Soc* **118:** 11309–11310.
13. Wang CR, Kai T, Tomiyama T, *et al.* (2000) C_{66} fullerene encaging a scandium dimer. *Nature* **408:** 426–427.
14. Stevenson S, Fowler PW, Heine T, *et al.* (2000) A stable non-classical metallofullerene family. *Nature* **408:** 427–428.
15. Xie SY, Gao F, Lu X, *et al.* (2004) Capturing the labile fullerene[50] as $C_{50}Cl_{10}$. *Science* **304:** 699.
16. Ioffe IN, Goryunkov AA, Tamm NB, *et al.* (2009) Fusing Pentagons in a Fullerene Cage by Chlorination: IPR D_2-C_{76} Rearranges into non-IPR $C_{76}Cl_{24}$. *Angew Chem Int Ed* **48:** 5904–5907.
17. Shvartsburg AA, Hudgins RR, Dugourd P, *et al.* (2000) Observation of "stick" and "handle" intermediates along the fullerene road. *Phys Rev Lett* **84:** 2421–2424.
18. Bates KR, Scuseria GE. (1997) Energetic analysis of pentagon road intermediates of C_{60}-buckminsterfullerene formation. *J Phys Chem A* **101:** 3038–3041.
19. Shinohara H. (2000) Endohedral metallofullerenes. *Rep Prog Phys* **63:** 843–892.
20. Akasaka T, Nagase S (eds). (2002) *Endofullerenes: A New Family of Carbon Clusters*, Springer.
21. Stevenson S, Burbank P, Harich K, *et al.* (1998) La_2@C_{72}: Metal-mediated stabilization of a carbon cage. *J Phys Chem A* **102:** 2833–2837.

22. Kato H, Taninaka A, Sugai T, Shinohara H. (2003) Structure of a missing caged metallofullerene: $La_2@C_{72}$. *J Am Chem Soc* **125:** 7782–7783.
23. Wan TSM, Zhang HW, Nakane T, *et al.* (1998) Production, isolation, and electronic properties of missing fullerenes: $Ca@C_{72}$ and $Ca@C_{74}$. *J Am Chem Soc* **120:** 6806–6807.
24. Yamada M, Wakahara T, Tsuchiya T, *et al.* (2008) Spectroscopic and theoretical study of endohedral dimetallofullerene having a Non-IPR fullerene cage: $Ce_2@C_{72}$. *J Phys Chem A* **112:** 7627–7631.
25. Shinohara H, Yamaguchi H, Hayashi N, *et al.* (1993) Isolation and spectroscopic properties of $Sc_2@C_{74}$, $Sc_2@C_{82}$, and $Sc_2@C_{84}$. *J Phys Chem* **97:** 4259–4261.
26. Kobayashi K, Nagase S. (2002) A stable unconventional structure of $Sc_2@C_{66}$ found by density functional calculations. *Chem Phys Lett* **362:** 373–379.
27. Wakahara T, Nikawa H, Kikuchi T, *et al.* (2006) $La@C_{72}$ having a non-IPR carbon cage. *J Am Chem Soc* **128:** 14228–14229.
28. Lu X, Nikawa H, Nakahodo T, *et al.* (2008) Chemical understanding of a non-IPR metallofullerene: Stabilization of encaged metals on fused-pentagon bonds in $La_2@C_{72}$. *J Am Chem Soc* **130:** 9129–9136
29. Lu X, Nikawa H, Tsuchiya T, *et al.* (2008) Bis-carbene adducts of non-IPR $La_2@C_{72}$: Localization of high reactivity around fused pentagons and electrochemical properties. *Angew Chem Int Ed* **47:** 8642–8645.
30. Stevenson S, Rice G, Glass T, *et al.* (1999) Small-bandgap endohedral metallofullerenes in high yield and purity. *Nature* **401:** 55–57.
31. Aihara J. (2001) Kinetic stability of carbon cages in non-classical metallofullerenes. *Chem Phys Lett* **343:** 465–469
32. Olmstead MM, Lee HM, Duchamp JC, *et al.* (2003) $Sc_3N@C_{68}$: Folded pentalenecoordination in an endohedral fullerene that does not obey the isolated pentagon rule. *Angew Chem Int Ed* **42:** 900–903.
33. Reveles JU, Heine T, Koster AM. (2005) C-13 NMR pattern of $Sc_3N@C_{68}$. Structural assignment of the first fullerene with adjacent pentagons. *J Phys Chem A* **109:** 7068–7072.
34. Cai T, Xu LS, Shu CY, *et al.* (2008) Synthesis and characterization of a non-IPR fullerene derivative: $Sc_3N@C_{68}[C(COOC_2H_5)_2]$. *J Phys Chem C* **112:** 19203–19208.
35. Kroto HW, Allaf AW, Balm SP. (1991) C_{60}: Buckminsterfullerene. *Chem Rev* **91:** 1213–1235.
36. Yang SF, Popov AA, Dunsch L. (2007) Violating the isolated pentagon rule(IPR): The endohedral non-IPR C_{70} cage of $Sc_3N@C_{70}$. *Angew Chem Int Ed* **46:** 1256–1259.

37. Yang S, Popov AA, Dunsch L. (2007) The role of an asymmetric nitride cluster on a fullerene cage: The Non-IPR endohedral $DySc_2N@C_{76}$. *J Phys Chem B* **111:** 13659–13663.
38. Beavers CM, Chaur MN, Olmstead MM, *et al.* (2009) Large Metal Ions in a Relatively Small Fullerene Cage: The Structure of $Gd_3N@C_2(22010)\text{-}C_{78}$ Departs from the Isolated Pentagon Rule. *J Am Chem Soc* **131:** 11519–11524.
39. Krause M, Wong J, Dunsch L. (2005) Expanding the world of endohedral fullerenes — The $Tm_3N@C_{2n}$ (39 <= n <= 43) clusterfullerene family. *Chem Eur J* **11:** 706–711.
40. Yang SF, Dunsch L. (2005) A large family of dysprosium-based trimetallic nitride endohedral fullerenes: $Dy_3N@C_{2n}$ (39 <= n <= 44). *J Phys Chem B* **109:** 12320–12328.
41. Popov AA, Krause M, Yang SF, *et al.* (2007) C_{78} cage isomerism defined by trimetallic nitride cluster size: A computational and vibrational spectroscopic study. *J Phys Chem B* **111:** 3363–3369.
42. Ma Y H, Wang T S, Wu JY, *et al.* (2011) Size effect of endohedral cluster on fullerene cage: Preparation and structural studies of $Y_3N@C_{78}\text{-}C_2$. *Nanoscale* **3:** 4955–4597.
43. Zhag JY, Bearden DW, Fuhrer T, *et al.* (2013) Metallofullerene with a Distorted Shape: Structural Investigation of $Y_3N@C_{78}$. *J. Am. Chem. Soc.* **135:** 3351–3354 .
44. Campanera JM, Bo C, Poblet JM. (2005) General rule for the stabilization of fullerene cages encapsulating trimetallic nitride templates. *Angew Chem Int Ed* **44:** 7230–7233.
45. Popov AA, Dunsch L. (2007) Structure, stability, and cluster-cage interactions in nitride clusterfullerenes $M_3N@C_{2n}$ (M = Sc, Y; 2n = 68–98): a density functional theory study. *J Am Chem Soc* **129:** 11835–11849.
46. Mercado BQ, Beavers CM, Olmstead MM, *et al.* (2008) Is the isolated pentagon rule merely a suggestion for endohedral fullerenes? The structure of a second egg-shaped endohedral fullerene-$Gd_3N@C_s(39663)\text{-}C_{82}$. *J Am Chem Soc* **130:** 7854–7855.
47. Beavers CM, Zuo TM, Duchamp JC, *et al.* (2006) An improbable, eggshaped endohedral fullerene that violates the isolated pentagon rule. *J Am Chem Soc* **128:** 11352–11353.
48. Zuo T, Walker K, Olmstead MM, *et al.* (2008) New egg-shaped fullerenes: non-isolated pentagon structures of $Tm_3N@C_s(51365)\text{-}C_{84}$ and $Gd_3N@C_s(51365)\text{-}C_{84}$. *Chem Commun* 1067–1069.
49. Wang CR, Kai T, Tomiyama T, *et al.* (2001) A scandium carbide endohedral metallofullerene: $(Sc_2C_2)@C_{84}$. *Angew Chem Int Ed* **40:** 397–399.

50. Krause M, Hulman M, Kuzmany H, *et al.* (2004) Fullerene quantum gyroscope. *Phys Rev Lett* **93:** 137403.
51. Shi ZQ, Wu X, Wang CR, *et al.* (2006) Isolation and characterization of $Sc_2C_2@C_{68}$: A metal-carbide endofullerene with a non-IPR carbon cage. *Angew Chem Int Ed* **45:** 2107–2111.
52. Dunsch L, Yang S, Zhang L, *et al.* (2010) Metal sulfide in a C_{82} fullerene cage: a new form of endohdral clusterfullerenes. *J Am Chem Soc* **132:** 5413–5421.
53. Chen N, Chaur MN, Moore C, *et al.* (2010) Synthesis of a new endohedral fullerene family, $Sc_2S@C_{2n}$ (n = 40–50) by the introduction of SO_2. *Chem Commun* **46:** 4818–4820.
54. Chen N, Beavers CM, Mulet-Gas M, *et al.* (2012) $Sc_2S@C_s(10528)$-C_{72}: A Dimetallic sulfide endohedral fullerene with a non isolated pentagon rule cage. *J Am Chem Soc* **134:** 7851–7860.
55. Chen N, Mulet-Gas M, Li YY, *et al.* (2013) $Sc_2S@C_2(7892)$–C_{70}: a metallic sulfide cluster inside a non-IPR C_{70} cage. *Chem Sci* **4:** 180–186.
56. Tan YZ, Liao ZJ, Qian ZZ, *et al.* (2008) Two I_h-symmetry-breaking C_{60} isomers stabilized by chlorination. *Nature Mater* **7:** 790–794.
57. Wang CR, Shi ZQ, Wan LJ, *et al.* (2006) $C_{64}H_4$: Production, isolation, and structural characterizations of a stable unconventional fulleride. *J Am Chem Soc* **128:** 6605–6610.
58. Loffler D, Bajales N, Cudaj M, *et al.* (2009) Non-IPR C_{60} solids. *J Chem Phys* **130:** 164705.
59. Lu X, Chen ZF, Thiel W, *et al.* (2004) Properties of fullerene[50] and D_{5h} decachlorofullerene[50]: A computational study. *J Am Chem Soc* **126:** 14871–14878.
60. Han X, Zhou SJ, Tan YZ, *et al.* (2008) Crystal structures of Saturn-like $C_{50}Cl_{10}$ and pineapple-shaped $C_{64}Cl_4$: Geometric implications of double and triple-pentagon-fused chlorofullerenes. *Angew Chem Int Ed* **47:** 5340–5343.
61. Zettergren H, Alcami M, Martin F. (2008) Stable non-IPR C_{60} and C_{70} fullerenes containing a uniform distribution of pyrenes and adjacent pentagons. *ChemPhysChem* **9:** 861–866.
62. Troshin PA, Avent AG, Darwish AD, *et al.* (2005) Isolation of two sevenmembered ring C58 fullerene derivatives: $C_{58}F_{17}CF_3$ and $C_{58}F_{18}$. *Science* **309:** 278–281.
63. Qian W, Chuang S-C, Amador RB, *et al.* (2003) Synthesis of Stable Derivatives of C_{62}: The First Nonclassical Fullerene Incorporating a Four-Membered Ring. *J Am Chem Soc* **125:** 2066–2067.
64. Tan YZ, Li J, Zhu F, *et al.* (2010) Chlorofullerenes featuring triple sequentially fused pentagons. *Nat Chem* **2:** 269–273.

65. Tan YZ, Li J, Zhou T, *et al.* (2010) Pentagon-fused hollow fullerene in C78family retrieved by chlorination. *J Am Chem Soc* **132:** 12648–12652.
66. Tan YZ, Zhou T, Bao J, *et al.* (2010) C72Cl4: A pristine fullerene with favorable pentagon-adjacent structure. *J Am Chem Soc* **132:** 17102–17104.
67. Ziegler K, Mueller A, Amsharov KY, Jansen M. (2010) Disclosure of the elusive C_{2v}-C_{72} carbon cage. *J Am Chem Soc* **132:** 17099–17101.
68. Ioffe IN, Chen CB, Yang SF, *et al.* (2010) Chlorination of C86 to C84Cl32 with nonclassical heptagon-containing fullerene cage formed by cage shrinkage. *Angew Chem Int Ed* **49:** 4784–4787.
69. Ziegler K, Mueller A, Amsharov KY, Jansen M. (2011) Capturing the moststable C_{56} fullerene cage by *in situ* chlorination. *Chem Asian J* **6:** 2412–2418.
70. Tan YZ, Han X, Wu X, *et al.* (2008) An entrant of smaller fullerene: C_{56} captured by chlorines and aligned in linear chains. *J Am Chem Soc* **130:** 15240–15241.
71. Amsharov KY, Ziegler K, Mueller A, Jansen M. (2012) Capturing the antiaromatic $_{\#6094}C_{68}$ carbon cage in the radio-frequency furnace. *Chem Eur J* **18:** 9289–9293.
72. Ross RB, Cardona CM, Guldi DM, *et al.* (2009) Endohedral fullerenes for organic photovoltaic device. *Nat Mater* **8:** 208–212.
73. Pinzón JR, Plonska-Brzezinska ME, Cardona CM, *et al.* (2008) Sc_3N@C-80-ferrocene electrondonor/acceptor conjugates as promising materials for photovoltaic applications. *Angew Chem Int Ed* **47:** 4173–4176.
74. Feng L, Slanina Z, Sato S, *et al.* (2011) Covalently linked porphyrin-La@C-82 hybrids: structural elucidation and investigation of intramolecular interactions. *Angew Chem Int Ed* **50:** 5909–5912.
75. Feng L, Radhakrishnan SG, Mizorogi N, *et al.* (2011) Synthesis and charge-transfer chemistry of $La_2@I_h$-C_{80}/$Sc_3N@I_h$-C_{80}–Zinc porphyrin conjugates: impact of endohedral cluster. *J Am Chem Soc* **133:** 7608–7618.
76. Rudolf M, Wolfrum S, Guldi DM, *et al.* (2012) Endohedral metallo-fullerenes — filled fullerene ferivatives towards multifunctional reaction center mimics. *Chem Eur J* **18:** 5136–5148.
77. Feng L, Rudolf M, Wolfrum S, *et al.* (2012) A paradigmatic change: linking fullerenes to electron acceptors. *J Am Chem Soc* **134:** 12190–12197.

CHAPTER 7

Metal Oxide Clusterfullerenes

*Steven Stevenson**

The recent discovery of our metal oxide clusterfullerenes is generating much excitement among scientists. The first reported endohedral metallofullerenes with metal oxide clusters inside fullerene cages contain scandium. $Sc_4O_2@C_{80}$ is interesting due to its mixed valence oxidation states of Sc^{2+}, Sc^{3+} and its electrochemical properties. $Sc_4O_2@C_{80}$ is amenable to ESR spectroelectrochemical investigations of its oxidized and reduced radical ions. $Sc_4O_2@C_{80}$ is the dominant species of the scandium metal oxide clusterfullerene family. $Sc_4O_3@C_{80}$ is the world record holder for the most atoms (seven) encapsulated inside a fullerene. The next scandium oxide clusterfullerene we isolated was $Sc_2O@C_{82}$, which has one predominant structural isomer as its preferred cage versus the existence of two structural C_{82} cage isomers for its Group VIA analogue, $Sc_2S@C_{82}$. In this chapter we present an expanded family of metal oxide clusterfullerenes to include other transition metal Group IIIB and 4f-block rare earth metal oxide clusters. We also discuss their synthesis, purification, characterization and theoretical calculations.

* Department of Chemistry, Indiana University-Purdue University at Fort Wayne IPFW, 2101 E. Coliseum Blvd, Fort Wayne, IN, 46805 USA E-mail: stevenss@ipfw.edu

1. INTRODUCTION

The entrapment of atoms inside the void volume of fullerene cages has created much excitement in the endohedral metallofullerene community. As a historical perspective, the early 1990s were represented by seminal reports of endohedrals with a metal atom(s) on the inside.[1–8] Enthusiasm was generated as progressively more metal atoms were entrapped. In 1999 there was a breakthrough[9] in the field of endohedral metallofullerenes when Stevenson and Dorn successfully added a Group VA element (i.e. N) to the metals to create blended clusters containing mixtures of metals with nitrogen. From 1999 to the present time, hundreds of papers have been published on these metal nitride clusterfullerenes, their properties, synthesis, characterization and applications. The 2000s were a decade of searching for new types of clusters (e.g. other metal nitrides with $M_3N@C_{2n}$,[10–18] metal carbides with $M_{2-4}C_2@C_{2n}$,[19,20] metal cyanos with $Sc_3NC@C_{78}$,$_{80}$,[21,22] hydrocarbons with $Sc_3CH@C_{80}$,[23] metal oxides with $M_{2,4}O_{2,3}@C_{80}$,[24,25] $M_2O@C_{82}$,[26] and metal sulfides with $M_2S@C_{2n}$ [27–30]).

There are now several excellent reviews on metal clusterfullerenes,[31–34] metal nitride clusterfullerenes,[30,35,36] metal sulfide clusterfullerenes,[30] and metal oxide clusterfullerenes.[37] For chemical nomenclature of clusters in endohedral metallofullerenes, elements inside the fullerene cage (C_{2n}) are placed left of the "@" symbol. Hence, $Sc_4O_2@I_h$-C_{80} has a Sc_4O_2 cluster entrapped within an icosahedral, I_h, C_{80} cage.

The metal oxide family was born in 2008 with a breakthrough in the Stevenson research group when he discovered a new way to create fullerene clusters with metals bonded to a Group VIA element (i.e. O).[24] The incorporation of metal oxide clusters into fullerenes has resulted in a new class of compounds not just for scandium, but also for the first time reported herein, other transition Group IIIB and rare earth (e.g. 4f block) metals. The X-ray crystal structures (Fig. 1) for these first metal oxide clusterfullerenes and their publication year are as follows: $Sc_4O_2@C_{80}$ (2008),[24] $Sc_4O_3@C_{80}$ (2010)[25] and $Sc_2O@C_{82}$ (2010).[26]

When comparing the ionic radius of scandium with other Group IIIB and rare earth metals, the Sc^{3+} ion is the smallest. The advantage of mixing the Sc^{3+} ion with oxygen is the ability to form metal oxide clusters with many atoms inside the cage. For example, $Sc_4O_3@C_{80}$ holds a world-record for the most atoms crammed inside a fullerene cage. Other distinguishing

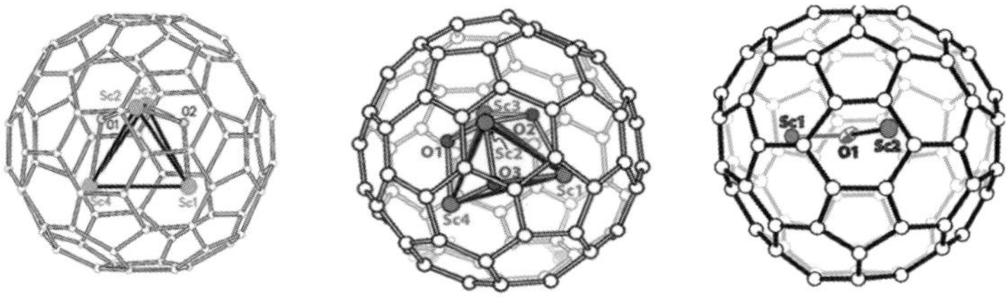

$Sc_4O_2@C_{80}$ $Sc_4O_3@C_{80}$ $Sc_2O@C_{82}$

Fig. 1 X-ray crystal structures for purified $Sc_4(\mu_3\text{-}O)_2@I_h\text{-}C_{80}$, $Sc_4(\mu_3\text{-}O)_3@I_h\text{-}C_{80}$, and $Sc_2(\mu_2\text{-}O)@C_s(6)\text{-}C_{80}$. (Adapted with permission from Refs. 24–26, Copyright 2008, 2010 American Chemical Society and Royal Society of Chemistry.)

features of metal oxide clusterfullerenes include the ability to "force" a metal (e.g. Sc) into oxidation states that are not preferred by an atom. In $Sc_4O_2@C_{80}$, two scandium atoms are in the preferred oxidation states of Sc^{3+}, but the other two scandium atoms are coerced into the less favorable Sc^{2+} oxidation state.[38,39] Electronically, $Sc_4O_2@C_{80}$ is described as $[Sc_4O_2]^{6+}@[C_{80}]^{6-}$.[38,39] The cluster, with an overall net charge of 6+, then fits nicely with the icosahedral C_{80} cage, which is well known to prefer a 6- charge when forming metallofullerenes.

As the field of endohedral metallofullerenes is gaining momentum, now is a good time to review metal oxide clusterfullerenes. The goal of this chapter is to provide a single document to serve as a place for newly interested scientists to quickly become acquainted with these new molecules. In this chapter, we will discuss many aspects of metal oxide clusterfullerene research. Topics that will be reviewed include their synthesis, purification, and characterization. Theoretical calculations and computational work will also be discussed.

2. ARC-PLASMA SYNTHESIS

The discovery of metal oxide clusterfullerenes was a result of our research into the feasibility of producing endohedral metallofullerenes in oxidizing atmospheres.[40] Since the 1990s, electric-arc synthesis of metallofullerenes was primarily performed in helium atmospheres. There was special attention to sealing reactors well to prevent air from entering the chamber. The rationale was that air in the chamber would be deleterious to the synthesis of both empty-cage fullerenes and metallofullerenes because the oxygen source from ambient air leaks would disrupt the synthetic mechanism of fullerene production and thereby decrease the fullerene yield. This belief was correct, as the presence of oxygen and air does decrease fullerene production.[41]

In 1999, the presence of added nitrogen in the reactor resulted in the discovery of the metal nitride clusterfullerenes.[9] In the early 2000s, there was a movement away from this inert He/N_2 atmosphere (Dorn method) as the Dunsch research group discovered a novel method for making metal nitride clusterfullerenes using NH_3 gas for metallofullerene formation in a reducing atmosphere (Dunsch method).[11,36,42]

In the mid-2000s, the Stevenson group began work on the feasibility of making endohedrals in oxidizing atmospheres.[40] Though people said the Stevenson group was crazy for intentionally adding oxygen and air into the reactor, we did, as predicted, achieve a much lower yield of fullerene product, we were able to burn more rods to offset the loss of fullerene production. The fear other scientists had of adding oxygen and air to their fullerene reactors in essence prevented their ability to discover these new metal oxide clusterfullerene molecules, which are readily formed in such a chemically harsh and oxidizing environment. A schematic of one of our arc-plasma reactors is shown in Fig. 2.

In 2007 we began to add copper,[43] air and copper nitrate hydrate[40] to scandium oxide packed rods for the creation of an oxidizing environment in the reactor to see which fullerenes and metallofullerenes could survive these harsh conditions. The hope was that in this Darwinistic approach of

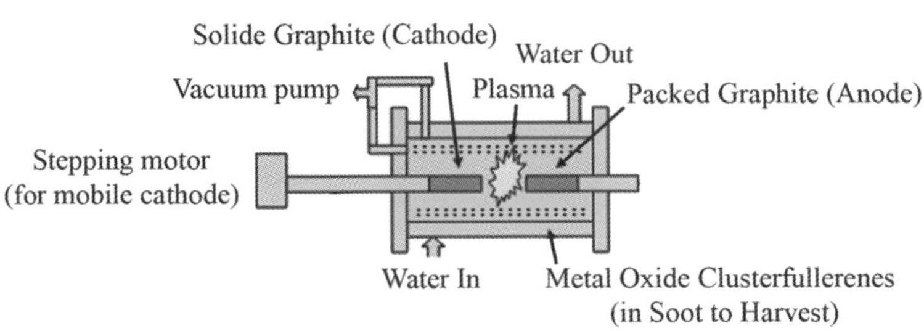

Fig. 2 Our electric-arc plasma reactor used to generate soot containing metal oxide, metal nitride, and other metal clusterfullerenes. (Adapted with permission from Ref. 43, Copyright 2007, Royal Society of Chemistry.)

the "survival of the fittest" of the most inert molecules, perhaps the more chemically resistant $Sc_3N@C_{80}$ clusterfullerenes would be selectively produced by suppressing the formation of empty cage fullerenes (e.g. the plethora of C_{60}, C_{70}). If empty cage fullerenes were formed in such reactive conditions, the hypothesis was that the more reactive empty-cage and metallofullerenes would not survive, undergo oxidative decomposition, and be chemically transformed into insoluble compounds and removed in the extraction and filtration process.

This Stevenson method of Chemically Adjusting Plasma Temperature, Energy and Reactivity (CAPTEAR) was reported in 2007.[40] Soot extracts and corresponding MALDI mass spectral data revealed optimal conditions for the selective synthesis of metal nitride clusterfullerenes. In the study, it should be noted that the synthesis of C_{60} and $Sc_3N@C_{80}$ were inversely proportional.[40]

Upon further study of mass spectral data obtained for soot extracts prepared under the Stevenson Method, we noticed that this CAPTEAR synthesis resulted in the emergence of at least two new families of molecules. These two new classes of molecules are the metallic nitride azaclusterfullerenes[44–46] and metal oxide clusterfullerenes.[24–26,38]

2.1. Scandium Oxide Clusterfullerenes

In 2008, we isolated the first of these new molecules, $Sc_4O_2@C_{80}$,[24] which is by far the most abundant of the metal oxide clusterfullerenes, followed by $Sc_2O@C_{82}$,[26] with the $Sc_4O_3@C_{80}$[25] being much lower in yield than other metal oxide endohedrals by an order of magnitude. It should be noted that optimizing reactor conditions for all three of the metal oxide endohedrals simultaneously is difficult. For example, when the synthesis is optimized for an improved yield of $Sc_4O_2@C_{80}$, other metal oxide clusterfullerenes (e.g. $Sc_2O@C_{82}$) can decrease, and vice versa. A typical MALDI mass spectrum showing the presence of scandium oxide endohedrals relative to other metallofullerenes, such as the metal nitride clusterfullerenes, is shown in Fig. 3.

2.2. ^{13}C-labeled Metal Oxide Clusterfullerenes

It can be advantageous to prepare ^{13}C labeled samples of metal oxide clusterfullerenes for NMR experiments. This is especially useful in NMR

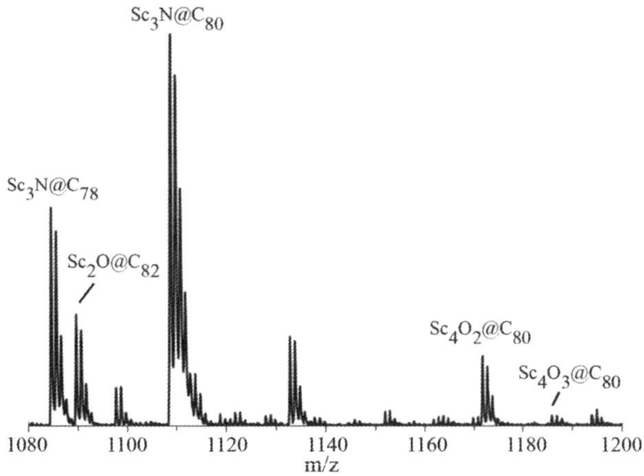

Fig. 3 MALDI mass spectrum of scandium oxide clusterfullerenes.

studies of carbon atoms on the fullerene cage. NMR data is useful for monitoring chemical shift behavior not only for pristine samples of purified metal oxide clusterfullerenes, but also for their derivatives upon functionalizing the carbon cage surface.

To demonstrate the feasibility of making ^{13}C labeled metal oxide metallofullerenes, our goal was to prepare ^{13}C isotope enriched samples of the two dominant scandium oxide clusterfullerenes, $Sc_4O_2@^{13}C_{80}$ and $Sc_2O@^{13}C_{82}$. We were also interested in their yield relative to the metal nitride clusterfullerene production. The experimental procedure was similar to our previously described[24,25] synthesis of $Sc_4O_2@C_{80}$ and $Sc_2O@C_{82}$, with the difference being added amorphous ^{13}C carbon powder to the typical packing material (e.g. scandium oxide, copper, copper nitrate). By adjusting the stoichiometric amounts of ^{13}C powder versus Sc_2O_3, we have prepared samples of various amounts of ^{13}C labeling.

Shown in Fig. 4 is the MALDI mass spectrum of extracted soot that shows extensive ^{13}C incorporation of labeled carbon atoms into the carbon cage. Note the shift of the m/z values from 1090 to a much broader isotopic distribution of peaks whose new maximum is 1095 for $Sc_2O@^{13}C_{82}$ and the shift of the dominant peak from the m/z value of naturally abundant $Sc_4O_2@C_{80}$ at 1172 to $Sc_4O_2@^{13}C_{80}$ with a much broader isotope distribution whose maximum is centered about 1177. The extent of isotopic enrichment corresponds to about 7% ^{13}C labeling. In addition to

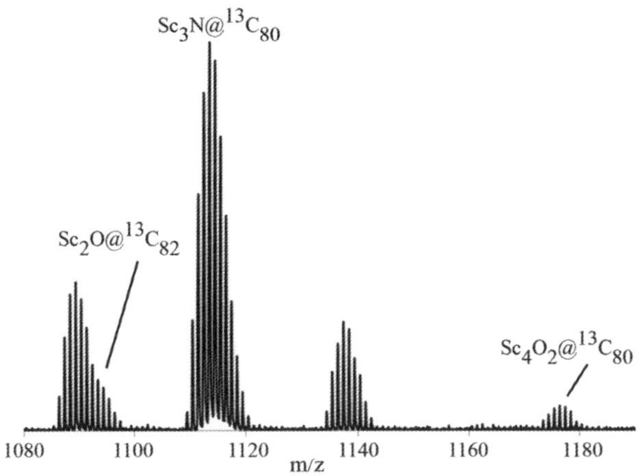

Fig. 4 MALDI mass spectrum demonstrating the successful ^{13}C isotopic labeling of metal oxide clusterfullerenes.

the successful synthesis of ^{13}C enriched metal oxide clusterfullerenes, ^{13}C labeled metal nitride clusterfullerenes (e.g. $Sc_3N@^{13}C_{80}$, $Sc_3N@^{13}C_{78}$) are also synthesized with a comparable extent of ^{13}C isotopic labeling.

2.3. Yttrium Oxide Clusterfullerenes

After the discovery of the scandium oxide endohedral metallofullerenes, we became curious to determine if we could substitute other metals for scandium. If so, then we could expand the family of metal oxide clusterfullerenes. The obvious choice for these investigations is other Group IIIB transition metals (e.g. Y) and 4f-block rare earth elements. If these non-scandium metals also could form metal oxide clusters, then the question would be the effect of these larger clusters on the carbon cage selections. Specifically, would the carbon cage have to expand to larger fullerene cages in order to encapsulate these larger metal oxide clusters? Would there be a new set of favored fullerene cages? To address these questions, we performed experiments in which we packed non-scandium oxides (e.g. yttrium, erbium, and lutetium) and subsequently vaporized them in our electric arc reactor as we previously described with our CAPTEAR synthesis.[40]

After soot harvesting and extraction with carbon disulfide, these yttrium soot extracts were characterized by MALDI mass spectrometry (Fig. 5a). The mass spectral data shows the presence of at least three sets of molecules, whose internal clusters vary and are comprised of metals (e.g. $Y_2@C_{2n}$), trimetallic nitrides (e.g. $Y_3N@C_{2n}$), and dimetallic oxides (e.g. $Y_2O@C_{2n}$). Upon comparison of yttrium with scandium endohedral clusterfullerenes, we note their similarities and differences.

For the trimetallic nitride endohedrals, the dominant metallofullerenes for this class of compounds are $Sc_3N@C_{80}$ and $Y_3N@C_{80}$. It should be noted that $Y_3N@C_{88}$ is the second most dominant member of the trimetallic nitride endohedrals and is almost as abundant as $Y_3N@C_{80}$ for these reactor conditions. In contrast, $Sc_3N@C_{88}$ is a very minor species in scandium soot extracts. This shift of 8 carbons in mass has also been noticed by Echegoyen[14,47] in their $M_3N@C_{80}$ studies, in which they varied the metal selection and noticed the formation of a series of dominant $M_3N@C_{2n}$ species (e.g. $M_3N@C_{80}$, $M_3N@C_{88}$, $M_3N@C_{96}$), for which there was also a +8 shift of "favored cages" (i.e. C_{80}, C_{88}, C_{96} cages differing by 8 carbon atoms).[14,47]

Next, we compare the clusters of scandium oxide versus yttrium oxide clusterfullerenes. Similar to scandium, the dominant species of the M_2O type of compounds are both in C_{82} cages, $Sc_2O@C_{82}$ and $Y_2O@C_{82}$. However, the substitution of Y for Sc has created a more rich variety of dimetallic oxide endohedrals. Whereas the $Sc_2O@C_{2n}$ family is limited primarily to $Sc_2O@C_{82}$, with yttrium there is a series of peaks in the MALDI mass spectrum (Fig. 5a) from C_{78} through C_{96} cages and beyond with peak maxima at $Y_2O@C_{82}$ and $Y_2O@C_{90}$. Note that these two dominant $Y_2O@C_{2n}$ species also differ by +8 carbons, the same difference as the two dominant $Y_3N@C_{80}$ and $Y_3N@C_{88}$ species.

Another significant difference between scandium oxide and yttrium oxide endohedrals is the absence of the larger and bulkier Y_4O_2 and Y_4O_3 clusters for C_{80} cages. Whereas $Sc_4O_2@C_{80}$ is the dominant metal oxide clusterfullerene, $Y_4O_2@C_{80}$ is either not produced or below the detection limits. For the 7-atom metal oxide (i.e. M_4O_3) clusterfullerenes, the dominant species is $Sc_4O_3@C_{80}$, which is formed in extremely low yields and is difficult to synthesize. In comparison, $Y_4O_3@C_{80}$, if made, is also below the mass spectral detection limits for our samples.

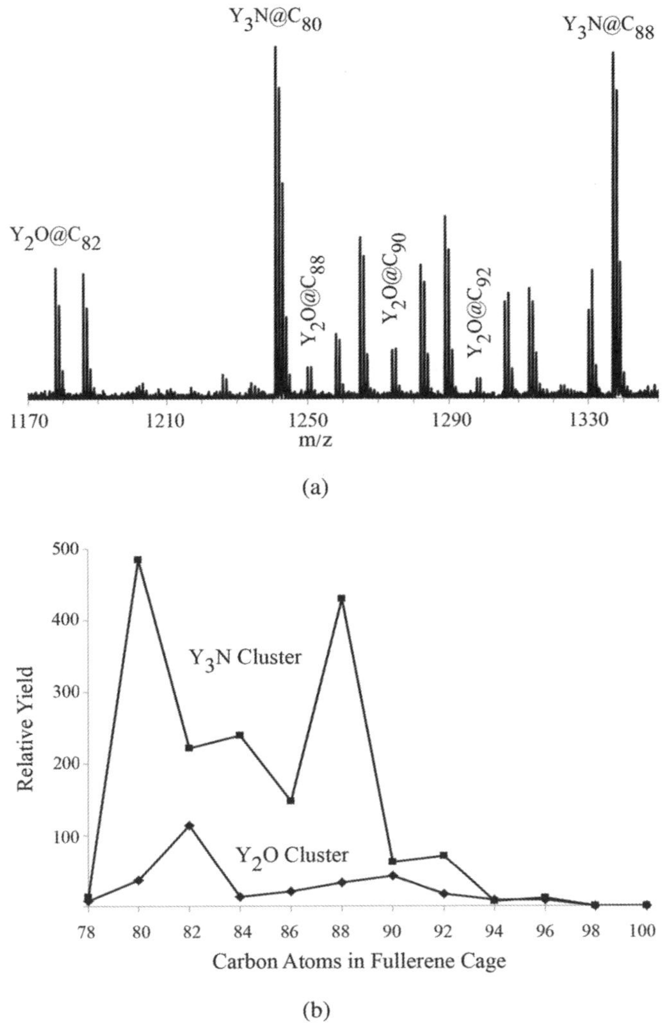

Fig. 5 MALDI mass spectrum of soot extract containing yttrium oxide clusterfullerenes and (b) correlation of carbon cage to cluster and relative yield.

The comparison of yttrium oxide clusterfullerenes with other yttrium endohedral metallofullerenes is summarized in Fig. 5b. The shift of 8 carbons for yttrium nitride clusters ($Y_3N@C_{80}$, $Y_3N@C_{88}$) is clearly observed and demonstrates these two species are the most favored members of the metal nitride family. Likewise, the two dominant dimetallic oxide structures, which also differ by +8 carbons, have shifted to $Y_2O@C_{82}$ and $Y_2O@C_{90}$ as their dominant family members.

2.4. Rare Earth Oxide Clusterfullerenes

To demonstrate the feasibility of making rare earth oxide clusterfullerenes, we selected lutetium as a representative of the 4f-block elements because of its similar ionic radius to yttrium. The produced lutetium-based endohedral metallofullerenes include cages with clusters that are dimetallic (e.g. $Lu_2@C_{2n}$), trimetallic nitrides (e.g. $Lu_3N@C_{2n}$) and dimetallic oxides (e.g. $Lu_2O@C_{2n}$). Similar to results for yttrium metal oxide endohedrals, the Lu_4O_2 and Lu_4O_3 clusters in C_{80} cages were also not observed in our mass spectra.

Analysis of MALDI mass spectra (Fig. 6a) for lutetium soot extracts also revealed the presence of many families of compounds. Similar to $Y_2O@C_{82}$, the dominant dimetallic oxide clusterfullerene also had a C_{82} cage and was $Lu_2O@C_{82}$. Analogous to Y_2O clusterfullerenes, the Lu_2O endohedrals are also created with an extensive range of carbon cages (e.g. $Lu_2O@C_{78}$ thru $Lu_2O@C_{116}$ and beyond).

In comparison with yttrium nitride clusterfullerenes, the dominant trimetallic nitride endohedral was $Lu_3N@C_{80}$ followed by $Lu_3N@C_{88}$ as the next dominant lutetium nitride clusterfullerene. These mass spectral results are summarized in Fig. 6b. We have investigated the ability of other rare earth elements to also form metal oxide clusterfullerenes (e.g. Er), and these species are detected in MALDI mass spectrometry.

3. PURIFICATION

Due to a low production yield of metal oxide clusterfullerenes, their separation and purification can be challenging. Another difficulty is extracts can have a complex mixture of many types of fullerenes. A typical extract from a CAPTEAR synthesis has empty-cage fullerenes mixed with several families of endohedral metallofullerenes, which contain clusters such as M_2, M_3, M_2C_2, M_3N, M_2O and M_4O_2. Because soot extracts often contain <3% of metal oxide clusterfullerenes, the removal of the remaining 97% of fullerenes and other endohedrals is difficult.

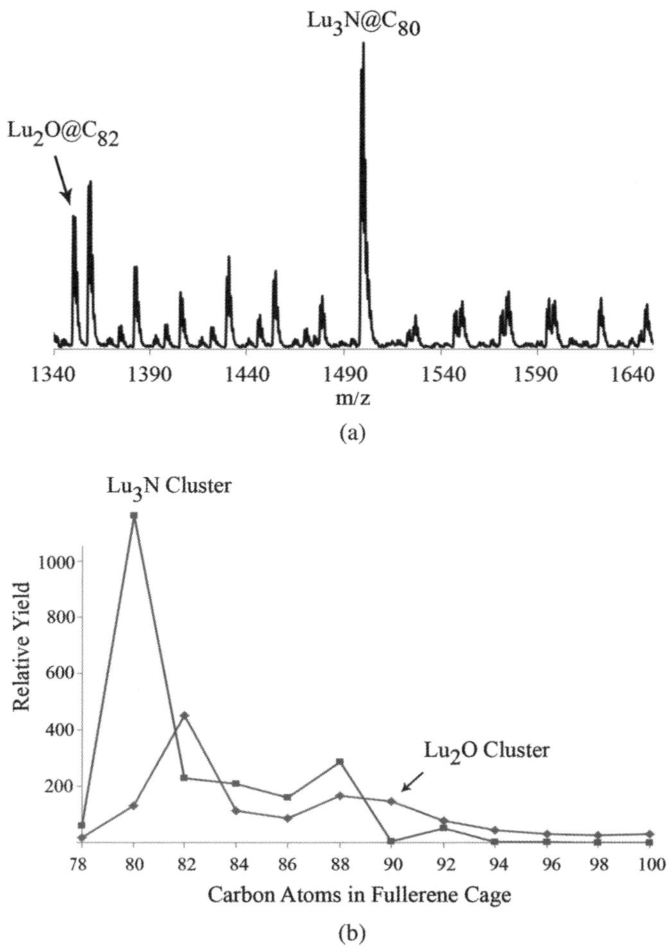

Fig. 6 MALDI mass spectrum of soot extract containing lutetium oxide clusterfullerenes and (b) correlation of carbon cage to cluster and relative yield.

To address these challenges for isolating metallic oxide endohedrals, we have used reactivity differences to obtain samples enriched in M_2O, M_4O_2 and M_4O_3 clusterfullerenes. Once contaminant empty-cage fullerenes are removed, then HPLC fraction collection can be utilized to isolate these metallic oxide clusterfullerenes. As a pre-cleanup step we manipulate reactivity differences between empty cage fullerenes and metallofullerenes by selective reaction with aminosilica gel[48,49] or by selective precipitation with Lewis acids.[50,51]

In the Stir and Filter Approach (SAFA),[48,49] extracts are dissolved in solvent (e.g. toluene, *o*-dichlorobenzene) and stirred in a slurry of pre-dried diaminosilica. Over time, the more reactive empty-cage fullerenes bond with amino groups and become immobilized onto the functionalized silica. Remaining in solution, the metal oxide clusterfullerenes are present in the filtrate. HPLC fraction collection on the filtrate obtained from this SAFA pre-cleanup step has led to the isolation of $Sc_4O_2@C_{80}$[24] and $Sc_4O_3@C_{80}$.[25]

In the Lewis acid approach, soot extracts are stirred in carbon disulfide, and Lewis acids (e.g. $AlCl_3$, $FeCl_3$) selectively complex with metal oxide clusterfullerenes. The reacted metal oxide clusterfullerenes are removed from solution via precipitation.[51] The empty cage fullerenes (e.g. C_{60}, C_{70}) remain in solution. After filtration, metallofullerenes in the precipitate are decomplexed and recovered via reaction with ice water to obtain endohedral enriched samples, including metal oxide and metal nitride clusterfullerenes. $Sc_2O@C_{82}$ was purified using this Lewis acid pre-cleanup step.[26]

For the isolation of $Sc_2O@C_{82}$,[26] soot extract was stirred in CS_2. Scandium nitride and scandium oxide clusterfullerenes were selectively precipitated upon addition of the Lewis acid, $AlCl_3$. The first stage of precipitation and decomplexation provided a sample enriched in $Sc_4O_2@C_{80}$, $Sc_3N@C_{78}$, and $Sc_3N@D_{5h}$-C_{80}. Further addition of the $AlCl_3$ to the filtrate led to the selective precipitation of the $Sc_2O@C_{82}$. Upon decomplexation, this fraction contains $Sc_2O@C_{82}$ as the second most dominant species. Purified $Sc_2O@C_{82}$ is readily obtained via HPLC fraction collection (Fig. 7).[26]

To evaluate whether this Lewis acid approach of selective precipitation and subsequent decomplexation would be feasible for non-Sc based clusterfullerenes, we prepared a soot extract containing $Y_2O@C_{82}$ (<1% in mixture) with the remaining 99+% of fullerenes being a mixture of empty-cage fullerenes (e.g. C_{60}, C_{70}, C_{84}), classical metallofullerenes (e.g. Y_2C_{82}, Y_2C_{84}, Y_2C_{90}), and yttrium nitride clusterfullerenes ($Y_3N@C_{80}$, $Y_3N@C_{88}$). While stirring yttrium soot extract in CS_2, aluminum chloride was added. Upon reaction overnight, the sample was filtered. The precipitate was collected, and decomplexed metallofullerenes were obtained from the precipitate via reaction with ice water and subsequent workup with carbon disulfide in a separatory funnel. The recovered yttrium clusterfullerenes were analyzed by MALDI (Fig. 8).

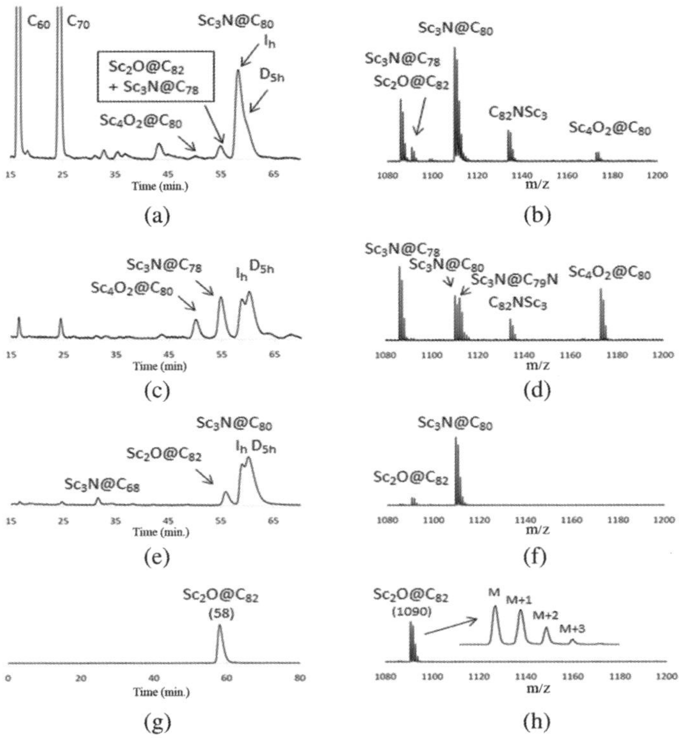

Fig. 7 (a–d) HPLC and (e–g) MALDI mass spectral data showing the fractionation of $Sc_4O_2@C_{80}$ (b, Stage 1) and $Sc_2O@C_{82}$ (c, Stage 2). HPLC and MALDI data for the purified sample of $Sc_2O@C_{82}$ is shown in (d) and (f). (Reproduced with permission from Ref. 26, Copyright 2010 American Chemical Society.)

These results demonstrate that $Y_2O@C_{82}$ clusterfullerenes selectively react with $AlCl_3$ Lewis acid and permit a sample enriched in $Y_2O@C_{82}$, which can be purified via HPLC fraction collection. Note that other endohedral metallofullerenes families (e.g. classical metallofullerenes such as $Y_2@C_{2n}$ and metal nitride clusterfullerenes, $Y_3N@C_{80}$, $Y_3N@C_{80}$) are also readily recovered in this decomplexation step, and hence can also be isolated by HPLC fraction collection.

4. CHARACTERIZATION

Due to their recent discovery, the characterization of metal oxide clusterfullerenes has been limited to $Sc_4O_2@C_{80}$, $Sc_4O_3@C_{80}$, and $Sc_2O@C_{82}$.

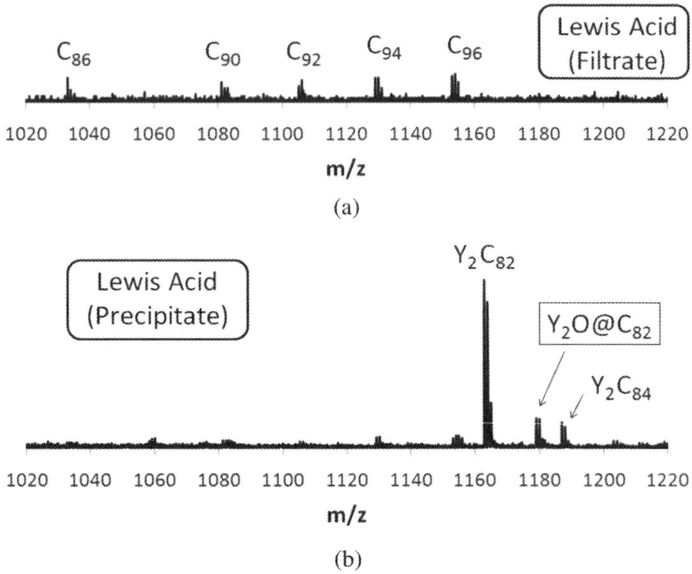

Fig. 8 MALDI mass spectra showing the successful separation of (a) empty-cage fullerenes from (b) $Y_2O@C_{82}$ using a selective precipitation method with Lewis acids as a pre-cleanup step.

Amounts of these purified samples are on the milligram scale. Hence, initial experiments use techniques which require small quantities, such as UV-Vis, IR, Raman, NMR, voltammetry, and ESR spectroelectrochemistry.

4.1. Spectroscopy

4.1.1 *UV-Vis*

To date, the only metal oxide clusterfullerenes that have been isolated are $Sc_4O_2@C_{80}$,[24] $Sc_4O_3@C_{80}$,[25] and $Sc_2O@C_{82}$.[26] We would expect the UV-Vis spectra (Fig. 9) for the $Sc_4O_2@C_{80}$[24] and the $Sc_4O_3@C_{80}$[25] species to be similar because their carbon cage (I_h-C_{80}) is identical, and they differ only by the internal cluster. Indeed their UV-Vis is quite similar. In contrast, $Sc_2O@C_{82}$ possesses both a different cluster, Sc_2O, and a different symmetry cage, $C_s(6)$-C_{82}. Hence, the UV-Vis spectrum for $Sc_2O@C_{82}$ is quite different and possesses peak maxima at 496 nm, 652 nm, 760 nm, and 808 nm.[26]

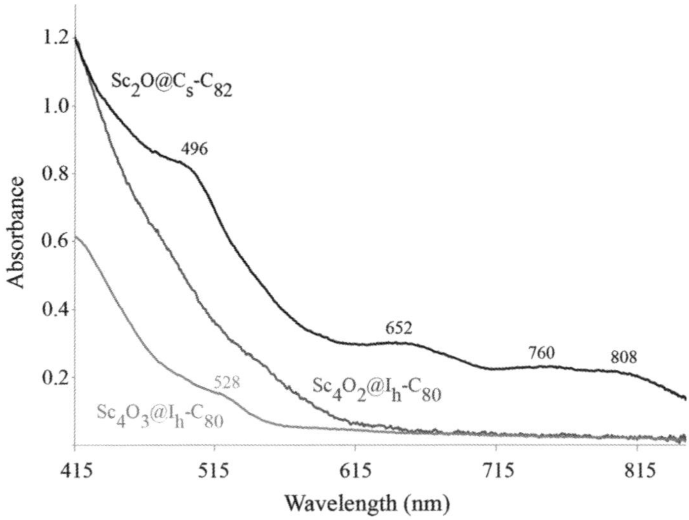

Fig. 9 UV-Vis spectra of purified $Sc_2(\mu_2\text{-}O)@C_s(6)\text{-}C_{82}$, $Sc_4(\mu_3\text{-}O)_2@I_h\text{-}C_{80}$, and $Sc_4(\mu_3\text{-}O)_3@I_h\text{-}C_{80}$ obtained in carbon disulfide. (Adapted with permission from Ref. 37, Copyright 2011 World Scientific Publishing.)

4.1.2 ^{13}C NMR

The only metal oxide clusterfullerene to be analyzed by NMR is $Sc_4O_2@C_{80}$.[39] In these seminal experiments, both ^{13}C NMR and ^{45}Sc NMR spectra were obtained (Fig. 10).[39] Knowing from X-ray crystal data that the cage has I_h symmetry, we would expect two carbons in the ^{13}C spectrum with a ratio of 3:1 (i.e. 60 carbons of one type and 20 carbons of another type). This expected ratio is experimentally observed for $Sc_4O_2@C_{80}$ with ^{13}C chemical shifts of 137.29 and 144.82 ppm.[39] This suggests a freely rotating Sc_4O_2 cluster inside the $C_{80}\text{-}I_h$ cage. For comparison with a Sc_3N cluster, also having an I_h cage, the ^{13}C chemical shifts of 136.87 and 144.18 ppm, are also similar to those from $Sc_4O_2@C_{80}$.[39]

4.1.3 ^{45}Sc NMR

The ^{45}Sc NMR spectrum (Fig. 10) of purified $Sc_4O_2@C_{80}$ was also obtained in two different solvents.[39] Measured in carbon disulfide, there were two peaks for the Sc_4O_2 cluster at 129 and 292 ppm. In *o*-dichlorobenzene,

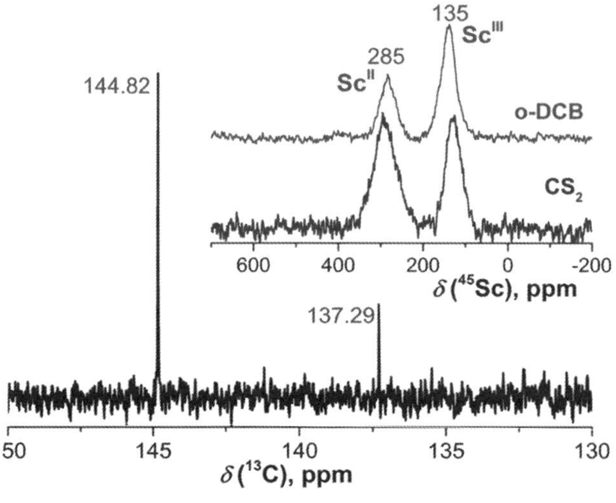

Fig. 10 ^{45}Sc and ^{13}C NMR of purified $Sc_4(\mu_3\text{-}O)_2@I_h\text{-}C_{80}$. (Reproduced with permission from Ref. 39, Copyright 2012 American Chemical Society.)

these two ^{45}Sc NMR peaks are at 138 and 285 ppm.[39] The data indicates a rather rigid cluster and an inability of the oxygen atoms to rearrange between the different faces of the Sc_4 tetrahedron based on the lack of coalescence of the two Sc signals.[39] The presence of two ^{45}Sc NMR peaks was also observed in $Sc_2C_2@C_{82}$,[52,53] $Sc_2C_2@C_{80}$,[54] and $Sc_3CN@C_{80}$.[23] Note the Sc_4O_2 cluster has four metal atoms in a mixed valence state, namely two atoms are at Sc^{2+} with the other two atoms at Sc^{3+}.[38,39]

4.1.4 *Vibrational*

The IR and Raman spectra for $Sc_4O_2@C_{80}$ have also been obtained.[39] Shown in Fig. 11 is a comparison of theoretical versus experimental spectra for $Sc_4O_2@I_h\text{-}C_{80}$. There is good agreement between the computed and experimental spectra. Also in Fig. 11 is a comparison of two species with the same I_h-cage but different clusters, Sc_3N and Sc_4O_2.[39] Overall, the spectra of $Sc_3N@C_{80}$ and $Sc_4O_2@C_{80}$ are similar, but with distinct differences in the 650–800 cm^{-1} range, a spectral region with no corresponding counterpart in $Sc_3N@C_{80}$. A conclusion is the Sc_4O_2 cluster causes a larger distortion in the C_{80} cage than does the Sc_3N cluster.[39]

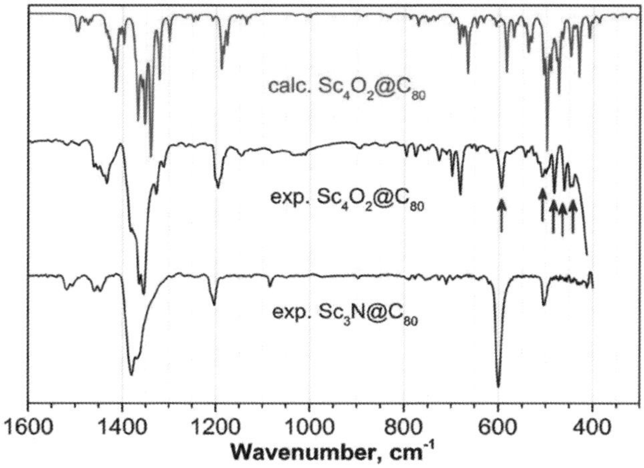

Fig. 11 Comparison of IR spectra for purified samples of $Sc_4(\mu_3\text{-}O)_2@I_h\text{-}C_{80}$ and $Sc_3N@I_h\text{-}C_{80}$. (Reproduced with permission from Ref. 39, Copyright 2012 American Chemical Society.)

Raman measurements were also obtained for $Sc_4O_2@C_{80}$ versus $Sc_3N@C_{80}$, with the results indicating similarities in spectra.[39] However, there were several distinct differences that were attributed to differences in the electronic structure of $Sc_4O_2@C_{80}$.

4.2. Voltammetry

Electrochemical measurements of metal oxide clusterfullerenes have begun.[30,39] Square wave and cyclic voltammetry data has been reported for $Sc_4O_2@C_{80}$ and is shown in Fig. 12.[39] At a scan rate of 20 mV/s, there are two reversible oxidation and two reversible reduction steps for $Sc_4O_2@C_{80}$.[39] The observed electrochemical gap of 1.10 V is compared to 0.88 V, the predicted DFT value. There was a third reduction step that was observed by higher scan rates or with square-wave voltammetry.[39] A remarkable observation was the reversible behavior of $Sc_4O_2@C_{80}$ on the cathodic scale when compared to other metallofullerenes, which usually show irreversible reductions.

Relative to $Fe(Cp)_2^{+/0}$, $Sc_4O_2@C_{80}$ has a first oxidation potential of 0.00 V, which is significantly more negative than most other clusterfullerenes. In contrast, the first reduction potential for $Sc_4O_2@C_{80}$ was determined to

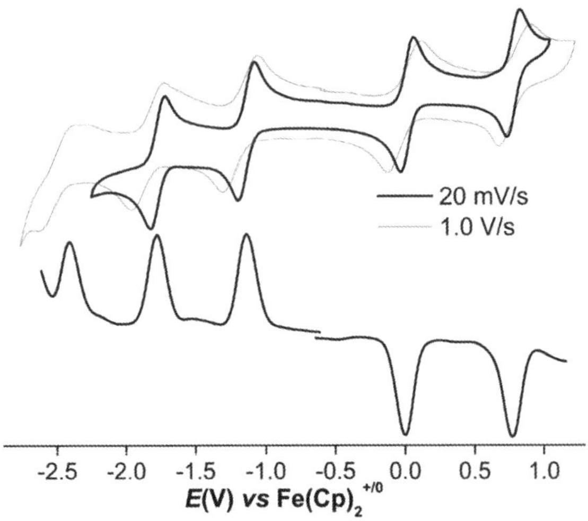

Fig. 12 Cyclic and square wave voltammetry of purified $Sc_4(\mu_3\text{-}O)_2@I_h\text{-}C_{80}$ in *o*-DCB. (Reproduced with permission from Ref. 39, Copyright 2012 American Chemical Society.)

be -1.10 V relative to $Fe(Cp)_2^{+/0}$.[39] When compared to yttrium and lanthanide metallic nitride clusterfullerenes, this value of −1.10 V for $Sc_4O_2@C_{80}$ is significantly more positive. It should also be noted that the electrochemical gap of 1.10 V is about twice as small as those for $M_3N@C_{80}$, which have their oxidation and reduction localized on the carbon cage.

Electrochemical results for another metal oxide clusterfullerene, $Sc_2O@C_s\text{-}C_{82}$ has also been obtained (Fig. 13) and compared with a Sc_2S cluster possessing the same $C_s(6)$ cage isomer ($Sc_2S@C_s\text{-}C_{82}$).[28] The electrochemical data for $Sc_2O@C_s(6)\text{-}C_{82}$ and $Sc_2S@C_s(6)\text{-}C_{82}$ are similar, as are their HOMO and LUMO orbitals.[28] A summary of electrochemical data for Sc_2O,[26] Sc_2S,[28] Sc_4C_2,[20] Sc_4O_2[30,39], Dy_3N,[55] Lu_3N,[56] and Sc_3N[57] clusterfullerenes is provided in Table 1.

4.3. ESR spectroelectrochemistry

Metal-metal bonding in endohedral metallofullerenes can be investigated with ESR spectroelectrochemistry. Indeed, $Sc_4O_2@C_{80}$ is reported to be a good candidate for *in situ* ESR spectroelectrochemistry.[39] This molecule has a reversible reduction and oxidation, and its anion and cation radicals

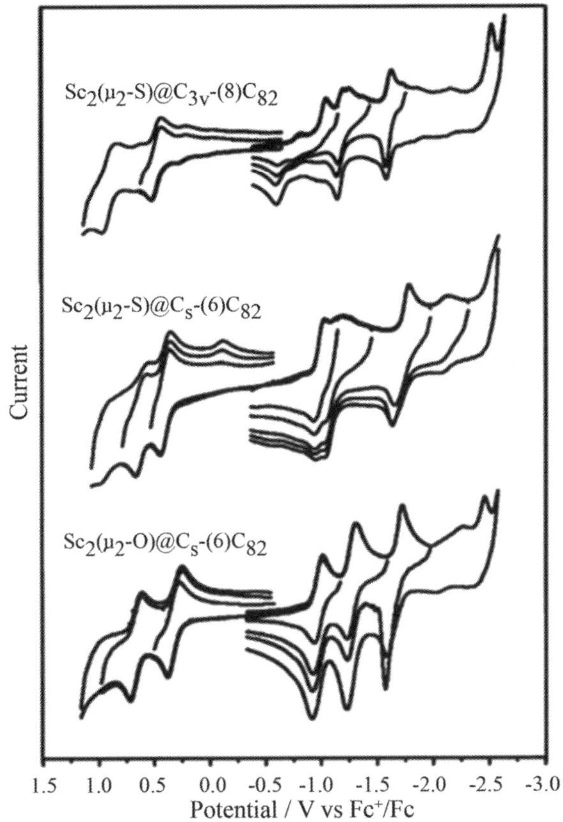

Fig. 13 Cyclic voltammograms comparison for purified $Sc_2(\mu_2\text{-}O)@C_{3v}(8)\text{-}C_{82}$, $Sc_2(\mu_2\text{-}O)@C_s(6)\text{-}C_{82}$, and $Sc_2(\mu_2\text{-}O)@C_s(6)\text{-}C_{82}$. (Reproduced with permission from Ref. 28, Copyright 2012 , Journal of Physical Organic Chemistry, John Wiley & Sons.)

of this clusterfullerene are stable in *o*-dichlorobenzene. The spin density distribution in its electrochemically generated anion and cation radical is recently investigated. Figure 14 shows the ESR spectra from the oxidation and reduction of $Sc_4O_2@C_{80}$ at the first oxidation/reduction voltammetric peaks.[39]

There is a rich hyperfine structure in the ESR spectra of the $Sc_4O_2@C_{80}$ ion radicals due to the two pairs of equivalent Sc atoms ($2Sc^{2+}$, $2Sc^{3+}$). Scandium is an element with a nuclear spin, I = 7/2, and therefore results in ESR spectra with signature patterns. The spectra show large hyperfine

Table 1 Comparison of Redox Properties of Purified $Sc_2O@C_{82}$ and $Sc_4O_2@C_{80}$ versus Metal Sulfide and Metal Nitride Clusterfullerenes. Peak Potentials in Square Brackets Denote Electrochemically Irreversible Processes. (Adapted with permission from Ref. 39, Copyright 2012 American Chemical Society.)

	II_{ox}	I_{ox}	I_{red}	II_{red}	III_{red}	EC gap (V)	ref
$Sc_4O_2@I_h$–C_{80}	0.79	0.00	−1.10	−1.73	−2.35	1.10	30,39
$Sc_4C_2@$ I_h–C_{80}	1.10	0.40	[−1.16]	−1.65		1.56	20
$Sc_3N@$ I_h–C_{80}	1.09	0.59	−1.26	−1.62	−2.37	1.85	57
$Dy_3N@$ I_h–C_{80}		0.70	[−1.37]	[−1.86]		2.07	55
$Lu_3N@$ I_h–C_{80}	[1.11]	[0.64]	[−1.42]	[−1.80]	[−2.26]	2.08	56
$Sc_2O@$ $C_s(6)$–C_{82}	0.72	0.35	−0.96	−1.28	−1.74	1.31	28
$Sc_2S@C_s(6)$–C_{82}	0.65	0.39	−0.98	−1.12	−1.73	1.37	28

coupling constants and indicate that both reduction and oxidation of $Sc_4O_2@C_{80}$ are *in cavea* redox processes. The Sc^{2+} atoms show a significantly large coupling constant (150.4 G) for the cation radical.[39] This large value is of significance as it is the largest recorded $a(^{45}Sc)$ value for scandium clusterfullerenes. The cause of this very large coupling constant was attributed to the nature of the HOMO localized on the six-atom Sc_4O_2 cluster.[39] This highest occupied molecular orbital is a Sc-Sc bonding molecular orbital and therefore has large contributions from the 4s atomic orbitals of Sc^{2+}.[39]

4.3.1 *Cation Radical*

Specific features of the ESR spectra obtained from the oxidation of $Sc_4O_2@C_{80}$ (Fig. 14a), show a nonsymmetric spectrum of relatively narrow lines (ca. 3.5 G) with a range spanning more than 2200 G.[39] There are 15-line multiplets having a hyperfine coupling constant of 18 G (right portion of spectrum), revealing the existence of two equivalent Sc atoms. Two other equivalent scandium atoms have a hyperfine coupling constant of 150.4 G. Note the *g* factor of the cation radical ($g = 1.9956$) is lower than the free electron value ($g = 2.0023$), suggesting the localization of the spin density on the scandium metal atoms.[39]

4.3.2 Anion Radical

Shown in Fig. 14b, the ESR spectrum of the anion radical also has a complex hyperfine pattern and spans of field of ca. 400 G.[39] The simulated ESR spectrum is obtained by having two pairs of equivalent Sc atoms with each pair having $a(^{45}\text{Sc})$ values of 2.6 and 27.4 G. The *g* factor is 1.9960 and also smaller than the free electron value. The assignment of hfc value to Sc^{2+} and Sc^{3+} is not straightforward for this ESR spectroelectrochemical

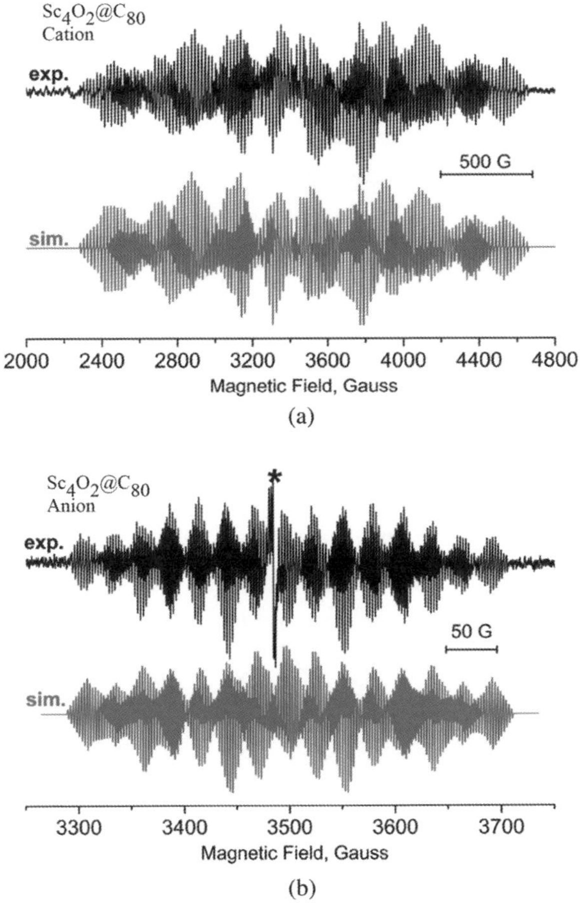

Fig. 14 ESR spectra obtained *in situ* during the electrochemical (a) oxidation and (b) reduction of $Sc_4(\mu_3\text{-}O)_2@I_h\text{-}C_{80}$. (Reproduced with permission from Ref. 39, Copyright 2012 American Chemical Society.)

data.[39] Extended DFT studies were performed to address the structure, dynamics and ^{45}Sc hyperfine coupling constants for the $Sc_4O_2@C_{80}$ anion and cation radicals.[39] Further theoretical details for $Sc_4O_2@C_{80}$ are also published.[58]

5. ELECTRONIC STRUCTURES AND HOMO-LUMOs

When we first discovered the metal oxide fullerenes and proposed new fullerene clusters of Sc_4O_2 and Sc_4O_3, there was no history of scandium and oxygen combining in nature with this stoichiometry. The stable form of combining scandium and oxygen is Sc_2O_3, but Sc_2O_3 is neutral. Fullerene scientists know that upon formation of metallofullerenes, the encapsulated clusters often possess a positive formal charge and electron transfer to a negatively charged fullerene cage for stabilization. For example, at the time of discovery, we knew the electronic structure of $Sc_3N@C_{80}$ was $[Sc_3N]^{6+}@[I_h\text{-}C_{80}]^{6-}$. Namely the Sc_3N cluster had a formal charge of +6 (i.e. each of the 3 Sc atoms was +3, the N was −3). Hence, the $[Sc_3N]^{6+}$ cluster was a perfect complement to a fullerene $I_h\text{-}C_{80}{}^{6-}$ cage, which needed six electrons for stability.[59] Naturally, when we determined our new molecules to have the chemical formulas of $Sc_4O_2C_{80}$ and $Sc_4O_3C_{80}$, we naturally looked at the $I_h\text{-}C_{80}{}^{6-}$ cage as the cage housing and set the formal charge of the clusters to be +6. In doing so, the Sc_4O_3 species would have all four Sc atoms at +3 and three O atoms at -2 for an overall formal charge of the Sc_4O_3 cluster to be +6, described as $(Sc_4O_3)^{6+}$. However, for the formal charge of the Sc_4O_2 cluster to add to +6, two Sc atoms could each be at +3, but the other two Sc atoms would be forced into the less preferred oxidation state of +2.

We needed the assistance of theoretical chemists to determine the following questions. Were we crazy during this discovery phase to suggest an endohedral metallofullerene with six ($Sc_4O_2@C_{80}$) and seven ($Sc_4O_3@C_{80}$) atoms inside the cage? After all, such a large number of atoms crammed inside a C_{80} cage would set a new world record for an isolated fullerene with the most atoms trapped inside. If there was a new family of metal oxide clusterfullerenes, what would be their theoretical stabilities? Is their formation energetically favorable? What would be their HOMO and LUMO diagrams? Where would the spin densities be upon addition or removal of an electron? To address these questions, scientists have investigated and now published these answers. Excellent theoretical results have now been published by several research groups.[30,31,45] We briefly discuss these results below.

5.1 $Sc_4(\mu3\text{-}O)_2@I_h\text{-}C_{80}$

An interesting HOMO-LUMO diagram is expected for $Sc_4O_2@C_{80}$ due to the two Sc atoms which are only partially oxidized at Sc^{2+}. In contrast, $Sc_3N@C_{80}$ and $Sc_4O_3@C_{80}$ all possess fully oxidized Sc^{3+} atoms. Shown in Fig. 15 is the orbital interaction diagram[38] for $Sc_4O_2@I_h\text{-}C_{80}$, whose structure was corroborated with X-Ray crystal data.[24]

In contrast to $Sc_3N@I_h\text{-}C_{80}$ and $Sc_4O_3@I_h\text{-}C_{80}$, both the highest occupied molecular orbital and the lowest unoccupied molecular orbital are primarily localized on the Sc atoms of $Sc_4(\mu_3\text{-}O)_2$.[38] Another difference between $Sc_4O_2@I_h\text{-}C_{80}$ and $Sc_3N@I_h\text{-}C_{80}$ is an absence of metal d electrons in the Sc_3N cluster.[38] Also shown in Fig. 15 is the formal transfer of six

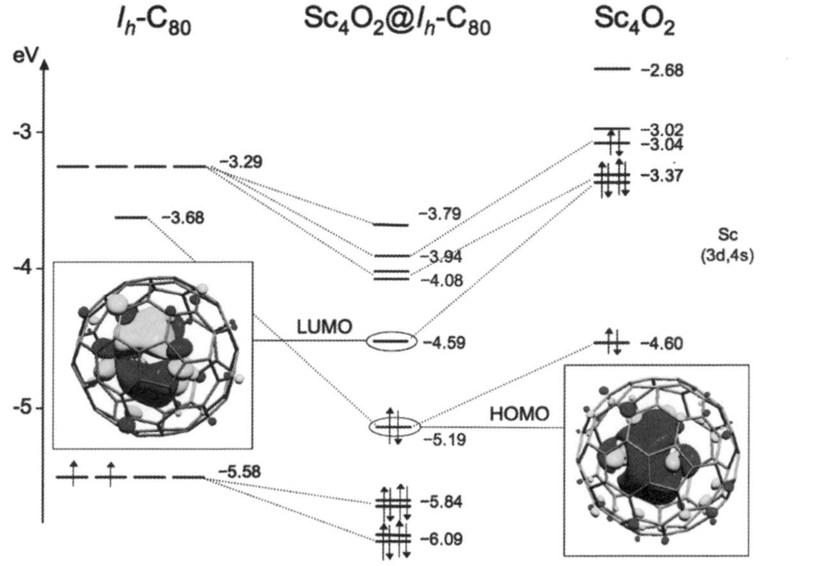

Fig. 15 Orbital interaction diagram for $Sc_4(\mu_3\text{-}O)_2@I_h\text{-}C_{80}$. (Reproduced with permission from Ref. 38, Copyright 2009 American Chemical Society.)

electrons from the three HOMOs of the Sc_4O_2 internal cluster to the icosahedral C_{80} cage.[38]

5.2 $Sc_4(\mu_3\text{-}O)_3@I_h\text{-}C_{80}$

When the first sample of $Sc_4O_3C_{80}$ was purified, we did not know the location of the additional oxygen atom. Was the correct structure an oxide of $Sc_4O_2@C_{80}$ (i.e. $Sc_4O_2@C_{80}O$), or was the extra oxygen atom on the inside of the cage (i.e. $Sc_4O_3@C_{80}$)? Theoretical calculations allowed the direct comparison of these two possible formulas. The seven-atom cluster with all three oxygen atoms within the cage's interior (i.e. Sc_4O_3) was 115 kcal/mol more stable than the structure having the extra atom located on the exterior to the cage surface (i.e. $Sc_4O_2@C_{80}O$).[38] Subsequent X-ray crystallography confirmed this theoretical prediction of the additional oxygen atom in the Sc_4O_3 cluster.[25] Shown in Fig. 16 is the orbital interaction diagram for $Sc_4(\mu_3\text{-}O)_3@I_h\text{-}C_{80}$.[38]

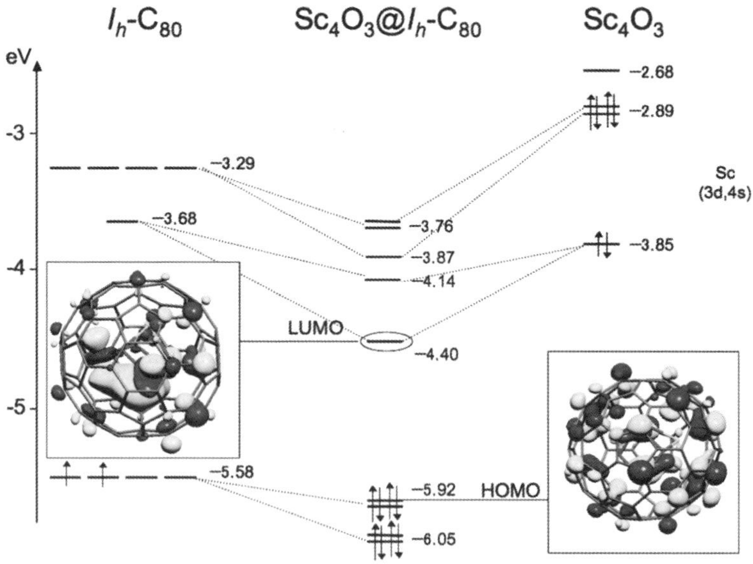

Fig. 16 Orbital interaction diagram for $Sc_4(\mu_3\text{-}O)_3@I_h\text{-}C_{80}$. (Reproduced with permission from Ref. 38, Copyright 2009 American Chemical Society.)

Analysis reveals the HOMO is localized on the C_{80} cage while the LUMO is on the Sc_4O_3 cluster, as similar to $Sc_3N@I_h$-C_{80}.[38] Similar to $Sc_4(\mu_3\text{-}O)_2@I_h$-$C_{80}$, three HOMOs of the free $Sc_4(\mu_3\text{-}O)_3$ cluster reveal delocalized Sc-Sc bonding interactions.[38] As a further comparison of the Sc_4O_2 and Sc_4O_3 cluster, Fig. 17 illustrates the molecular structure and atom-type descriptions for the Sc_4O_2 and Sc_4O_3 clusters in I_h-C_{80} cages.[38]

5.3 $Sc_2(\mu_2\text{-}O)@C_s(6)$-$C_{82}$

In contrast to Sc_4O_2 and Sc_4O_3 clusters, the Sc_2O cluster has been observed experimentally in a non-fullerene setting.[60] Indeed, the gaseous oxide, Sc_2O, has been detected by mass spectrometry from the reaction of scandium metal with scandium oxide.[60] X-ray crystallography of purified $Sc_2(\mu_2\text{-}O)@C_s(6)$-$C_{82}$ confirms the Sc_2O cluster is inside a $C_s(6)$-C_{82} cage with Sc-O bond distances of 1.934 Å and 1.867 Å. The Sc1-O-Sc2 angle is 156.6 Å.[26] The electronic structure can be described as $[Sc_2(\mu_2\text{-}O)]^{4+}@[C_s\text{-}C_{82}]^{4-}$.[26]

DFT calculations indicate the $Sc_2(\mu_2\text{-}O)$ cluster can rotate quite freely inside the $C_s(6)$-$C_{82}{}^{4-}$ cage.[26] As shown in Fig. 18, the transfer of four electrons from the Sc_2O cluster to the C_{82} cage stabilizes $Sc_2(\mu_2\text{-}O)@C_s(6)$-$C_{82}$. Upon encapsulation the Sc_2O cluster, the HOMO and LUMO orbitals are primarily delocalized over the carbon cage.[26]

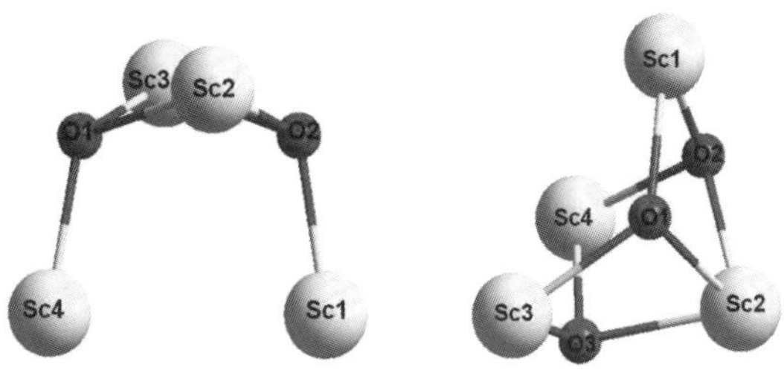

Sc_4O_2
(a)

Sc_4O_3
(b)

Fig. 17 Molecular structures and atom-type descriptions for the $Sc_4(\mu_3\text{-}O)_2$ and $Sc_4(\mu_3\text{-}O)_3$ clusters entrapped in I_h-C_{80} cages. (Reproduced with permission from Ref. 38, Copyright 2009 American Chemical Society.)

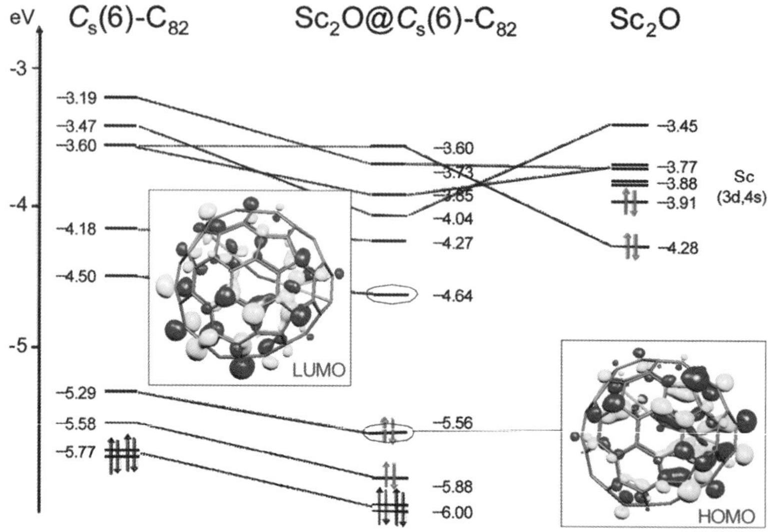

Fig. 18 Molecular orbital diagram for $Sc_2(\mu_2\text{-}O)@C_s(6)\text{-}C_{82}$. (Reproduced with permission from Ref. 26, Copyright 2010 American Chemical Society.)

6. CONCLUSIONS

The future outlook is exciting for metal oxide clusterfullerenes. With the replacement of scandium with other transition and rare earth metals, the metal oxide family of compounds is now extended. We look forward to experimental studies and theoretical investigations for new yttrium compounds such as $Y_2O@C_{82}$, $Lu_2O@C_{82}$ and metal oxide clusters of different M_xO_y stoichiometric compositions. New M_xO_y clusters could also have mixed valencies (M^{2+}, M^{3+}), have interesting ESR spectroelectrochemistry, and HOMO-LUMO theoretical calculations in comparison to the prototypical $Sc_4O_2@C_{80}$ metal oxide clusterfullerene. Given the remarkable electrochemical redox reactions that can be done *in cavea* with $Sc_4O_2@C_{80}$, perhaps these fundamental properties will be conducive toward application development. As we are in the discovery phase with the initial wave of experiments coming forth, it will be even more exciting to see these molecules give up their secrets as we continue to investigate their science and applications.

ACKNOWLEDGEMENTS

The author thanks the NSF CAREER grant (CHI-0547988) and NSF RUI grant (CHE-1151668) for financial support.

REFERENCES

1. Alvarez MM, Gillan EG, Holczer K, Kaner RB, Min KS, Whetten RL. (1991) La_2C_{80} — A soluble dimetallofullerene. *J Phys Chem* **95:** 10561–10563.
2. Chai Y, Guo T, Jin CM, Haufler RE, Chibante LPF, Fure J, Wang LH, Alford JM, Smalley RE. (1991) Fullerenes with metals inside. *J Phys Chem* **95:** 7564–7568.
3. Gillan EG, Yeretzian C, Min KS, Alvarez MM, Whetten RL, Kaner RB. (1992) Endohedral rare-earth fullerene complexes. *J Phys Chem* **96:** 6869–6871.
4. Ross MM, Nelson HH, Callahan JH, McElvany SW. (1992) Production and characterization of metallofullerenes. *J Phys Chem* **96:** 5231–5234.
5. Shinohara H, Sato H, Saito Y, Ohkohchi M, Ando Y. (1992) Mass spectroscopic and ESR characterrization of soluble yttrium-containing metallofullerenes YC_{82} and Y_2C_{82}. *J Phys Chem* **96:** 3571–3573.
6. Yannoni CS, Hoinkis M, Devries MS, Bethune DS, Salem JR, Crowder MS, Johnson RD. (1992) Scandium clusters in fullerene cages. *Science* **256:** 1191–1192.
7. Shinohara H, Yamaguchi H, Hayashi N, Sato H, Ohkohchi M, Ando Y, Saito Y. (1993) Isolation and spectroscopic properties of $Sc_2@C_{74}$, $Sc_2@C_{82}$, and $Sc_2@C_{84}$. *J Phys Chem* **97:** 4259–4261.
8. Beyers R, Kiang CH, Johnson RD, Salem JR, Devries MS, Yannoni CS, Bethune DS, Dorn H, Burbank P, Harich K, Stevenson S. (1994) Preparation and structure of crystals of the metallofullerene $Sc_2@C_{84}$. *Nature* **370:** 196–199.
9. Stevenson S, Rice G, Glass T, Harich K, Cromer F, Jordan MR, Craft J, Hadju, E, Bible R, Olmstead MM, Maitra K, Fisher AJ, Balch AL, Dorn HC. (1999) Small-bandgap endohedral metallofullerenes in high yield and purity. *Nature* **401:** 55–57.
10. Krause M, Kuzmany H, Georgi P, Dunsch L, Vietze K, Seifert G. (2001) Structure and stability of endohedral fullerene $Sc_3N@C_{80}$: A Raman, infrared, and theoretical analysis. *J Chem Phys* **115:** 6596–6605.
11. Dunsch L, Georgi P, Krause M, Wang CR. (2003) New clusters in endohedral fullerenes: The metalnitrides. *Synthetic Met* **135:** 761–762.

12. Krause M, Wong J, Dunsch L. (2005) Expanding the world of endohedral fullerenes — The $Tm_3N@C_2n$ (39 <= n <= 43) clusterfullerene family. *Chem Eur J* **11:** 706–711.
13. Melin F, Chaur MN, Engmann S, Elliott B, Kumbhar A, Athans AJ, Echegoyen L. (2007) The large $Nd_3N@C_2n$ (40 <= n <= 49) cluster fullerene family: preferential templating of a C_{88} cage by a trimetallic nitride cluster. *Angew Chem Int Ed* **46:** 9032–9035.
14. Chaur MN, Melin F, Elliott B, Kumbhar A, Athans AJ, Echegoyen L. (2008) New $M_3N@C_2n$ endohedral metallofullerene families (M = Nd, Pr, Ce; n = 40–53): Expanding the preferential templating of the C_{88} cage and approaching the C_{96} cage. *Chem Eur J* **14:** 4594–4599.
15. Olmstead MM, de Bettencourt-Dias A, Duchamp JC, Stevenson S, Dorn HC, Balch AL. (2000) Isolation and crystallographic characterization of $ErSc_2N@C_{80}$: an endohedral fullerene which crystallizes with remarkable internal order. *J Am Chem Soc* **122:** 12220–12226.
16. Stevenson S, Fowler PW, Heine T, Duchamp JC, Rice G, Glass T, Harich K, Hajdu E, Bible R, Dorn HC. (2000) Materials science — A stable non-classical metallofullerene family. *Nature* **408:** 427–428.
17. Olmstead MH, de Bettencourt-Dias A, Duchamp JC, Stevenson S, Marciu D, Dorn HC, Balch AL. (2001) Isolation and structural characterization of the endohedral fullerene $Sc_3N@C_{78}$. *Angew Chem Int Ed* **40:** 1223–1225.
18. Stevenson S, Phillips JP, Reid JE, Olmstead MM, Rath SP, Balch AL. (2004) Pyramidalization of Gd_3N inside a C_{80} cage. The synthesis and structure of $Gd_3N@C_{80}$. *Chem Commun* 2814–2815.
19. Tan K, Lu X, Wang CR. (2006) Unprecedented mu(4)-C-2(6-) anion in $Sc_4C_2@C_{80}$. *J Phys Chem B* **110:** 11098–11102.
20. Wang TS, Chen N, Xiang JF, Li B, Wu JY, Xu W, Jiang L, Tan K, Shu CY, Lu X, Wang CR. (2009) Russian-doll-type metal carbide endofullerene: Synthesis, isolation, and characterization of $Sc_4C_2@C_{80}$. *J Am Chem Soc* **131:** 16646–16647.
21. Wang TS, Feng L, Wu JY, Xu W, Xiang JF, Tan K, Ma YH, Zheng JP, Jiang L, Lu X, Shu CY, Wang CR. (2010) Planar quinary cluster inside a fullerene cage: Synthesis and Structural Characterizations of $Sc_3NC@C_{80}$-I_h. *J Am Chem Soc* **132:** 16362–16364.
22. Wu JY, Wang TS, Ma YH, Jiang L, Shu CY, Wang CR. (2011) Synthesis, isolation, characterization, and theoretical studies of $Sc_3NC@C_{78}$-C_2. *J Phys Chem C* **115:** 23755–23759.
23. Krause M, Ziegs F, Popov AA, Dunsch L. (2007) Entrapped bonded hydrogen in a fullerene: The five-atom cluster Sc_3CH in C_{80}. *Chemphyschem* **8:** 537–540.

24. Stevenson S, Mackey MA, Stuart MA, Phillips JP, Easterling ML, Chancellor CJ, Olmstead MM, Balch AL. (2008) A distorted tetrahedral metal oxide cluster inside an lcosahedral carbon cage. Synthesis, isolation, and structural characterization of Sc_4(mu(3)-O)(2)@I_h-C_{80}. *J Am Chem Soc* **130:** 11844–11845.
25. Mercado BQ, Olmstead MM, Beavers CM, Easterling ML, Stevenson S, Mackey MA, Coumbe CE, Phillips JD, Phillips JP, Poblet JM, Balch AL. (2010) A seven atom cluster in a carbon cage, the crystallographically determined structure of Sc_4(mu(3)-O)(3)@I_h-C_{80}. *Chem Commun* **46:** 279–281.
26. Mercado BQ, Stuart MA, Mackey MA, Pickens JE, Confait BS, Stevenson S, Easterling ML, Valencia R, Rodriguez-Fortea A, Poblet JM, Olmstead MM, Balch AL. (2010) Sc_2(mu(2)-O) trapped in a fullerene cage: The isolation and structural characterization of Sc_2(mu(2)-O)@Cs(6)-C_{82} and the relevance of the thermal and entropic effects in fullerene isomer selection. *J Am Chem Soc* **132:** 12098–12105.
27. Chen N, Beavers CM, Mulet-Gas M, Rodriguez-Fortea A, Munoz EJ, Li YY, Olmstead MM, Balch AL, Poblet JM, Echegoyen L. (2012) Sc_2S@Cs(10528)-C_{72}: A dimetallic sulfide endohedral fullerene with a non isolated pentagon rule cage. *J Am Chem Soc* **134:** 7851–7860.
28. Mercado BQ, Chen N, Rodriguez-Fortea A, Mackey MA, Stevenson S, Echegoyen L, Poblet JM, Olmstead MM, Balch AL. (2011) The shape of the Sc_2(mu(2)-S) unit trapped in C_{82}: Crystallographic, computational, and electrochemical studies of the isomers, Sc_2(mu(2)-S)@Cs(6)-C_{82} and Sc_2(mu(2)-S)@C_3v(8)-C_{82}. *J Am Chem Soc* **133:** 6752–6760.
29. Dunsch L, Yang SF, Zhang L, Svitova A, Oswald S, Popov AA. (2010) Metal sulfide in a C_{82} fullerene cage: A new form of endohedral clusterfullerenes. *J Am Chem Soc* **132:** 5413–5421.
30. Rivera-Nazario DM, Pinzón JR, Stevenson S, Echegoyen LA. (2012) Buckyball maracas: exploring the inside and outside properties of endohedral fullerenes. *J Phys Org Chem* **26:** 194–205.
31. Yang SF. (2012) Synthesis, separation, and molecular structures of endohedral fullerenes. *Curr Org Chem* **16:** 1079–1094.
32. Shinohara H. (2000) Endohedral metallofullerenes. *Rep Prog Phys* **63:** 843–892.
33. Lu X, Akasaka T, Nagase S. (2011) Chemistry of endohedral metallofullerenes: the role of metals. *Chem Commun* **47:** 5942–5957.
34. Rodriguez-Fortea A, Irle S, Poblet JM. (2011) Fullerenes: formation, stability, and reactivity. *Wiley Interdisciplinary Reviews-Computational Molecular Science* **1:** 350–367.

35. Chaur MN, Athans AJ, Echegoyen L. (2008) Metallic nitride endohedral fullerenes: synthesis and electrochemical properties. *Tetrahedron* **64:** 11387–11393.
36. Dunsch L, Yang S. (2007) Metal nitride cluster fullerenes: Their current state and future prospects. *Small* **3:** 1298–1320.
37. Stevenson S. (2011) Metallic Oxide Clusters in Fullerene Cages. In *Handbook of Carbon Nano Materials*; D'Souza F D, Kadish KM, Eds, World Scientific: Singapore, Vol. 1; pp. 185–205.
38. Valencia R, Rodriguez-Fortea A, Stevenson S, Balch AL, Poblet JM. (2009) Electronic structures of scandium oxide endohedral metallofullerenes, Sc_4(mu(3)-O)(n)@I_h-C_{80} (n = 2, 3). *Inorg Chem* **48:** 5957–5961.
39. Popov A, Chen N, Pinzón JR, Stevenson S, Echegoyen L, Dunsch L. (2012) Redox-active scandium oxide cluster inside a fullerene cage: Spectroscopic, voltammetric, electron spin resonance spectroelectrochemical, and extended density functional theory study of $Sc_4O_2@C_{80}$ and its ion radicals. *J Am Chem. Soc* **134:** 19607–19618.
40. Stevenson S, Thompson MC, Coumbe HL, Mackey MA, Coumbe CE, Phillips JP. (2007) Chemically adjusting plasma temperature, energy, and reactivity (CAPTEAR) method using NOx and combustion for selective synthesis of $Sc_3N@C_{80}$ metallic nitride fullerenes. *J Am Chem Soc* **129:** 16257–16262.
41. Takikawa H, Imamura M, Kouchi M, Sakakibara T. (1998) Decrease in fullerene productivity due to air leakage in carbon arc method. *Fullerene Sci Techn* **6:** 339–349.
42. Dunsch L, Krause M, Noack J, Georgi P. (2004) Endohedral nitride cluster fullerenes — Formation and spectroscopic analysis of L_3-xMxN@C_{2n} ($0 \leq x \leq 3$; N = 39,40). *J Phys Chem Solids* **65:** 309–315.
43. Stevenson S, Mackey MA, Thompson MC, Coumbe HL, Madasu PK, Coumbe CE, Phillips JP. (2007) Effect of copper metal on the yield of $Sc_3N@C_{80}$ metallofullerenes. *Chem Commun* 4263–4265.
44. Ling Y, Stevenson S, Zhang Y. (2011) Metallic nitride azafullerenes: A novel class of fullerenes with interesting electronic properties. *Chem Phys Lett* **508:** 121–124.
45. Stevenson S, Ling Y, Coumbe CE, Mackey MA, Confait BS, Phillips JP, Dorn HC, Zhang Y. (2009) Preferential encapsulation and stability of La_3N cluster in 80 atom cages: Experimental synthesis and computational investigation of $La_3N@C_{79}N$. *J Amer Chem Soc* **131:** 17780–17782.
46. Zuo TM, Xu LS, Beavers CM, Olmstead MM, Fu WJ, Crawford D, Balch AL, Dorn HC. (2008) $M_2@C_{79}N$ (M = Y, Tb): Isolation and characterization of stable endohedral metallofullerenes exhibiting M-M bonding interactions inside aza 80 fullerene cages. *J Am Chem Soc* **130:** 12992–12997.

47. Chaur MN, Melin F, Ashby J, Elliott B, Kumbhar A, Rao AM, Echegoyen L. (2008) Lanthanum nitride endohedral fullerenes La3N@C2n (43≤ n ≤55): Preferential formation of $La_3N@C_{96}$. *Chem Eur J* **14:** 8213–8219.
48. Stevenson S, Mackey MA, Coumbe CE, Phillips JP, Elliott B, Echegoyen L. (2007) Rapid removal of D_{5h} isomer using the "stir and filter approach" and isolation of large quantities of isomerically pure $Sc_3N@C_{80}$ metallic nitride fullerenes. *J Am Chem Soc* **129:** 6072–+.
49. Stevenson S, Harich K, Yu H, Stephen RR, Heaps D, Coumbe C, Phillips JP. (2006) Nonchromatographic "stir and filter approach" (SAFA) for isolating $Sc_3N@C_{80}$ metallofullerenes. *J Am Chem Soc* **128:** 8829–8835.
50. Akiyama K, Hamano T, Nakanishi Y, Takeuchi E, Noda S, Wang ZY, Kubuki S, Shinohara H. (2012) Non-HPLC rapid separation of metallofullerenes and empty cages with $TiCl_4$ lewis acid. *J Am Chem Soc* **134:** 9762–9767.
51. Stevenson S, Mackey MA, Pickens JE, Stuart MA, Confait BS, Phillips JP. (2009) Selective complexation and reactivity of metallic nitride and oxometallic fullerenes with lewis acids and use as an effective purification method. *Inorg Chem* **48:** 11685–11690.
52. Lu X, Nakajima K, Iiduka Y, Nikawa H, Mizorogi N, Slanina Z, Tsuchiya T, Nagase S, Akasaka T. (2011) Structural elucidation and regioselective functionalization of an unexplored carbide cluster metallofullerene $Sc_2C_2@C_s(6)\text{-}C_{82}$. *J Am Chem Soc* **133:** 19553–19558.
53. Miyake Y, Suzuki S, Kojima Y, Kikuchi K, Kobayashi K, Nagase S, Kainosho M, Achiba Y, Maniwa Y, Fisher K. (1996) Motion of scandium ions in Sc_2C_{84} observed by Sc-45 solution NMR. *J Phys Chem* **100:** 9579–9581.
54. Kurihara H, Lu X, Iiduka Y, Mizorogi N, Slanina Z, Tsuchiya T, Akasaka T, Nagase S. (2011) $Sc_2C_2@C_{80}$ rather than $Sc_2@C_{82}$: Templated formation of unexpected $C_{2v}(5)\text{-}C_{80}$ and temperature-dependent dynamic motion of internal Sc_2C_2 cluster. *J Am Chem Soc* **133:** 2382–2385.
55. Yang SF, Zalibera M, Rapta P, Dunsch L. (2006) Charge-induced reversible rearrangement of endohedral fullerenes: Electrochemistry of tridysprosium nitride clusterfullerenes $Dy_3N@C_{2n}$ (2n = 78, 80). *Chem Eur J* **12:** 7848–7855.
56. Pinzon JR, Zuo TM, Echegoyen L. (2010) Synthesis and electrochemical studies of Bingel-Hirsch derivatives of $M_3N@I_h\text{-}C_{80}$ (M = Sc, Lu). *Chem Eur J.* **16:** 4864–4869.
57. Elliott B, Yu L, Echegoyen L. (2005) A simple isomeric separation of D_{5h} and I_h $Sc_3N@C_{80}$ by selective chemical oxidation. *J Am Chem Soc* **127:** 10885–10888.
58. Popov AA, Avdoshenko SM, Pendas AM, Dunsch L. (2012) Bonding between strongly repulsive metal atoms: an oxymoron made real in a confined space of endohedral metallofullerenes. *Chem Commun* **48:** 8031–8050.

59. Kobayashi K, Sano Y, Nagase S. (2001) Theoretical study of endohedral metallofullerenes: Sc_3-nLanN@C_{80} (n = 0–3). *J Comput Chem* **22:** 1353–1358.
60. Clemmer DE, Dalleska NF, Armentrout PB. (1992) Gas-phase thermochemistry of the group-3 dioxides = ScO_2, YO_2 and LaO_2. *Chem Phys Lett* **190:** 259–265.

CHAPTER 8

Nitrogen Atom-Based Endohedral Fullerenes and Potential Applications

*B.J. Farrington and K. Porfyrakis**

Endohedral fullerenes N@C_{60} and N@C_{70} and their chemical derivatives are some of the most exotic members of the fullerene family. In this chapter we focus on these remarkable molecules. We start by outlining methods of synthesis and purification of N@C_{60} and N@C_{70}, then move on to their physical properties with particular focus of their distinct electron spin resonance (ESR) characteristics. Much of the recent work on N@C_{60} has involved chemical modification of the fullerene carbon cage. We therefore highlight the main reaction schemes used to chemically modify N@C_{60} and discuss the effect that this chemical modification has on the physical and ESR properties of the molecules. One of the most promising applications for N@C_{60} and N@C_{70} is as quantum information storage elements (qubits) in a future quantum computer. We summarize the current state of work in this area and discuss the remaining challenges that need to be overcome to realize a N@C_{60} electron spin based quantum computer.

* Department of Materials Science, University of Oxford Oxford, OX1 3PH, United Kingdom E-mail: kyriakos.porfyrakis@materials.ox.ac.uk

1. INTRODUCTION

This chapter focuses on one of the most exotic classes of endohedral fullerenes, nitrogen atom-based endohedral fullerenes, $N@C_{60}$ and $N@C_{70}$. Nitrogen atom-based endohedral fullerenes are structurally similar to the noble gas endohedral fullerenes, in that the endohedral atom is located at the center of the fullerene cage. What makes the nitrogen analogue remarkable is the ability of the fullerene cage to contain an extremely reactive atom such as nitrogen without forming any chemical bonds to it, behaving as a chemical analogue of an ion trap. This results in paramagnetic molecules with very unusual electron spin properties that are ideal candidates for use in applications such as electron spin based quantum information processing.

2. SYNTHESIS

There have been a number of different techniques used to synthesize $N@C_{60}$ and $N@C_{70}$. A common requirement for all methods is the acceleration of N^+ ions towards a fullerene cage to achieve nitrogen implantation, with the optimum acceleration of the N^+ ions consistently found to be around 40 eV. All systems also use an effusion cell to sublime the empty fullerene starting materials either onto a target or into the path of the N^+ ions.

2.1. Commercial Ion Source Method

The first report of the synthesis of $N@C_{60}$ was made by Murphy and coworkers in 1996.[1] Their production method used a commercial ion source to bombard a thin film of deposited C_{60} with nitrogen ions. Using this method both $^{14}N@C_{60}$ and $^{15}N@C_{60}$ were detected and selectively produced using a modified N_2 isotope ratio feed gas source. The above method for ion implantation was subsequently modified to allow continuous deposition of C_{60} on to the target by utilizing an effusion cell to sublime C_{60} on to the target (see Fig. 1). The optimized parameters of the ion beam energy and beam current were found to be 40 eV and 50 μA respectively, producing a yield of $N@C_{60}$ in C_{60} of $10^{-4}-10^{-5}$ (10–100 ppm).[2] To date this method remains the most widely used technique to synthesize nitrogen encapsulated fullerenes due the combination of relatively good yields and

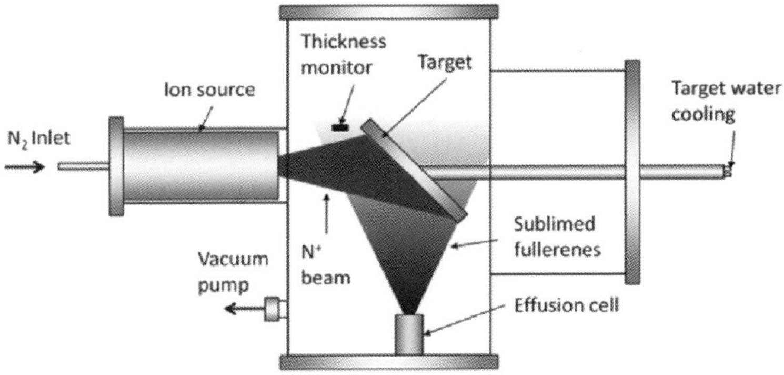

Fig. 1 Schematic view of nitrogen implantation apparatus using a commercial ion source.

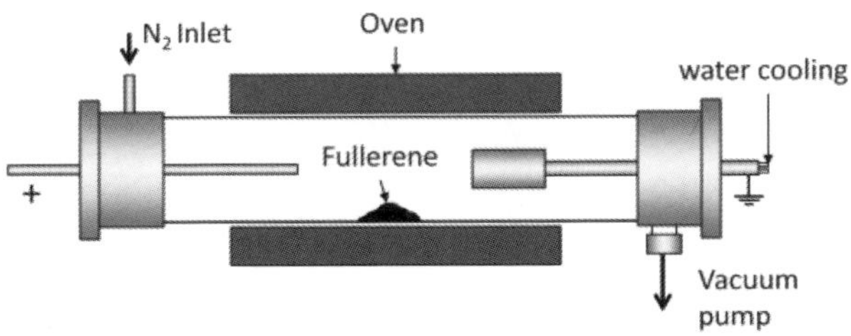

Fig. 2 Schematic view of a glow discharge reactor.

large mass of implanted material that can be produced at a time (up to 500 mg of soluble fullerenes).

2.2. Glow Discharge

Murphy and co-workers also demonstrated a second, lower cost method for the production of $N@C_{60}$ using a glow discharge reactor (see Fig. 2).[3] The drawback to this technique is the low purity of the produced $N@C_{60}$, about 5 ppm in C_{60}. Conditions for production of $N@C_{60}$ using the glow discharge technique were later optimized by Ito *et al.* to produce $N@C_{60}$ in a 15 ppm yield,[4] a significant improvement on Murphy's work, however it remains a lower production yield in comparison to the commercial ion source method.

2.3. R.F. Plasma Method

In 2002, Huang and co-workers reported the formation of N@C_{60} from a radio frequency (rf) plasma. The set-up uses a capacitively coupled plasma system, incorporating a 13.56 MHz plasma generator. The plasma is generated from within the evacuated chamber using the partial atmosphere of nitrogen as the source of N^+ while C_{60} is introduced by evaporation from an effusion cell (see Fig. 3). This technique was optimized by the authors to produce a yield of N@C_{60} in C_{60} of between 200–500 ppm[5]; this remains the highest yield reported to date. Interestingly, N@C_{60} was only reported to be recovered from the outer casing of the chamber with all of the material deposited on the cathode completely insoluble in CS_2, suggesting the cages were seriously deformed at the cathode surface.[6] To reduce the destruction of the fullerene species, an external ring electrode arrangement was trialed (see Fig. 3). However this method failed to yield any measurable amount of N@C_{60}. This was attributed to insufficient acceleration of the N^+ towards the fullerene in (b), while in (a) this acceleration is present from the natural negative self-bias of the cathode.

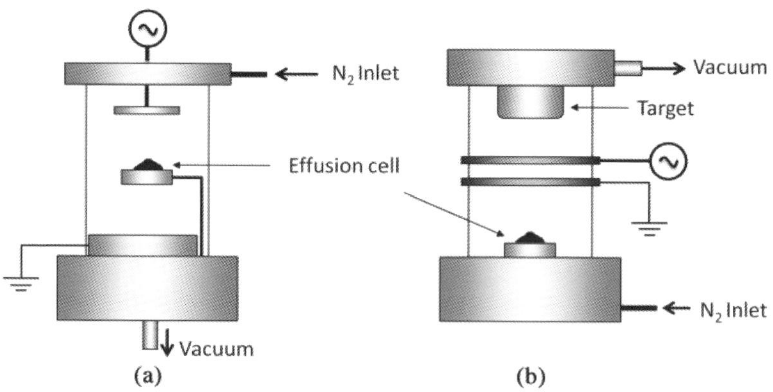

Fig. 3 (a) Schematic of the internal parallel plate electrode arrangement of the capacitively coupled RF plasma implantation method (b) external ring electrode arranged capacitively coupled plasma implantation system, equipped with two external ring electrodes.

2.4. Electron Cyclotron Resonance Method

Electron cyclotron resonance has also been used to effectively create a nitrogen plasma. As with the commercial ion source method, the energy of the incident N^+ on the C_{60} target can be optimized using an electric field gradient between the plasma and the target. The optimal beam energy of 40 eV is in agreement to the value identified by Murphy *et al.*[7] It was then subsequently shown that by introducing radial as well as axial magnetic confinement to the plasma, the electron temperature can be increased by a factor of three to 30 eV.[8] Using the relative intensities of optical emission spectra for the exited nitrogen species, Kaneko and co-workers showed that increasing the electron temperature significantly raised the proportion of N^+ and N^* species compared to N_2^* exiting the plasma generation region, leading to a maximum yield of $N@C_{60}$ in C_{60} of 300 ppm.[8]

3. PURIFICATION

Purification remains one of the biggest challenges to the synthesis of endohedrally-doped nitrogen fullerenes due to the almost complete chemical and physical indistinguishability of $N@C_{60}$ and $N@C_{70}$ with C_{60} and C_{70}, respectively. By far the most common method for purification uses HPLC (High Pressure Liquid Chromatography) to exploit the larger polarizability and slightly larger mass of nitrogen doped endohedral fullerenes compared to the empty fullerene analogue. This effect causes endohedral molecules to have a greater interaction with the HPLC column's stationary phase, causing it to elute at a slightly longer time compared to either C_{60} or C_{70}.

The enrichment of $N@C_{60}$ using HPLC, has been demonstrated using a range of HPLC columns such as Pyren,[2] Buckyprep,[9] PYE[9] and PBB,[10] with the separation of elution time between C_{60} and $N@C_{60}$ calculated from the ESR signal intensity distribution recorded from consecutive aliquots sampled from within the $N@C_{60}/C_{60}$ peak.[2] The most efficient protocol for the enrichment of both $N@C_{60}$ and $N@C_{70}$ uses a 5PBB column, for which the retention time of $N@C_{60}$ and $N@C_{70}$ is 3% longer than C_{60} and C_{70}, and a two-stage purification process.[10] The first stage consists of single pass injections in which the C_{60}/C_{70} peak is split into two at its apex. This process is repeated four further times, each time retaining and injecting the

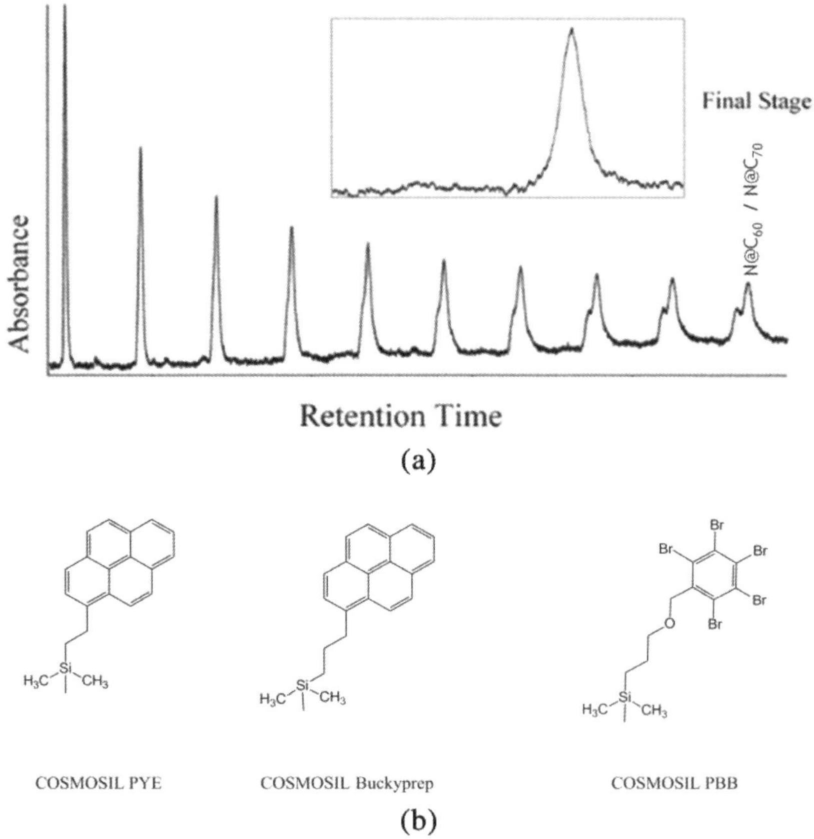

Fig. 4 a) The recycling HPLC profile for the second stage of $N@C_{60}$ purification using a 5PBB column, showing separation of C_{60} from $N@C_{60}$.[10] The inset shows the final cycle from the following stage containing pure $N@C_{60}$. The purification protocol for $N@C_{70}$ is identical to that shown above. b) Chemical structure of HPLC column stationary phase commonly used to purify $N@C_{60}$ and $N@C_{70}$.

second fraction collected on the previous run. In the second stage the enriched material is injected and recycled 12 times as shown in Fig. 4 with the $N@C_{60}/N@C_{70}$ peak collected. This recycling procedure is repeated a further two times on the $N@C_{60}/N@C_{70}$ peak to isolate pure endohedral fullerene material (see insert of Fig. 4). It has also been shown that the separation of $N@C_{60}$ from the parent C_{60} peak increases proportionally with the ratio of hexane to toluene used as the eluent.[9] Purification using an iterative HPLC approach has been used successfully by a handful of groups to isolate N@C_{60} with a purity of up to 99.5%.[10–12]

4. PHYSICAL PROPERTIES

N@C_{60} and N@C_{70} are both black solids, identical to empty fullerenes. There is however still some debate over the exact UV-vis absorption profiles as they have been reported as being both identical to the empty fullerenes[11] or exhibiting weaker absorption peaks in the 440–640 cm^{-1} region.[10] Either way the absorption profile is very similar to that found for the empty fullerenes, with no additional absorption peaks present.

4.1. Electron Spin Resonance

The most fascinating and unique feature of N@C_{60} and N@C_{70} is the combination of the highly symmetric central location of the nitrogen atom within the fullerene cage[13] together with negligible mixing between the nitrogen atom and carbon atom electronic wave functions. This environment imparts a degree of isolation to the nitrogen atom usually only attainable using an ion trap and permits the nitrogen center to retain the same $^4S_{3/2}$ electron spin ground state[1] present in gas phase *N*. The upshot of this is that the higher symmetry N@C_{60} molecules exhibit an extraordinarily long electron phase coherence time of up to 250 μs,[14] the longest coherence time reported for a molecular system.

The ESR spectrum of N@C_{60} is dominated by an isotropic hyperfine interaction between the $^4S_{3/2}$ electron spin ground state and the ^{14}N or ^{15}N nuclear spin of $I = 1$ or $\frac{1}{2}$, respectively. By considering only the first-order hyperfine interaction, the allowed electron transitions are triply degenerate, as illustrated in Fig. 5. This causes the observed continuous-wave (CW) EPR spectrum of ^{14}N@C_{60} and ^{15}N@C_{60} dissolved in toluene at room temperature to comprise of three or two sharp resonance peaks respectively (see Fig. 6), corresponding to each of the allowed nuclear spin states. Each of the EPR resonances is very narrow. Their intrinsic line width has been measured to be less than 0.3 μT, with the observed line width often limited by inhomogeneous broadening caused by the magnet stability and field homogeneity of the spectrometer.

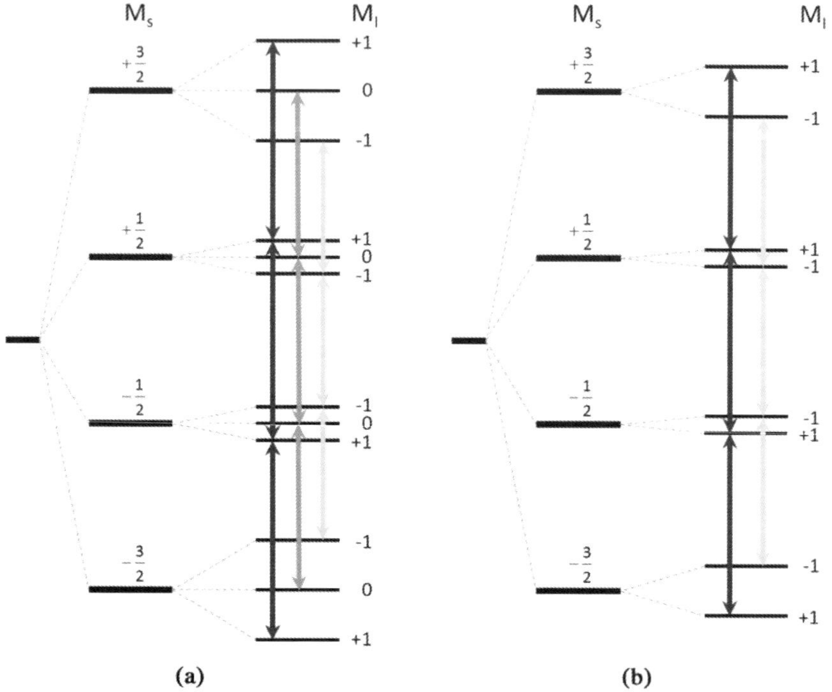

Fig. 5 Energy level diagrams of a) $^{14}N@C_{60}$ and b) $^{15}N@C_{60}$ in a magnetic field. $^{14}N@C_{60}$ and $^{15}N@C_{60}$ have a nuclear spin of I = 1 and I = 1/2 respectively. In a magnetic field, Zeeman splitting gives rise to a 12-level structure in $^{14}N@C_{60}$ and an 8-level structure in $^{15}N@C_{60}$. By considering only the first-order hyperfine interaction and the selection rules $\Delta M_S = 1$ and $\Delta M_I = 0$, the allowed transitions are triply degenerate (indicated as different shades of gray).

The solution-based CW ESR measurements of $N@C_{70}$ appear almost identical to that of $N@C_{60}$ except for the slight reduction in hyperfine interaction due to the larger volume inside the cage. However when the translational and rotational motion of the endohedral fullerenes is restricted through the use of a solid host matrix, the presence of the axial symmetry in $N@C_{70}$ lifts the degeneracy between the electronic transitions in the spectrum. This gives rise to additional angular dependent transitions caused by zero-field splitting (ZFS) that appear as shoulders on the principal hyperfine peaks. The magnitude of this ZFS term is ≈ 2 G.[14]

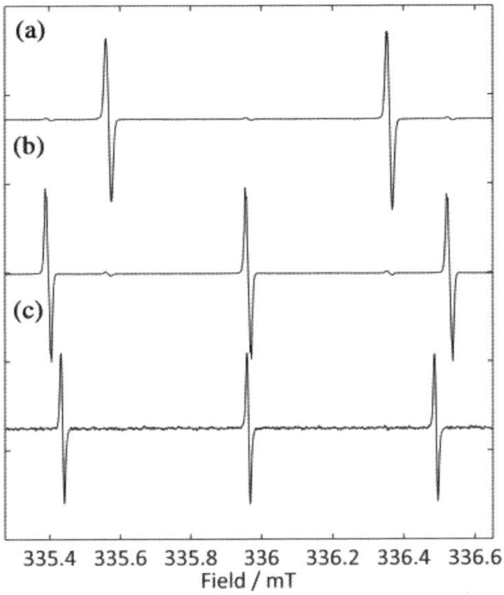

Fig. 6 Comparison of solution-based CW ESR spectra of a) $^{15}N@C_{60}$, b) $^{14}N@C_{60}$ and $^{14}N@C_{70}$.

4.2. Stability

The thermal stability of $N@C_{60}$ was the first decay parameter to be investigated. Weidinger *et al.* recorded the *in-situ* ESR spin signal dependence as a function of temperature of a low purity oxygen-free $N@C_{60}$ sample in both a solution of chloronapthalene and in a powder.[15] The results indicated that in both solid and solution, appreciable decomposition of $N@C_{60}$ occurred only at temperatures higher than 480 K with the decomposition rate increasing with temperature; this corresponds to a low escape energy of only 1.7 eV, far less than recorded for other endohedral fullerenes such as $He@C_{60}$. A proposed mechanism for the escape of the nitrogen via an internal azabridge is proposed and outlined in Fig. 7. The positive correlation between external pressure and thermal stability[16] also supports the proposed nitrogen escape mechanism.

The photo stability of $N@C_{60}$ has also been investigated by a number of groups with the exact decay mechanism(s) still yet to be determined. It has been shown that pulsing a low-purity $N@C_{60}$ and $N@C_{70}$ toluene solution

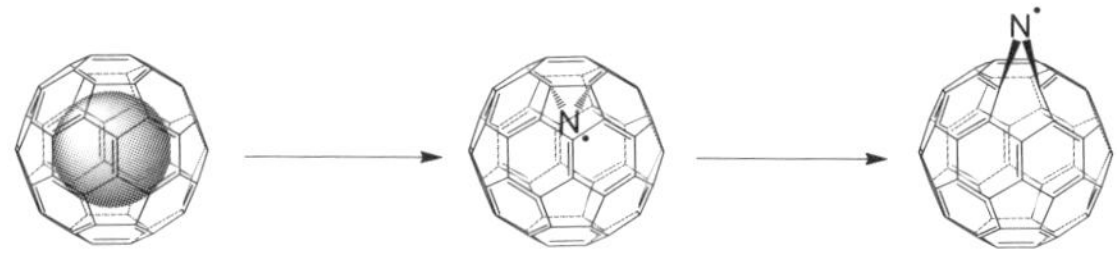

Fig. 7 Weidinger and co-workers proposed mechanism for the escape of nitrogen from the fullerene cage. Initially an internal azabridge is formed, which then inverts and bonds exohedrally to the cage.[17]

with a 355 nm Nd:YAG laser results in the deterioration of the ESR signal.[18] The authors used the observed relative decay rates of N@C_{60} and N@C_{70} as evidence that the decay was caused by vibrational excitation initiated by non-radiative internal conversion of the exited triplet molecular species to the electronic ground state. This mechanism would also explain the observation that when oxygen (known to efficiently quench the triplet exited state of fullerenes) is present in the solvent of the endohedral fullerene solution the photo stability is significantly enhanced.[19] The photo stability of N@C_{60} when irradiated with 254 nm radiation also depends on the solvent used.[20] A larger decay rate of N@C_{60} has been found in aromatic solvents such as benzene or toluene compared to non-aromatic solvents such as carbon disulphide or decalin. This has been rationalized by a combination of aromatic pi interaction of the solvent molecules with the fullerene cage and the formation of radical adduct with the fullerene cages.

5. CHEMICAL DERIVATIVES OF N@C_{60} AND N@C_{70}

5.1. Comparison with C_{60}/C_{70}

Due to the isolation of the fullerene and nitrogen electronic orbitals in N@C_{60} and N@C_{70}, the chemical reactivity of both are ostensibly identical to that of C_{60} and C_{70}. The principal consideration that must be made when attempting chemical functionalization of nitrogen-doped fullerenes is their reduced thermal and photostability compared to the empty fullerene analogues. This often means that modification of the synthetic protocols used for empty fullerenes is required to minimize the exposure of N@C_{60} to elevated temperatures. Additionally the restricted availability of highly enriched nitrogen-doped endohedral fullerenes makes the reaction scale that can be used of the order of a few tens of micrograms rather than the multi-gram scale that is possible for empty fullerenes. This makes purification and characterization non trivial. To date, relativity few examples of chemical derivatives of endohedral nitrogen fullerenes exist. The majority of the work has so far focused on N@C_{60} rather than N@C_{70} presumably due to the added complication of isolating isomerically pure N@C_{70} derivatives caused by the non-equivalent reaction sites of the molecule.

5.2. Exo Ring-Adducts

5.2.1. *Cyclopropane Exohedral Adducts*

The first example of successful functionalization of N@C_{60} was published by Hirsch and co-workers in 1997 using a Bingel type nucleophilic cyclopropanation reaction (see Fig. 8d).[3] Interestingly, the reactivity of N@C_{60} is reported to be identical to C_{60} along with no measureable spin loss after the functionalization reaction. As mentioned earlier, the ESR spectra of the mono-adduct shows the presence of extra hyperfine coupling features due to the lowering of the cage symmetry. Hirsch and co-workers went on to synthesize the hexa-adduct analogue of the above N@C_{60} derivative (see Fig. 8c) using an analogous protocol.[21] An ESR study of the hexa-adduct shows that increasing the cage symmetry compared to the mono adduct raises the T_1 and T_2 relaxation times by a factor of two,[21] illustrating

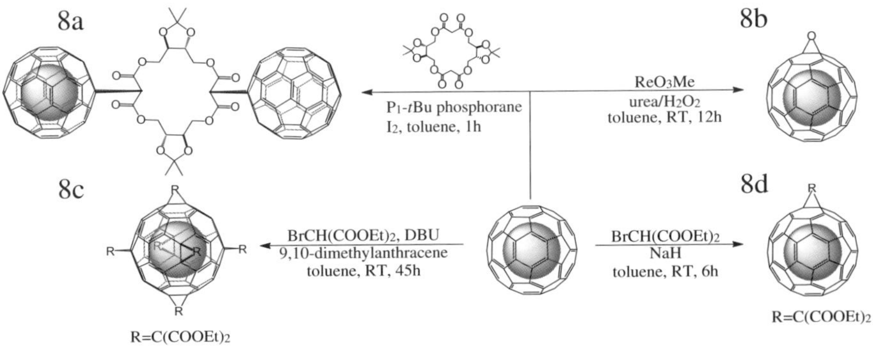

Fig. 8 Examples of reaction schemes used to synthesize cyclopropane derivatized N@C_{60} molecules.

a potential benefit of using a multiply substituted N@C_{60} derivative to create N@C_{60} dimers for use in quantum computation systems.

A bis-cyclopropane adduct has also been used to form a semi-ridged N@C_{60}-C_{60} dimer molecule using hindered phosphazene base *tert*-butyliminotris(dimethylamino)phosphorane (P1-*t*Bu) and iodine to reduce both the formation of fullerene salts and the reaction time[22] (see Fig. 8a). Despite using these optimized conditions, the N@C_{60} purity was reduced by 50% after the reaction. This result combined with the rapid decomposition of N@C_{60}O **8b** seen at temperatures sufficient for spontaneous opening of the epoxide ring to form a biradical suggest an instability of N@C_{60} to radical formation on the fullerene cage.[23]

5.2.2. *Pyrrolidine Exohedral Adducts*

The most widely used method to functionalize N@C_{60} is using a Prato 1,3-dipolar cycloaddition to create an exohedral 5 membered pyrrolidine ring adduct.[24] This method has gained popularity for two reasons. The reaction is a "one-pot" reaction and is compatible with a wide variety of aldehyde and amino acid precursors allowing specific functionality to be introduced easily. Additionally the reaction has been proven to conserve a relatively high proportion of N@C_{60} spin centers, ranging between 70 and 80%, in the functionalized product with good long-term stability when kept in the dark and at room temperature. Figure 9 gives an overview of the type of N@C_{60} pyrrolidine derivatives that have been synthesized to date, most of which use the readily available methyl-glycine amino acid as a reaction precursor. Recently however, the use of a N-(4-(hexyloxy)benzyl)glycine amino acid derivative has been shown to enhance the pyrrolidine formation reaction rate allowing a 45% yield of **9f** using a 5 minute reaction time.[25]

Using this amino acid derivative multi paramagnetic center molecules have been synthesized to probe the effect of two center electron dipole-dipole interactions on the ESR spectrum of N@C_{60} (see Fig. 10). Molecule **10a** is based on the coupling between an N@C_{60} molecule and a Cu^{2+} ion in a porphyrin.[26] Due to a combination of the strength of the coupling and the fast relaxation of the Cu^{2+} center, the presence of the Cu^{2+} ion was found to induce significant relaxation broadening of the N@C_{60} transition, acting as a chemical switch that was reversible on demetallation of the

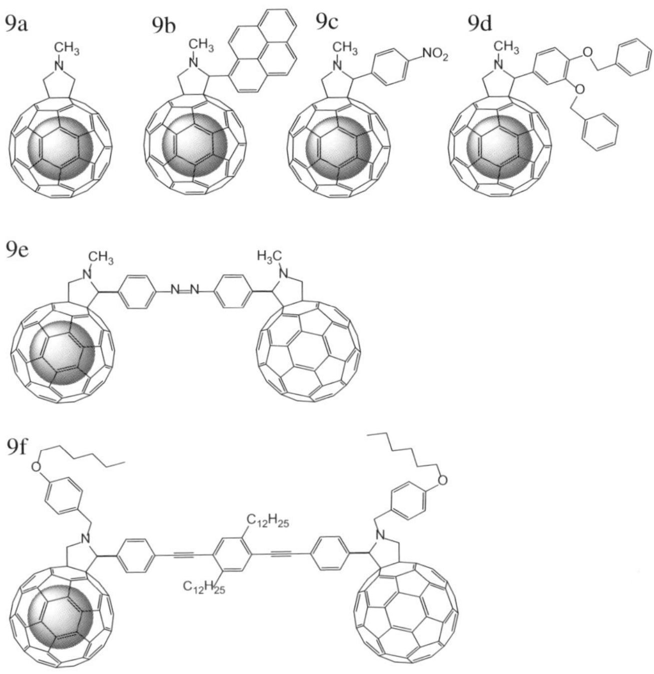

Fig. 9 Examples of the variety of N@C_{60} pyrrolidine derivatives that have been synthesized using a Prato 1,3-dipolar cycloaddition.

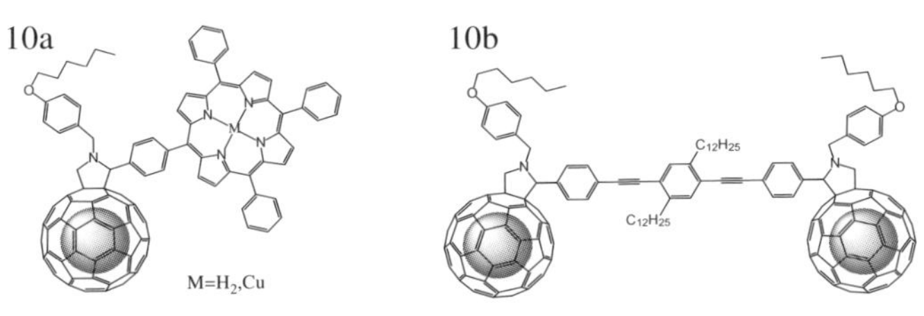

Fig. 10 Chemical structures of two spin center molecules containing N@C_{60}.

porphyrin moiety. By optimizing the synthesis and purification protocols to be compatible with the micro gram scale that highly enriched N@C_{60} can currently be produced in, molecule **10b** was synthesized in sufficient abundance to be characterized using CW ESR.[25] The ESR spectrum showed additional electron dipole–dipole transitions caused by the presence of the second electron spin center. This result shows that the chemistry of $N@C_{60}$ is now sufficiently mature to permit the functional modification of pure $N@C_{60}$ without the need to be diluted with empty C_{60} to provide a sufficient molar scale for a reaction.

5.2.3. *ESR Studies*

Solution-state CW ESR spectra of functionalized $N@C_{60}$ look almost identical to pristine $N@C_{60}$ due to the rapid tumbling of the molecule on the time-scale of the measurement. However given a suitably high resolution spectrometer, it is possible to observe a slight broadening of the hyperfine lines caused by electron-nuclear coupling to the protons attached to the adduct.[21] As in the case of $N@C_{70}$, when the motion of the fullerenes is restricted by either measuring the CW ESR in the solid state or a frozen solution, the effect of the distortion of the fullerene cage can be seen by the appearance of satellite peaks adjacent to each of the hyperfine lines caused by the presence of additional transitions due to ZFS (see Fig. 11). The size of the axial ZFS component (D) has been shown to lie within the range of 6–17 MHz with the rhombohedral distortion (E) much smaller at between 0.4 and 0.9 Mhz.

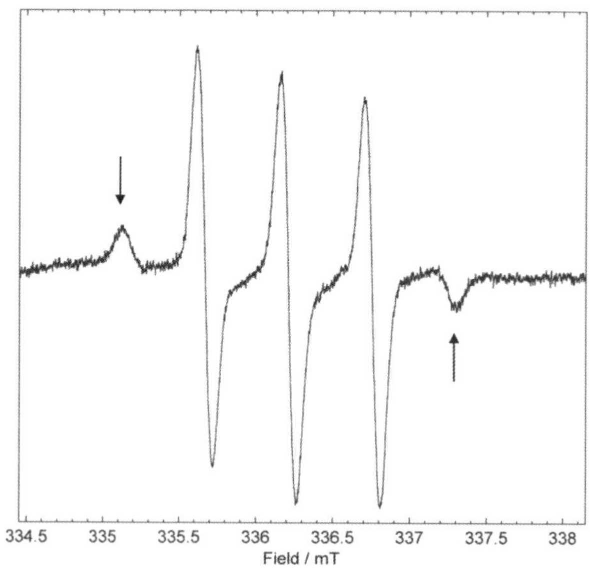

Fig. 11 Frozen-solution CW ESR spectrum recorded from a $^{14}N@C_{60}$ pyrrolidine adduct. The additional peaks due to zero field spitting (ZFS) are clearly visible (indicated by arrows).

Table 1 Electron Spin Environment of Chemically Functionalized $N@C_{60}$ Derivatives

Molecule	Exohedral ring size	A_{iso}/MHz	D/MHz	E/Mhz
$^{14}N@C_{60}$[1]		15.73	—	—
$^{15}N@C_{60}$		22.35[28]		
$^{14}N@C_{70}$		15.00[21]	5.5[14]	—
8a[22]	3	15.73	8.6	0.45
8b[23]	3	15.45	6.6	0.5
8d[21]	3	—	6.1	0.45
$N@C_{60}$–C_{60}[29]	4	15.55	14	0.56
9a[27]	5	15.95	16.4	0.4
9b[19]	5	$A_{yy/xx}$ = 16.44, A_{zz} = 14.7	16.75	0.93
9c[30]	5	$A_{yy/xx}$ = 14.2, A_{zz} = 18.5	17	0.8
9d[27]	5	15.95	17.2	0.6
9e[31]	5	$A_{yy/xx}$ = 14.6, A_{zz} = 18.6	16.1	0.7
9f[25]	5	15.736	14.19	0.74
10a[26]	5	—	16.3	0.4

A clear correlation between the size of ring formed between $N@C_{60}$ and different adducts on the ESR ZFS magnitude is shown in Table 1. The trend shows an increase in ZFS going from 3-4-5 membered rings. This observation was rationalized by considering the overlap of the exohedral bonds with the radial *p* orbital of the fullerene.[27] As the overlap increases the proportion of fullerene electron density shifted to the substituent also increases, leading to a greater perturbation of the environment of the nitrogen atom and so a larger ZFS parameter.

5.2.4. *Stability*

The thermal stability of $N@C_{60}$ adducts has been shown to vary for different derivatives. The cyclopropane linked adducts of the form of **8d** have a

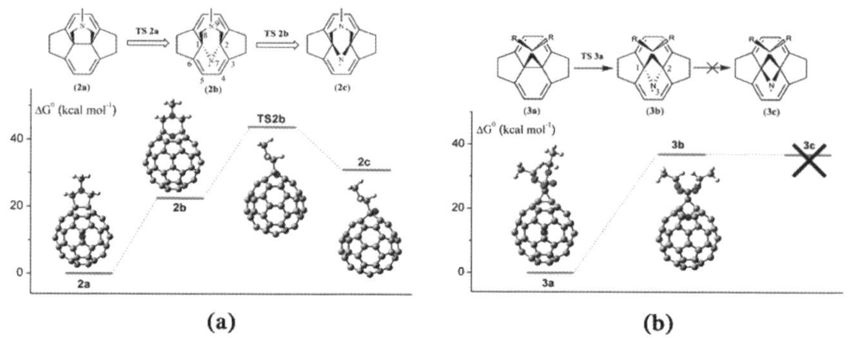

Fig. 12 DFT simulated reaction coordinates for the escape of nitrogen from the fullerene cage when attached to either a) a pyrrolidine ring or b) a cyclopropane ring.[33]

similar stability compared to pristine N@C_{60}, decomposing at around 500K[32] with the stability then increasing when multiple adducts are present as in **8c** to provide stability up to 570 K.[32] Pyrrolidine derivatives such as **9a–d** however decompose rapidly above 410 K.[27,30] DFT simulations of the reaction coordinates for the escape of nitrogen from both ring sizes (see Fig. 12) show that the energy for formation of an adduct with the inside of the cage and subsequent flip from endo-nitrogen to exo-nitrogen has a significantly lower activation barrier for the pyrrolidine derivative compared to a cyclopropane analogue owing to the difference in ring size formed in the transition state.

The photo stability of N@C_{60} derivatives shows similar behavior to their thermal stability. Pyrrolidine groups accelerate the photo decomposition by more than an order of magnitude compared to pristine N@C_{60}, while cyclopropane derivatives show comparable stability to N@C_{60}.[19]

6. POTENTIAL APPLICATIONS

6.1. Quantum Information Processing

Classical information theory governs the limits of the computational speed, or more precisely the number of steps that are required to perform any given calculation. These laws are based on classical mechanics and are used by every commercial computer chip that is currently in existence. If however a computer architecture that can take advantage of quantum mechanical phenomena such as superposition and entanglement is built, a different set of laws governed by quantum information theory would be applicable. This fundamentally different method of carrying out calculations offers significant computational speed ups for certain tasks such as factorizing very large numbers.[34]

The long electron spin T_1 and T_2 coherence times achievable in N@C_{60} make it an attractive system to use in a quantum computer architecture for the following reasons. Information can be stored and read from an ensemble of electron spins using short microwave pulses of the order of a few tens of ns long for each operation. Consequently the 250 μs phase coherence time achievable in N@C_{60} would permit over 10,000 sequential operations to be performed before electron spin decoherence becomes overwhelming. It has also been shown, using N@C_{60}, that by employing composite microwave pulses, large systematic errors in single qubit operations can be corrected for to retain a fidelity of between 0.999 and 0.9999.[35] N@C_{60} therefore meets all of the basic criteria for fault-tolerant quantum computation and consequently has been proposed as the quantum storage element of a number of quantum information processing architectures.[36–37] The nuclear spin of N@C_{60} has also been used to store quantum information, enhancing the storage time that is possible when purely using the electron spin, via a technique known as *bang-bang* decoupling.[38] To further explore the potential of N@C_{60} as a quantum computing element, the creation of controlled entanglement between electron spins in an array of at least two magnetically distinct N@C_{60} centers with fixed electron–electron coupling is required.[39]

6.1.1. *2 Qubit Assemblies*

The simplest array consists of two qubits, and this is already sufficiently complex to test the implementation of key quantum logic elements such as to create a universal CNOT gate. It has already been mentioned that 98% of the spin density of the unpaired electrons resides inside the fullerene cage, hence the most significant coupling interaction occurs via electron–electron dipolar coupling rather than exchange coupling processes.[25] Since the magnitude of dipolar coupling is both distance and angle dependent with respect to the applied magnetic field,[40] both of these parameters must be accurately controlled to fix the electron coupling interaction energy. Control of the separation between two N@C_{60} centers has now been realized by chemically attaching two N@C_{60} molecules to a molecular wire **3b**.[25] However macroscopic alignment has only been achieved so far on isolated N@C_{60} and N@C_{70} molecules using liquid crystal matrices.[41] The challenge now centers on exerting macroscopic alignment on arrays of N@C_{60} or N@C_{70}.

6.1.2. *1D and 2D arrays*

Fullerene molecules are known to self-assemble into 1-dimensional ordered arrays inside single-walled carbon nanotubes (SWNTs) using high-temperature conditions. By applying a low-temperature method based on refluxing n-hexane it has been shown to be possible to encapsulate N@C_{60} within SWCNTs.[42]

Two-dimensional structures can be formed on surfaces by exploiting non-covalent interactions between the constituent molecules. Self-assembled molecular networks have been shown to form porous structures that can act as hosts for fullerenes or related molecules, with the spatial arrangement of the guest fullerene molecules principally controlled by the size and shape of the voids formed by the network.[43–44] To date, these methods have not yet been transferred to work with endohedral nitrogen fullerenes, however this should be achievable using low-temperature solution phase deposition techniques. Once a successful architecture has been found, the last challenge is to find a suitably scalable method to address individual qubits, for which a number of approaches have been suggested.[37,45–46] These include using local magnetic field gates beneath each fullerene to control the magnetic environment of N@C_{60} element in a two-dimensional solid state array. The use of very large applied field gradients to provide spatial selectivity to a solid state N@C_{60} thin film has been trialed, similar to an MRI scanner. However to achieve the required 1 nm spatial resolution significantly larger field gradients will be required than are currently achievable. Possibly the easiest way to implement this approach is based on utilizing the difference in transitional frequencies found in ^{14}N@C_{60}, ^{15}N@C_{60} and ^{31}P@C_{60} to create an hetero-endohedral fullerene array of individually addressable qubits.

7. SUMMARY

This chapter explores the physics and chemistry of the exotic nitrogen atom endohedral fullerenes $^{14}N@C_{60}$, $^{15}N@C_{60}$, $^{14}N@C_{70}$ and $^{15}N@C_{70}$. We outline the different method of synthesis using nitrogen implantation, the most widely used of which uses a commercial ion source, yielding up to 100 ppm purity of nitrogen-doped fullerenes in empty fullerenes. The purification and isolation of the pure endohedral fullerenes using HPLC chromatography are outlined.

The highly symmetry environment of the paramagnetic center results in some very unusual electron spin properties, including the ultra-narrow transitional line widths < 0.3 μT, the presence of a pure $^4S_{3/2}$ electron spin ground state and the longest electron T_2 relaxation lifetime achievable in a molecular system of 250 μs .

The main schemes used to chemically modify $N@C_{60}$ are Prato cycloaddition and Bingel cycloaddition. The effect that chemical modification has on the physical and ESR properties of these molecules can be seen by the introduction of a non-zero ZFS term in the electron spin Hamiltonian and is observable as additional features in the solid state ESR spectra.

One of the most promising applications for $N@C_{60}$ and $N@C_{70}$ is as quantum information elements (qubit) in a future quantum computer. We summarize the current state of work in this area and discuss the remaining challenges that need to be overcome to realize a $N@C_{60}$ electron spin based quantum computer.

REFERENCES

1. Almeida Murphy T, Pawlik T, Weidinger A, *et al.* (1996) Observation of Atomlike Nitrogen in Nitrogen-Implanted Solid C_{60}. *Phys Rev Lett* **77:** 1075.
2. Weidinger A, Waiblinger M, Pietzak B, *et al.* (1998) Atomic nitrogen in C_{60}: $N@C_{60}$. *Appl Phys A: Mater Sci Process* **66:** 287–292.
3. Pietzak B, Waiblinger M, Murphy TA, *et al.* (1997) Buckminsterfullerene C_{60}: a chemical Faraday cage for atomic nitrogen. *Chem Phys Lett* **279:** 259–263.
4. Ito S, Shimotani H, Takagi H, *et al.* (2008) On the synthesis conditions of N and N_2 endohedral fullerenes. *Fullerenes, Nanotubes, Carbon Nanostruct* **16:** 206–213.
5. Ata M, Huang H, Akasaka T. (2004) Nitrogen radio-frequency plasma processing of fullerenes. *J Phys Chem B* **108:** 4640–4646.

6. Takahashi N, Dock H, Matsuzawa N, *et al.* (1993) Plasma-polymerized C_{60}/C_{70} mixture films: Electric conductivity and structure. *J Appl Phys* **74:** 5790–5798.
7. Abe S, Sato G, Kaneko T, *et al.* (2006) Effects of ion energy control on production of nitrogen-C_{60} compounds by ion implantation. *Jpn J Appl Phys, Part 1* **45:** 8340–8343.
8. Kaneko T, Abe S, Ishida H, *et al.* (2007) An electron cyclotron resonance plasma configuration for increasing the efficiency in the yield of nitrogen endohedral fullerenes. *Phys Plasmas* **14:** 1107051–1107053.
9. Suetsuna T, Dragoe N, Harneit W, *et al.* (2003) Separation of N_2@C_{60} and N@C_{60}. *Chem Eur J* **9:** 598.
10. Kanai M, Porfyrakis K, Briggs GAD, *et al.* (2004) Purification by HPLC and the UV/Vis absorption spectra of the nitrogen-containing incar-fullerenes N@C_{60}, and N@C_{70}. *Chem Commun* 210–211.
11. Jakes P, Dinse KP, Meyer C, *et al.* (2003) Purification and optical spectroscopy of N@C_{60}. *Phys Chem Chem Phys* **5:** 4080–4083.
12. Nikawa H, Araki Y, Slanina Z, *et al.* (2010) The effect of atomic nitrogen on the C_{60} cage. *Chem Commun* **46:** 631–633.
13. Weidinger A, Waiblinger M, Pietzak B, *et al.* (1998) Atomic nitrogen in C_{60}:N@C_{60}. *Appl Phys A* **66:** 287–292.
14. Morton JJL, Tyryshkin AM, Ardavan A, *et al.* (2006) Electron spin relaxation of N@C_{60} in CS_2. *J Chem Phys* **124:** 014501–014508.
15. Waiblinger M, Pietzak B, Lips K, *et al.* (1998) Thermal stability of N@C_{60}. *AIP Conf Proc* **442:** 388–391.
16. Iwasiewicz-Wabnig A, Porfyrakis K, Briggs GAD, *et al.* (2009) Investigations of N@C_{60} and N@C_{70} stability under high pressure and high temperature conditions. *Phys Status Solidi B* **246:** 2767–2770.
17. Waiblinger M, Lips K, Harneit W, *et al.* (2001) Thermal stability of the endohedral fullerenes N@C_{60}, N@C_{70}, and P@C_{60}. *Phys Rev B* **63:** 045421–045425.
18. Goedde B, Weiden N, Dinse K-P. (1999) Light-induced release of nitrogen from fullerene cages. *AIP Conf Proc* **486:** 144–147.
19. Liu G, Khlobystov AN, Ardavan A, *et al.* (2011) Photochemical stability of N@C_{60} and its pyrrolidine derivatives. *Chem Phys Lett* **508:** 187–190.
20. Farrington BJ, Hingston TJ, Briggs GA, *et al.* (2011) Photostability of N@C_{60} in Common Solvents. *ECS Transactions* **35:** 113–117.
21. Dietel E, Hirsch A, Pietzak B, *et al.* (1999) Atomic Nitrogen Encapsulated in Fullerenes: Effects of Cage Variations. *J Am Chem Soc* **121:** 2432–2437.
22. Hörmann F, Hirsch A, Porfyrakis K, *et al.* (2011) Synthesis and Magnetic Properties of a Nitrogen-Containing Fullerene Dimer. *Eur J Org Chem* **2011:** 117–121.

23. Jones MAG, Britz DA, Morton JJL, *et al.* (2006) Synthesis and reactivity of $N@C_{60}O$. *Phys Chem Chem Phys* **8:** 2083–2088.
24. Maggini M, Scorrano G, Prato M. (1993) Addition Of Azomethine Ylides To C_{60}: Synthesis, Characterization, And Functionalization Of Fullerene Pyrrolidines. *J Am Chem Soc* **115:** 9798–9799.
25. Farrington BJ, Jevric M, Rance GA, *et al.* (2012) Chemistry at the Nanoscale: Synthesis of an $N@C_{60}$–$N@C_{60}$ Endohedral Fullerene Dimer. *Angew Chem Int Ed* **51:** 3587–3590.
26. Liu G, Khlobystov AN, Charalambidis G, *et al.* (2012) $N@C_{60}$–Porphyrin: A Dyad of Two Radical Centers. *J Am Chem Soc* **134:** 1938–1941.
27. Franco L, Ceola S, Corvaja C, *et al.* (2006) Synthesis and magnetic properties of $N@C_{60}$ derivatives. *Chem Phys Lett* **422:** 100–105.
28. Pietzak B, Waiblinger M, Murphy TA, *et al.* (1998) Properties of endohedral $N@C_{60}$. *Carbon* **36:** 613–615.
29. Goedde B, Waiblinger M, Jakes P, *et al.* (2001) Nitrogen doped C_{60} dimers ($N@C_{60}$-C_{60}). *Chem Phys Lett* **334:** 12–17.
30. Zhang J, Morton JJL, Sambrook MR, *et al.* (2006) The effects of a pyrrolidine functional group on the magnetic properties of $N@C_{60}$. *Chem Phys Lett* **432:** 523–527.
31. Zhang J, Porfyrakis K, Morton JJL, *et al.* (2008) Photoisomerization of a Fullerene Dimer. *J Phys Chem C* **112:** 2802–2804.
32. Waiblinger M, Lips K, Harneit W, *et al.* (2001) Thermal stability of the endohedral fullerenes $N@C_{60}$, $N@C_{70}$, and $P@C_{60}$. *Phys Rev B* **63:** 045421.
33. Liu G, Khlobystov AN, Briggs GAD, *et al.* (2012) Catalytic and non-catalytic roles of pendant groups in the decomposition of $N@C_{60}$: a DFT investigation. *Chem Commun* **48:** 5148–5150.
34. Shor PW. (1994) Algorithms for quantum computation: discrete logarithms and factoring. *1994 Proceedings, 35th Annual Symposium on Foundations of Computer Science*, pp 124–134.
35. Morton JJL, Tyryshkin AM, Ardavan A, *et al.* (2005) High Fidelity Single Qubit Operations Using Pulsed Electron Paramagnetic Resonance. *Phys Rev Lett* **95:** 200501.
36. Benjamin SC, Browne DE, Fitzsimons J, *et al.* (2006) Brokered graph-state quantum computation. *New J Phys* **8:** 9.
37. Harneit W. (2002) Fullerene-based electron-spin quantum computer. *Phys Rev A* **65:** 032322.
38. Morton JJL, Tyryshkin AM, Ardavan A, *et al.* (2006) Bang-bang control of fullerene qubits using ultrafast phase gates. *Nat Phys* **2:** 40–43.
39. Ju C, Suter D, Du J. (2007) Two-qubit gates between noninteracting qubits in endohedral-fullerene-based quantum computation. *Phys Rev A* **75:** 012318.

40. Levitt MH. (2008) *Spin Dynamics: Basics of Nuclear Magnetic Resonance.* Wiley
41. Meyer C, Harneit W, Lips K, *et al.* (2002) Alignment of the endohedral fullerenes N@C_{60} and N@C_{70} in a liquid-crystal matrix. *Phys Rev A* **65:** 0612011–0612014.
42. Simon F, Kuzmany H, Rauf H, *et al.* (2004) Low temperature fullerene encapsulation in single wall carbon nanotubes: synthesis of N@C_{60}@SWCNT. *Chem Phys Lett* **383:** 362–367.
43. Silly F, Shaw AQ, Porfyrakis K, *et al.* (2007) Pairs and heptamers of C_{70} molecules ordered via PTCDI-melamine supramolecular networks. *Appl Phys Lett* **91:** 253109.
44. Theobald JA, Oxtoby NS, Phillips MA, *et al.* (2003) Controlling molecular deposition and layer structure with supramolecular surface assemblies. *Nature* **424:** 1029–1031.
45. Simon CB, *et al.* (2006) Towards a fullerene-based quantum computer. *J Phys: Condens Matter* **18:** S867-S883.
46. Suhovoy E, Blank A. (2008) High-resolution ESR imaging of N@C_{60} radicals on a surface. *Isr J Chem* **48:** 45–51.

CHAPTER 9

Noble-Gas Fullerenes

*R. James Cross, Jr.**

Noble-gas fullerene compounds have no chemical bond between the gas atom and the carbon atoms of the cage, yet they are remarkably stable, since several C-C bonds must be broken to let the gas atom escape. Because of this lack of bonding, their chemistry is nearly identical to that of empty fullerenes, so that the noble gas atom can be used as a tracer to follow fullerene reactions, using NMR and mass spectrometry to detect the noble gas atom. These compounds have been found in nature in sites related to meteor impacts.

1. INTRODUCTION

The noble-gas fullerene compounds are unusual in that they are van der Waals molecules, since there is no chemical bond between the noble-gas atom and the carbon atoms of the cage. Quantum calculations, X-ray studies[1] and NMR measurements all indicate that the noble gas atom is located at the center of the carbon cage, so that, in C_{60}, there are 60 simultaneous van der Waals bonds to the cage. Depending on the noble gas, this can result in a substantial binding energy. Most van der Waals molecules

*Department of Chemistry, Yale University P.O. Box 208107, New Haven, CT 06529-8107, USA E-mail: james.cross@yale.edu

are so unstable that they last for only a few molecular collisions. In the fullerene case, several carbon-carbon bonds must be broken to allow the gas atom to escape, and so these molecules are quite stable. Indeed, in one experiment, He@C_{60} was sublimed into an argon stream at 650°C, passed through an oven heated to 1100°C, and then condensed on the other side.[2] It passed through undecomposed, without loss of the helium atom. This test shows that noble-gas fullerenes are more stable than almost all other organic species.

Unlike the metallofulerenes, there is no charge transfer with the carbon cage. Thus we would expect that the chemistry of the noble-gas fullerenes would be very similar to that of the unlabeled fullerene. While this might seem boring, it means that one can study the chemistry of ordinary fullerenes by using the endohedral noble gas atom as an inert label. Below we describe experiments using ^{3}He NMR to study fullerene chemistry. Each product gives a separate, narrow peak in the NMR spectrum. These experiments would be largely meaningless if the chemistry of ^{3}He@C_{60} differed significantly from that of empty C_{60}.

Noble-gas fullerene compounds are found in nature at sites related to meteor impacts. The implication is that the fullerenes were formed in some sort of astrochemical process with the noble gases trapped inside.

2. MAKING NOBLE-GAS FULLERENES

The first noble-gas fullerenes were made by Helmut Schwartz[3,4] by crossing beams of C_{60}^+ with ^{3}He and ^{4}He. Neutralization and reionization showed that the helium was inside the C_{60} cage. No macroscopic quantities could be recovered.

A method for bulk synthesis was developed at Yale in a collaboration between Prof. Martin Saunders, the present author, and their groups of students and postdocs. It started with the simple observation that fullerenes are made by the Krätchmer–Huffman process in an electric arc in an atmosphere of helium gas.[5,6] Since there is an empty volume inside the fullerene molecules, it seems reasonable that some of the time a helium atom must be trapped there. Using crude van der Waals radii and the ideal gas law, one can estimate that one fullerene in a million will contain a helium atom. But how does one detect this? Clearly ordinary mass spectroscopy will not work. The main peak for C_{60} is at m/e = 720, but there are peaks at 721, 722, … due to varying numbers of ^{13}C's in the molecule. The signal at 724 due to four ^{13}C's is orders of magnitude larger than that due to $^4He@C_{60}$. A solution was suggested by Prof. Karl Turekian of Yale's Geology department. He mentioned that a former student of his, Prof. Robert Poreda at the University of Rochester, had a sensitive mass spectrometer for the detection of very low-level amounts of noble gases in minerals. In this instrument, the sample is thoroughly outgassed in an oven. A valve between the oven and the rest of the vacuum system is closed and the oven is heated at a constant temperature for a period of time. Then, a gate valve between the mass spectrometer and the pumps is closed, and the valve to the oven is opened. The mass spectrometer then sees a stepwise increase in the signal for the desired mass. The gate valve is then opened and the valve to the oven closed. The process is repeated at a succession of higher temperatures. By using an unpumped mass spectrometer, the sensitivity is greatly increased over the conventional method of flowing a stream of gas past an ionizing filament into the pump. By measuring the helium released from the fullerene, the problem of competing mass peaks due to ^{13}C is avoided. By using high mass resolution and gettering pumps, which don't pump noble gases, he can achieve backgrounds of 10^4 atoms of ^3He. Poreda reported that a sample of C_{60} contained one part in 850,00 ^4He and one part in 10^{12} of ^3He.[7] This isotope ratio is expected from tank helium produced from natural gas wells. The helium was released at temperatures around 650°C.

If helium gets out at 650°C, perhaps it will go in at 650°C. A sample of C_{60} was heated to 650°C in a fused silica tube along with ~3 atm of ^3He. After a few hours, the sample was cooled, dissolved, filtered, and dried. Prof. Poreda found that one C_{60} in 10^6 contained ^3He — an increase of six orders of magnitude over the natural incorporation. This was quickly followed by experiments showing similar incorporation of Ne, Ar, Kr, and Xe.[8] We then devised a method to greatly increase the pressure of noble gas. The fused silica ampoule was replaced by a tube made of OFHC (oxygen-free high conductivity) copper. This is cleaned and annealed. One end is crimped to form a high-vacuum seal. The fullerene sample is put in, and the tube is attached to a vacuum line. The air is pumped out and the noble gas is added, while the tube is cooled with liquid nitrogen. Finally, the top of the tube is crimped. At room temperature the pressure of noble gas is about 3 atm. Two of these tubes are put into a high-pressure vessel, which is then partially filled with water and sealed with a metal o-ring. The high-pressure vessel is then heated to 650°C for about 8 hrs. The water cannot boil, as there is nowhere for the steam to go, so the pressure inside the vessel rises to 3500 atm. This flattens the copper tubes, compressing the gas inside. After the run, the tubes are broken open, and the contents are dissolved in CS_2. About 15% is insoluble and removed by filtration. The soluble portion is then dried under vacuum. Analysis by conventional mass spectrometry shows incorporation at about 0.1% for He, Ne, Ar, and Kr and at about 0.03% for Xe. Since the samples are dissolved and then recrystallized, any noble gas on the outside of the fullerenes will be lost. The noble gas atoms must therefore be on the inside of the fullerene.

The mechanism for this reaction is problematic. Patchkovskii and Thiel did a series of *ab initio* calculations.[9] They showed that it required more than 200 kcal/mol to force a helium atom through one of the six-membered rings in C_{60}. This is too much energy for a reaction occurring at 650°C. Furthermore, the energy would be considerably higher for the heavier noble gases. They then tried mechanisms where one or more C-C bonds were broken to make larger rings. While this decreased the energy required to get a helium atom into the C_{60} cage, when the energy required to break the bond(s) was added to the energy required to push the helium atom through the ring, the total remained at ~200 kcal/mol. These calculations agreed with some perplexing experimental observations. In the incorporation reaction most of the C_{60} and most of the noble gas is not used up in the reaction. If the reaction is allowed to continue for much longer than 8 hrs, very little additional $X@C_{60}$ is formed. However, if the product from one reaction is purified and used as the starting material for a second run, the yield is doubled. This can be done at least six times to get an incorporation of 0.6%. It is obvious that the equilibrium yield is very much higher than this. Indeed, calculations give the equilibrium yield as ~10% under the experimental conditions of the reaction.[10,11] It is evident that some sort of catalyst or, more properly, promoter is present in the initial sample and that this is used up during the reaction. When the C_{60} was sublimed and then handled in a nitrogen-filled glove bag, the yield of $^3He@C_{60}$ dropped by a factor of two. The situation bears some similarity to the sublimation of C_{60}. C_{60} sublimes under vacuum between 600° and 700°C, but usually only part of the sample sublimes. The rest remains as an insoluble black solid. If a sample of C_{60} is put into the bottom of a glass tube and pumped out and then immediately heated, most of the C_{60} decomposes. If the sample is warmed slowly under vacuum for many hours, then a much larger amount sublimes. If the sublimed C_{60} is heated further without exposing it to air, almost all of it resublimes. The usual explanation is that when C_{60} is crystalized from solution, molecules of the solvent get trapped in the interstitial spaces in the crystals. While C_{60} is stable at 700°, the solvent is not, and the decomposition of the solvent creates free radicals that add to and decompose the C_{60}. An alternate explanation is that C_{60} reacts slowly with oxygen in the air to form an epoxide or simply to adsorb oxygen, and this causes the decomposition. Probably both explanations are correct.

We examined many possible catalysts and eventually discovered potassium cyanide. Komatsu had discovered that grinding C_{60} and KCN in a ball mill at room temperature produced a dimer C_{120}.[12] Basically, CN^- adds to one of the 30 double bonds in C_{60} to produce an anion radical. This then adds to a double bond in a neighboring C_{60} to produce a radical anion on that molecule. This site then causes the elimination of the CN^- to produce the dimer, two C_{60} molecules bound together in a four-membered ring. We did a high-pressure run using a finely ground mixture of C_{60} and KCN. The amount of decomposed fullerene increased to 90% but the incorporation fraction increased by an order of magnitude to 1%.[13] One convenient feature of this method is that it uses only a small amount of noble gas. This makes it practical to use expensive, isotopically separated gases such as ^{3}He, ^{129}Xe, and ^{22}Ne.

Early on, the Yale group formed a collaboration with the late Prof. Chava Lifshitz at the Hebrew University in Jerusalem. She took a mass spectrum of $Ne@C_{70}$ and reported that she had evidence that some C_{70} molecules contained two neon atoms. We were skeptical. If the chance of getting one neon inside was 0.1%, the chance of getting two inside would be one in 10^6 — far too small to see. We repeated the experiment using neon enriched in ^{21}Ne, since the dominant isotope of neon is ^{20}Ne which has exactly half the mass of ^{40}Ar, a common impurity in neon. Back came a gorgeous mass spectrum showing three sets of peaks: one due to C_{60}

containing various numbers of ^{13}C, a second due to $Ne@C_{60}$ with the various carbon and neon isotopes, and a third with two neon atoms inside.[14] The ratio of C_{60}:$Ne@C_{60}$:$Ne_2@C_{60}$ is 1000:1:0.02. Evidently, only a small fraction of the C_{60} molecules open up, but then they are filled with one or two neon atoms.

The high-pressure method will eventually be superseded by the molecular surgery method[15] (see Chapter 10). Molecular surgery can give nearly 100% incorporation — far higher than the 1% achieved by the high-pressure method. However, the high-pressure method is a "one size fits all" method working for all the noble gases. In molecular surgery, the size of the hole must be carefully tailored to fit the noble gas. If the hole opened in the fullerene is too small, the atom cannot get inside. If it is too large, the atom will get in at high temperatures and then immediately leave at low temperatures before the hole can be chemically closed. This requires different chemistry for different atoms and molecules.

A second method to make $He@C_{60}$ and $Ne@C_{60}$ is to shoot a beam of He^+ or Ne^+ at solid C_{60}.[16] Helium requires at least 30 eV of kinetic energy in the beam; neon requires at least 200 eV. Basically, the ion is shot through one of the rings in C_{60}, where it is quickly neutralized by electron transfer. Stopping and neutralizing the ion releases an enormous amount of energy (tens of eV), and this must be rapidly dissipated to the surrounding solid to avoid decomposing the nascent $He@C_{60}$ molecule. Because the ions can penetrate to at most one or two molecules deep into the surface, the surface must be continuously renewed. The apparatus consists of a slowly rotating cylinder. On one side a molecular beam of C_{60} deposits a continuous layer of C_{60} on the surface. The other side is struck by a beam of ions. After several hours, the experiment is stopped, and the fullerene is dissolved off the cylinder. We could get about 0.1% incorporation. In principle, one might think that by increasing the intensity of the ion beam or decreasing the intensity of the C_{60} beam, that one could get higher incorporations. However, doing this results in much more decomposition of the C_{60}, so that no soluble portion remains. The high-pressure method gives more product with less work and is therefore preferred. The beam method, however, is used to make $N@C_{60}$, one of the most intriguing endohedral fullerenes[17] (see Chapter 8). It has also been used to make $Li@C_{60}$.[18–25]

The heavier noble-gas fullerene compounds can be separated from the empty C_{60} by using HPLC. This was first done for Ar@C_{60} by DiCanillo *et al.*[26] and then later at Yale for Kr@C_{60}[27] and Xe@C_{60}.[28] The separations are difficult, since the starting material is only 1% or less X@C_{60} and because the peak for X@C_{60} is very close to the much larger peak for empty C_{60}. Because the column is easily saturated, only a few micrograms can be injected each time. By collecting the tail end of the C_{60} peak for many injections, one can get a smaller amount of enriched product. This can then be reinjected in a second pass to get further enrichment. After three passes, the X@C_{60} is almost pure. One problem is that C_{60} slowly oxidizes in air to give the epoxide $C_{60}O$. This elutes just after the peak for X@C_{60}, so the process also enriches the epoxide. This problem can be eliminated by periodically reducing the epoxide with triphenyl phosphine to give C_{60}.

Kr@C_{60} shows small changes in the UV-vis spectrum and a slightly smaller lifetime of the lowest triplet electronic state,[27] presumable due to the presence of the heavy atom inside. ^{129}Xe@C_{60} has been used for a variety of NMR studies as described below.

3. THEORETICAL CALCULATIONS

At first, it might seem easy to do *ab initio* calculations on noble-gas fullerene compounds. They are closed-shell systems with high symmetry. Indeed, there have been many calculations.[10,11,29–33] However, the most important result is the binding energy, the energy difference between X@C_{60} and X+C_{60}. As mentioned above, the binding is due entirely to the van der Waals interaction and this requires electron correlation. It is completely absent in a Hartree–Fock calculation. Most density-functional methods either ignore correlation or treat it very approximately. This means that one must use the Møller–Plesset perturbation method or the coupled-cluster method. With the large number of electrons, such a calculation gets very expensive. As an example, consider the calculation done by Frunzi *et al.* on Xe@C_{60}.[34] He used a large basis set [6-31G(d)] and a relativistic core potential. As expected, Hartree–Foch and DFT (B3LYP) gave negative binding energies, −32 kcal/mol. These results also directly contradict the experiment, since the equilibrium yields would be far less than what was measured. MP2 gave +40 kcal/mol. The sign is correct, but the magnitude seems large. So, Frunzi put in a counterpoise correction, and this brought the binding energy down to +13 kcal/mol.

An *ab initio* calculation gives the total energy of the molecule, i.e. the energy difference between C_{60} and 60 bare carbon nuclei and 720 free electrons. This energy is huge. The binding energy is then the small difference between two large numbers. Since the errors in the calculations are much higher than the binding energy, the only hope of getting a reasonable answer is if the errors largely cancel. For that to happen, one must use the same basis set of orbitals for both the reactants and products. In the case of Xe@C_{60}, however, there are many unfilled carbon orbitals that can be used for the xenon atom, and, conversely, many unfilled xenon orbitals that can be used for the carbon cage. Thus, the basis set for Xe@C_{60} is larger than that for Xe and C_{60} taken separately. With the counterpoise correction one calculates the energy of the free xenon atom with all the basis functions of the carbon cage but without the carbon atoms, and for the carbon cage, all the xenon orbitals are included without the xenon atom. Although the final answer looks reasonable, it requires a lot of faith to believe it.

Patchkovskii and Thiel[10] did an accurate calculation on He@C_{60}, using MP2 with a large basis set and a counterpoise correction. Since helium is much smaller than xenon, there is little overlap of the basis sets for helium and the carbon cage, so their results are probably accurate. They got a binding energy of 2.0 kcal/mol. When they moved the helium away from the center of the cage, they lost the I_h symmetry, and they had to use a smaller basis set to be able to do the calculations in a reasonable amount of computer time. To a high degree of accuracy the potential was spherically symmetric and was given by

$$V(r) = [-2.001+1.905r^2 + 0.8428r^4 + 0.8098r^6]\text{kcal/mol}. \qquad (1)$$

As expected, the potential is very anharmonic. It is largely flat at the bottom of the well and rises rapidly as the helium approaches the carbon cage.

The trend going from helium to xenon is that the binding energy increases, at least through argon or krypton and then gets smaller for xenon. The larger the noble gas atom is, the stronger the van der Waals potential becomes. Xenon is sufficiently large that its outer orbitals start to overlap with the inside electron density of the carbon cage. Also, the width of the potential decreases; the larger the noble-gas atom becomes, the less it can move off center before bumping into the carbon atoms. A simple model is to treat the noble-gas-carbon interaction as a Lennard-Jones (12,6) potential.[11,29] As one goes from helium to xenon, both the well depth ε and the position of the minimum r_e increase. For helium the radius of the cage R is much larger than r_e, so that the system is way outside the minimum. For argon and krypton, $R \approx r_e$ to give a strong binding energy. For xenon, $R < r_e$, and the system has moved partially up the repulsive wall.

4. NMR MEASUREMENTS

The most important use of noble-gas fullerene compounds is as tracers to study the chemical reactions of fullerenes. Two of the many noble-gas nuclei are useful for NMR: ^{3}He and ^{129}Xe. Both have a nuclear spin of ½. ^{3}He has a high gyromagnetic ratio, which means that it can be detected easily. However, it requires a special NMR probe. One can use a proton probe for a 400 MHz instrument for ^{3}He on a 500 MHz spectrometer with a little retuning, or one can buy a specially made probe for ^{3}He. ^{129}Xe has a lower gyromagnetic ratio, and so a multinuclear probe can be used. The lower sensitivity for ^{129}Xe can be compensated for by separating the ^{129}Xe containing fullerene from the unlabeled fullerene using HPLC, as described above.

Putting a ^{3}He atom inside C_{60} shifts the NMR resonance upfield by 6.3 ppm, −6.3 ppm using the usual convention where downfield shifts are positive. A simplified explanation for this is that when C_{60} is placed in a magnetic field, the π electrons in the C_{60} form a ring current like the secondary of a transformer, partially shielding the magnetic field on the inside. This means that one must go to a higher field (lower frequency) to bring the ^{3}He nucleus into resonance. A similar effect is well known in proton NMR when a hydrogen atom is located near the face of an aromatic ring. The effect is very sensitive to the electronic structure of the fullerene. C_{70} gives an upfield shift of 28.8 ppm. Higher fullerenes generally fall between these two limits. Figure 1 shows an NMR spectrum of a mixture of higher fullerenes.

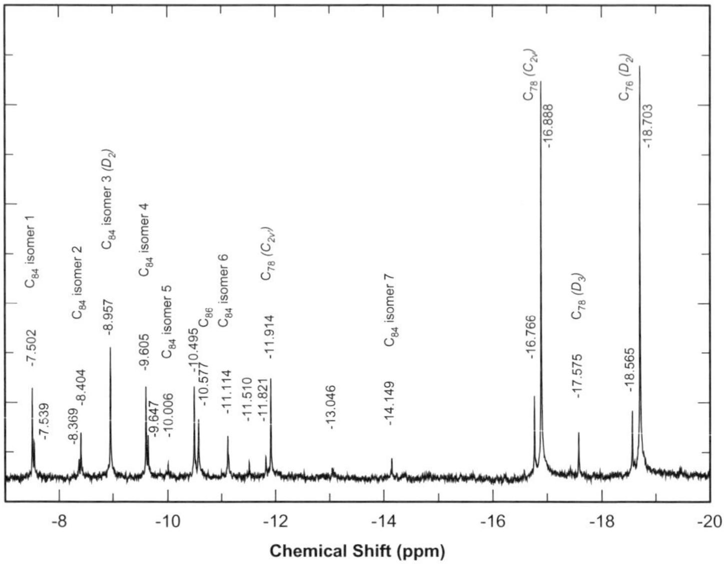

Fig. 1 ^{3}He NMR of a mixture of higher fullerenes.

Note that the lines are all narrow and well resolved. Beginning with C_{76} each fullerene generally has multiple isomers, and each isomer gives a separate peak. Quantum calculations can be used to assign the various peaks to specific isomers.[35,36] Many of the peaks show small satellite peaks nearby. These are due to the presence of two ^{3}He atoms inside. The chemical shift is an average of the shift over the positions inside the fullerene cage. With two ^{3}He atoms inside, the average is slightly different than when only one atom is inside.

When six electrons are added to C_{60} or to C_{70}, stable, closed-shell hexa-anions are formed. This results in very large changes in the chemical shifts.[37] The resonance for C_{60} moves from −6.3 ppm to −48.7 ppm. In contrast, the resonance for C_{70} moves in the opposite direction, from −28.8 ppm to +8.3 ppm. The three isomers of C_{76} also show large differences between the neutral and anionic species; the peaks move in different directions for the different isomers.[38] In the case of C_{60}, there is no satellite peak corresponding to $^3He_2@C_{60}$. Either C_{60} is too small to fit two helium atoms, or the two peaks are too close together to be resolved. The spectrum of the C_{60} hexa-anion, however, shows a very small satellite peak,[39] showing that the second explanation is the correct one.

When chemical groups are added to the outside of C_{60}, the electronic structure is altered, and the ^3He resonance shifts. Generally, adding one group across one of the 30 double bonds gives an upfield shift of 3–4 ppm; the exact amount depends on the group added. Further additions cause larger changes. The 30 double bonds in C_{60} are all equivalent, so there is only one monoadduct. However, once one group has been added, the symmetry is broken, so there are multiple isomers for the higher adducts. Each gives a separate peak in the NMR. The situation is quite different than for proton or ^{13}C NMR. There is a single resonance for the ^{13}C NMR for C_{60}. But once a group has been added, this splits into a large number of closely spaced peaks. Multiple additions further complicate the spectrum. Similarly, the proton NMR of the adducts often gives a complicated spectrum that is difficult or impossible to interpret.

As an example of the power of NMR to study fullerene reactions, consider the Diels–Alder addition of 9,10 dimethylanthacene (DMA),

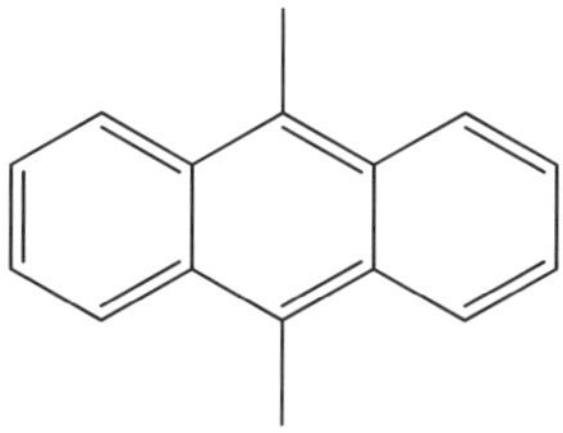

to C_{60}. Hirsch has shown that the reaction is reversible.[40] Figure 2 shows the ^3He NMR spectrum for various concentrations of DMA at 25°C.[41] The top panel shows peaks for unreacted C_{60} and for the monoadduct. There are small peaks for two bis adducts. As the concentration of DMA increases, we see the C_{60} peak decrease and higher adducts appearing. We see 6 bis adducts, 11 tris adducts, and 10 tetrakis adducts. From the peak intensities we can get the various equilibrium constants. By repeating the experiment at various temperatures we can get ΔH and ΔS for the reaction. ^3He NMR was used to study many other types of fullerene reactions.[42–55]

The NMR and reactivity of $^{129}Xe@C^{60}$ might be expected to be different from those of $^3He@C_{60}$. The helium atom is small and basically rattles around inside the carbon cage. Its electrons are tightly bound to the nucleus and don't overlap the electrons on the cage. In contrast, a xenon atom is a tight fit, and its outer 5p electrons overlap with the inside π

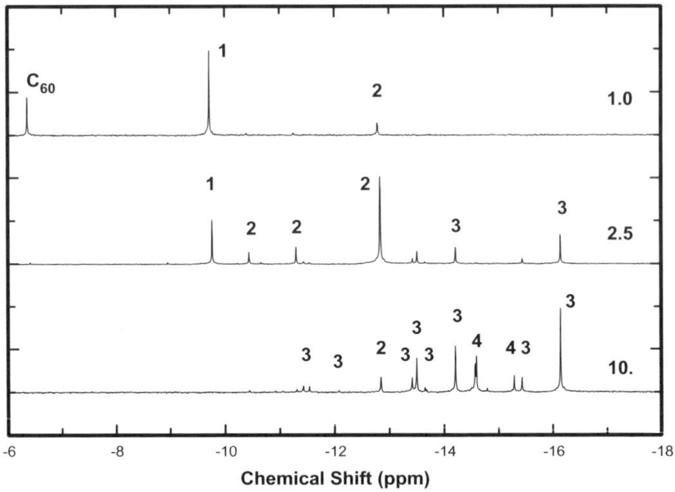

Fig. 2 ^{3}He NMR of mixtures of C_{60} and DMA, showing the various adducts. The number near each peak gives the number of DMA molecules bound to C_{60}. The number on the far right is the ratio of the concentrations of DMA to C_{60}.

electrons of the cage. Frunzi *et al.*[34] made solutions containing $^{129}Xe@C_{60}$, $^3He@C_{60}$, and varying amounts of DMA and took both the ^{3}He and ^{129}Xe NMR spectra at the same temperature. Adding one DMA molecule to $^3He@C_{60}$ caused an upfield shift of 3.53 ppm. Adding a DMA molecule to $^{129}Xe@C_{60}$ caused a downfield shift of 10.8 ppm. By measuring the ratios of peak areas, he got the ratio of the equilibrium constants for the reaction of DMA with $^{129}Xe@C_{60}$ and $^3He@C_{60}$. This ratio went from 1.4 at 45°C to 0.3 at 10°. This means that putting xenon inside changes both ΔH and ΔS for the reaction with DMA.

The molecular surgery technique (see Chapter 10) has made it possible to use $H_2@C_{60}$. Frunzi *et al.*[56] have shown that the chemical shift differences and the equilibrium constants for the addition of DMA to $H_2@C_{60}$ are identical to those for $^3He@C_{60}$. This is as expected, since H_2, like ^{3}He, is small and basically rattles around inside the cage. The NMR resonances for the endohedral protons are all upfield from TMS. Almost all the resonances for the exohedral protons are downfield from TMS, and thus one proton NMR spectrum will give both regions with no problems in assigning the spectrum. There are two identical protons in $H_2@C_{60}$ and only one ^3He in $^3He@C_{60}$. One can get almost 100% incorporation in $H_2@C_{60}$ as compared to 1% for $^3He@C_{60}$. Finally, $H_2@C_{60}$ requires only a standard proton probe, which every NMR center has. One note of caution must be added. The region upfield from TMS may not be included in many standard parameter sets. In digital, Fourier-transform NMR there is a window of frequencies. Lines outside this window are folded back into the window, an artifact of digital sampling. The effect is the same as the wheels of cars in movies, which often appear to be rotating backwards.

5. OTHER SPECTROSCOPIC STUDIES

We have discussed spectroscopy of noble-gas fullerenes in the radio region of the spectrum. Noble gas atoms don't absorb again until well into the UV region. In the IR, the presence of an endohedral atom or molecule is expected to change the IR lines of the fullerene by small amounts, typically a few wave numbers.[57] In the case of C_{60} there are 46 normal modes, four of which have T_{1u} symmetry and are IR allowed. For $X@C_{60}$ there is a fifth T_{1u} mode due to the vibration of the noble gas atom inside the cage. The frequencies, however, are only a few wave numbers, an inconvenient range for most spectrometers. As we saw previously, the force constant for helium is very low. For xenon the force constant is much larger, but the mass is higher, so the frequency remains low.

Measurements of the UV-vis spectrum of $Kr@C_{60}$ showed small changes in the absorption bands.[27] Deep in the UV, where the noble-gas atom starts to absorb, the fullerene cage absorbs very strongly, so this region is unlikely to provide any useful information. However, in the X-ray region, the fullerene cage is transparent, but the heavier noble gas can absorb and scatter. X-ray diffraction on $Kr@C_{60}$ showed that the Kr is in the center of the cage.[1] Measurement of the X-ray absorption edge shows structure relating to the fullerene cage.[58] Kilcoyne *et al.*[59] made $Xe@C_{60}$ using the molecular-beam method. This was then ionized, mass separated, and irradiated with synchrotron radiation in the range of 60–100 eV, producing $Xe@C_{60}^{3+}$. The excess ion signal over background showed oscillations due to the confinement of the xenon in the C_{60} cage.

6. GEOCHEMICAL APPLICATIONS

Noble-gas fullerene compounds are found in nature at sites related to meteor impacts. The oldest site and the first one studied is in Sudbury, Canada,[60] dated at 2.8 Gyr. Rocks were dissolved in hydrofluoric acid and extracted with toluene. The extract contained both C_{60} and C_{70}. The fullerenes were found to contain minute amounts of ^{3}He and ^{4}He.[61] However, the ^{3}He/^{4}He isotope ratio was larger than that for terrestrial and atmospheric helium and was typical of interstellar helium. This strongly implies that the fullerenes were present in the meteor and survived the impact and that the noble-gas fullerenes were formed by some as-yet-unknown astrophysical process. It also shows that the noble-gas fullerenes have lifetimes exceeding billions of years.

Noble-gas fullerenes were also found in the K-T boundary layer, associated with the extinction of the dinosaurs, and in three carbonaceous chondrite meteors,[62,63] The authors found helium inside with an extraterrestrial isotope ratio, and, in one meteor also extraterrestrial argon. Most of the helium and argon on earth is produced by radioactive decay: ^{4}He as alpha particles from the decay of heavy atoms and ^{40}Ar from the decay of ^{40}K. This accounts for the very different isotope ratios found on earth and in interstellar space. Extraterrestrial helium was also found inside fullerenes in the Permian–Triassic boundary,[64] an even larger extinction, and in sediments associated with the Younger-Dryas cooling event[63] that occurred 12,900 years ago, just as the last ice age was ending.

There is now abundant evidence from many sources that fullerenes exist as a minor component of interstellar dust.[65–68] The mechanism of formation is currently unknown, and, of course, it is not clear how noble gases get inside. One possible mechanism might be in the stellar wind of a carbon-rich, red-giant star. Toward the end of its life, such a star would emit huge amounts of carbon in its wind, as the outer layers of the star are blown off. The carbon would cool down as it expanded similar to the carbon vapor in an arc in the Krätschmer–Huffman process.[5] Any fullerene produced would then condense, along with other material into dust particles, thus shielding the fullerene from the UV light present in interstellar space. It is an interesting conjecture that fullerenes, some containing noble gases, might have been present in the dust that originally formed the earth. These would have been destroyed long ago by geological processes.

REFERENCES

1. Lee HM, Olmstead MM, Suetsuna T, Shimotani H, Dragoe N, Cross RJ, Kitazawa K, Balch AL (2002) Crystallographic characterization of Kr@C_{60} in (0.09Kr@C_{60}/0.91C_{60})··{NiII(OEP)}··2C_6H_6. *Chem. Com.*: 1352–1353.
2. Cross RJ, Saunders M (2005) Transmutation of fullerenes. *J. Am. Chem. Soc.* **127:** 3044–3047.
3. Weiske T, Hrusák J, Böhme DK, Schwarz H (1992) Endohedral Fullerene-Noble Gas Clusters Formed with High-Energy Bimolecular Reactions of C_X^{N+} (X = 60, 70 – N = 1,2,3). *Helv. Chim. Acta* **75:** 79–89.
4. Weiske T, Wong T, Krätschmer W, Terlouw JK, Schwarz H (1992) The Neutralization of He@$_{60}^{+}$ in the Gas-Phase — Compelling Evidence for the Existence of an Endohedral Structure for He@C_{60}. *Angewandte Chemie-International Edition in English* **31:** 183–185.
5. Krätschmer W, Lamb LD, Fostiropoulos K, Huffman DR (1990) Solid C_{60} — a New Form of Carbon. *Nature* **347:** 354–358.
6. Krätschmer W, Fostiropoulos K, Huffman DR (1990) The Infrared and Ultraviolet-Absorption Spectra of Laboratory-Produced Carbon Dust — Evidence For the Presence of the C_{60} Molecule. *Chem. Phys. Lett.* **170:** 167–170.
7. Saunders M, Jiménez-Vázquez HA, Cross RJ, Poreda RJ (1993) Stable compounds of Helium and Neon: He@C_{60} and Ne@C_{60}. *Science* **259:** 1428–1430.
8. Saunders M, Jiménez-Vázquez HA, Cross RJ, Mroczkowski S, Gross ML, Giblin DE, Poreda RJ (1994) Incorportion of Helium, Neon, Argon, Krypton, and Xenon into Fullerenes using High Pressure. *J. Am. Chem. Soc.* **116:** 2193–2194.
9. Patchkovskii S, Thiel W (1996) How Does Helium Get into Buckminsterfullerene? *J. Am. Chem. Soc.* **118:** 7164.
10. Patchkovskii S, Thiel W (1997) Equilibrium Yield for Helium Incorporation into Buckminsterfullerene: Quantum Chemical Evaluation. *J. Chem. Phys.* **106:** 1796.
11. Jiménez-Vázquez HA, Cross RJ (1996) Equilibrium Constants for Noble-Gas Fullerene Compounds. *J. Chem. Phys.* **104:** 5589.
12. Komatsu K, Wang GW, Murata Y, Tanaka T, Fujiwara K, Yamamoto K, Saunders M (1998) Mechanochemical Synthesis and Characterization of the Fullerene Dimer C_{120}. *J. Org. Chem.* **63:** 9358–9366.
13. Cross RJ, Khong A, Saunders M (2003) Using Cyanide to put Noble Gases inside C_{60}. *J. Org. Chem.* **68:** 8281–8283.
14. Laskin J, Peres T, Lifshitz C, Saunders M, Cross RJ, Khong A (1998) An Artificial Molecule of Ne_2 inside C_{70}. *Chem. Phys. Lett.* **285:** 7–9.

15. Komatsu K, Murata M, Murata Y (2005) Encapsulation of Molecular Hydrogen in Fullerene C_{60} by Organic Synthesis. *Science* **307:** 238–240.
16. Shimshi R, Cross RJ, Saunders M (1997) Beam implantation: A new method for preparing cage molecules containing atoms at high incorporation levels. *J. Am. Chem. Soc.* **119:** 1163–1164.
17. Murphy TA, Pawlik T, Weidinger A, Höhne M, Alcala R, Spaeth J-M (1996) Observation of Atomlike Nitrogen in Nitrogen-Implanted Solid C_{60}. *Phys. Rev. Lett.* **77:** 1075.
18. Tellgmann R, Krawez N, Lin SH, Hertel IV, Campbell EEB (1996) Endohedral fullerene production. *Nature* **382:** 407–408.
19. Campbell EEB, Tellgmann R, Krawez N, Hertel IV (1997) Production and LDMS characterisation of endohedral alkali- fullerene films. *Journal of Physics and Chemistry of Solids* **58:** 1763–1769.
20. Gromov A, Krätschmer W, Krawez N, Tellgmann R, Campbell EEB (1997) Extraction and HPLC purification of Li@$C_{60/70}$. *Chem. Com.*: 2003–2004.
21. Campbell EEB (1998) Endohedral fullerenes. *Applied Physics a-Materials Science & Processing* **66:** 241–241.
22. Campbell EEB, Ehlich R, Heusler G, Knospe O, Sprang H (1998) Capture dynamics in collisions between fullerene ions and rare gas atoms. *Chem. Phys.* **239:** 299–308.
23. Campbell EEB, Fanti M, Hertel IV, Mitzner R, Zerbetto F (1998) The hyperpolarisability of an endohedral fullerene: Li@C_{60}. *Chem. Phys. Lett.* **288:** 131–137.
24. Campbell EEB, in *Endofullerenes: A New Family of Carbon Clusters*, edited by T. Akasaka and S. Nagase (Klewer, Dordrecht, 2002), Vol. 3, p. 67–97.
25. Aoyagi S, et al. (2010) A layered ionic crystal of polar Li@C_{60} superatoms. *Nat. Chem.* **2:** 678–683.
26. DiCamillo BA, Hettich RL, Guiochon G, Compton RN, Saunders M, Jiménez-Vázquez HA, Khong A, Cross RJ (1996) Enrichment and Characterization of a Noble Gas Fullerene: Ar@C_{60}. *J. Phys. Chem.* **100:** 9197.
27. Yamamoto K, Saunders M, Khong A, Cross RJ, Grayson M, Gross ML, Benedetto AF, Weisman RB (1999) Isolation and Spectral Properties of Kr@ C_{60}, a Stable van der Waals Molecule. *J. Am. Chem. Soc.* **121:** 1591–1596.
28. Syamala MS, Cross RJ, Saunders M (2002) ^{129}Xe NMR Spectrum of Xenon Inside C_{60}. *J. Am. Chem. Soc.* **124:** 6216–6219.
29. Pang L, Brisse F (1993) Endohedral Energies and Translation of Fullerene-Noble Gas Clusters G@C_n (G = He, Ne, Ar, Kr, and Xe – N = 60 and 70). *J. Phys. Chem.* **97:** 8562–8563.
30. Cislowski J, *Electronic Structure Calculations on Fullerenes and their Derivatives* (Oxford University Press, New York, 1995).

31. Son MS, Sung YK (1995) The Atom-Atom Potential — Exohedral and Endohedral Complexation Energies of Complexes of X@C_{60} between Fullerene and Rare-Gas Atoms (X = He, Ne, Ar, Kr, and Xe). *Chem. Phys. Lett.* **245:** 113–118.
32. Darzynkiewicz RB, Scuseria GE (1997) Noble gas endohedral complexes of C_{60} buckminsterfullerene. *J. Phys. Chem. A* **101:** 7141–7144.
33. Darzynkiewicz RB, Scuseria GE (1998) Noble gas endohedral complexes of C_{60} buckminsterfullerene (vol 101, pg 7143, 1997). *J. Phys. Chem. A* **102:** 3458–3458.
34. Frunzi M, Cross RJ, Saunders M (2007) Effect of xenon on fullerene reactions. *J. Am. Chem. Soc.* **129:** 13343–13346.
35. Wang GW, Wu P (2009) Calculations and assignments of endohedral ^{3}He chemical shifts of open-cage fullerenes and higher fullerenes. *Theor Chem Acc* **123:** 375–381.
36. Stepanek P, Bour P, Straka M (2010) Assignment of the He@C_{84} isomers in experimental NMR spectra using density functional calculations. *Chem. Phys. Lett.* **500:** 54–58.
37. Shabtai E, Weitz A, Haddon RC, Hoffman RE, Rabinovitz M, Khong A, Cross RJ, Saunders M, Cheng PC, Scott LT (1998) ^{3}He NMR of He@C_{60}^{6-} and He@ C_{70}^{6-}, New Records for the Most Shielded and the Most Deshielded ^{3}He inside a Fullerene. *J. Am. Chem. Soc.* **120:** 6389–6393.
38. Sternfeld T, Saunders M, Cross RJ, Rabinovitz M (2003) The Inside Story of Fullerene Anions: A ^{3}He NMR Aromaticity Probe. *Angew. Chem., Int. Ed. Engl.* **42:** 3136–3139.
39. Sternfeld T, Hoffman RE, Saunders M, Cross RJ, Syamala MS, Rabinovitz M (2002) Two helium atoms inside fullerenes: Probing the internal magnetic field in C_{60}^{6-} and C_{70}^{6-}. *J. Am. Chem. Soc.* **124:** 8786–8787.
40. Lamparth I, Maichle-Mössmer C, Hirsch A (1995) Reversible Template-Directed Activation of Equatorial Double-Bonds of the Fullerene Framework — Regioselective Direct Synthesis, Crystal-Structure, and Aromatic Properties of T_h-$C_{66}(COOEt)_{12}$. *Angew. Chem., Int. Ed. Engl.* **34:** 1607–1609.
41. Wang GW, Saunders M, Cross RJ (2001) Reversible Diels-Alder addition to fullerenes: A study of equilibria using ^{3}He NMR spectroscopy. *J. Am. Chem. Soc.* **123:** 256–259.
42. Cross RJ, Jiménez-Vázquez HA, Lu Q, Saunders M, Schuster DI, Wilson SR, Zhao H (1996) Differentiation of isomers resulting from bisaddition to C_{60} using ^{3}He NMR spectrometry. *J. Am. Chem. Soc.* **118:** 11454–11459.
43. Saunders M, Cross RJ, Jiménez-Vázquez HA, Shimshi R, Khong A (1996) Noble Gas Atoms Inside Fullerenes. *Science* **271:** 1693–1697.

44. Smith AB, Strongin RM, Brard L, Furst GT, Atkins JH, Romanow WJ, Saunders M, Jiménez-Vázquez HA, Owens KG, Goldschmidt RJ (1996) Synthesis and Characterization of the First $C_{70}O$ Epoxides. Utilization of ^3He NMR in Analysis of Fullerene Derivatives. *J. Org. Chem.* **61:** 1904–1905.
45. Billups WE, Gonzalez A, Gesenberg C, Luo WM, Marriott T, Alemany LB, Saunders M, Jimenez-Vazquez HA, Khong A (1997) ^3He NMR spectra of highly reduced C_{60}. *Tetrahedron Letters* **38:** 175–178.
46. Billups WE, Luo WM, Gonzalez A, Arguello D, Alemany LB, Marriott T, Saunders M, Jimenez-Vazquez HA, Khong A (1997) Reduction of C_{60} using Anhydrous Hydrazine. *Tetrahedron Lett.* **38:** 171–174.
47. Jensen AW, Khong A, Saunders M, Wilson SR, Schuster DI (1997) Photocycloaddition of Cyclic 1,3-Diones to C_{60}. *J. Am. Chem. Soc.* **119:** 7303–7307.
48. Ruttimann M, Haldimann RF, Isaacs L, Diederich F, Khong A, Jiménez-Vázquez H, Cross RJ, Saunders M (1997) π-Electron Ring Current Effects in Multiple Adducts of $^3He@C_{60}$ and $^3He@C_{70}$: A ^3He NMR Study. *Chem. Euro. J.* **3:** 1071–1076.
49. Birkett PR, Bühl M, Khong A, Saunders M, Taylor R (1999) The ^3He NMR spectra of a [60]fullerene cation and some arylated [60]fullerenes. *J. Chem. Soc. Perk. T* **2:** 2037–2039.
50. Boltalina OV, Bühl M, Khong A, Saunders M, Street JM, Taylor R (1999) The ^3He NMR spectra of $C_{60}F_{18}$ and $C_{60}F_{36}$; the parallel between hydrogenation and fluorination. *J. Chem. Soc. Perk. T* **2:** 1475–1479.
51. Wang GW, Weedon BR, Meier MS, Saunders M, Cross RJ (2000) ^3He NMR Study of $^3He@C_{60}H_6$ and $^3He@C_{70}H_{2-10}$. *Org. Lett.* **2:** 2241–2243.
52. Wilson SR, Yurchenko ME, Schuster DI, Khong A, Saunders M (2000) Synthesis of Fluorous Fullerene Adducts: Reversible Solubilization of Fullerenes in Perfluorinated Solvents. *J. Org. Chem.* **65:** 2619–2623.
53. Fujiwara K, Komatsu K, Wang GW, Tanaka T, Hirata K, Yamamoto K, Saunders M (2001) Derivatization of fullerene dimer C_{120} by the Bingel reaction and a ^3He NMR study of $^3He@C_{120}$ monoadducts. *J. Am. Chem. Soc.* **123:** 10715–10720.
54. Nossal J, Saini RK, Sadana AK, Bettinger HF, Alemany LB, Scuseria GE, Billups WE, Saunders M, Khong A, Weisemann R (2001) Formation, Isolation, Spectroscopic Properties, and Calculated Properties of Some Isomers of $C_{60}H_{36}$. *J. Am. Chem. Soc.* **123:** 8482–8495.
55. Peera A, Saini RK, Alemany LB, Billups WE, Saunders M, Khong A, Syamala MS, Cross RJ (2003) Formation, isolation, and spectroscopic properties of some isomers of $C_{60}H_{38}$, $C_{60}H_{40}$, $C_{60}H_{42}$, and $C_{60}H_{44}$ — Analysis of the effect of

the different shapes of various helium-containing hydrogenated fullerenes on their ^3He chemical shifts. *Eur. J. Org. Chem.*: 4140–4145.

56. Frunzi M, Xu H, Cross RJ, Saunders M (2009) NMR Temperature-Jump Method for Measuring Reaction Rates: Reaction of Dimethylanthracene with $H_2@C_{60}$. *J. Phys. Chem. A* **113:** 4996–4999.
57. Brown S, Cao J, Musfeldt JL, Dragoe N, Cimpoesu F, Ito S, Takagi H, Cross RJ (2006) Search for microscopic evidence for molecular level negative thermal expansion in fullerenes. *Phys. Rev. B* **73:** -.
58. Ito S, Takeda A, Miyazaki T, Yokoyama Y, Saunders M, Cross RJ, Takagi H, Berthet P, Dragoe N (2004) Kr Extended X-ray Absorption Fine Structure Study of Endohedral $Kr@C_{60}$. *J. of Phys. Chem. B* **108:** 3191–3195.
59. Kilcoyne ALD, Aguilar A, Muller A, Schippers S, Cisneros C, Alna'Washi G, Aryal NB, Baral KK, Esteves DA, Thomas CM, Phaneuf RA (2010) Confinement Resonances in Photoionization of $Xe@C_{60}^+$. *Phys. Rev. Lett.* **105:** 213001.
60. Becker L, Bada JL, Winans RE, Hunt JE, Bunch TE, French BM (1994) Fullerenes in the 1.85-Billion-Year-Old Sudbury Impact Structure. *Science* **265:** 642–645.
61. Becker L, Poreda RJ, Bada JL (1996) Extraterrestrial helium trapped in fullerenes in the Sudbury impact structure. *Science* **272:** 249–252.
62. Becker L, Poreda RJ, Bunch TE (2000) Fullerenes: An extraterrestrial carbon carrier phase for noble gases. *P. Natl. Acad. Sci. USA* **97:** 2979–2983.
63. Firestone RB, et al. (2007) Evidence for an extraterrestrial impact 12,900 years ago that contributed to the megafaunal extinctions and the Younger Dryas cooling. *P. Natl. Acad. Sci. USA* **104:** 16016–16021.
64. Becker L, Poreda RJ, Hunt AG, Bunch TE, Rampino M (2001) Impact event at the Permian-Triassic boundary: Evidence from extraterrestrial noble gases in fullerenes. *Science* **291:** 1530–1533.
65. Dibrozolo FR, Bunch TE, Fleming RH, Macklin J (1994) Fullerenes in an Impact Crater on the Ldef Spacecraft. *Nature* **369:** 37–40.
66. Garcia-Hernandez DA, Rao NK, Lambert DL (2011) Are C_{60} Molecules Detectable in Circumstellar Shells of R Coronae Borealis STars? *Astrophys. J.* **729:** 126–126.
67. Zhang Y, Kwok S (2011) Detection of C_{60} in the Protoplanetary Nebula IRAS 01005+7910. *Astrophys. J.* **730:** 126–126.
68. Roberts KRG, Smith KT, Sarre PJ (2012) Detection of C_{60} in embedded young stellar objects, a Herbig Ae/Be star and an unusual post-asymptotic giant branch star. *Monthly Notices of the Royal Astronomical Society* **421:** 3277–3285.

CHAPTER 10

Electrochemical Properties of Endohedral Metallofullerenes

Luis Echegoyen,[*,a] Frederic Melin[b] and Manuel N. Chaur[c]*

1. INTRODUCTION

Endohedral metallofullerenes (EMFs) are fullerenes that encapsulate a metal or a metallic cluster. Since the discovery of fullerenes,[1] many attempts to encapsulate different moieties into the carbon allotropes have resulted in the preparation of a variety of compounds with interesting structural, electronic and chemical properties, suitable for many applications such as in optoelectronic devices and in medical and biological diagnostics and therapeutics.[2,9] Electrochemistry has been a very useful tool for the characterization of the electronic properties of these compounds and when coupled with ESR and DFT calculations, the HOMO-LUMO energy levels can be estimated thus providing a much better understanding of the physical and chemical properties of these endohedral compounds.

* Corresponding author.
[a] Chemistry Department, University of Texas at El Paso, El Paso TX, USA. E-mail: echegoyen@utep.edu
[b] Faculté Chimie, Université de Strasbourg, Strasbourg, France. E-mail: fmelin@unistra.fr
[c] Departamento de Química, Universidad del Valle, Cali, Colombia.

In general, the electrochemical properties of EMFs have shown to be dependent on the entrapped moiety and on the size and symmetry of the fullerene cage, and as a result, electrochemistry has been used as a tool for the characterization and even for the separation of some of these compounds. Herein we have compiled some of the most remarkable electrochemical data of EMFs classified on the basis of the entrapped moiety, and we also review theoretical calculations reported for these compounds.

2. ELECTROCHEMISTRY OF CLASSICAL EMFs AND THEIR DERIVATIVES

2.1 The $M@C_{2n}$ Families

Electrochemistry of these compounds has extensively been reported by Suzuki, Akasaka, Gu, Dunsch and Wang *et al.*, specially for the $M@C_{82}$ compounds[10,24] (see Table 1), and these have been reviewed elsewhere.[6,9]

Ce@C_{82}, Gd@C_{82}, Y@C_{82} and the major [C_{2v}] and minor [C_s] isomers of La@C_{82} and Pr@C_{82} exhibit very low electrochemical potential gaps (defined as $\Delta E_{gap}=E_{ox1}-E_{red1}<0.50V$). Electrochemistry reveals a reversible first oxidation step, close to that of the ferrocene/ferrocenium couple, and

Table 1 Redox Potentials (V vs Fc^+/Fc) of the $M@C_{82}$ EMFs

Metallofullerene	$E ox_2$	$E_{1/2} ox_1$	$E_{1/2} red_1$	$E_{1/2} red_2$	$E_{1/2} red_3$	$E_{1/2} red_4$	$E_{1/2} red_5$	$E_{1/2} red_6$
La@C_{82} [C_{2v}]	1.07	0.07	−0.42	−1.37	−1.53	−2.26	−2.46	
La@C_{82} [C_s]	1.08	−0.07	−0.47	−1.40	−2.01	−2.40		
Pr@C_{82} [C_{2v}]	1.08	0.07	−0.39	−1.35	−1.46	−2.21		
Pr@C_{82} [C_s]	1.05	−0.07	−0.48	−1.39	−1.99	−1.99		
Ce@C_{82} [C_{2v}]	1.08	0.08	−0.41	−1.41	−1.53	−1.79	−2.25	−2.50
Gd@C_{82} [C_{2v}]	1.08	0.09	−0.39	−1.38	−2.22			
Y@C_{82} [C_{2v}]	1.07	0.10	−0.37	−1.34	−2.22	−2.47		
Sm@C_{82} [C_{2v}]			−0.28	−0.63	−1.52	−1.88	−2.03	−2.32
Yb@C_{82} [C_s]			−0.33	−0.65	−1.58	−1.81		
Yb@C_{82} [C_2]			−0.60	−0.76	−1.33	−1.73		
Yb@C_{82} [C_{2v}]			−0.33	−0.67	−1.56	−1.90		
Ca@C_{82} [C_2]			−0.59	−0.74	−1.30	−1.70		
Ca@C_{82} [II]			−0.65	−0.96	−1.55	−1.90		

the observation of six reversible reduction steps indicates that these compounds are very good electron donors and acceptors, a significantly different behavior from that observed for the empty cage C_{82}, which exhibits four reversible reductions and no oxidation processes.[18] These results not only reflect the strong influence of the metal on the electronic properties of these compounds but also indicate a very active redox behavior.

When studying the electrochemical behavior of Sm@C_{82} and all the isolated isomers of Ca@C_{82}, Yb@C_{82} and Tm@C_{82}, no oxidation steps were detected, therefore it was assumed that these compounds possess larger electrochemical gaps. However, recently Akasaka and Nagase reported the preparation and electrochemical characterization of all the isomers of Yb@C_{2n} (n = 40–42).[25] These compounds exhibit one or two reversible oxidation steps along with four reduction steps as measured in 1,2-dichlorobenzene. It has been assumed from the diamagnetism that the metal adopts a 2+ formal oxidation state in these compounds, leading to a M^{2+}@C_{82}^{2-} electronic structure. From these studies, we can conclude that the formal oxidation state of the entrapped metal, and its size, have a clear influence on the redox properties of the EMFs.

In the case of Yb@C_{2n} and Ca@C_{2n} EMFs, the electrochemistry is highly dependent on the structure of the carbon cage, as observed by Gu *et al.* when studying the Yb@C_{2n}[23] and Ca@C_{2n}[24] families (Table 2).

Noteworthy are the dramatic differences in electron accepting abilities of isomeric endohedral fullerenes. This difference reached 0.22 V for the isomers of Yb@C_{76} and even 0.3 V between two isomers of Yb@C_{82}. The cage size also influences the reduction potentials. The C_s and C_{2v} isomers of Yb@C_{82} are the best electron acceptors of the family and the second isomer of Yb@C_{76} was the worst.

2.2. Endohedral Dimetallofullerenes

The electrochemical properties of M_2 moieties encapsulated in C_{72}, C_{80} and C_{82} cages have been reported,[6–9,26,3] and the results of these studies are summarized in Table 3.

The La_2 and Ce_2 moieties encapsulated in a non-IPR C_{72} cage exhibit very similar redox properties,[27,30] indicating that they share a very similar electronic structure. So do the La_2 and Ce_2 encapsulated in a C_{80} cage, as

Table 2 Redox Potentials (V vs Fc^+/Fc) of the $M@C_{2n}$ (M = Yb, Ca) EMFs

Metallofullerene	$E_{1/2}\,red_1$	$E_{1/2}\,red_2$	$E_{1/2}\,red_3$	$E_{1/2}\,red_4$
$Yb@C_{74}$ [II]	−0.52	−0.96	−1.55	−1.99
$Yb@C_{76}$ [I]	−0.46	−0.83	−1.46	−1.89
$Yb@C_{76}$ [II]	−0.68	−1.02	−1.59	−2.01
$Yb@C_{78}$	−0.48	−0.79	−1.46	−1.83
$Yb@C_{80}$	−0.57	−0.95	−1.55	−1.90
$Yb@C_{82}$ [C_s]	−0.33	−0.65	−1.58	−1.81
$Yb@C_{82}$ [C_2]	−0.60	−0.76	−1.33	−1.73
$Yb@C_{82}$ [C_{2v}]	−0.33	−0.67	−1.56	−1.90
$Yb@C_{84}$ [II]	−0.63	−0.88	−1.26	−1.64
$Yb@C_{84}$ [III]	−0.49	−0.68	−1.57	−1.79
$Yb@C_{84}$ [IV]	−0.46	−0.72	−1.34	−1.54
$Ca@C_{76}$	−0.61	−0.99	−1.57	−1.97
$Ca@C_{82}$ [C_2]	−0.65	−0.96	−1.55	−1.90
$Ca@C_{82}$ [II]	−0.59	−0.74	−1.30	−1.70
$Ca@C_{84}$ [II]	−0.64	−0.90	−1.27	−1.65

Table 3 Redox Potentials (V vs Fc^+/Fc) of $M_2@C_{2n}$ Compounds; Comparison with $La@C_{82}$ and $Ce@C_{82}$

Metallofullerene	$E\,ox_2$	$E\,ox_1$	$E\,red_1$	$E\,red_2$	$E\,red_3$
$La_2@C_{72}$ [D_2]	0.75	0.24	−0.68	−1.92	
$Ce_2@C_{72}$ [D_2]	0.82	0.18	−0.81	−1.86	
$La_2@C_{78}$[D_{3h}]	0.62	0.26	−0.40	−1.84	−2.28
$La_2@C_{80}$ [I_h]	0.95	0.56	−0.31	−1.72	−2.13
$Ce_2@C_{80}$ [I_h]	0.95	0.57	−0.39	−1.71	
$Sc_2@C_{82}$ [I]	0.10	−0.12	−1.26	−1.88	
$Sc_2@C_{82}$ [III]		0.07	−0.87	−1.29	−1.85
$Er_2@C_{82}$		0.19	−0.87	−1.26	

well as the Sc_2 and Er_2 entrapped in a C_{82} cage. Since a higher degree of charge transfer between the encapsulated moiety and the cage is believed to occur in these structures (six electrons), it was expected that these compounds would be less prone to reduction. Surprisingly, it was observed that the electron accepting abilities were better than anticipated. In particular, $La_2@C_{80}$ and $Ce_2@C_{80}$, described by Nagase *et al.* were even easier to reduce than $La@C_{82}$ and $Ce@C_{82}$. This is probably the result of the LUMO being localized on the bimetallic cluster, as predicted by calculations in the case of $La_2@C_{80}$. In contrast, the electron donating ability of these two compounds was significantly weaker.

EMFs show a very large range of redox properties: some of them can be very easily oxidized and reduced, such as the historically important $La@C_{82}$, whereas others are difficult to oxidize, such as $Ca@C_{82}$. These dramatic differences have been exploited to efficiently purify large quantities of endohedral fullerenes using either one or a combination of chemical and electrochemical oxidative or reductive treatments. The first report of such an interesting method of purification was published by Alford *et al.* in 1998, who described the separation of the empty cages from the small band-gap gadolinium endohedral fullerenes in the arcing soot, using an electrochemical reduction followed by a selective chemical reoxidation.[34] The procedure was later improved by the same group to obtain easily and quickly three different fractions of $Gd@C_{2n}$ compounds.[35] This is obviously a fast, efficient and economical alternative to the standard purification methods used in this field, HPLC.

3. ELECTROCHEMISTRY OF NON-CLASSICAL METALLOFULLERENES

The so-called non-classical EMFs involve metallic nitrides, sulfides, oxides, carbides[36–39] and more recently cyanide cluster metallofullerenes.[40–42] All of these compounds have been extensively studied in recent years.[4–9]

3.1 Metallic Nitride Endohedral Fullerenes

One of the earliest electrochemical studies by Echegoyen and co-workers corresponds to the mixture of isomers of $Sc_3@C_{80}$, the predominant $[I_h]$ isomer and the less abundant $[D_{5h}]$.[43] These two compounds exhibit a pronounced difference in the first oxidation potential and, therefore, they can be electrochemically separated (see Fig. 1). This feature clearly shows the influence of the carbon cage symmetry on the electrochemical properties of metallic nitride fullerenes.

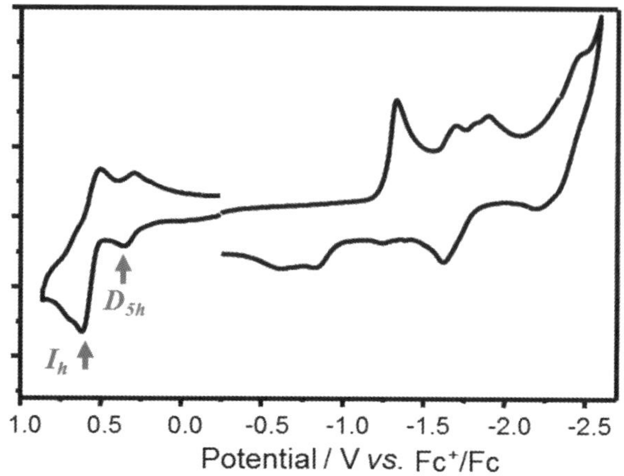

Fig. 1 Cyclic voltammogram of the isomeric mixture ([I_h] and [D_{5h}]) of $Sc_3N@C_{80}$ obtained in *o*-DCB + 0.05 M (*n*-Bu)$_4$NPF$_6$ (scan rate 100 mV.s^{-1}).

Interestingly, all the $M_3N@C_{80}$ [I_h] compounds studied so far in *o*-DCB have shown an electrochemical behavior quite similar to that of $Sc_3N@C_{80}$, with irreversible reduction steps and reversible oxidation processes.[4–9,44–53]

The nature of the metal has a more pronounced influence on the reduction processes than on the oxidation potentials (see Table 4). The case of $Sc_3N@C_{80}$ seems unique since this compound has a significantly lower electrochemical gap than the others. The scandium hyperfine splitting in the EPR spectrum of the radical anion of $Sc_3N@C_{80}$,[57] together with quantum calculations, suggested a major contribution of the cluster to the LUMO (see Section 4.2.). For the other metallic nitride EMFs, it is now believed that there is no significant contribution from the cluster to either the LUMO or the HOMO,[57–60] except when Ti is included, as in $TiSc_2N@C_{80}$ [61–62] and $TiY_2N@C_{80}$,[63] where Yang and coworkers determined that the first reduction and oxidation processes, being both reversible, are located in the cluster, changing the valence state of Ti from Ti^{2+} in the anion to Ti^{3+} in the neutral state and to Ti^{4+} in the cation.[63]

The electrochemical properties of metallic nitride fullerenes with cages larger than C_{80} have also been reported by Echegoyen and co-workers when studying the families of Gd, Nd, Ce, Pr and La nitride fullerenes.[6,45–48,57,60] (See Fig. 2 and Table 5).

Table 4 Relevant Redox Potential (in V vs Fc^+/Fc) of the $M_3N@C_{80}$ Compounds Obtained in *o*-DCB + 0.05 M $(n\text{-}Bu)_4NPF_6$, Metal Electronegativities (χ) and Effective Valencies; [a] Ref. 54–56

Metallofullerene	χ	Effective valency[a]	$Ep_c\ red_1$	$Ep_c\ red_2$	$E_{1/2}\ ox_1$	ΔE_{gap}
$Sc_3N@C_{80}$ $[I_h]$	1.36	2.4	−1.29	−1.56	0.59	1.88
$Sc_3N@C_{80}$ $[I_h]$	1.36	2.4	−1.24	−1.62	0.62	1.86
$Sc_3N@C_{80}$ $[I_h]$	1.36	2.4	−1.22	−1.59	0.62	1.84
$Sc_3N@C_{80}$ $[I_h]$	1.36	2.4	−1.27		0.57	1.84
$Lu_3N@C_{80}$ $[I_h]$	1.27		−1.40		0.64	2.04
$Tm_3N@C_{80}$ $[I_h]$	1.25	2.9	−1.43	−1.78	0.65	2.08
$Er_3N@C_{80}$ $[I_h]$	1.24		−1.42	−1.80	0.63	2.05
$Y_3N@C_{80}$ $[I_h]$	1.22		−1.41	−1.83	0.64	2.05
$Dy_3N@C_{80}$ $[I_h]$	1.22	2.8	−1.37	−1.86	0.70	2.07
$Gd_3N@C_{80}$ $[I_h]$	1.20		−1.44	−1.86	0.58	2.02
$Nd_3N@C_{80}$ $[I_h]$	1.14		−1.42	−1.89	0.63	2.05
$Pr_3N@C_{80}$ $[I_h]$	1.13		−1.41	−1.84	0.59	2.00
$ScYErN@C_{80}$ $[I_h]$			−1.55	−1.97	0.64	2.19
$TiSc_2N@C_{80}$ $[I_h]$			−0.94	−1.58	0.16	1.10
$TiY_2N@C_{80}$ $[I_h]$			−1.11	−1.79	0.00	1.11
$Sc_3NC@C_{80}$ $[I_h]$			−1.05	−1.68	0.60	1.65
$Sc_3N@C_{80}$ $[D_{5h}]$	1.36		−1.33		0.34	1.67
$Lu_3N@C_{80}$ $[D_{5h}]$	1.27		−1.41		0.45	1.86
$Tm_3N@C_{80}$ $[D_{5h}]$	1.25		−1.45		0.39	1.84
$Dy_3N@C_{80}$ $[D_{5h}]$	1.22		−1.40	−1.85	0.40	1.80

$Gd_3N@C_{86}$ and the two non-IPR metallic nitride EMFs $Gd_3N@C_{82}$ [C_s(39663)][64] and $Gd_3N@C_{84}$ [C_s(51365)][65] showed three irreversible reduction and one reversible oxidation processes, a behavior qualitatively similar to that of the IPR $Gd_3N@C_{80}$ [I_h]. In contrast, for $Gd_3N@C_{88}$, both reductions and oxidations are reversible. Remarkably, changing the cage size does not affect significantly the reduction potential of these compounds, which displayed very close first reduction potentials (see Table 5), but dramatically influenced their oxidation potentials. These results

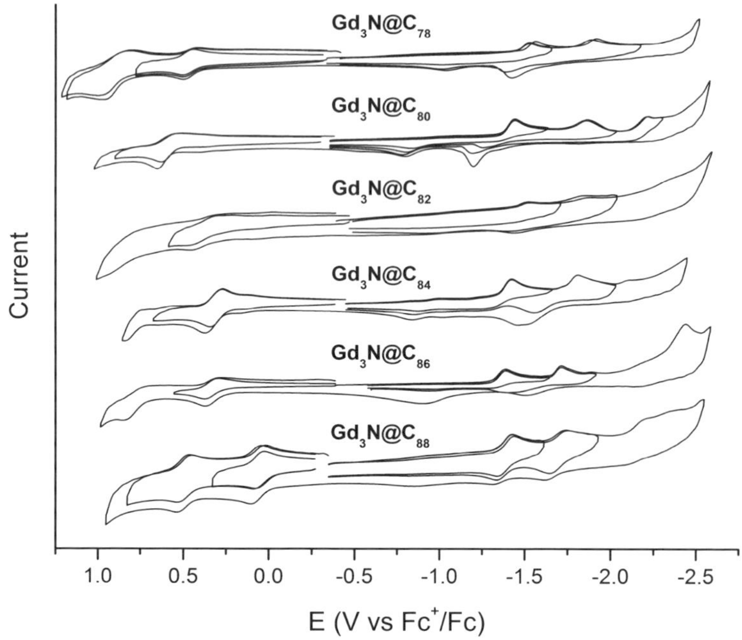

Fig. 2 Cyclic voltammograms of the $Gd_3N@C_{2n}$ ($40 \leq n \leq 44$) compounds obtained in *o*-DCB + 0.05 M $(n\text{-Bu})_4NPF_6$ (scan rate 0.1 V.s^{-1}).

strongly suggest that the HOMO of the metallic nitride EMFs are probably cage-centered. A significant lowering of the electrochemical bandgap was observed when increasing the cage size, from C_{80} to C_{88}. Similar observations of the reversibility of the redox steps and on the electrochemical bandgap (see Table 5) were made in the case of the $Nd_3N@C_{2n}$ and $Pr_3N@C_{2n}$ families.[6,57]

The redox behavior of $Gd_3N@C_{88}$, i.e. the reversibility of both oxidation and reduction processes and the very low HOMO-LUMO gap, was later found to be characteristic of all of the metallic nitride clusters encapsulated in a C_{88} cage, since the Nd, Pr, and Ce counterparts showed very similar behavior.[45–48,52] This family of $M_3N@C_{88}$ compounds did not exhibit many differences in the oxidation and reduction potentials, in contrast to the $M_3N@C_{80}$ family. This was presumably due to the very similar electronic properties of Gd, Nd, Pr, and Ce (including their electronegativity) and a negligible contribution of the cluster to both the HOMO and

Table 5 Redox Potential (in V vs Fc+/Fc) of the $M_3N@C_{2n}$ (M=Nd, Pr, Ce and La and 44≤n≤48) Compounds Obtained in o-DCB + 0.05 M $(n\text{-}Bu)_4NPF_6$.; [a]ΔEgap = $E_{1/2}$ ox_1– Ep_c red_1; Inside Square Brackets are the Cage Isomer Labels

Metallofullerene	Ep_c red_1	Ep_c red_2	$E_{1/2}$ ox_1	$E_{1/2}$ ox_2	ΔE_{gap} [a]
$Gd_3N@C_{80}$ [I_h]	−1.44	−1.86	0.58		2.02
$Gd_3N@C_{82}$	−1.53	−1.87	0.38		1.91
$Gd_3N@C_{84}$ [C_s]	−1.37	−1.76	0.32		1.69
$Gd_3N@C_{86}$	−1.39	−1.72	0.33		1.72
$Gd_3N@C_{88}$	−1.43	−1.74	0.06	0.49	1.49
$Nd_3N@C_{80}$ [I_h]	−1.42	−1.90	0.63		2.05
$Nd_3N@C_{84}$	−1.44	−2.02	0.31		1.75
$Nd_3N@C_{86}$	−1.46	−1.79	0.36		1.82
$Nd_3N@C_{88}$	−1.36	−1.75	0.07	0.53	1.43
$Pr_3N@C_{80}$ [I_h]	−1.41	−1.81	0.59		2.00
$Pr_3N@C_{86}$	−1.48	−1.80	0.31		1.79
$Pr_3N@C_{88}$	−1.34	−1.72	0.09	0.54	1.43
$Pr_3N@C_{92}$	−1.46		0.35		1.81
$Pr_3N@C_{96}$	−1.51	−1.86	0.14	0.53	1.65
$Ce_3N@C_{88}$	−1.30	−1.57	0.08	0.63	1.38
$Ce_3N@C_{92}$	−1.48		0.32		1.81
$Ce_3N@C_{96}$	−1.50	−1.84	0.18	0.67	1.68
$La_3N@C_{88}$	−1.36	−1.67	0.21	0.66	1.57
$La_3N@C_{92}$	−1.44	−1.64	0.36		1.80
$La_3N@C_{96}$	−1.54	−1.77	0.14	0.53	1.68

LUMO.[60] It was very surprising that such low bandgap fullerenes were the most abundant species obtained in the soluble extract of the arcing soot in the case of Nd, Pr, and Ce. Only $La_3N@C_{88}$ showed different behavior with irreversible reduction steps and a slightly larger electrochemical bandgap, probably an indication of a different symmetry for this carbon cage although this has not been firmly established yet.

The electrochemistry and EPR of some trimetallic nitride EMFs with cages smaller than C_{80} have also been reported (Table 6).[50,66–68] It was expected that $Sc_3N@C_{68}$, which possesses a non-IPR cage of [D_3]

Table 6 Cathodic Peak Potentials and Anodic Half Wave Potentials (in V vs Fc^+/Fc) of $Sc_3N@C_{68}$, $Sc_3N@C_{78}$ and $Dy_3N@C_{78}$; Comparison with $M_3N@C_{80}$;[a] $\Delta Egap = E_{1/2}\,ox_1 - Ep_c\,red_1$

Metallofullerene	Ep_c red$_1$	Ep_c red$_2$	$E_{1/2}$ ox$_1$	$E_{1/2}$ ox$_2$	ΔE_{gap} [a]
$Sc_3N@C_{80}$ [I_h]	−1.24	−1.62	0.62		1.86
$Sc_3N@C_{68}$ [D_3]	−1.45	−2.05	0.33	0.85	1.88
$Sc_3N@C_{78}$ [D_3h]	−1.54		0.12		1.66
$Dy_3N@C_{80}$ [I_h]	−1.37	−1.86	0.70		2.07
$Dy_3N@C_{78}$ [22010-C_2]	−1.54	−1.93	0.47		2.01
$Gd_3N@C_{78}$ [22010-C_2]	−1.53	−1.89	0.47	1.00	2.00

Table 7 Relevant Redox Potentials (in V vs Fc^+/Fc) of Metal Carbide Fullerenes; Inside Square Brackets are the Cage Isomer Labels

Metallofullerene	E ox$_1$	E red$_1$	E red$_2$	E red$_3$
$Sc_4C_2@C_{80}$ [I_h]	0.40	−1.16	−1.65	
$Sc_3C_2@C_{80}$ [I_h]	−0.06	−0.50	−1.64	−1.82
$Sc_2C_2@C_{80}$ [C_{2v} (5)]	0.41	−0.74	−1.33	
$Sc_2C_2@C_{82}$ [C_{3v} (8)]	0.47	−0.94		
$Sc_2C_2@C_{82}$ [C_s (6)]	0.42	−0.93	−1.30	
$Sc_2C_2@C_{82}$ [C_{2v} (9)]	0.25	−0.74	−0.96	
$Sc_2@C_{82}$ [III]	0.07	−0.87	−1.29	−1.85
$Er_2@C_{82}$	0.19	−0.87	−1.26	

symmetry would show a very different behavior than the IPR $Sc_3N@C_{80}$ [I_h], due to their very different cage structures. Dunsch *et al.* observed two irreversible reduction steps for $Sc_3N@C_{68}$,[66] cathodically shifted compared to those of $Sc_3N@C_{80}$ (see Table 6). The main difference between the two compounds was observed in the oxidative scan. The donating ability of $Sc_3N@C_{68}$ was found to be significantly stronger than that of $Sc_3N@C_{80}$ [I_h]. The radical cation[66] $[Sc_3N@C_{68}]^{+\bullet}$ and radical anion[69] $[Sc_3N@C_{68}]^{-\bullet}$ were generated by electrolysis and both characterized by EPR. The EPR spectra consisted of 22 lines consistent with three equivalent scandiums with a small hyperfine splitting of 1.28 gauss for the cation and 1.75 gauss for the anion. Based on the smaller hyperfine splitting, as compared to that of $[Sc_3N@C_{80}]^{-\bullet}$,[70] predominant delocalization of the unpaired spin on the cage was proposed, which confirmed the contribution of the cage to the HOMO of these compounds. Dunsch *et al.* also reported the redox behavior of $Dy_3N@C_{78}$[50] and Echegoyen *et al.* reported that of $Sc_3N@C_{78}$ and the non-IPR $Gd_3N@C_{78}$.[68] These EMFs showed two irreversible reduction steps and one reversible oxidation step. Their electron accepting ability was also significantly weaker and their electron donating ability stronger than that of the parent metallic nitride EMF with an $[I_h]$-C_{80} cage.

3.2. Metallic Carbide Endohedral Fullerenes

The metallic nitride EMFs are not the only non-classical EMFs studied by electrochemistry. Nagase and co-workers reported the electrochemistry of some metallic carbide endohedral fullerenes (Table 7).[71–74]

Some similarities were observed between the voltammograms of $Sc_2C_2@C_{82}$ and $Er_2@C_{82}$ and for isomer [III] of $Sc_2@C_{82}$ (see Table 7).[29,71,74] These similarities are now well understood by assuming that these species share the same C_{82} carbon cage. The electronic structure of this metallic carbide endohedral fullerene was established by DFT calculations[71] to be $(Sc_2C_2)^{4+}C_{82}^{4-}$. $Sc_3C_2@C_{80}$ is paramagnetic and thus has a very small electrochemical bandgap in *o*-DCB.[72] The charge state of this compound was established by DFT calculations to be $(Sc^{3+})_3C_2^{3-}C_{80}^{6-}$,[73] and interestingly the redox state of the molecule does not seem to influence the formal oxidation state of the Sc atoms or the charge on the cage, but affects directly the formal charge on the C_2 moiety. Therefore the C_2 moiety displayed a remarkable range of charge, since it varied sequentially in this structure from 2+ (first oxidation) to 3– (third reduction). A very interesting case is $Sc_3C_2@C_{80}$ $[I_h]$ where the Sc_3C_2 cluster transfers seven electrons to the carbon cage[72]; as a result, this metal carbide metallofullerene exhibits an open shell structure with lower oxidation and reduction potentials than other metal carbides.

3.3. Metallic Sulfide and Metallic Oxide Endohedral Fullerenes

The encapsulation of oxometallic and metallic sulfide clusterfullerenes has also been recently reported. These compounds exhibit very interesting redox properties. For instance, $Sc_2(\mu_2\text{-}S)@C_{72}$ and $La_2@C_{72}$ show different HOMO-LUMO gaps, which is the result of the difference in their cage symmetries, which is a remarkable example of how not only the size but also the geometry of the cluster has a large influence in the fullerene cage selection during the synthesis process. Although Popov *et al.* calculate a much more stable non-IPR $C_{72}^{6-} - D_2$ (10611) fullerene,[58] the fixed trigonal geometry of the Sc_2S moiety is not able to stabilize the two adjacent pentalene units present in the 10611 isomer, whereas in the $C_{72}^{6-} - C_s$ (10528) the Sc_2S cluster can geometrically interact easily with the two pentalene units.[75] So it is a combination of electronic correspondence between the cluster and the cage, along with the corresponding size and geometric complementarity.

In the case of $Sc_2(\mu_2\text{-}S)@C_{82}$, the electrochemistry of the two isolated isomers (C_{3v} and C_s) is another example of the influence of the cage symmetry on the redox properties.[76] Changing the non-metallic atom of the cluster, O for S, results in similar reduction/oxidation potentials, indicating localization of the HOMO and LUMO on the cage as described in Table 8.

On the other hand, the electrochemistry of $Sc_4(\mu_3\text{-}O)_2@C_{80}$ considerably differs from those of the previously mentioned metallic sulfide and oxometallic clusterfullerenes,[77] which was explained on the basis of quantum chemical calculations (see Section 4.3.). The CV exhibits two reversible oxidation and reduction steps with an electrochemical gap of 1.10 V, and, the first oxidation potential is observed at 0.00 V. Similar behavior has also been observed in clusterfullerenes such as $TiY_2N@C_{80}$ and other

Table 8 Relevant Redox Potential (in V vs Fc^+/Fc) of Metallic Sulfide Fullerenes and Oxometallic Clusterfullerenes; Inside Square Brackets are the Cage Isomer Labels

Metallofullerene	E ox_1	E red_1	E red_2	E red_3
$Sc_2(\mu_2\text{-}S)@C_{72}$ [C_s (10528)]	0.64	−1.14	−1.53	−2.24
$Sc_2(\mu_2\text{-}S)@C_{82}$ [C_{3v} (8)]	0.52	−1.04	−1.19	−1.63
$Sc_2(\mu_2\text{-}S)@C_{82}$ [C_s (6)]	0.39	−0.98	−1.12	−1.73
$Sc_2(\mu_2\text{-}O)@C_{82}$ [C_s (6)]	0.35	−0.96	−1.28	−1.74
$Sc_4(\mu_3\text{-}O)_2@C_{80}$ [I_h]	0.00	−1.10	−1.73	−2.35

mixed metallic nitride metallofullerenes[63] where the oxidation occurs in the cluster.

4. INSIGHTS FROM QUANTUM CHEMICAL CALCULATIONS

The redox properties of endohedral fullerenes thus show somewhat unexpected features, such as the low dependence of the potentials of the metallic nitrides on the nature of the metal. In contrast, in monometallofullerenes, the influence of the encapsulated metals is more pronounced.[18] To understand these experimental findings, the orbitals of several endohedral fullerenes, including their highest occupied molecular orbital (HOMO) and lowest unoccupied molecular orbital (LUMO) have been computed by DFT in recent years. This work is not trivial, in particular for endohedral fullerenes with encapsulated lanthanide metals, which often show problems of convergence at the DFT level due to their quasi-degenerate 4f orbitals. It is also necessary to take into consideration different orientational isomers when the encapsulated moiety is able to rotate within the carbon cage. Despite these inherent difficulties, recent calculations were able to predict the redox potentials and the electrochemical gaps in solution with very good accuracy for a variety of endohedral compounds. Comprehensive studies of the bonding in endohedral fullerenes are also available now.[78]

4.1. Mono and Dimetallofullerenes

Soon after the discovery of La@C_{2n} in 1985,[79] and the isolation of La@C_{82} in 1991 by Smalley *et al.*,[80] the electronic structures of monometallofullerenes attracted considerable attention. In particular, the charge transfer between the encapsulated metal and the carbon cage in the relatively abundant M@C_{82} (C_{2v}) family has been the subject of numerous theoretical studies.[81–89] It is now well established that in these compounds, the metal is off-center and interacts specifically with a hexagonal ring of the cage. Sc, Y, La, Ce, Pr, Nd and Gd favor a +3 oxidation state whereas Ca, Sm, Eu and Yb prefer a +2 oxidation state. However, after encapsulation inside the C_{82} cage, the effective charge on the metals can deviate significantly from these absolute values due to a certain degree of hybridization between the orbitals of the metal and those of the carbon cage.[87] Trivalent metals form open-shell endohedral fullerenes, with very small electrochemical gaps.[18] In the case of Sc, Y and La, it was established that the unpaired electron is mostly delocalized on the carbon cage.[87] The d-orbitals of the metal, however, contribute significantly to the fully occupied valence orbitals. In contrast, Ca, Sm, Yb and Eu lead to closed-shell endohedral fullerenes with larger electrochemical gaps.[23–25,86,90] These metals can be encapsulated in a wide range of carbon cages.

Theoretical calculations of dimetallofullerenes are still very scarce. Most studies have been concerned with the relatively abundant $La_2@C_{80}$ (I_h) compound[28,91] and related $M_2@C_{80}$ (M=Sc, Y, La, Ce, Pr, Eu, Gd, Yb and Lu) species, although most of them have never been isolated.[59,92–93] C_{80} (I_h) has fourfold degenerate HOMOs with only two electrons,[94–95] and is thus highly kinetically unstable. Encapsulated moieties which can donate six electrons can fill the HOMOs of C_{80} and thus stabilize this fullerene cage. This ideal situation is achieved in the case of $La_2@C_{80}$, which is consistent with the relative abundance of this dimetallofullerene.[28,91] As expected, its HOMO is mostly localized on the cage, so oxidation occurs there. Its LUMO, in contrast, corresponds to a La-La bonding orbital. This orbital has a relatively low energy, so the first reduction potential of this molecule is one of the highest among endohedral fullerene families.[28] The separation between the first two reduction potentials is considerably larger than the separation between the first two oxidation potentials, presumably because there is more repulsion between electrons localized on the La-La moiety than between electrons delocalized on the whole C_{80} cage. In another isolated compound, $Ce_2@C_{80}$, the C_{80} cage is also stabilized by six electrons. The compound, however, is in a triplet state due to the presence of one f electron on each Ce atom. The HOMO is localized on both cerium atoms.[92–93] Despite their predicted different electronic structures, the redox properties of $La_2@C_{80}$ and $Ce_2@C_{80}$ (I_h) are very similar, a situation which is not fully understood at the present time. Interestingly, the electronic structures of $La_2@C_{72}$ (D_2) and $La_2@C_{78}$ (D_{3h}) exhibit similar features to those of $La_2@C_{80}$ (I_h).[59] In the same way, the electronic structures of $Ce_2@C_{72}$ (D_2), $Ce_2@C_{78}$ (D_{3h}) and $Ce_2@C_{80}$ (I_h) share similar characteristics,[96] which are also observed in their redox properties.

4.2. Metallic Nitride Endohedral Fullerenes

Calculations for this family of compounds started soon after their discovery in 1999 for those containing scandium,[70] and were then extended to those containing yttrium in 2006–2007[58,97] and finally to those with lanthanide metals since 2009.[57,98–100] These studies suggested that the lowest energy conformers of both I_h and D_{5h} isomers of $Sc_3N@C_{80}$ have a cage-based HOMO whereas their LUMOs exhibit a significant contribution from the encapsulated cluster (see Fig. 3). The LUMO of $Sc_3N@C_{78}$ is believed to be localized on the cluster as well.[101] In contrast, the HOMO and LUMO of $Y_3N@C_{80}$ are both mainly localized on the carbon cage (Fig. 3). The Gadolinium compounds ($Gd_3N@C_{80}$, $Gd_3N@C_{84}$ and $Gd_3N@C_{88}$) again exhibit a LUMO with a considerable metal contribution. Finally the HOMO and LUMO of the lanthanum compounds ($La_3N@C_{92}$ and

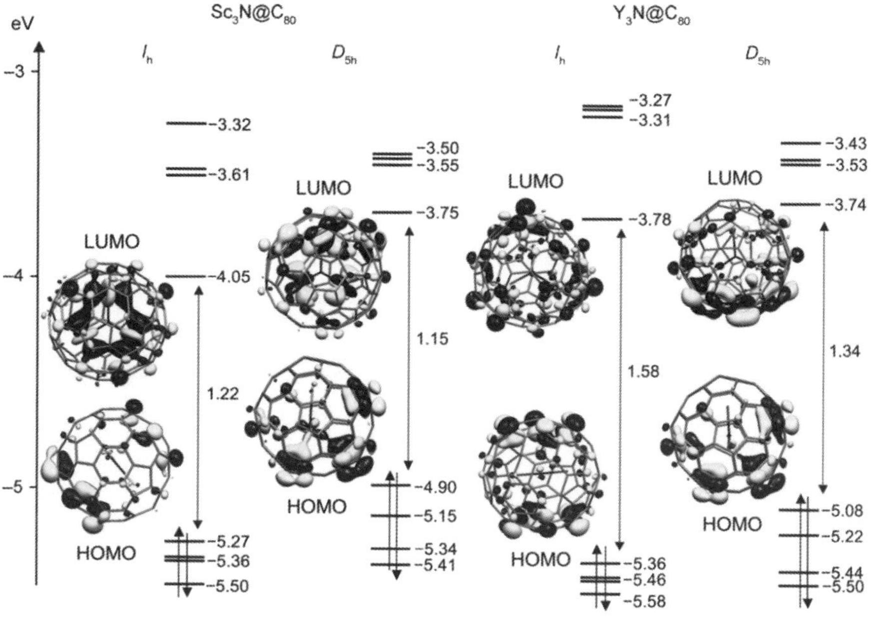

Fig. 3 Computed HOMO and LUMO levels of $Sc_3N@C_{80}$ and $Y_3N@C_{80}$ both (I_h) and (D_{5h}) isomers. Reprinted by permission of John Wiley & Sons, Inc. from R. Valencia, A. Rodriguez-Fortea, A. Clotet, C. de Graaf, M. N. Chaur, L. Echegoyen, J. M. Poblet, *Chem. Eur. J.* **15**, 10997 (2009). Copyright 2009 by John Wiley & Sons, Inc."

$La_3N@C_{96}$) are essentially cage orbitals. The frontier orbital of the neutral species, however, are not sufficient to describe accurately where the first oxidation and reduction takes place, due to significant reorganization of the electronic density upon electron transfer. It is necessary to analyze the orbitals of the $M_3N@C_{2n}^-$ radical anions and $M_3N@C_{2n}^+$ radical cations as well. From these studies it appears that only in the case of the $Sc_3N@C_{80}^-$ isomers, the extra electron resides mainly on the encapsulated cluster.[59,69,98] Interestingly, the HOMO of the closed-shell $Sc_3N@C_{80}^{2-}$ dianion exhibits a major contribution from the cluster as well,[99] and in the $Sc_3N@C_{80}^{3-}$ trianion radical, up to 41% of the spin density can still be found on the cluster.[100] The electronic structure of $Sc_3N@C_{80}$ is thus unique among metallic nitride endohedral fullerenes and explains why this compound exhibits redox properties significantly different from those of the others.[44]

For most of the metallic nitride endohedral fullerenes, oxidation and reduction involves essentially the carbon cage, and the electrochemical bandgaps are thus highly dependent on the nature of the fullerene host. Interestingly, an excellent correlation between the electrochemical bandgaps of these endohedral fullerenes and the energy gap between the LUMO-4 and LUMO-3 orbitals of the corresponding free cages was reported.[57] This correlation nicely confirmed the ionic model of interaction proposed since 2001 for these compounds, i.e. a transfer of six electrons from the encapsulated cluster to the carbon cage. This donation of electrons allows the stabilization of kinetically unstable fullerenes, such as the C_{80} (I_h) cage, and even non-classical fullerenes including several which do not obey the isolated pentagon rule (non-IPR fullerenes).[58,102] This correlation also supported a simple orbital rule proposed by Poblet *et al.* in 2005 to understand how cage selection occurs in this family of compounds: "a suitable fullerene host must have three low-lying unoccupied molecular orbitals and a sizeable energy gap between the LUMO-4 and the LUMO-3."[103–104] The high dependence of the electrochemical bandgaps on the nature of the cage can in turn be used to identify the fullerene host, when not known by complementary X-Ray studies. For example, the carbon cage for $La_3N@C_{92}$ has been proposed to be C_{92} (T) by comparison between computed electrochemical bandgaps and experimental values.[57]

DFT calculations can also serve to explain the nucleophilic reactivity of endohedral fullerene anions prepared by electrochemical or chemical methods. $Lu_3N@C_{80}^{2-}$, for instance, reacts easily with benzal bromide ($PhCHBr_2$) and leads to a methano-[6,6] derivative. Under the same conditions, $Sc_3N@C_{80}^{2-}$ does not react.[99] Poblet *et al.* showed that the HOMO of $Sc_3N@C_{80}^{2-}$ is mainly localized on the cluster, whereas the HOMO of $Lu_3N@C_{80}^{2-}$ is mainly localized on the cage. The C_{80} host cage is thus expected to be more nucleophilic for $Lu_3N@C_{80}^{2-}$ than $Sc_3N@C_{80}^{2-}$. Interestingly, the contribution of the cage is predominant in the singly occupied molecular orbital (SOMO) of both corresponding trianions. As a consequence these latter species are both decent nucleophiles which react with benzal bromide.[100]

4.3. Metallic Carbide, Metallic Oxide and Metallic Sulfide Endohedral Fullerenes

Similarly to that of metallic nitride endohedral fullerenes, the stability of the other cluster fullerenes is linked to a donation of electrons by the encapsulated moiety to the fullerene host. In the case of metallic carbide endohedral fullerenes, clusters which transfer four electrons to the cage (M_2C_2, M=Sc, Y or Er)[71,105–106] will select carbon cages with large (LUMO-3)-(LUMO-2) energy gaps (in particular C_{82} (C_{3v})), whereas clusters which transfer six electrons (such as Sc_3C_2 and Sc_4C_2)[73–74] will select carbon cages with large (LUMO-4)-(LUMO-3) energy gaps (in particular C_{80} (I_h)). Interestingly, for $Sc_2C_2@C_{82}$, $Sc_3C_2@C_{80}$ and $Sc_4C_2@C_{80}$, the scandium atoms are all in the +3 oxidation state, but the C_2 moiety respectively carries a −2, −3 or −6 charge and thus displays a remarkable charge-flexibility. With the exception of $Sc_3C_2@C_{80}$, these endohedral fullerenes have a closed shell electronic configuration, and their HOMO is mostly localized on the carbon cage. In contrast, their LUMO often exhibits a large contribution from the cluster. Oxidation and reduction are thus predicted to occur respectively on the carbon cage and on the encapsulated cluster.[59,105] The spin localization computations for the corresponding mono oxidized and mono reduced species usually confirm these predictions. Interestingly, in the open shell $Sc_3C_2@C_{80}$ endofullerene, oxidation and reduction apparently involves only the cluster, and affects specifically the formal charge of

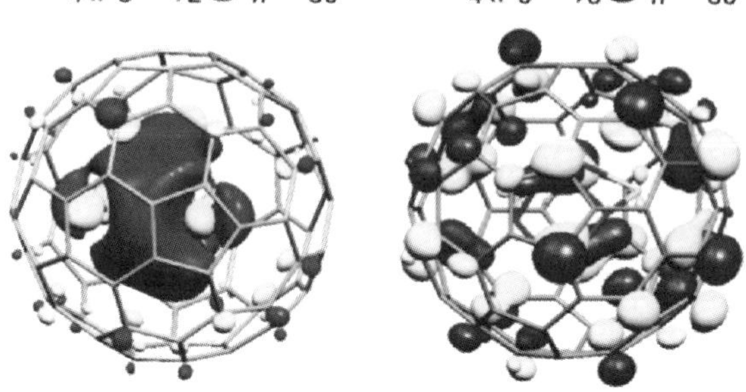

Fig. 4 Computed HOMOs of $Sc_4O_2@C_{80}$ (I_h) and $Sc_4O_3@C_{80}$ (I_h). Reprinted with permission from R. Valencia, A. Rodríguez-Fortea, S. Stevenson, A. L. Balch, J. M. Poblet, *Inorg. Chem.*, **48**, 5957 (2009). Copyright 2009 American Chemical Society."

the C_2 unit: from −2 in $Sc_3C_2@C_{80}^+$ to −6 in $Sc_3C_2@C_{80}$.[3–73] These results emphasize again the remarkable charge flexibility of the C_2 unit.

Several metallic oxide cluster fullerenes have been characterized recently, including $Sc_4O_2@C_{80}$ (I_h) and $Sc_4O_3@C_{80}$ (I_h). In these species, the C_{80} cage receives six electrons from the cluster and the oxygen atoms carry a −2 charge. Two scandium atoms are thus formally in the +3 oxidation state and two in the +2 oxidation state in $Sc_4O_2@C_{80}$ (I_h).[107] The HOMO (Fig. 4) and LUMO of this compound are mostly delocalized on the metal atoms, which suggest that both reduction and oxidation concerns the cluster. Spin density analysis of the mono reduced and mono oxidized compound confirmed this fact. The computations also predicted a small electrochemical gap for this endohedral fullerene. In $Sc_4O_3@C_{80}$ (I_h), in contrast, the four scandium atoms are in the +3 oxidation state.[107] Its HOMO is mainly a cage orbital (Fig. 4), whereas its LUMO is still localized on the cluster. A larger electrochemical gap was predicted for this compound.

Metallic sulfide clusters with the general formula M_2S (M= Sc, Y, Dy and Lu) have been characterized recently for several fullerenes, including C_{82} (C_s) and (C_{3v}),[39,108] and non-IPR C_{70} (C_2)[109] and C_{72} (C_s).[75] DFT studies have been very helpful to identify the cages, and to establish that the cluster formally transfers four electrons to the carbon cages in these structures. The calculations also showed that the HOMO (Fig. 5) and LUMO of these compounds are mainly localized on the cage. It is thus not surprising that

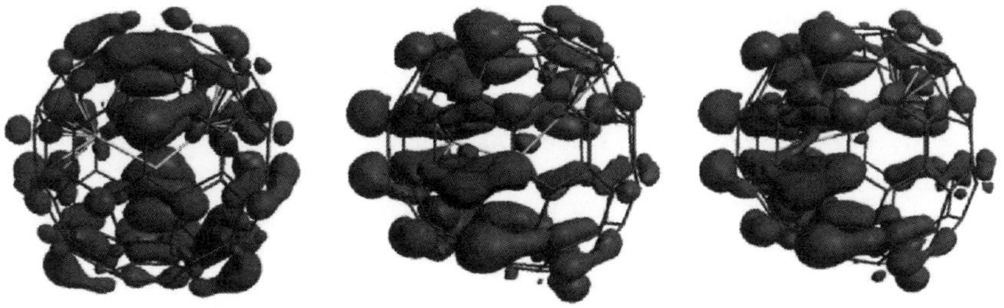

Fig. 5 Computed HOMOs of $Sc_2S@C_{82}$ (C_{3v}), $Sc_2S@C_{82}$ (C_s) and $Sc_2O@C_{82}$ (C_s). Reprinted with permission from B. Q. Mercado, N. Chen, A. Rodríguez-Fortea, M. A. Mackey, S. Stevenson, L. Echegoyen, J. M. Poblet, M.M. Olmstead, A. L. Balch, *J. Am. Chem. Soc.*, 133, 6752 (2011). Copyright 2011 American Society."

$Sc_2S@C_{82}$ (C_s) and (C_{3v}) exhibit quite different electrochemical gaps. Interestingly, the frontier orbitals of $Sc_2O@C_{82}$ (C_s) are localized on the cage as well (Fig. 5), which explains why the electrochemical behavior of the isoelectronic $Sc_2S@C_{82}$ (C_s) and $Sc_2O@C_{82}$ (C_s) species is very similar despite the different nature of the clusters.[108]

ACKNOWLEDGMENTS

L.E. thanks the National Science Foundation, Grants CHE-1110967 and CHE-1124075 for financial support and the Robert A. Welch Foundation for an endowed chair, Grant AH-0033. M.N.C. thanks the Centro de Excelencia en Nuevos Materiales (CENM) and the Vicerrectoría de Investigaciones of Universidad del Valle for financial support.

REFERENCES

1. Kroto HW, Heath JR, O'Brien SC, *et al.* (1985) C_{60}: Buckminsterfullerene. *Nature* **318:** 162–163.
2. Shinohara H. (2000) Endohedral metallofullerenes. *Rep Prog Phys* **63:** 843–892.
3. Akasaka T, Nagase S (eds). (2002) *Endofullerenes: A New Family of Carbon Clusters.* Springer.

4. Dunsch L, Yang S. (2007) Metal nitride cluster fullerenes: their current state and future prospects. *Small* **3:** 1298–1320.
5. Dunsch L, Yang S. (2007) Endohedral clusterfullerenes-playing with cluster and cage sizes. *Phys Chem Chem Phys* **9:** 3067–3081.
6. Chaur MN, Melin F, Ortiz AL, Echegoyen L. (2009) Chemical, electrochemical, and structural properties of Endohedral Metallofullerenes. *Angew Chem Int Ed* **48:** 7514–7538.
7. Yang S, Liu F, Chen C, *et al.* (2011) Fullerenes encaging metal clusters — clusterfullerenes. *Chem Commun* **47:** 11822–11839.
8. Rodriguez-Fortea A, Balch AL, Poblet JM. (2012) Endohedral metallofullerenes: A unique host–guest association. *Chem Soc Rev* **40:** 3551–3563.
9. Lu X, Feng L, Akasaka T, Nagase S. (2012) Current status and future developments of endohedral metallofullerenes. *Chem Soc Rev* **41:** 7723–7760.
10. Akasaka T, Okubo S, Kondo M, *et al.* (2000) Isolation and characterization of two Pr@C_{82} isomers. *Chem Phys Lett* **319:** 153–156.
11. Suzuki T, Maruyama Y, Kato T, *et al.* (1993) Electrochemical properties of La@C_{82}. *J Am Chem Soc* **115:** 11006–11007.
12. Suzuki T, Maruyama Y, Kato T, *et al.* (1994) Electrochemical properties of lanthanofullerenes. *Proc Electrochem Soc* **94–24:** 1077–1086.
13. Yamamoto K, Funasaka H, Takahashi T, Akasaka T. (1994) Isolation of an ESR-Active Metallofullerene of La@C_{82}. *J Phys Chem* **98:** 2008–2011.
14. Kikuchi K, Nakao Y, Suzuki S, *et al.* (1994) Characterization of the Isolated Y@C_{82}. *J Am Chem Soc* **116:** 9367–9368.
15. Suzuki T, Kikuchi K, Nakao Y, *et al.* (1995) Electrochemical studies of fullerenolanthanides. *Proc Electrochem Soc* **95–10:** 259–266.
16. Suzuki T, Maruyama Y, Kato T. (1995) Electrochemical properties of fullerenolanthanides. *Synth Met* **70:** 1443–1446.
17. Kirbach U, Dunsch L. (1996) The existence of stable Tm@C_{82} isomers. *Angew Chem Int Ed* **35:** 2380–2383.
18. Suzuki T, Kikuchi K, Oguri F, *et al.* (1996) Electrochemical properties of fullerenolanthanides. *Tetrahedron* **52:** 4973–4982.
19. Wang W, Ding J, Yang S, Li X-Y. (1997) Electrochemical properties of 4f-block metallofullerenes. *Proc Electrochem Soc* **97–14:** 417–428.
20. Dunsch L, Kirbach U, Kuran P. (1997) Electrochemistry of mono- and dimetallofullerenes. *Proc Electrochem Soc* **97–42:** 761–771.
21. Akasaka T, Okubo S, Wakahara T, *et al.* (1999) Isolation and characterization of a Pr@C_{82} isomer. *Proc Electrochem Soc* **99–12:** 771–778.
22. Sun B, Li M, Luo H, *et al.* (2002) Electrochemical properties of metallofullerenes and their anions. *Electrochim Acta* **47:** 3545–3549.

23. Xu J, Li M, Shi Z, Gu Z. (2006) Electrochemical survey: The effect of the cage size and structure on the electronic structures of a series of ytterbium metallofullerenes. *Chem Eur J* **12:** 562–567.
24. Zhang Y, Xu J, Hao C, *et al.* (2006) Synthesis, isolation, spectroscopic and electrochemical characterization of some calcium-containing metallofullerenes. *Carbon* **44:** 475–479.
25. Lu X, Slanina Z, Akasaka T, *et al.* (2010) Yb@C_{2n} (n = 40, 41, 42): New fullerene allotropes with unexplored electrochemical properties. *J Am Chem Soc* **132:** 5896–5905.
26. Iiduka Y, Ikenaga O, Sakuraba A, *et al.* (2005) Chemical reactivity of Sc_3N@C_{80} and La_2@C_{80}. *J Am Chem Soc* **127:** 9956–9957.
27. Lu X, Nikawa H, Nakahodo T, *et al.* (2008) Chemical understanding of a non-IPR metallofullerene: stabilization of encaged metals on fused-pentagon bonds in La_2@C_{72}. *J Am Chem Soc* **130:** 9129–9136.
28. Suzuki T, Maruyama Y, Kato T, *et al.* (1995) Electrochemistry and ab initio study of the dimetallofullerene La_2@C_{80}. *Angew Chem Int Ed* **34:** 1094–1096.
29. Anderson MR, Dorn HC, Stevenson SA. (2000) Making connections between metallofullerenes and fullerenes: Electrochemical investigations. *Carbon* **38:** 1663–1670.
30. Kato H, Taninaka A, Sugai T, Shinohara H. (2003) Structure of a missing-caged metallofullerene: La_2@C_{72}. *J Am Chem Soc* **125:** 7782–7783.
31. Cao B, Wakahara T, Tsuchiya T, *et al.* (2004) Isolation, characterization, and theoretical study of La_2@C_{78}. *J Am Chem Soc* **126:** 9164–9165.
32. Yamada M, Nakahodo T, Wakahara T, *et al.* (2005) Positional control of encapsulated atoms inside a fullerene cage by exohedral addition. *J Am Chem Soc* **127:** 14570–14571.
33. Yamada M, Wakahara T, Tsuchiya T, *et al.* (2008) Spectroscopic and theoretical study of endohedral dimetallofullerene having a non-IPR fullerene cage: Ce_2@C_{72}. *J Phys Chem A* **112:** 7627–7631.
34. Diener MD, Alford JM. (1998) Isolation and properties of small-bandgap fullerenes. *Nature* **393:** 668–671.
35. Bolskar RD, Alford JM. (2003) Chemical oxidation of endohedral metallofullerenes: Identification and separation of distinct classes. *Chem. Commun* 1292–1293.
36. Stevenson S, Rice G, Glass T, *et al.* (1999) Small-bandgap endohedral metallofullerenes in high yield and purity. *Nature* **401:** 55–57.
37. Wang C-R, Kai T, Tomiyama T, *et al.* (2001) A scandium carbide endohedral metallofullerene: (Sc_2C_2)@C_{84}. *Angew Chem Int Ed* **40:** 397–399.

38. Stevenson S, Mackey MA, Stuart MA, *et al.* (2008) A distorted tetrahedral metal oxide cluster inside an icosahedral carbon cage. Synthesis, isolation, and structural characterization of $Sc_4(\mu_3\text{-}O)_2$@*Ih*-C_{80}. *J Am Chem Soc* **130:** 11844–11845.
39. Dunsch L, Yang S, Zhang L, *et al.* (2010) Metal sulfide in a C_{82} fullerene cage: A new form of endohedral clusterfullerenes. *J Am Chem Soc* **132:** 5413–5421.
40. Wang T-S, Feng L, Wu J-Y, *et al.* (2010) Planar quinary cluster inside a fullerene cage: Synthesis and structural characterizations of $Sc_3NC@C_{80}$-I_h. *J Am Chem Soc* **132:** 16362–16364.
41. Wu JY, Wang TS, Ma YH, *et al.* (2011) Synthesis, isolation, characterization, and theoretical studies of $Sc_3NC@C_{78}$-C_2. *J Phys Chem C* **155:** 23755–23759.
42. Feng Y, Wang T, Wu J, *et al.* (2013) Spin-active metallofullerene stabilized by the core of an NC moiety. *Chem Commun* **49:** 2148–2150.
43. Elliott B, Yu L, Echegoyen L. (2005) A simple isomeric separation of D_{5h} and I_h $Sc_3N@C_{80}$ by selective chemical oxidation. *J Am Chem Soc* **127:** 10885–10888.
44. Cardona CM, Elliott B, Echegoyen L. (2006) Unexpected chemical and electrochemical properties of $M_3N@C_{80}$ (M = Sc, Y, Er). *J Am Chem Soc* **128:** 6480–6485.
45. Melin F, Chaur MN, Engmann S, *et al.* (2007) The large $Nd_3N@C_{2n}$ ($40 \le n \le 49$) cluster fullerene family: Preferential templating of a C_{88} cage by a trimetallic nitride cluster. *Angew Chem Int Ed* **46:** 9032–9035.
46. Chaur MN, Melin F, Elliott B, *et al.* (2008) New $M_3N@C_{2n}$ endohedral metallofullerene families (M = N, Pr, Ce; n = 40–53): Expanding the preferential templating of the C_{88} cage and approaching the C_{96} cage. *Chem Eur J* **14:** 4594–4599.
47. Chaur MN, Melin F, Ashby J, *et al.* (2008) Lanthanum nitride endohedral fullerenes $La_3N@C_{2n}$ ($43 \le n \le 55$): Preferential Formation of $La_3N@C_{96}$. *Chem Eur J* **14:** 8213–8219.
48. Chaur MN, Athans AJ, Echegoyen L. (2008) Metallic nitride endohedral fullerenes: Synthesis and electrochemical properties. *Tetrahedron* **64:** 11387–11393.
49. Cai T, Xu L, Anderson MR, *et al.* (2006) Structure and enhanced reactivity rates of the D_{5h} $Sc_3N@C_{80}$ and $Lu_3N@C_{80}$ metallofullerene isomers: The importance of the pyracylene motif. *J Am Chem Soc* **128:** 8581–8589.
50. Yang S, Zalibera M, Rapta P, Dunsch L. (2006) Charge-induced reversible rearrangement of endohedral fullerenes: Electrochemistry of tridysprosium

nitride clusterfullerenes $Dy_3N@C_{2n}$ (2n = 78, 80). *Chem Eur J* **12:** 7848–7855.

51. Chen N, Zhang EY, Wang CR. (2006) C_{80} encaging four different atoms: The synthesis, isolation, and characterizations of $ScYErN@C_{80}$. *J Phys Chem B* **110:** 13322–13325.
52. Chaur MN, Melin F, Elliott B, *et al.* (2007) $Gd_3N@C_{2n}$ (n = 40, 42, and 44): remarkably low HOMO-LUMO gap and unusual electrochemical reversibility of $Gd_3N@C_{88}$. *J Am Chem Soc* **129:** 14826–14829.
53. Zuo T, Olmstead MM, Beavers CM, *et al.* (2008) Preparation and structural characterization of the I_h and the D_{5h} isomers of the endohedral fullerenes $Tm_3N@C_{80}$: Icosahedral C_{80} cage encapsulation of a trimetallic nitride magnetic cluster with three uncoupled Tm^{3+} ions. *Inorg Chem* **47:** 5234–5244.
54. Alvarez L, Pichler T, Georgi P, *et al.* (2002) Electronic structure of pristine and intercalated $Sc_3N@C_{80}$ metallofullerene. *Phys Rev B* **66:** 035107/1–035107/7.
55. Krause M, Liu X, Wong J, *et al.* (2005) The electronic and vibrational structure of endohedral $Tm_3N@C_{80}$ (I) fullerene — Proof of an encaged Tm^{3+}. *J Phys Chem A* **109:** 7088–7093.
56. Shiozawa H, Rauf H, Pichler T, *et al.* (2005) Electronic structure of the trimetal nitride fullerene $Dy_3N@C_{80}$. *Phys Rev B* **72:** 195409/1–195409/5.
57. Chaur MN, Valencia R, Rodríguez-Fortea A, *et al.* (2009) Trimetallic nitride endohedral fullerenes: Experimental and theoretical evidence for the $M_3N^{6+}@C_{2n}^{\ 6-}$ model. *Angew Chem Int Ed* **48:** 1425–1428.
58. Popov AA, Dunsch L. (2007) Structure, stability, and cluster-cage interactions in nitride clusterfullerenes $M_3N@C_{2n}$ (M = Sc, Y; 2n = 68–98): A density functional theory study. *J Am Chem Soc* **129:** 11835–11849.
59. Popov AA, Dunsch L. (2008) Hindered cluster rotation and 45Sc hyperfine splitting constant in distonoid anion radical $Sc_3N@C_{80}^{\ -}$, and spatial spin–charge separation as a general principle for anions of endohedral fullerenes with metal-localized lowest unoccupied molecular orbitals. *J Am Chem Soc* **130:** 17726–17742.
60. Chaur MN, Aparicio-Anglès X, Mercado BQ, *et al.* (2010) Structural and electrochemical property correlations of metallic nitride endohedral metallofullerenes. *J Phys Chem C* **114:** 13003–13009.
61. Yang S, Chen C, Popov AA, *et al.* (2009) An endohedral titanium(III) in a clusterfullerene: putting a non-group-III metal nitride into the C_{80}-I_h fullerene cage. *Chem Commun* 6391–6393.

62. Popov AA, Chen C, Yang S, *et al.* (2010) Spin-flow vibrational spectroscopy of molecules with flexible spin density: Electrochemistry, ESR, cluster and spin dynamics, and bonding in $TiSc_2N@C_{80}$. *ACS Nano* **4:** 4857–4871.
63. Chen C, Liu F, Li S, *et al.* (2012) Titanium/Yttrium Mixed Metal Nitride Clusterfullerene $TiY_2N@C_{80}$: Synthesis, Isolation, and Effect of the Group-III Metal. *Inorg Chem* **51:** 3039–3045.
64. Mercado BQ, Beavers CM, Olmstead MM, *et al.* (2008) Is the isolated pentagon rule merely a suggestion for endohedral fullerenes? The structure of a second egg-shaped endohedral fullerene-$Gd_3N@C_s$(39663)-C_{82}. *J Am Chem Soc* **130:** 7854–7855.
65. Zuo T, Walker K, Olmstead MM, *et al.* (2008) New egg-shaped fullerenes: non-isolated pentagon structures of $Tm_3N@C_s$(51 365)-C_{84} and $Gd_3N@C_s$(51 365)-C_{84}. *Chem Commun* 1067–1069.
66. Yang S, Rapta P, Dunsch L. (2007) The spin state of a charged non-IPR fullerene: The stable radical cation of $Sc_3N@C_{68}$. *Chem Commun* 189–191.
67. Cai T, Xu L, Shu C, *et al.* (2008) Selective formation of a symmetric $Sc_3N@C_{78}$ bisadduct: Adduct docking controlled by an internal trimetallic nitride cluster. *J Am Chem Soc* **130:** 2136–2137.
68. Beavers CM, Chaur MN, Olmstead MM, *et al.* (2009) Large metal ions in a relatively small fullerene cage: the structure of $Gd_3N@C_2$(22010)-C_{78} departs from the isolated pentagon Rule. *J Am Chem Soc* **131:** 11519–11524.
69. Rapta P, Popov AA, Yang S, Dunsch L. (2008) Charged states of $Sc_3N@C_{68}$: an in situ spectroelectrochemical study of the radical cation and radical anion of a non-IPR fullerene. *J Phys Chem A* **112:** 5858–5865.
70. Jakes P, Dinse K-P. (2001) Chemically induced spin transfer to an encased molecular cluster. An EPR study of $Sc_3N@C_{80}$ radical anions. *J Am Chem Soc* **123:** 8854–8855.
71. Iiduka Y, Wakahara T, Nakajima K, *et al.* (2007) Experimental and theoretical studies of the scandium carbide endohedral metallofullerene $Sc_2C_2@C_{82}$ and its carbene derivative. *Angew Chem Int Ed* **46:** 5562–5564.
72. Wakahara T, Sakuraba A, Iiduka Y, *et al.* (2004) Chemical reactivity and redox property of $Sc_3@C_{82}$. *Chem Phys Lett* **398:** 553–556.
73. Tan K, Lu X. (2006) Electronic structure and redox properties of the open-shell metal–carbide endofullerene $Sc_3C_2@C_{80}$: A density functional theory investigation. *J Phys Chem A* **110:** 1171.
74. Wang TS, Chen N, Xiang JF, *et al.* (2009) Russian-doll-type metal carbide endofullerene: Synthesis, isolation, and characterization of $Sc_4C_2@C_{80}$. *J Am Chem Soc* **131:** 16646–16647.

75. Chen N, Beavers CM, Mulet-Gas M, *et al.* (2012) $Sc_2S@C_s(10528)-C_{72}$: A dimetallic sulfide endohedral fullerene with a non isolated pentagon rule cage. *J Am Chem Soc* **134:** 7851–7860.
76. Chen N, Chaur MN, Moore C, *et al.* (2010) Synthesis of a new endohedral fullerene family, $Sc_2S@C_{2n}$ (n = 40–50) by the introduction of SO_2. *Chem Commun* **46:** 4818–4820.
77. Popov AA, Chen N, Pinzon JR, *et al.* (2012) Redox-active scandium oxide cluster inside a fullerene cage: Spectroscopic, voltammetric, electron spin resonance spectroelectrochemical, and extended density functional theory study of $Sc_4O_2@C_{80}$ and its ion radicals. *J Am Chem Soc* **134:** 19607.
78. Popov AA, Dunsch L. (2009) Bonding in endohedral metallofullerenes as studied by quantum theory of atoms in molecules. *Chem Eur J* **15:** 9707–9729.
79. Heath JR, O'Brien SC, Zhang Q, *et al.* (1985) Lanthanum complexes of spheroidal carbon shells. *J Am Chem Soc* **107:** 7779–7780.
80. Chai Y, Guo T, Jin C, *et al.* (1991) Fullerenes with metals inside. *J Phys Chem* **95:** 7564–7568.
81. Laasonen K, Andreoni W, Parrinello M. (1992) Structural and electronic properties of $La@C_{82}$. *Science* **258:** 1916–1919.
82. Nagase S, Kobayashi K, Kato T, Achiba Y. (1993) A theoretical approach to C_{82} and LaC_{82}. *Chem Phys Lett* **201:** 475–480.
83. Nagase S, Kobayashi K. (1993) Metallofullerenes MC_{82} (M = Sc, Y, and La). A theoretical study of the electronic and structural aspects. *Chem Phys Lett* **214:** 57–63.
84. Nagase S, Kobayashi K. (1994) Theoretical study of the lanthanide fullerene CeC_{82}. Comparison with ScC_{82}, YC_{82} and LaC_{82}. *Chem Phys Lett* **228:** 106–110.
85. Nagase S, Kobayashi K. (1994) The ionization energies and electron affinities of endohedral metallofullerenes MC_{82} (M = Sc, Y, La): density functional calculations. *J Chem Soc Chem Commun* 1837–1838.
86. Kobayashi K, Nagase S. (1997) Structures of the $Ca@C_{82}$ isomers: A theoretical prediction. *Chem Phys Lett* **274:** 226–230.
87. Lu J, Zhang X, Zhao X, *et al.* (2000) Strong metal-cage hybridization in endohedral $La@C_{82}$, $Y@C_{82}$ and $Sc@C_{82}$. *Chem Phys Lett* **332:** 219–224.
88. Amamiya S, Okada S, Suzuki S, Nakao K. (2001) Relativistic calculation on electronic structure of $La@C_{82}$ molecule and crystal. *Synthetic Met* **121:** 1137–1138.

89. Slanina Z, Uhlík F, Lee S-L, *et al.* (2011) Computed stabilities in metallofullerene series: Al@C_{82}, Sc@C_{82}, Y@C_{82}, and La@C_{82}. *J Quantum Chem* **111:** 2712–2718.
90. Liu J, Shi Z, Gu Z. (2009) The cage and metal effect: Spectroscopy and electrochemical survey of a series of Sm-containing high metallofullerenes. *Chem Asian J* **4:** 1703–1711.
91. Kaoru K, Shigeru N, Takeshi A. (1995) A theoretical study of C_{80} and La_2@C_{80}. *Chem Phys Lett* **245:** 230–236.
92. Kobayashi K, Nagase S. (1996) Structures and electronic states of endohedral dimetallofullerenes: M_2@C_{80} (M = Sc, Y, La, Ce, Pr, Eu, Gd, Yb and Lu). *Chem Phys Lett* **262:** 227–232.
93. Muthukumar K, Larsson JA. (2008) Explanation of the different preferential binding sites for Ce and La in M_2@C_{80} (M = Ce, La)-a density functional theory prediction. *J Mater Chem* **18:** 3347–3351.
94. Fowler PW. (1986) How unusual is C_{60}? Magic numbers for carbon clusters. *Chem Phys Lett* **131:** 444–450.
95. Fowler PW, Zerbetto F. (1995) Charging and equilibration of fullerene isomers. *Chem Phys Lett* **243:** 36–41.
96. Muthukumar K, Larsson JA. (2010) The role of ellipticity on the preferential binding site of Ce and La in C_{78}-D_{3h}-A density functional theory study. *Nanoscale* **2:** 1250–1255.
97. Rodriguez-Fortea A, Campanera JM, Cardona CM, *et al.* (2006) Dancing on a fullerene surface: Isomerization of Y_3N@(N-ethylpyrrolidino-C_{80}) from the 6,6 to the 5,6 regioisomer. *Angew Chem Int Ed* **45:** 8176–8180.
98. Valencia R, Rodríguez-Fortea A, Clotet A, *et al.* (2009) Electronic structure and redox properties of metal nitride endohedral fullerenes M_3N@C_{2n} (M = Sc, Y, La, and Gd; 2n = 80, 84, 88, 92, 96). *Chem Eur J* **15:** 10997–11009.
99. Li F-F, Rodríguez-Fortea A, Poblet JM, Echegoyen L. (2011) Reactivity of metallic nitride endohedral metallofullerene anions: Electrochemical synthesis of a Lu_3N@I_h-C_{80} derivative. *J Am Chem Soc* **133:** 2760–2765.
100. Li F–F, Rodríguez-Fortea A, Peng P, *et al.* (2012) Electrosynthesis of a Sc_3N@Ih-C_{80} methano derivative from trianionic Sc_3N@I_h-C_{80}. *J Am Chem Soc* **134:** 7480–7487.
101. Krause M, Popov A, Dunsch L. (2006) Vibrational structure of endohedral fullerene Sc_3N@C_{78} (D3h′): Evidence for a strong coupling between the Sc_3N cluster and C_{78} cage. *Chem Phys Chem* **7:** 1734–1740.
102. Rodríguez-Fortea A, Alegret N, Balch AL, Poblet JM. (2010) The maximum pentagon separation rule provides a guideline for the structures of endohedral metallofullerenes. *Nat Chem* **2:** 955–961.

103. Campanera JM, Bo C, Poblet JM. (2005) General rule for the stabilization of fullerene cages encapsulating trimetallic nitride templates. *Angew Chem Int Ed* **44:** 7230–7233.
104. Valencia R, Rodríguez-Fortea A, Poblet JM. (2007) Large fullerenes stabilized by encapsulation of metallic clusters. *Chem Commun* 4161–4163.
105. Valencia R, Rodríguez-Fortea A, Poblet JM. (2008) Understanding the stabilization of metal carbide endohedral fullerenes $M_2C_2@C_{82}$ and related systems. *J Phys Chem A* **112:** 4550–4555.
106. Nishimoto Y, Wang Z, Morokuma K, Irle S. (2012) Molecular and electronic structures of endohedral fullerenes, $Sc_2C_2@C_{3v}–C_{82}$ and $Sc_2@C_{3v}–C_{82}$: benchmark for SCC-DFTB and proposal of new inner cluster structures. *Phys Status Solidi B* **249:** 324–334.
107. Valencia R, Rodríguez-Fortea A, Stevenson S, *et al.* (2009) Electronic structures of scandium oxide endohedral metallofullerenes, $Sc_4(\mu_3\text{-}O)_n@I_h\text{-}C_{80}$ (n = 2, 3). *Inorg Chem* **48:** 5957–5961.
108. Mercado BQ, Chen N, Rodríguez-Fortea A, *et al.* (2011) The shape of the $Sc_2(\mu_2\text{-}S)$ unit trapped in C_{82}: Crystallographic, computational, and electrochemical studies of the isomers, $Sc_2(\mu_2\text{-}S)@C_s(6)\text{-}C_{82}$ and $Sc_2(\mu_2\text{-}S)@C_{3v}(8)\text{-}C_{82}$. *J Am Chem Soc* **133:** 6752–6760.
109. Chen N, Mulet-Gas M, Li YY, Stene RE, Atherton CW, Rodriguez-Fortea A, Poblet J M., Echegoyen L. (2013) $Sc_2S@C_2(7892)–C_{70}$: A metallic sulfide cluster inside a non-IPR C_{70} cage. *Chem. Sci.* **4:** 180–186.

CHAPTER 11

Chemical Functionalization of Endohedral Metallofullerenes

*Yutaka Maeda**

Endohedral metallofullerenes, which encapsulate one or more metal atoms inside a hollow fullerene cage, have attracted special attention because they engender new spherical molecules with unique electronic properties and structures that are unexpected for empty fullerenes. This chapter describes the chemical functionalization of endohedral metallofullerenes.

1. INTRODUCTION

Since the first proposal of the spherical molecular structure of C_{60} in 1985, encapsulation of metal atoms inside a hollow fullerene cage has been attempted.[1,2] In 1991, Smalley *et al.* reported the synthesis and extraction of a lanthanum-encapsulated fullerene, La@C_{82}.[3] Triggered by extraction of La@C_{82} as a stable molecule, great efforts have been made for the synthesis and characterization of novel metallofullerenes. Understanding their unprecedented structure and how the chemical and electronic properties of empty fullerenes change upon endohedral metal doping are of great

*Department of Chemistry, Tokyo Gakugei University 4-1-1 Nukuikitamachi, Koganei Tokyo, Japan. E-mail: ymaeda@u-gakugei.ac.jp

interest.[4–13] Theoretical and experimental studies of metallofullerenes have revealed that remarkable metallofullerene characteristics originate in the electron transfer that takes place from the encapsulated metal atoms to the carbon cages. For example, the formal electronic structure of La@C_{82} is described as $La^{3+}C_{82}^{3-}$. A pioneering study of metallofullerenes revealed that not only the characteristic redox properties, but also the chemical reactivity of La@C_{82}, are significantly different from that of an empty fullerene.[14]

Metallofullerenes of various types, which are distinguished by the kind and the number of metal atoms and structure of fullerene carbon cages, are produced and characterized. The recent successful production and isolation of endohedral metallofullerenes has encouraged the chemical functionalization of endohedral metallofullerenes, which facilitates the clarification of their physical and chemical properties. This review presents a summary of recent advances in the functionalization of endohedral metallofullerenes. The combination of endohedral and exohedral functionalization of fullerenes will introduce new concepts of chemistry, and will be of great help in fullerene-based applications.

2. FUNCTIONALIZATION OF ENDOHEDRAL METALLOFULLERENES

Various types of functionalization of endohedral metallofullerenes have been performed based on the reaction of empty fullerenes. The effect of metal doping into the fullerene cages was revealed in diverse studies of exohedral functionalization. As described later, regioselective functionalization of metallofullerenes and structural analyses of the derivatives have become important topics because regioselectivity provides fundamental information related to the chemical properties of metallofullerenes and leads to improvements in the mass production of derivatives for practical applications.

Because La@C_{2n} is known as a prototype of metallofullerenes, the reactivity and regioselectivity of La@C_{2v}-C_{82} and La_2@I_h-C_{80} have been well investigated. Although La@C_{2v}-C_{82} has nonequivalent C–C bonds of 35 types and 24 nonequivalent carbon atoms in the C_{2v}-C_{82} cage, high regioselectivity has been observed in several reactions. Structural determination of La@C_{2v}-C_{82} derivatives indicates that charge density, spin density, and local strain are important factors for regioselectivity depending on the reaction (Table 1).

Table 1 Mulliken Charge Density, Spin Density, and POAV of La@C_{2v}-C_{82}

Carbon number	Charge density	Spin density	POAV
1	−0.136	−0.012	11.2
2	−0.170	0.023	11.3
3	−0.099	0.031	8.8
4	−0.071	0.035	9.6
5	−0.045	−0.019	9.2
6	−0.037	−0.004	10.5
7	−0.092	−0.242	9.1
8	−0.061	0.008	8.9
9	−0.022	0.037	9.9
10	−0.020	0.066	10.7
11	−0.021	−0.016	10.6
12	−0.047	−0.003	7.7
13	−0.006	−0.001	10.9
14	0.000	0.057	11.0
15	−0.027	0.014	7.4
16	−0.012	−0.001	10.6
17	−0.036	−0.031	8.2
18	0.004	0.045	11.0
19	−0.006	0.046	10.9
20	−0.026	0.011	8.4
21	0.002	−0.029	10.5
22	−0.006	−0.012	10.7
23	0.006	0.063	10.7
24	−0.025	−0.026	8.3

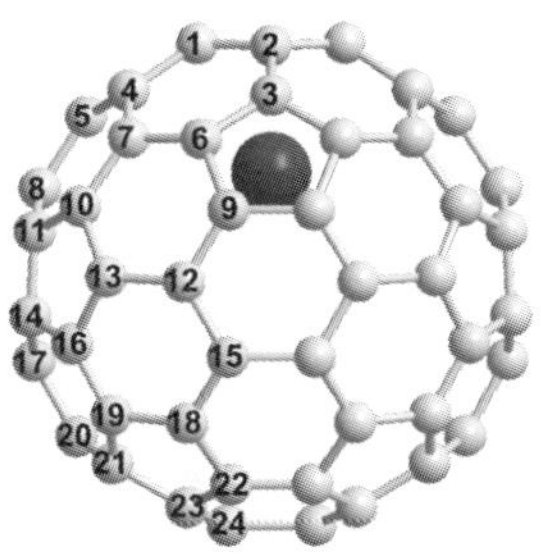

The I_h-C_{80} cage has advantages related to regioselectivity because of its high symmetrical structure, which has only nonequivalent C–C bonds of two types and nonequivalent carbons (Fig. 1). The chemistry of M_3N@I_h-C_{80} is also well investigated not only because of its symmetrical I_h-C_{80} cage, but also because of its high production yields. Regioselectivity

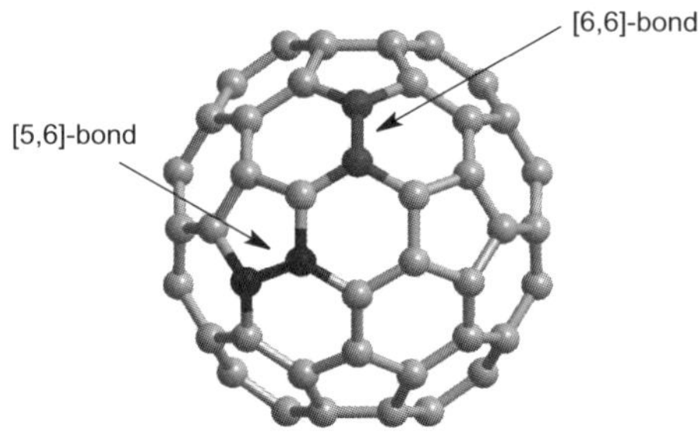

Fig. 1 Structure of I_h-C_{80}

depending on the metal atoms of $M_3N@C_{80}$ has been well studied and remains an interesting topic.

2.1. Nucleophilic Additions

The nucleophilic addition reaction of fullerenes using malonate anion, the Bingel–Hirsch reaction, has been widely used to produce functional fullerene-based materials. The Bingel–Hirsch reaction of $M_3N@C_n$ (M = Y, Sc, Lu: n = 68, 78, 80, 82, 84, 88) has been investigated under various conditions (Scheme 1). The reaction of $Y_3N@I_h$-C_{80} with bromomalonate in the presence of 1,8-diazabicyclo[5.4.0]undec-7-ene (DBU) afforded the corresponding [6,6]-open mono-adduct.[15] The structure of the mono-adduct was determined through crystallographic analysis.[16] It is particularly interesting that the Bingel–Hirsch reaction of $M_3N@I_h$-C_{80} (M = Sc, Lu) did not take place under the same reaction conditions.[15] This result revealed a change in the chemical reactivity of metallofullerenes depending on encapsulated metals. The mono-adduct of $M_3N@I_h$-C_{80} (M = Sc, Lu) was obtained when NaH was chosen as the base instead of DBU.[17] The addition of DMF with DBU also affords the corresponding mono-adduct. The theoretical study suggests that the presence of a polar solvent, such as DMF, lowers the energy of the transition state at the first step of the Bingel–Hirsch reaction, which might explain why the presence of DMF favors the reaction.[18]

The reaction of $Sc_3N@D_{3h}$-C_{78} with excess diethyl bromomalonate in the presence of DBU caused the formation of only one mono-adduct and one bis-adduct predominantly (>70%) with the formation of minor

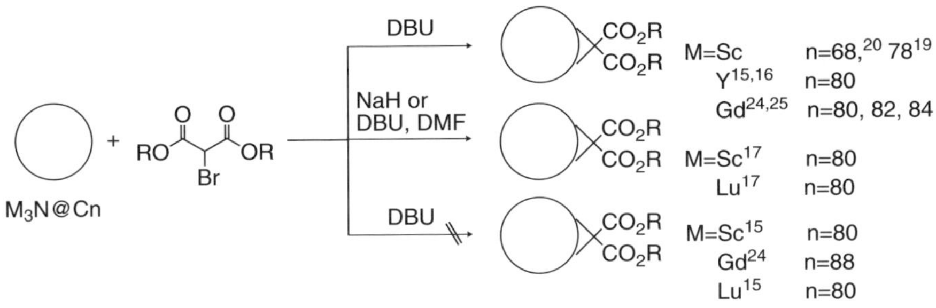

Scheme 1

amounts of the multi-adducts.[19] In contrast, $Sc_3N@I_h$-C_{80} does not react under the same reaction conditions, which indicates the high reactivity of $Sc_3N@D_{3h}$-C_{78} toward malonate anion. From the symmetrical ^{13}C-NMR spectra of mono-adduct and bis-adduct, the structures of these adducts were proposed. The regioselective addition of malonate anion was confirmed theoretically. It was supported by the surface electron density value of LUMO of mono-adduct, which corresponds to the kinetically preferred site for nucleophilic attack.

Non-IPR $Sc_3N@C_{68}$ derivatives $Sc_3N@C_{68}(C(CO_2Et)_2)_n$ (n = 1–5) were synthesized and characterized.[20] The HPLC analysis of the reaction mixture revealed only one fraction of mono-adduct with a fraction of multi-adducts. Four nonequivalent methylene hydrogens in the ^{1}H-NMR and 66 sp^2 carbon peaks in the aromatic region in ^{13}C-NMR of mono-adduct indicated that the highly regioselectively obtained mono-adduct has C_1 symmetry. The addition site of substituent at the adjacent carbons of the 5,5-ring junction was proposed by calculated LUMO electron density, absorption spectra, and comparison of the experimental and calculated ^{13}C-NMR. The experimental and theoretical results suggest that fused pentagons increase the reactivity of adjacent carbons of the 5,5-ring junction. In the reaction of non-IPR $La@C_{72}$ and $La_2@C_{72}$ with radical or carbene, high reactivity of the adjacent carbons of the 5,5-ring junction was also observed, as described later.[21–23]

The Bingel–Hirsch reaction of $Gd_3N@C_n$ (n = 80, 82, 84, 88) was also conducted.[24,25] Mono-adducts and bis-adducts were obtained from $Gd_3N@C_{80}$ and mono-adducts were obtained from $Gd_3N@C_{82}$ and $Gd_3N@C_{84}$. Furthermore, $Gd_3N@C_{88}$ did not react at all, which indicates that the fullerene cage size strongly influences the metallofullerene reactivity. A comprehensive computational study demonstrated that functionalization of Gd_3N non-IPR fullerenes via the Bingel–Hirsch reaction occurs at C–C bonds of which the distance is greater than the average C–C distance and of which the pyramidalization angles are greater than the average angle.

Singly bonded La@C_{2v}-C_{82} derivatives were prepared from the Bingel–Hirsch reaction.[26,27] In the reaction of La@C_{2v}-C_{82} with diethyl bromomalonate in the presence of DBU, four isomers of singly bonded mono-adducts as diamagnetic molecules were obtained along with one methanofullerene. Structures of singly bonded La@C_{2v}-C_{82} derivatives were analyzed using ^{13}C-NMR. The structure of the major isomer of La@C_{2v}-C_{82}(CBr(CO_2Et)$_2$) was determined using X-ray structure analysis. Theoretical studies revealed that the addition position carbon atom (C_{23}, see Table 1) is highly strained, and that it is the most positively charged. The carbon atom also has high spin density. The bis-adduct La@C_{2v}-C_{82}(CH(CO_2Et)$_2$)$_2$ was obtained when diethyl malonate was used instead of diethyl bromomalonate.[28] The structure of La@C_{2v}-C_{82}(CH(CO_2Et)$_2$)$_2$ was revealed by X-ray structure analysis. The addition positions of substituents are C21 and C23. It is particularly interesting that a pair of La@C_{2v}-C_{82}(CH(CO_2Et)$_2$)$_2$ formed a singly bonded dimer in the single crystal.

The highly regioselective and chemoselective reaction of Dy@C_{2v}-C_{82} with phosphonium ylide was reported (Scheme 3).[29] The mono-adduct was obtained in a 90% yield. X-ray structure analysis revealed that the substituent is bonded at the strained carbon atoms, which are close to the encapsulated Dy atom. One addition position carbon atom (addition position, C2 and C3, see Table 1) is highly strained and is the most negatively charged.

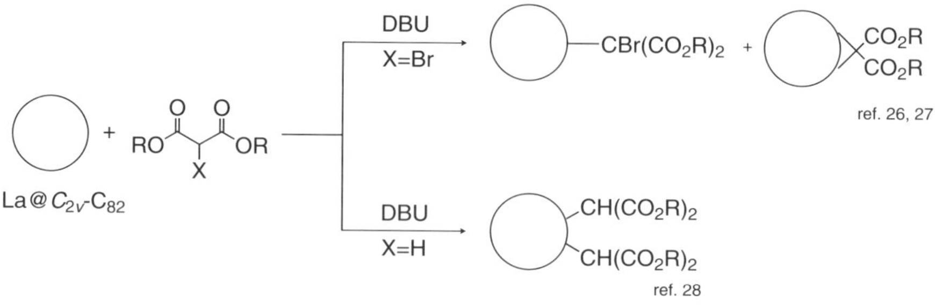

Scheme 2

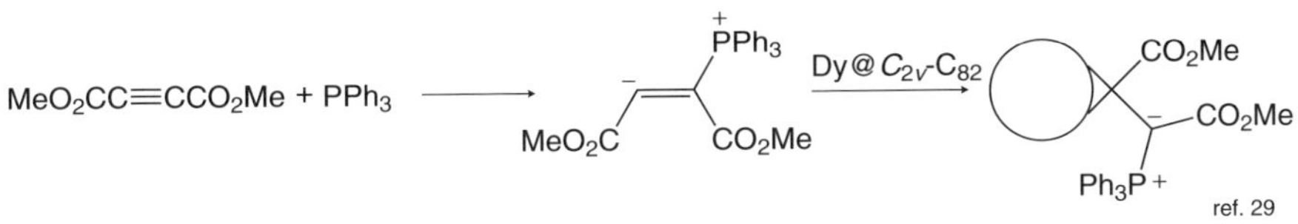

Scheme 3

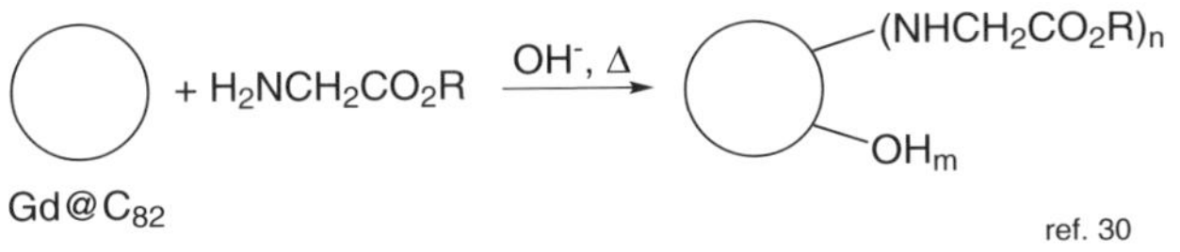

Scheme 4

Nucleophilic addition of glycine esters to $Gd@C_{82}$ was conducted.[30] The multiple addition products, $Gd@C_{82}(NHCH_2COOR)_n(OH)_m$, were detected using mass spectrometric analysis (Scheme 4).

2.2. 1,3-Dipole Cycloaddition

The Prato reaction of $M@C_{82}$ (M = Y,[31] La,[32] Gd[31,33]) was achieved in 2004 (Scheme 5). Reaction of La@C_{2v}-C_{82} with azomethine ylide, which was generated from *N*-methylglycine and paraformaldehyde, afforded pyrrolidino-La@C_{2v}-C_{82}. The HPLC, ESR, and MS analyses suggest production of two regioisomers of mono-adducts with the formation of bis-adducts. Similar results were obtained for the reaction of $M@C_{82}$ (M = Y, Gd). MS analysis of these pyrrolidino-$M@C_{82}$ revealed the formation of the mono-, bis-, tris-, hexa-, and penta-adducts.[31] It is particularly interesting that $Gd@C_{82}((CH(CO_2H))_2NCH_3)$ was formed in the reaction of $Gd@C_{82}$ with *N*-methylglycine in the presence of O_2 (Scheme 6).[33] As an intermediate for the production of $Gd@C_{82}((CH(CO_2H))_2NCH_3)$, generation of imine ($HO_2CHC = N^+(CH_3)C^-HCO_2H$) from *N*-methylglycine via oxidation by O_2 was proposed. Recently, two plausible reaction mechanisms were proposed for the reaction of C_{60} under similar reaction conditions.[34] One is a similar reaction mechanism. The other involves the addition of glycine radical via hydrogen abstraction by $^1O_2$, which is produced by energy transfer from excited C_{60}.

Recently, the Prato reaction was applied to the construction of a photoactive donor–acceptor hybrid based on metallofullerenes. 2-formyl-9,10-bis(1,3-dithiol-2-ylidene)-9,10-dihydroanthracene (exTTF-CHO) was used to synthesize La@C_{2v}-C_{82}/ex-TTF congugate.[35]

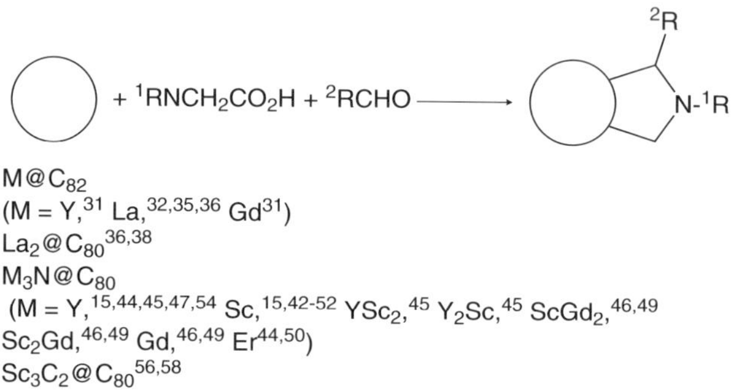

Scheme 5

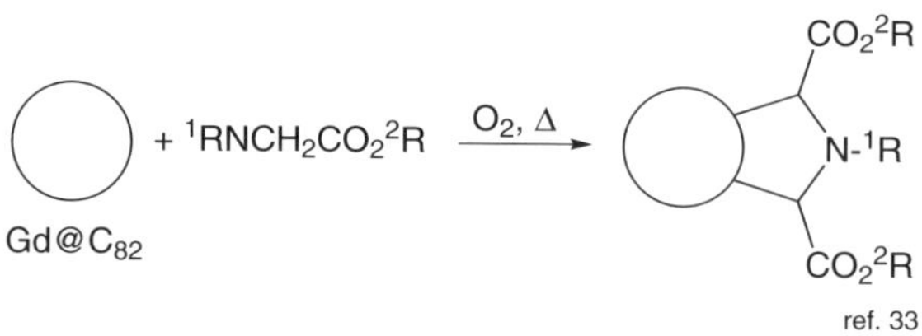

Scheme 6

Although 24 nonequivalent cage carbons exist in La@C_{2v}-C_{82}, only two isomers of mono-adduct (molar ratio 1:1) were formed, which was analyzed based on ESR of mono-adducts and NMR of mono-adduct anions. Highly regioselective Prato reaction of La@C_{2v}-C_{82} using pyridine-4-carbaldehyde was also reported.[36] Supramolecular donor–acceptor complex (La@C_{2v}-C_{82}(Py)/Por) was formed by the addition of zinc tetraphenylporphyrin to the pyridino-pyrrolidino-La@C_{2v}-C_{82} (La@C_{2v}-C_{82}(Py)). The addition positions of ylide (C21-C23, see Table 1) were proposed at the same position of La@C_{2v}-C_{82}(Cp*),[37] (Cp*: 1,2,3,4,5-pentamethyl-1,3-cyclopentadiene) based on the charge density, local strain, and molecular orbital of pristine fullerene and thermodynamic stability of adducts by DFT calculation. Laser flash photolysis (LFP) experiments of the donor–acceptor dyads (La@C_{2v}-C_{82}/ex-TTF and La@C_{2v}-C_{82}(Py)/Por) were conducted. The fast charge separation and charge recombination process were observed in both covalently and non-covalently bonded donor–acceptor conjugated La@C_{2v}-C_{82} compared to the corresponding C_{60} analogues.

Differences in the addition pattern of azomethine ylide at M_2@I_h-C_{80} were reported. The reaction with the azomethine ylide produced from the reaction of glycine and exTTF-CHO or pyridine-4-carbaldehyde in toluene under reflux for 10–60 min caused the formation of [5,6]-addition products of La_2@I_h-C_{80} (La_2@I_h-C_{80}/ex-TTF and La_2@I_h-C_{80}(Py)).[36,38] In contrast, the reaction of M_2@I_h-C_{80} (M=La, Ce) with *N*-triphenylmethyl-5-oxazolidinone, as a precursor of azomethine ylide, in toluene under reflux for 15 hr afforded not only [5,6]-, but also [6,6]-adduct (scheme 7).[39,40] It is particularly interesting that the molar ratio of [5,6]-adduct and [6,6]-adduct (La 1:4, Ce 1:1) depends on the internal metal, which suggests the regioselectivity of fullerenes or thermodynamic stability of adducts is tunable by the encapsulated metal atoms. Dorn *et al.* reported the isomerization of Sc_3N@I_h-C_{80}(C_2H_4NTrt) (Trt: triphenylmethyl) from [6,6]- to [5,6]-isomer at 130°C, which indicated that the [6,6]-adduct of Sc_3N@I_h-C_{80}(C_2H_4NTrt) is the kinetic product and the [5,6]-adduct of Sc_3N@I_h-C_{80}(C_2H_4NTrt) is the thermodynamic product.[41] In the thermal treatment of [6,6]-La_2@I_h-C_{80}(C_2H_4NTrt) at 60°C for one day, pristine La_2@I_h-C_{80} and the bis-adducts were detected instead of isomerization. The authors presumed that different formation ratios are attributable to the difference in stability of the isomers.

An LFP experiment of La_2@I_h-C_{80}/ex-TTF conjugate and La_2@I_h-C_{80}(Py)/porphyrin supramolecular was conducted. It was revealed that the lifetime of La_2@I_h-$C_{80}^{\bullet -}$/ex-TTF$^{\bullet +}$ is about one order of magnitude shorter than that of $C_{60}^{\bullet -}$/ex-TTF$^{\bullet +}$. La_2@I_h-C_{80}(Py)/porphyrin showed that fast charge

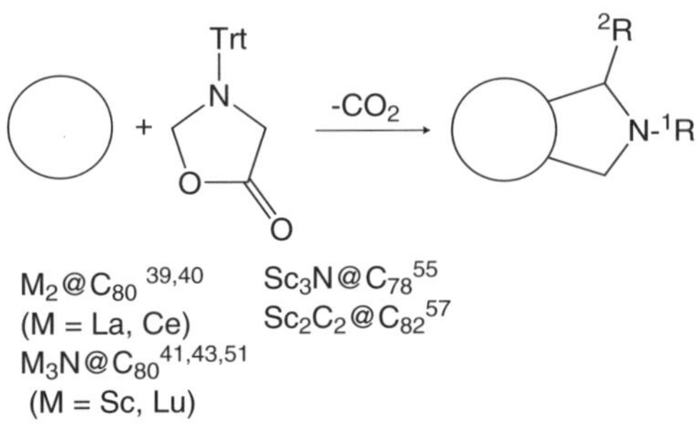

Scheme 7

separation and slow charge recombination process takes place compared to the corresponding C_{60} analogues. The fast charge separation process is explained by the low reduction potential of metallofullerenes. However, a slow charge recombination process was explained by the SOMO of $La_2@I_h\text{-}C_{80}^{\bullet-}$, which is localized between the encapsulated La atoms.

The Prato reaction of $M_3N@I_h\text{-}C_{80}$ (M = Sc,[15,42–53] Y,[15,44,45,47,54] Gd,[46,49] Er,[42,44,50] Lu[43,51]), $Sc_3N@C_{78}$,[55] $Sc_nC_2@C_m$ (n = 2,3; m = 80, 82)[56–58] has been widely investigated, revealing encapsulated metal effects on the reactivity and regioselectivity of fullerenes. Glycine derivatives and aldehyde (Scheme 5)[15,42,44–46,54] and *N*-triphenylmethyl-5-oxazolidinone (Scheme 7)[41,47,51] were used respectively as precursors of azomethine ylide. It is particularly interesting that the addition pattern of azomethine ylide, [5,6]-addition or [6,6]-addition, depends on the encapsulated metal clusters. Furthermore, encapsulated-cluster-dependent isomerization took place. Thermal isomerization of [5,6]-adduct to [6,6]-adduct was observed for $Gd_3N@I_h\text{-}C_{80}$.[46] However, thermal isomerization of [6,6]-adducts to [5,6]-adducts was observed for $M_3N@I_h\text{-}C_{80}$ (M = Sc,[41] Y,[44] Er,[44]) and $ScGd_2@I_h\text{-}C_{80}$.[46]

Systematic studies of the Prato reaction of $Sc_nY_{3-n}N@I_h\text{-}C_{80}$[45] and $Sc_nGd_{3-n}N@I_h\text{-}C_{80}$[46] (n = 0–3) revealed clearly that the regioselectivity of the $M_3N@I_h\text{-}C_{80}$ in the Prato reaction depends remarkably on the encaged metal atom. In the reaction of $Sc_3N@I_h\text{-}C_{80}$, [5,6]-adduct was produced predominantly; the yield of [6,6]-adduct increased with the decrease in the number of Sc atoms. From these systematic experimentally obtained results, the metal size in $I_h\text{-}C_{80}$ is proposed as an important factor in regioselectivity, as supported by results of theoretical studies.

It is particularly interesting that a cyclic voltammogram of [5,6]-*N*-alkyl-pyrrolidino-$M_3N@I_h\text{-}C_{80}$ (M = Sc,[48,52] Y[44], Er[44]) shows a reversible wave, whereas those of [6,6]-*N*-Trt-pyrrolidino-$Lu_3N@I_h\text{-}C_{80}$[51] and [6,6]-*N*-alkyl-pyrrolidino-$M_3N@I_h\text{-}C_{80}$ (M = Sc,[17] Y[44,47]) show an irreversible wave. As a notable exception, a reversible voltammogram result for [6,6]-*N*-Trt-pyrrolidino-$Sc_3N@I_h\text{-}C_{80}$, was reported in 2011.[51]

In the reaction of $Sc_3N@D_{5h}\text{-}C_{80}$ with *N*-triphenylmethyl-5-oxazolidinone, a mono-adduct and bis-adduct were obtained.[43] Comparison with $Sc_3N@I_h\text{-}C_{80}$, D_{5h} isomer showed high reactivity under the same conditions. Higher reactivity of D_{5h} isomer than I_h isomer was also observed in the reaction with cyclopentadiene derivatives, as described later.[43]

2.3. Carbene Addition

Reaction of diazirine to C_{60} was studied widely because C_{60} acts as a chemical probe for the photolysis of diazirine. Photoreaction of metallofullerenes with diazirine was widely studied.[23,37,59–69] Particularly, 2-adamantane-2,3'-[3H]-diazirine (AdN_2) was used for the precursor of electrophilic carbene because electrophilic carbene addition takes place with high regioselectivity (Scheme 8). Furthermore, high crystallinity of the Ad group facilitates the structure determination of metallofullerenes. X-ray structure analyses of the corresponding adduct were achieved for $M@C_{2v}$-C_{82} (M = Sc,[59] Y,[60] La,[61] Ce,[62] Gd[63]), $Ce@C_s$-C_{82}, $La_2@C_n$ (n = 72,[22,23] 78,[64] 80[65–67]), $Sc_2C_2@C_{2v}$-C_{80},[68,69] and $Sc_3C_2@C_{80}$.[70]

Only two isomers (ratio for two isomers: *ca.* 4:1) were obtained in the reaction of $M@C_{2v}$-C_{82} (M = Y, La, Ce, Gd) with AdN_2. X-ray structure analysis revealed that the addition position of major isomers and that of minor isomers were, respectively, C1-C2 and C2-C3 (Fig. 2). Results of theoretical studies indicated that electrophilic carbene selectively attacks

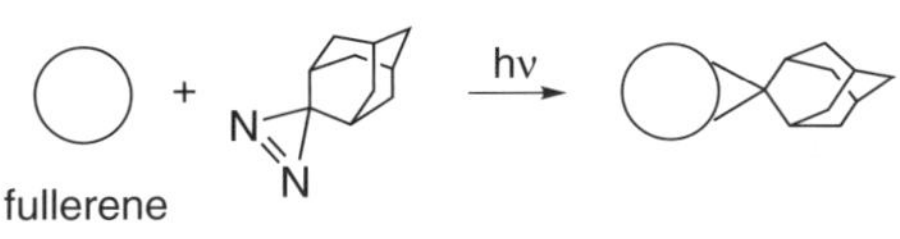

Scheme 8

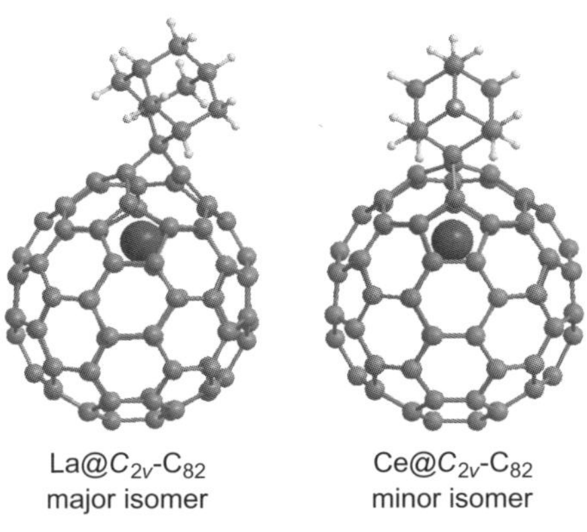

Fig. 2 Structure of $M@C_{2v}$-C_{82}(Ad)

Table 2 Mulliken Charge Density (Upper) and POAV (Lower) of Selected Atoms in $M@C_{2v}$-C_{82}

Metallofullerene	C1	C2	C3	C4	C6	C7
$Sc@C_{2v}$-C_{82} [59]	−0.188	−0.215	−0.110	−0.089	−0.014	−0.066
	11.7	11.2	8.4	9.0	10.5	9.2
$Y@C_{2v}$-C_{82} [60]	−0.135	−0.172	−0.096	−0.076	−0.004	−0.004
	11.7	11.5	8.2	9.2	10.5	9.1
$La@C_{2v}$-C_{82} [61]	−0.136	−0.170	−0.099	−0.071	−0.037	−0.092
	11.2	11.3	8.8	9.6	10.5	9.1
$Ce@C_{2v}$-C_{82} [62]	−0.152	−0.206	−0.136	−0.102	−0.011	−0.072
	11.7	11.5	8.4	9.4	10.3	8.9
$Gd@C_{2v}$-C_{82} [63]	−0.123	−0.175	−0.086	−0.060	−0.002	−0.058
	11.2	11.3	8.8	9.6	10.5	9.1

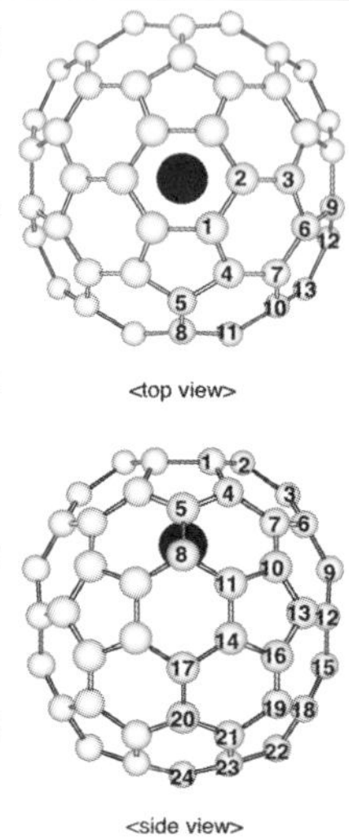

C2 atom, which has high charge density and local strain. In the reaction of $Sc@C_{2v}$-C_{82} with AdN_2, however, four isomers of mono-adduct were obtained under similar reaction conditions. Production of four regioisomers of $Sc@C_{2v}$-C_{82} was explained by high charge density of C1 atom compared to that in $M@C_{2v}$-C_{82} (M = Y, La, Ce, Gd) (Table 2).

The reaction of non-IPR fullerene, $La_2@D_2$-C_{72}, with AdN_2 afforded six mono-adduct isomers.[22] As expected, the local strain and charge density in the fused-pentagon areas are markedly higher than the other areas. X-ray structure analyses of three isomers of $La_2@D_2$-C_{72}(Ad) (ratio in six isomers (addition position): 40%(c–e), 36%(e–f), 9%(b–f)) were performed. The addition positions of the Ad group are the 5,6-bond in the fused pentagon, which has high local strain and high charge density (Table 3). These results show that the local strain and charge density control the reaction of metallofullerene with electrophilic carbene on the fullerene cage carbons and that they afford the corresponding mono-adducts with high regioselectivity.

Table 3 Mulliken Charge Density (Upper) and POAV (Lower) of Selected Atoms in $La_2@D_2$-C_{72}

	a	**b**	**c**	**d**	**e**	**f**	**g**
Charge density	−0.131	−0.160	−0.135	−0.177	−0.124	−0.139	−0.130
POAV	8.7	11.5	13.1	9.3	14.9	13.1	10.8

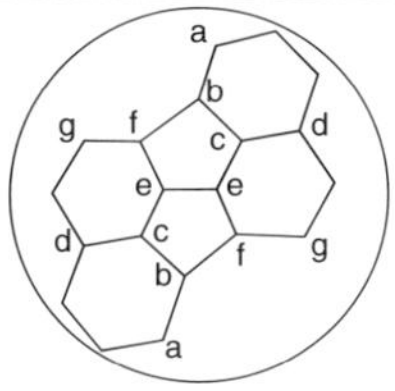

2.4. Reaction of Diazo Compound or Tosylhydrazone

In 1995, Suzuki *et al.* reported reaction of La@$C_{2}v$-C_{82} with diphenyldiazomethane.[71] Four new sets of octet signals were observed using ESR measurements. Mass spectra revealed the formation of corresponding adducts: La@$C_{2}v$-$C_{82}(CPh_2)_n$ (n = 1–3). The reaction of Tb@C_{82} with an α-diazocarbonyl compound was conducted in the presence of $Cu(MeCN)_4PF_6$ as a catalyst.[72] The mass spectra displayed a series of molecular ion peaks of the corresponding multi-adducts: Tb@$C_{82}(CR_2)_n$ (n = 1–6). It is particularly interesting that C_{60} did not afford the corresponding multi-adduct under the same conditions.

Tosylhydrazone was used to prepare metallofullerene-based donor–acceptor conjugates and semiconducting materials for photovoltaic cells. The structures of mono-adducts of La@C_{2v}-C_{82},[73] $M_2@I_h$-C_{80} (M = La[74], Ce[75]), and $M_3N@I_h$-C_{80} (M = Sc, Y)[76] were determined using X-ray structure analysis. Three isomeric mono-adducts of La@C_{2v}-C_{82} were isolated as major products. The addition position of addendum is the same position of Ad carbene (C2-C3, see Fig. 2, Table 2). This result probably indicates their similar formation pathway, i.e. carbene addition. The [6,6]-open adducts were obtained in the reaction of M@I_h-C_{80} (M = Ce_2,[75] Sc_3N,[76] Y_3N,[76] Lu_3N[77]) with tosylhydrazone.

2.5. Radical Addition

Photoreaction of La@C_s-C_{82} with perfluoroalkyl iodide (1.8 equiv.) was conducted to afford seven La@C_s-$C_{82}(C_8F_{17})_2$ isomers in high yield (>80%)

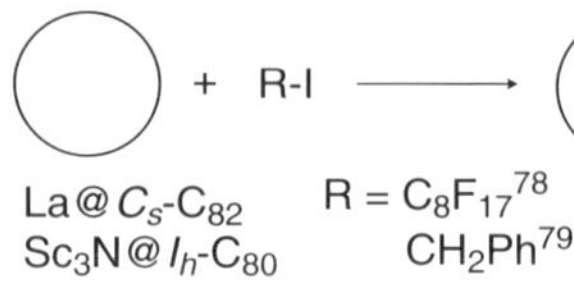

Scheme 9

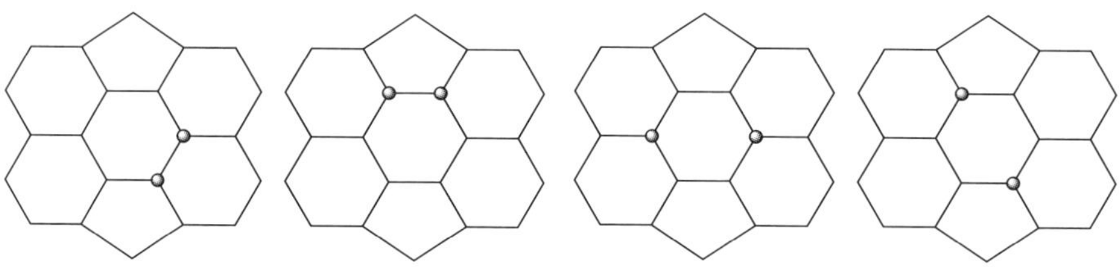

Fig. 3 Addition patterns of I_h-C_{80}

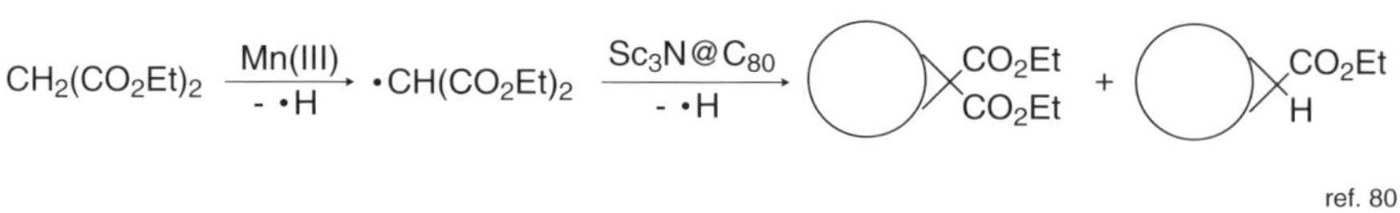

Scheme 10

(Scheme 9).[78] Perfluoroalkylation of fullerenes drastically increased the solubility of these adducts in perfluorocarbon solvents, which caused an easy and fast separation and purification of La@C_s-$C_{82}(C_8F_{17})_2$. Photochemical reaction of Sc_3N@I_h-C_{80} with benzyl bromide (1000 equiv.) was reported.[79] Sc_3N@I_h-$C_{80}(CH_2Ph)_2$ was produced in 82% yield and high regioselectivity. X-ray crystallographic analysis revealed that the addition of benzyl radicals takes place at 1,4-position on [566] ring junctions (Fig. 3). It is noteworthy that both photoreactions caused formation of corresponding bis-adducts in high yields.

Manganese(III)-catalyzed free radical addition of diethyl malonate with Sc_3N@I_h-C_{80} was conducted (Scheme 10).[80] The NMR analyses of adducts revealed that [6,6]-open adducts, Sc_3N@I_h-$C_{80}(C(CO_2Et)_2)$ and Sc_3N@I_h-$C_{80}(CH(CO_2Et))$, were obtained in the reaction. Ion peaks for the octa-adduct of Sc_3N@I_h-C_{80} and deca-adduct of Lu_3N@I_h-C_{80} were detected

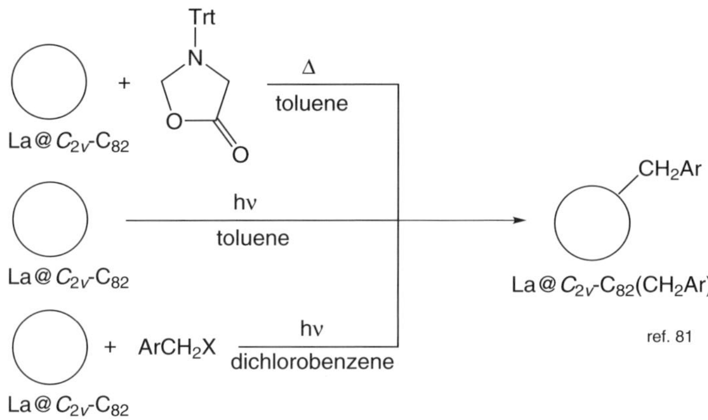

Scheme 11

respectively based on mass analysis. This result shows the high reactivity of $Sc_3N@I_h$-C_{80} toward the radical compared to $Lu_3N@I_h$-C_{80} because of the encapsulated metal cluster effect. As described later, $Sc_3N@C_{80}$ is more highly reactive to diene than $Lu_3N@C_{80}$ is.[43]

Although only pyrrolidino derivatives were found for the diamagnetic metallofullerenes, four benzyl adducts of La@C_{2v}-C_{82} were formed in the presence of 3-triphenylmethyl-5-oxazolidinone in toluene under reflux (Scheme 11).[81] The same mono-adducts were also obtained from a toluene solution of La@C_{2v}-C_{82} under photoirradiation. An open question remains as to why ESR signals of paramagnetic metallofullerenes disappear when they are stored in toluene for a long period. The photoreaction of La@C_{2v}-C_{82} in toluene furnished the answer. The addition position of radical species on La@C_{2v}-C_{82} was revealed by X-ray crystallographic analysis of La@C_{2v}-C_{82}($CHClC_6H_3Cl_2$), which was produced in the photoreaction of La@C_{2v}-C_{82} with a,a,2,4-tetrachlorotoluene in 1,2-dichlorobenzene. The addition site carbon of the substituent (C10, see Table 1) has the highest spin density in La@C_{2v}-C_{82}. This result suggests that La@C_{2v}-C_{82}($CHClC_6H_3Cl_2$) is formed as the kinetically preferred mono-adduct. The thermal retro-reaction of the adduct afforded the pristine metallofullerene in high yield.[82]

In 1991, Smalley *et al.* reported that although La@C_{60} and La@C_{74} were also detected in mass spectra of raw soot, only La@C_{82} was extracted in toluene. Recently, these insoluble metallofullerenes, La@C_n (n = 72,[21]

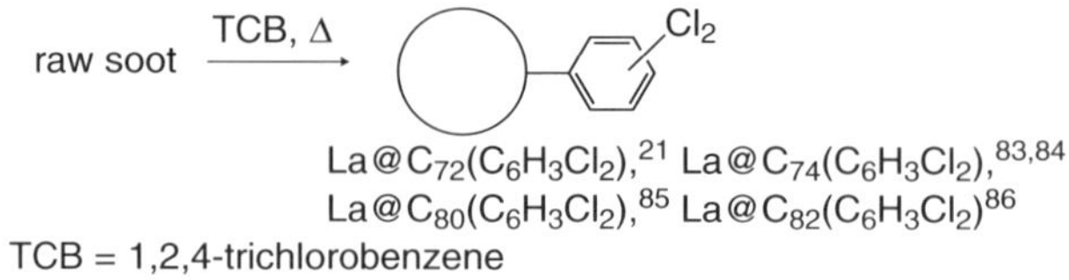

Scheme 12

74,[83,84] 80,[85] 82[86]) were extracted in 1,2,4-trichlorobenzene and isolated in forms of their dichlorophenyl derivatives (Scheme 12). X-ray crystallographic analyses and DFT calculations of these adducts suggest that the addition site carbon of the addendum has high SOMO spin density and a high POAV value on fullerene cages, which indicate that these adducts are produced by the addition of dichlorophenyl radical to fullerenes. The failure of extraction of missing fullerenes as a pristine form results from their high reactivity, which is stabilized by the addition of substituent at the reactive carbon on the fullerene cage. The first highly enantioselective metallofullerene synthesis was reported recently using La@C_{72}($C_6H_3Cl_2$).[87]

Addition of an odd number of substituents to paramagnetic metallofullerenes was also reported. Two Y@C_{2v}-$C_{82}(CF_3)_5$ isomers were isolated and characterized with ^{19}F-NMR, as diamagnetic stable isomers (Scheme 13).[88] DFT calculation proposed a 1,4-addition pattern of the five CF_3 groups, of which isomers are 14 kJmol^{-1} more stable than the third isomer and at least 50 kJmol^{-1} more stable than the other 53 isomers.

Trifluoromethylation of diamagnetic metallofullerenes was investigated in detail, revealing addition-pattern guidelines of multi-addition of Sc_3N@C_{80} according to results of detailed structural analyses of Sc_3N@$C_{80}(CF_3)_n$ (n = 2, 4, 8, 10, 12, 14, 16, 18), including X-ray crystallographic analyses of the adducts (I_h: n = 10, 12, 14, 16; D_{5h}: n = 14, 18).[89–93] Although Sc_3@D_{5h}-C_{80} is much less reactive than the I_h cage isomer toward dienes,[94] the reaction rate of Sc_3@D_{5h}-C_{80} with CF_3 radical was the same rate as that found in the case of Sc_3@I_h-C_{80}.[89] ^{19}F-NMR and DFT calculation suggests that the two CF_3 additions take place at 1,4-addition. Local strain of the fullerene cage is one important factor of chemical reactivity of fullerenes. Experimental and theoretical studies of the Sc_3N@I_h-$C_{80}(CF_3)_n$ (n < 8) revealed that the addition of CF_3 to Sc_3N@I_h-C_{80} engenders avoidance of pyrene-type carbon atoms (triple-hexagon junction: THJ), which have

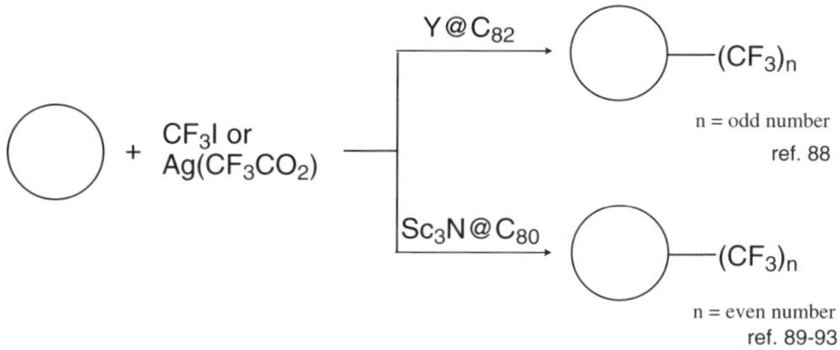

Scheme 13

lower local strain than cage carbons of other types (e.g. pentagon–hexagon–hexagon junction or pentagon–pentagon–hexagon junction).[95] However, four and eight THJs are substituted for $Sc_3N@I_h$-$C_{80}(CF_3)_n$ (n = 14, 16). Results suggest that the formation of isolated aromaticity and their coordination with the internal cluster are the main factors for such addition patterns. Electrochemical studies of $Sc_3N@I_h$-$C_{80}(CF_3)_2$ revealed that trifluoromethylation markedly decreases the HOMO/LUMO gap of $Sc_3N@I_h$-C_{80} from 1.85 V to 1.63 V. That result was also supported by the DFT calculation.[89] The increase of the HOMO/LOMO gap (1.59–1.90 V) with the increase of the number of CF_3 group ($Sc_3N@I_h$-$C_{80}(CF_3)_n$: n = 2, 4, 10, 12) was also clarified using cyclic voltammetry.[90,91,96] Trifluoromethylation of $Sc_3N@I_h$-C_{80} might become useful as an important practical application, perhaps as a photovoltaic material, because of its tunable molecular orbital levels.

2.6. Addition of Organosilicon Compound

Disilylation is known as the first example of exohedral functionalization of metallofullerenes (Scheme 14).[14] The disilylation reaction of fullerenes is used for chemical probes and the tuning of electronic properties. Although empty fullerenes react only photochemically with disilirane and digermirane,[97–106] it is noteworthy that specific metallofullerenes can be functionalized with disilirane and digermirane both thermally and photochemically.[14,107–119] The high reactivity of metallofullerenes toward disilirane and digermirane is attributable to their electronic properties. As Table 4 shows, highly reactive fullerenes show lower reduction

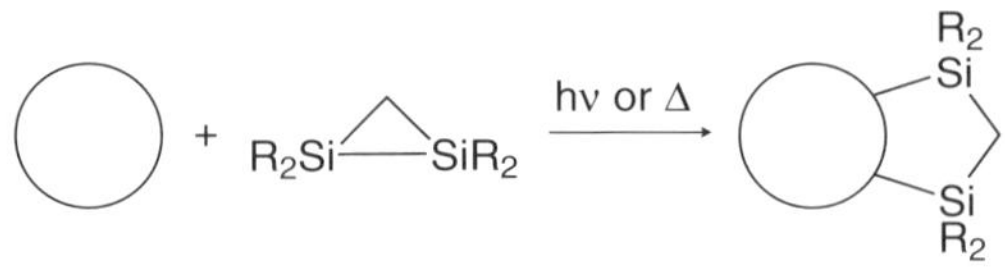

$M@C_{82}$
(M = Y[114,115], La[14,110,114,115,] Ce[114,115], Pr[111], Gd[108])
$M_2@C_{80}$[107,116,118]
(M = La, Ce)
$Sc_3N@C_{80}$[113,117]
$Sc_3N@C_{80}$ reacts photochemically, not thermally.

Scheme 14

Table 4 Redox Potentials of Fullerenes and its Reactivity toward Disilirane

Compound	E_{ox} V vs. Fc/Fc^+	E_{red} V vs. Fc/Fc^+	Thermal	Photo
C_{60}	1.21	−1.12	×	○
C_{70}	1.19	−1.09	×	○
C_{76}	0.81	−0.94	×	○
C_{78}	0.85	−0.73	×	○
C_{82}	0.72	−0.69	×	○
C_{84}	1.04	−0.75	×	○
$Y@C_{2v}$-C_{82}	0.1	−0.37	○	○
$La@C_{2v}$-C_{82}	0.07	−0.42	○	○
$Ce@C_{2v}$-C_{82}	0.08	−0.41	○	○
$Gd@C_{2v}$-C_{82}	0.09	−0.39	○	○
$La_2@I_h$-C_{80}	0.56	−0.31	○	○
$Ce_2@I_h$-C_{80}	0.56	−0.31	○	○
$Sc_3N@$ I_h-C_{80}	0.57	−0.39	×	○

and oxidation potential than low-reactive fullerenes do. This difference in reactivity is noteworthy as the first example attributable to encapsulated species. These findings suggest that fullerene reactivity is controllable through oxidation and reduction of fullerenes. Reactions of the anions and cations of $M@C_{82}$ (M = Y, La, Ce) were conducted, with results showing that oxidation activates metallofullerene reactivity toward disilirane and that reduction suppresses metallofullerene reactivity toward disilirane.[114] Recently, reductive addition of $M_3N@I_h$-C_{80} (M = Sc, Lu) with benzyl bromide was reported.[120,121] The reactivity of the metallofullerenes toward benzyl bromide was enhanced by the reduction. The corresponding adduct was obtained from $Lu_3N@I_h$-C_{80} dianion, whereas the corresponding adduct was not obtained from $Sc_3N@I_h$-C_{80} dianion. Interestingly, the corresponding adduct was obtained from $Sc_3N@I_h$-C_{80} trianion. The difference on the reactivity was studied theoretically, which showed notable difference in their electronic structures, such as a molecular orbital and spin densities of anionic fullerenes. These experimental results suggest that oxidation and reduction of metallofullerene is effective for mechanistic studies and tuning of its reactivity.

Addition of silylene, a divalent silicon species, on $Lu_3N@I_h$-C_{80} has been reported.[122] The thermal reaction of silylene with $Lu_3N@I_h$-C_{80} afforded the [6,6]-silacyclopropane derivative. The [6,6]-silacyclopropane derivative readily photoisomerized to [5,6]-silafulleroid, which is unprecedented, with the exception of a [5,6]-azafulleroid formed by reaction with azide.[123]

Tuning of the electrochemical property of fullerenes is important not only for reactivity control but also for practical application. Chemical functionalization of empty fullerenes to control their integration and electronic properties is now necessary for fullerene-based photovoltaic cells. Electrochemical studies of C_{60} derivatives revealed that silylated fullerenes have greater electropositive oxidation potentials than their carbon or oxygen addition products do.[124–126] Redox data of metallofullerenes and their derivatives, bis-silylation[118] and carbosilylation[127] of $M_2@I_h$-C_{80} (M = La, Ce) and bis-silylation of $Sc_3N@I_h$-C_{80},[113,117] show that the silyl metallofullerenes are more electron-negative than pristine metallofullerenes, depending on the number of silyl groups in the same manner as silylation of C_{60}.[126] In contrast to silylation, the enhancement of the electron-accepting character of the metallofullerene was achieved by the addition of tetracyanoethylene oxide.[128] It is noteworthy that the electronic properties of metallofullerenes can be tuned by chemical functionalization.

2.7. [2+2] and [4+2] Cycloaddition

Benzyne, generated by the diazotization of anthranilic acid with isoamyl nitrite, can be added to $Gd@C_{82}$[129] and $Sc_3N@I_h$-C_{80}[130] respectively, forming

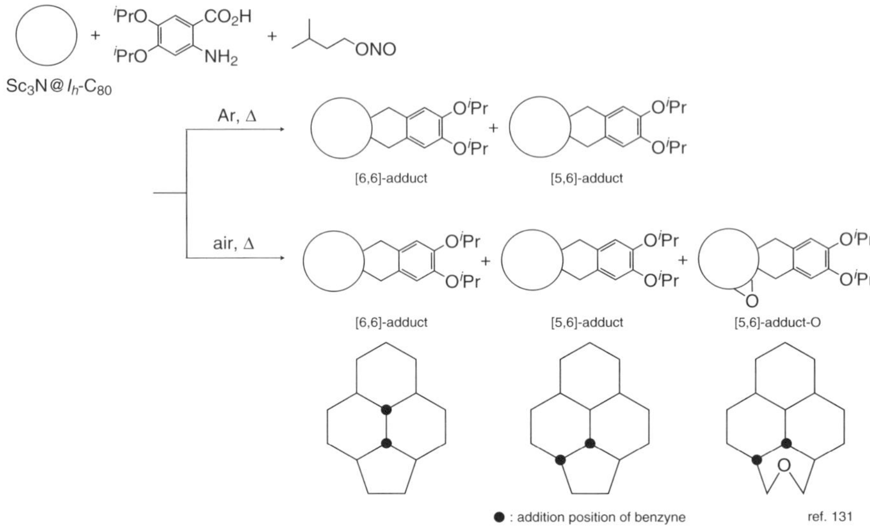

Scheme 15

two isolable isomers of mono-adducts under an Ar atmosphere. Isolated benzyne adducts of Gd@C_{82} were characterized by MS, UV, and CV. It is particularly interesting that MS analysis of benzyne adducts of Gd@C_{82} showed an ion peak assigned to mono-adduct (Gd@C_{82}(C_6H_4)) accompanied by the ion peak assigned to Gd@C_{82}(C_6H_4O). A report on the reaction of Sc_3N@I_h-C_{80} with benzyne clarified why the ion peak of Gd@C_{82}(C_6H_4O) was detected (Scheme 15). X-ray structure analyses of mono-adducts of Sc_3N@I_h-C_{80} revealed the structures of major and minor isomers, which are the [5,6]-regioisomer and the [6,6]-regioisomer, Sc_3N@I_h-C_{80}(C_6H_2($O^iPr)_2$), respectively. It is particularly interesting that in the presence of H_2O in air, oxygen atom addition product of [5,6]-Sc_3N@I_h-C_{80}(C_6H_2($O^iPr)_2$)(O) was obtained.[131] The structure was clearly determined by X-ray structure analysis. The addition of a benzyne unit and an oxygen atom occurred at [5,6]-ring junctions in a single pentagon. These results suggest that the addition of benzyne activates the pentagon.

Reaction of La@C_{2v}-C_{82} with benzyne was conducted, affording La@C_{2v}-C_{82}((C_6H_4)$_2$$NO_2$), which has been characterized using X-ray structure analysis (Scheme 16).[132] The addition of two benzyne units took place at the [5,6]-ring junctions. The mechanism, with two benzynes added after addition of NO_2, was proposed based on theoretical studies.

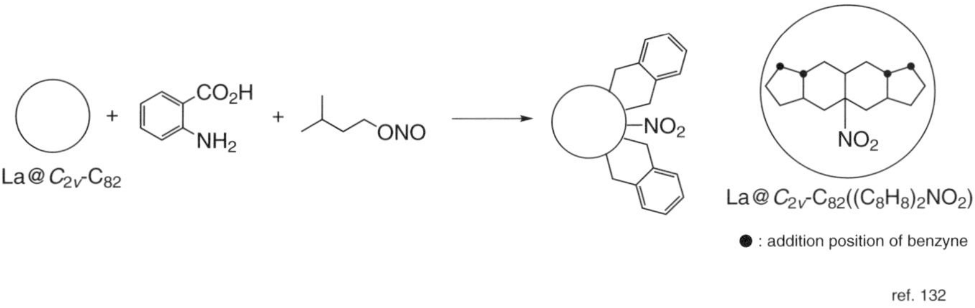

Scheme 16

The fact that fullerenes such as C_{60} and C_{70} act as polyalkenes and possess a reactive dienophile character derived from their electron-deficient properties has prompted the development of numerous efficient [4+2] cycloadditions of fullerenes. The [4+2] cycloaddition of metallofullerenes was achieved on $M_3N@I_h$-C_{80} (M = Sc, Gd) with 6,7-dimethoxyisochroman-3-one.[133,134] The addition site at $Sc_3N@I_h$-C_{80} was revealed by X-ray structure analysis, representing the first example for metallofullerene derivatives.[135] The addition takes place at the [5,6]-bond on the C_{80} cage.

The reaction of La@C_{2v}-C_{82} with cyclopentadiene derivatives (Cp: cyclopentadiene) caused formation of the stereospecific mono-adduct selectively.[37,136] The addition site of Cp* was clarified by results of X-ray structure analysis. The theoretical studies revealed that α-LUMO and β-LUMO+1 of La@C_{2v}-C_{82} present a favorable orbital at addition site carbons to react with Cp and Cp*. In comparison with C_{60}, La@C_{2v}-C_{82} shows lower chemical reactivity toward Cp and Cp*. Activation energies of the retro-reaction of the La@C_{2v}-C_{82}(Cp) and La@C_{2v}-C_{82}(Cp*) were estimated experimentally as lower than that of the C_{60} adduct. The concerted mechanism of the reaction of La@C_{2v}-C_{82} with Cp* was confirmed experimentally and theoretically.[137]

Based on the lower reactivity of metallofullerenes than that of empty fullerenes, advanced separation techniques have been developed using cyclopentadiene-immobilized stationary phase. These separation methods increase efficiency and accelerate the isolation of metallofullerenes.[138,139] This chemical separation method was applied to the separation of cage isomers $Sc_3N@I_h$-C_{80} and $Sc_3N@D_{5h}$-C_{80}. The chemical separation method is more effective for the separation of these isomers than HPLC separation is because these isomers show similar retention times in HPLC separation.[43,94]

3. CONCLUSIONS AND OUTLOOK

The chemical functionalization of metallofullerenes has progressed rapidly during the last decade. The development of characteristic novel reactions and the clarification of metallofullerenes' structural chemistry were achieved. Roles of encapsulated metal atoms on chemical reactions and the control of the metal motion inside of the fullerene cage have also been elucidated. Not only the basic science, but also investigations of charge-transport properties and properties in an excited state of metallofullerene derivatives are studied, which accelerates the application of metallofullerenes for photovoltaic devices, field effect transistors, and so on.[13,38,47,48,52,74,140–149] The numerous combinations of exohedral functional group, endohedral metal atoms, and fullerene cages promise the establishment of a new field of chemistry.

ACKNOWLEDGMENTS

This work was supported in part by a Grant-in-Aid for Young Scientists (B)(No. 23750035) from the Ministry of Education, Culture, Sports, Science, and Technology of Japan.

REFERENCES

1. Heath JR, O'Brien SC, Zhang Q, *et al.* (1985) Lanthanum Complexes of Spheroidal Carbon Shells. *J Am Chem Soc* **107:** 7779–7780.
2. Kroto HW, Heath JR, O'Brien SC, *et al.* (1985) C_{60}: Buckminsterfullerene. *Nature* **318:** 162–163.
3. Chai Y, Guo T, Jin C, *et al.* (1991) Fullerenes with metals inside. *J Phys Chem* **95:** 7564–7568.
4. Akasaka T, Nagase S. (2002) Endofullerenes: A New Family of Carbon Clusters. *Kluwer Academic Publishers, Dordrecht.*
5. Dunsch L, Yang S. (2007) Metal nitride cluster fullerenes: Their current state and future prospects. *Small* **3:** 1298–1320.
6. Chaur MN, Melin F, Ortiz AL, *et al.* (2009) Chemical, electrochemical, and structural properties of endohedral metallofullerenes. *Angew Chem Int Ed* **48:** 7514–7538.

7. Yamada M, Akasaka T, Nagase S. (2010) Endohedral Metal Atoms in Pristine and Functionalized Fullerene Cages. *Acc Chem Res* **43:** 92–102.
8. Lu X, Akasaka T, Nagase S. (2011) Chemistry of endohedral metallofullerenes: The role of metals. *Chem Commun* **47:** 5942–5957.
9. Maeda Y, Tsuchiya T, Lu X, *et al.* (2011) Current progress on the chemical functionalization and supramolecular chemistry of M@C_{82}. *Nanoscale* **3:** 2421–2429.
10. Rodríguez-Fortea A, Balch AL, Poblet JM. (2011) Endohedral metallofullerenes: A unique host-guest association. *Chem Soc Rev* **40:** 3551–3563.
11. Akasaka T, Lu X. (2012) Structural and electronic properties of endohedral metallofullerenes. *Chem Rec* **12:** 256–269.
12. Lu X, Feng L, Akasaka T, *et al.* (2012) Current status and future developments of endohedral metallofullerenes. *Chem Soc Rev* **41:** 7723–7760.
13. Rudolf M, Wolfrum S, Guldi DM, *et al.* (2012) Endohedral Metallofullerenes — Filled Fullerene Derivatives towards Multifunctional Reaction Center Mimics. *Chem Eur J* **18:** 5136–5148.
14. Akasaka T, Kato T, Kobayashi K, *et al.* (1995) Exohedral adducts of La@C_{82}. *Nature* **374:** 600–601.
15. Cardona CM, Kitaygorodskiy A, Echegoyen L. (2005) Trimetallic Nitride Endohedral Metallofullerenes: Reactivity Dictated by the Encapsulated Metal Cluster. *J Am Chem Soc* **127:** 10448–10453.
16. Lukoyanova O, Cardona CM, Rivera J, *et al.* (2007) Open Rather than Closed" Malonate Methano-Fullerene Derivatives. The Formation of Methanofulleroid Adducts of Y_3N@C_{80}. *J Am Chem Soc* **129:** 10423–10430.
17. Pinzón JR, Zuo T, Echegoyen L. (2010) Synthesis and Electrochemical Studies of Bingel–Hirsch Derivatives of M_3N@*Ih*-C_{80} (M = Sc, Lu). *Chem Eur J* **16:** 4864–4869.
18. Gao X, Ishimura K, Nagase S, *et al.* (2009) Dichlorocarbene Addition onto C_{60} from the Trichloromethyl Anion: Carbene Mechanism or Bingel Mechanism? *J Phys Chem A* **113:** 3673–3676.
19. Cai T, Xu L, Shu C, *et al.* (2008) Selective Formation of a Symmetric Sc_3N@C_{78} Bisadduct: Adduct Docking Controlled by an Internal Trimetallic Nitride Cluster. *J Am Chem Soc* **130:** 2136–2137.
20. Cai T, Xu L, Shu C, *et al.* (2008) Synthesis and characterization of a non-IPR fullerene derivative: Sc_3N@$C_{68}[C(COOC_2H_5)_2]$. *J Phys Chem C* **112:** 19203–19208.
21. Wakahara T, Nikawa H, Kikuchi T, *et al.* (2006) La@C_{72} having a non-IPR carbon cage. *J Am Chem Soc* **128:** 14228–14229.

22. Lu X, Nikawa H, Nakahodo T, *et al.* (2008) Chemical understanding of a non-IPR metallofullerene: Stabilization of encaged metals on fused-pentagon bonds in $La_2@C_{72}$. *J Am Chem Soc* **130:** 9129–9136.
23. Lu X, Nikawa H, Tsuchiya T, *et al.* (2008) Bis-carbene adducts of non-IPR $La_2@C_{72}$: Localization of high reactivity around fused pentagons and electrochemical properties. *Angew Chem Int Ed* **47:** 8642–8645.
24. Chaur MN, Melin F, Athans AJ, *et al.* (2008) The influence of cage size on the reactivity of trimetallic nitride metallofullerenes: A mono- and bis-methanoadduct of $Gd_3N@C_{80}$ and a monoadduct of $Gd_3N@C_{84}$. *Chem Commun:* 2665–2667.
25. Alegret N, Chaur MN, Santos E, *et al.* (2010) Bingel–Hirsch Reactions on Non-IPR $Gd_3N@C_{2n}$ (2n = 82 and 84). *J Org Chem* **75:** 8299–8302.
26. Feng L, Nakahodo T, Wakahara T, *et al.* (2005) A singly bonded derivative of endohedral metallofullerene: $La@C_{82}CBr(COOC_2H_5)_2$. *J Am Chem Soc* **127:** 17136–17137.
27. Feng L, Wakahara T, Nakahodo T, *et al.* (2006) The Bingel monoadducts of $La@C_{82}$: Synthesis, characterization, and electrochemistry. *Chem Eur J* **12:** 5578–5586.
28. Feng L, Tsuchiya T, Wakahara T, *et al.* (2006) Synthesis and characterization of a bisadduct of $La@C_{82}$. *J Am Chem Soc* **128:** 5990–5991.
29. Li X, Fan L, Liu D, *et al.* (2007) Synthesis of a $Dy@C_{82}$ Derivative Bearing a Single Phosphorus Substituent via a Zwitterion Approach. *J Am Chem Soc* **129:** 10636–10637.
30. Lu X, Zhou X, Shi Z, *et al.* (2004) Nucleophilic addition of glycine esters to $Gd@C_{82}$. *Inorg Chim Acta* **357:** 2397–2400.
31. Feng L, Lu X, He X, *et al.* (2004) Reactions of endohedral metallofullerenes with azomethine ylides: An efficient route toward metallofullerene-pyrrolidines. *Inorg Chem Commun* **7:** 1010.
32. Cao B, Wakahara T, Maeda Y, *et al.* (2004) Lanthanum Endohedral Metallofulleropyrrolidines: Synthesis, Isolation, and EPR Characterization. *Chem Eur J* **10:** 716–720.
33. Lu X, He X, Feng L, *et al.* (2004) Synthesis of pyrrolidine ring-fused metallofullerene derivatives. *Tetrahedron* **60:** 3713–3716.
34. Skanji R, Messaouda MB, Zhang Y, *et al.* (2012) Sequential photo-addition of glycine methyl-ester to [60]fullerene. *Tetrahedron* **68:** 2713–2718.
35. Takano Y, Obuchi S, Mizorogi N, *et al.* (2012) Stabilizing Ion and Radical Ion Pair States in a Paramagnetic Endohedral Metallofullerene/Extended Tetrathiafulvalene Conjugate. *J Am Chem Soc* **134:** 16103–16106.
36. Tsuchiya T, Rudolf M, Wolfrum S, *et al.* (2013) Coordinative interactions between proohyrins and C_{60}, $La@C_{82}$, and $La_2@C_{80}$. *Chem Eur J* **19:** 558–565.

37. Maeda Y, Sato S, Inada K, *et al.* (2010) Regioselective exohedral functionalization of $La@C_{82}$ and its 1,2,3,4,5-pentamethylcyclopentadiene and adamantylidene adducts. *Chem Eur J* **16:** 2193–2197.
38. Takano Y, Herranz MA, Martín N, *et al.* (2010) Donor-acceptor conjugates of lanthanum endohedral metallofullerene and π-extended tetrathiafulvalene. *J Am Chem Soc* **132:** 8048–8055.
39. Yamada M, Wakahara T, Nakahodo T, *et al.* (2006) Synthesis and structural characterization of endohedral pyrrolidinodimetallofullerene: $La_2@C_{80}(CH_2)_2NTrt$. *J Am Chem Soc* **128:** 1402–1403.
40. Yamada M, Okamura M, Sato S, *et al.* (2009) Two regioisomers of endohedral pyrrolidinodimetallofullerenes $M_2@I_h\text{-}C_{80}(CH_2)_2NTrt$ (M = La, Ce; Trt = trityl): Control of metal atom positions by addition positions. *Chem Eur J* **15:** 10533–10542.
41. Cai T, Slebodnick C, Xu L, *et al.* (2006) A pirouette on a metallofullerene sphere: Interconversion of isomers of *N*-tritylpyrrolidino I_h $Sc_3N@C_{80}$. *J Am Chem Soc* **128:** 6486–6492.
42. Cai T, Ge Z, Iezzi EB, *et al.* (2005) Synthesis and characterization of the first trimetallic nitride templated pyrrolidino endohedral metallofullerenes. *Chem Commun* 3594–3596.
43. Cai T, Xu L, Anderson MR, *et al.* (2006) Structure and Enhanced Reactivity Rates of the D_{5h} $Sc_3N@C_{80}$ and $Lu_3N@C_{80}$ Metallofullerene Isomers: The Importance of the Pyracylene Motif. *J Am Chem Soc* **128:** 8581–8589.
44. Cardona CM, Elliott B, Echegoyen L. (2006) Unexpected Chemical and Electrochemical Properties of $M_3N@C_{80}$ (M = Sc, Y, Er). *J Am Chem Soc* **128:** 6480–6485.
45. Chen N, Fan L, Tan K, *et al.* (2007) Comparative Spectroscopic and Reactivity Studies of $Sc_{3-x}Y_xN@C_{80}$ (x = 0–3). *J Phys Chem C* **111:** 11823–11828.
46. Chen N, Zhang E-Y, Tan K, *et al.* (2007) Size Effect of Encaged Clusters on the Exohedral Chemistry of Endohedral Fullerenes: A Case Study on the Pyrrolidino Reaction of $Sc_xGd_{3-x}N@C_{80}$ (x = 0–3). *Org Lett* **9:** 2011–2013.
47. Pinzón JR, Cardona CM, Herranz MÁ, *et al.* (2009) Metal Nitride Cluster Fullerene $M_3N@C_{80}$ (M = Y, Sc) Based Dyads: Synthesis, and Electrochemical, Theoretical and Photophysical Studies. *Chem Eur J* **15:** 864–877.
48. Pinzón JR, Gasca DC, Sankaranarayanan SG, *et al.* (2009) Photoinduced Charge Transfer and Electrochemical Properties of Triphenylamine $I_h\text{-}Sc_3N@C_{80}$ Donor-Acceptor Conjugates. *J Am Chem Soc* **131:** 7727–7734.
49. Braun K, Dunsch L, Pipkorn R, *et al.* (2010) Gain of a 500-fold sensitivity on an intravital MR Contrast Agent based on an endohedral Gadolinium-Cluster-Fullerene-Conjugate: A new chance in cancer diagnostics. *Int J Med Sci* **7:** 136–146.

50. Gimenez-Lopez MC, Gardener JA, Shaw AQ, *et al.* (2010) Endohedral metallofullerenes in self-assembled monolayers. *Phys Chem Chem Phys* **12:** 123–131.
51. Chen N, Pinzón JR, Echegoyen L. (2011) Influence of the Encapsulated Clusters on the Electrochemical Behaviour of Endohedral Fullerene Derivatives: Comparative Study of *N*-Tritylpyrrolidino Derivatives of $Sc_3N@I_h$-C_{80} and $Lu_3N@I_h$-C_{80}. *Chem Phys Chem* **12:** 1422–1425.
52. Wolfrum S, Pinzon JR, Molina-Ontoria A, *et al.* (2011) Utilization of $Sc_3N@C_{80}$ in long-range charge transfer reactions. *Chem Commun* **47:** 2270–2272.
53. Pinzón JR, Plonska-Brzezinska ME, Cardona CM, *et al.* (2008) $Sc_3N@C_{80}$-Ferrocene Electron-Donor/Acceptor Conjugates as Promising Materials for Photovoltaic Applications. *Angew Chem Int Ed* **47:** 4173–4176.
54. Echegoyen L, Chancellor CJ, Cardona CM, *et al.* (2006) X-Ray crystallographic and EPR spectroscopic characterization of a pyrrolidine adduct of $Y_3N@C_{80}$. *Chem Commun* 2653–2655.
55. Cai T, Xu L, Gibson HW, *et al.* (2007) $Sc_3N@C_{78}$: Encapsulated Cluster Regiocontrol of Adduct Docking on an Ellipsoidal Metallofullerene Sphere. *J Am Chem Soc* **129:** 10795–10800.
56. Wang T, Wu J, Xu W, *et al.* (2010) Spin Divergence Induced by Exohedral Modification: ESR Study of $Sc_3C_2@C_{80}$ Fulleropyrrolidine. *Angew Chem Int Ed* **49:** 1786–1789.
57. Lu X, Nakajima K, Iiduka Y, *et al.* (2011) Structural Elucidation and Regioselective Functionalization of An Unexplored Carbide Cluster Metallofullerene $Sc_2C_2@C_s(6)$-C_{82}. *J Am Chem Soc* **133:** 19553–19558.
58. Wang T, Wu J, Feng Y, *et al.* (2012) Preparation and ESR study of $Sc_3C_2@C_{80}$ bis-addition fulleropyrrolidines. *Dalton Trans* **41:** 2567–2570.
59. Hachiya M, Nikawa H, Mizorogi N, *et al.* (2012) Experimental Chemical Properties of $Sc@C_{2v}(9)$-C_{82} Probed with Adamantylidene Carbene. *J Am Chem Soc* **134:** 15550–15555.
60. Lu X, Nikawa H, Feng L, *et al.* (2009) Location of the yttrium atom in $Y@C_{82}$ and its influence on the reactivity of cage carbons. *J Am Chem Soc* **131:** 12066–12067.
61. Maeda Y, Matsunaga Y, Wakahara T, *et al.* (2004) Isolation and characterization of a carbene derivative of $La@C_{82}$. *J Am Chem Soc* **126:** 6858–6859.
62. Takano Y, Aoyagi M, Yamada M, *et al.* (2009) Anisotropic magnetic behavior of anionic $Ce@C_{82}$ carbene adducts. *J Am Chem Soc* **131:** 9340–9346.
63. Akasaka T, Kono T, Takematsu Y, *et al.* (2008) Does $Gd@C_{82}$ have an anomalous endohedral structure? Synthesis and single crystal X-ray structure of the carbene adduct. *J Am Chem Soc* **130:** 12840–12841.

64. Cao B, Nikawa H, Nakahodo T, *et al.* (2008) Addition of adamantylidene to $La_2@C_{78}$: Isolation and single-crystal X-ray structural determination of the monoadducts. *J Am Chem Soc* **130:** 983–989.
65. Yamada M, Someya C, Wakahara T, *et al.* (2008) Metal atoms collinear with the spiro carbon of 6,6-open adducts, $M_2@C_{80}$(Ad) (M = La and Ce, Ad = adamantylidene). *J Am Chem Soc* **130:** 1171–1176.
66. Ishitsuka MO, Sano S, Enoki H, *et al.* (2011) Regioselective bis-functionalization of endohedral dimetallofullerene, $La_2@C_{80}$: Extremal La-La distance. *J Am Chem Soc* **133:** 7128–7134.
67. Yamada M, Someya CI, Nakahodo T, *et al.* (2011) Synthesis of endohedral metallofullerene glycoconjugates by carbene addition. *Molecules* **16:** 9495–9504.
68. Kurihara H, Lu X, Iiduka Y, *et al.* (2011) $Sc_2C_2@C_{80}$ rather than $Sc_2@C_{82}$: Templated formation of unexpected C_{2v}(5)-C_{80} and temperature-dependent dynamic motion of internal Sc_2C_2 cluster. *J Am Chem Soc* **133:** 2382–2385.
69. Kurihara H, Lu X, Iiduka Y, *et al.* (2012) Chemical understanding of carbide cluster metallofullerenes: A case study on $Sc_2C_2@C_{2v}$(5)-C_{80} with complete X-ray crystallographic characterizations. *J Am Chem Soc* **134:** 3139–3144.
70. Iiduka Y, Wakahara T, Nakahodo T, *et al.* (2005) Structural determination of metallofullerene Sc_3C_{82} revisited: A surprising finding. *J Am Chem Soc* **127:** 12500–12501.
71. Suzuki T, Maruyama Y, Kato T, *et al.* (1995) Chemical Reactivity of a Metallofullerene: EPR Study of Diphenylmethano-La@C_{82} Radicals. *J Am Chem Soc* **117:** 9606–9607.
72. Feng L, Zhang X, Yu Z, *et al.* (2002) Chemical Modification of Tb@C_{82} by Copper(I)-Catalyzed Cycloadditions. *Chem Mater* **14:** 4021–4022.
73. Feng L, Slanina Z, Sato S, *et al.* (2011) Covalently Linked Porphyrin-La@C_{82} Hybrids: Structural Elucidation and Investigation of Intramolecular Interactions. *Angew Chem Int Ed* **50:** 5909–5912.
74. Feng L, Gayathri Radhakrishnan S, Mizorogi N, *et al.* (2011) Synthesis and Charge-Transfer Chemistry of $La_2@I_h$-C_{80}/$Sc_3N@I_h$-C_{80} Zinc Porphyrin Conjugates: Impact of Endohedral Cluster. *J Am Chem Soc* **133:** 7608–7618.
75. Guldi DM, Feng L, Radhakrishan SG, *et al.* (2010) A molecular $Ce_2@I_h$-C_{80} Switch-unprecedented oxidative pathway in photo induced charge transfer reactivity. I Am Chem Soc 132: 9078–9080.
76. Shu C, Xu W, Slebodnick C, *et al.* (2009) Syntheses and Structures of Phenyl-C_{81}-Butyric Acid Methyl Esters (PCBMs) from $M_3N@C_{80}$. *Org Lett* **11:** 1753–1756.

77. Ross RB, Cardona CM, Guldi DM, *et al.* (2009) Endohedral fullerenes for organic photovoltaic devices. *Nat Mater* **8:** 208–212.
78. Tagmatarchis N, Taninaka A, Shinohara H. (2002) Production and EPR characterization of exohedrally perfluoroalkylated paramagnetic lanthanum metallofullerenes: (La@C_{82})–$(C_8F_{17})_2$. *Chem. Phys. Lett.* **355:** 226–232.
79. Shu C, Slebodnick C, Xu L, *et al.* (2008) Highly Regioselective Derivatization of Trimetallic Nitride Templated Endohedral Metallofullerenes via a Facile Photochemical Reaction. *J Am Chem Soc* **130:** 17755–17760.
80. Shu C, Cai T, Xu L, *et al.* (2007) Manganese(III)-Catalyzed Free Radical Reactions on Trimetallic Nitride Endohedral Metallofullerenes. *J Am Chem Soc* **129:** 15710–15717.
81. Takano Y, Yomogida A, Nikawa H, *et al.* (2008) Radical coupling reaction of paramagnetic endohedral metallofullerene La@C_{82}. *J Am Chem Soc* **130:** 16224–16230.
82. Takano Y, Ishitsuka MO, Tsuchiya T, *et al.* (2010) Retro-reaction of singly bonded La@C_{82} derivatives. *Chem Commun* **46:** 8035–8036.
83. Nikawa H, Kikuchi T, Wakahara T, *et al.* (2005) Missing metallofullerene La@C_{74}. *J Am Chem Soc* **127:** 9684–9685.
84. Lu X, Nikawa H, Kikuchi T, *et al.* (2011) Radical derivatives of insoluble La@ C_{74}: X-ray structures, metal positions, and isomerization. *Angew Chem Int Ed* **50:** 6356–6359.
85. Nikawa H, Yamada T, Cao B, *et al.* (2009) Missing metallofullerene with C_{80} cage. *J Am Chem Soc* **131:** 10950–10954.
86. Akasaka T, Lu X, Kuga H, *et al.* (2010) Dichlorophenyl derivatives of La@ C3*v*(7)-C_{82}: Endohedral metal induced localization of pyramidalization and spin on a triple-hexagon junction. *Angew Chem Int Ed* **49:** 9715–9719.
87. Sawai K, Takano Y, Izquierdo M, *et al.* (2011) Enantioselective synthesis of endohedral metallofullerenes. *J Am Chem Soc* **133:** 17746–17752.
88. Kareev IE, Lebedkin SF, Bubnov VP, *et al.* (2005) Trifluoromethylated endohedral metallofullerenes: Synthesis and characterization of Y@$C_{82}(CF_3)_5$. *Angew Chem Int Ed* **44:** 1846–1849.
89. Shustova NB, Popov AA, Mackey MA, *et al.* (2007) Radical Trifluoromethylation of Sc_3N@C_{80}. *J Am Chem Soc* **129:** 11676–11677.
90. Shustova NB, Chen Y-S, Mackey MA, *et al.* (2009) Sc_3N@(C_{80}-I_h(7))$(CF_3)_{14}$ and Sc_3N@(C_{80}-I_h(7))$(CF_3)_{16}$. Endohedral Metallofullerene Derivatives with Exohedral Addends on Four and Eight Triple-Hexagon Junctions. Does the Sc_3N Cluster Control the Addition Pattern or Vice Versa? *J Am Chem Soc* **131:** 17630–17637.

91. Shustova NB, Peryshkov DV, Kuvychko IV, *et al.* (2011) Poly(perfluoroalkylation) of Metallic Nitride Fullerenes Reveals Addition-Pattern Guidelines: Synthesis and Characterization of a Family of $Sc_3N@C_{80}(CF_3)_n$ (n = 2–16) and Their Radical Anions. *J Am Chem Soc* **133:** 2672–2690.
92. Yang SF, Chen C, Jiao M, *et al.* (2011) Trifluoromethylated D_{5h} and I_h Isomers of $Sc_3N@C_{80}$: Synthesis, Isolation, and Addition Patterns of $Sc_3N@D_{5h}$-$C_{80}(CF_3)_{18}$ and $Sc_3N@I_h$-$C_{80}(CF_3)_{14}$. *Inorg Chem* **50:** 3766–3771.
93. Yang SF, Chen C, Lanskikh MA, *et al.* (2011) New isomers of trifluoromethylated derivatives of metal nitride clusterfullerene: $Sc_3N@C_{80}(CF_3)_n$ (n = 14 and 16). *Chem Asian J* **6:** 505–509.
94. Stevenson S, Mackey MA, Coumbe C, *et al.* (2007) Rapid Removal of D_{5h} Isomer Using the "Stir and Filter Approach" and Isolation of Large Quantities of Isomerically Pure $Sc_3N@C_{80}$ Metallic Nitride Fullerenes. *J Am Chem Soc* **129:** 6072–6073.
95. Yang SF, Popov AA, Dunsch L. (2008) Carbon Pyramidalization in Fullerene Cages Induced by the Endohedral Cluster: Non-Scandium Mixed Metal Nitride Clusterfullerenes. *Angew Chem Int Ed* **47:** 8196–8200.
96. Popov AA, Shustova NB, Svitova AL, *et al.* (2010) Redox-Tuning Endohedral Fullerene Spin States: From the Dication to the Trianion Radical of $Sc_3N@C_{80}(CF_3)_2$ in Five Reversible Single-Electron Steps. *Chem Eur J* **16:** 4721–4724.
97. Akasaka T, Ando W, Kobayashi K, *et al.* (1993) Photochemical [2 + 3] cycloaddition of C_{60} with disiliranes. *J Am Chem Soc* **115:** 10364–10365.
98. Akasaka T, Mitsuhida E, Ando W, *et al.* (1994) Adduct of C_{70} at the equatorial belt: Photochemical cycloaddition with disilirane. *J Am Chem Soc* **116:** 2627–2628.
99. Akasaka T, Maeda Y, Wakahara T, *et al.* (1999) Novel metal-free bis-silylation: C_{60}-sensitized reaction of disilirane with benzonitrile. *Org Lett* **1:** 1509–1512.
100. Akasaka T, Maeda Y, Wakahara T, *et al.* (2000) A first photochemical bis-germylation of C_{60} with digermirane. *Org Lett* **2:** 2671–2674.
101. Maeda Y, Sato R, Wakahara T, *et al.* (2000) C_{60}-sensitized bis-silylation of nitrile and carbonyl compounds with disilirane. *J Organomet Chem* **611:** 414–419.
102. Han A, Wakahara T, Maeda Y, *et al.* (2001) Photochemical cycloaddition of C_{78} with disilirane. *Chem Lett* 974–975.
103. Sasaki Y, Fujitsuka M, Ito O, *et al.* (2001) Photoinduced electron-transfer reactions between C_{60} and cyclic disiliranes (c-R_2Si-X-SiR_2; X = SiR_2, CH_2, O, NPh, S). *Heterocycles* **54:** 777–787.

104. Wakahara T, Han A, Niino Y, *et al.* (2002) Silylation of higher fullerenes. *J Mater Chem* **12:** 2061–2064.
105. Wakahara T, Kako M, Maeda Y, *et al.* (2003) Synthesis and characterization of cyclic silicon compounds of fullerenes. *Curr Org Chem* **7:** 927–943.
106. Han AH, Wakahara T, Maeda Y, *et al.* (2009) A new method for separating the D_3 and C_{2v} isomers of C_{78}. *New J Chem* **33:** 497–500.
107. Akasaka T, Nagase S, Kobayashi K, *et al.* (1995) Synthesis of the first adducts of the dimetallofullerenes $La_2@C_{80}$ and $Sc_2@C_{84}$ by addition of a disilirane. *Angew Chem Int Ed* **34:** 2139–2141.
108. Akasaka T, Nagase S, Kobayashi K, *et al.* (1995) Exohedral derivatization of an endohedral metallofullerene $Gd@C_{82}$. *J Chem Soc Chem Commun* 1343–1344.
109. Akasaka T, Kato T, Nagase S, *et al.* (1996) Chemical derivatization of endohedral metallofullerene $La@C_{82}$ with digermirane. *Tetrahedron* **52:** 5015–5020.
110. Kato T, Akasaka T, Kobayashi K, *et al.* (1996) ESR study on the reactivity of two isomers of LaC_{82} with disilirane. *Appl Magn Reson* **11:** 293–300.
111. Akasaka T, Okubo S, Kondo M, *et al.* (2000) Isolation and characterization of two $Pr@C_{82}$ isomers. *Chem Phys Lett* **319:** 153–156.
112. Wakahara T, Sakuraba A, Iiduka Y, *et al.* (2004) Chemical reactivity and redox property of Sc_3C_{82}. *Chem Phys Lett* **398:** 553–556.
113. Iiduka Y, Ikenaga O, Sakuraba A, *et al.* (2005) Chemical reactivity of $Sc_3N@C_{80}$ and $La_2@C_{80}$. *J Am Chem Soc* **127:** 9956–9957.
114. Maeda Y, Miyashita J, Hasegawa T, *et al.* (2005) Chemical reactivities of the cation and anion of $M@C_{82}$ (M = Y, La, and Ce). *J Am Chem Soc* **127:** 2143–2146.
115. Yamada M, Feng L, Wakahara T, *et al.* (2005) Synthesis and characterization of exohedrally silylated $M@C_{82}$ (M = Y and La). *J Phys Chem B* **109:** 6049–6051.
116. Yamada M, Nakahodo T, Wakahara T, *et al.* (2005) Positional control of encapsulated atoms inside a fullerene cage by exohedral addition. *J Am Chem Soc* **127:** 14570–14571.
117. Wakahara T, Iiduka Y, Ikenaga O, *et al.* (2006) Characterization of the bis-silylated endofullerene $Sc_3N@C_{80}$. *J Am Chem Soc* **128:** 9919–9925.
118. Wakahara T, Yamada M, Takahashi S, *et al.* (2007) Two-dimensional hopping motion of encapsulated La atoms in silylated $La_2@C_{80}$. *Chem Commun* 2680–2682.
119. Yamada M, Wakahara T, Tsuchiya T, *et al.* (2008) Location of the metal atoms in $Ce_2@C_{78}$ and its bis-silylated derivative. *Chem Commun* **8:** 558–560.

120. Li FF, Rodríguez-Fortea A, Poblet JM, *et al.* (2011) Reactivity of Metallic Nitride Endohedral Metallofullerene Anions: Electrochemical Synthesis of a $Lu_3N@I_h$-C_{80} Derivative. *J Am Chem Soc* **133:** 2760–2765.
121. Li FF, Rodríguez-Fortea A, Peng P, *et al.* (2012) Electrosynthesis of a $Sc_3N@I_h$-C_{80} Methano Derivative from Trianionic $Sc_3N@I_h$-C_{80}. *J Am Chem Soc* **134:** 7480–7487.
122. Sato K, Kako M, Suzuki M, *et al.* (2012) Synthesis of Silylene-Bridged Endohedral Metallofullerene $Lu_3N@I_h$-C_{80}. *J Am Chem Soc* **134:** 16033.
123. Liu TX, Wei T, Zhu S-E, *et al.* (2012) Azide Addition to an Endohedral Metallofullerene: Formation of Azafulleroids of $Sc_3N@I_h$-C_{80}. *J Am Chem Soc* **134:** 11956–11959.
124. Suzuki T, Maruyama Y, Akasaka T, *et al.* (1994) Redox properties of organofullerenes. *J Am Chem Soc* **116:** 1359–1363.
125. Akasaka T, Suzuki T, Maeda Y, *et al.* (1999) Photochemical bissilylation of C_{60} with disilane. *J Org Chem* **64:** 566–569.
126. Maeda Y, Aminur Rahman GM, Wakahara T, *et al.* (2003) Synthesis and characterization of tetrakis-silylated C_{60} isomers. *J Org Chem* **68:** 6791–6794.
127. Yamada M, Minowa M, Sato S, *et al.* (2010) Thermal carbosilylation of endohedral dimetallofullerene $La_2@I_h$-C_{80} with silirane. *J Am Chem Soc* **132:** 17953–17960.
128. Yamada M, Minowa M, Sato S, *et al.* (2011) Regioselective cycloaddition of $La_2@I_h$-C_{80} with tetracyanoethylene oxide: Formation of an endohedral dimetallofullerene adduct featuring enhanced electron-accepting character. *J Am Chem Soc* **133:** 3796–3799.
129. Lu X, Xu J, He X, *et al.* (2004) Addition of Benzyne to $Gd@C_{82}$. *Chem Mater* **16:** 953–955.
130. Li FF, Pinzón JR, Mercado BQ, *et al.* (2011) [2 + 2] Cycloaddition Reaction to $Sc_3N@I_h$-C_{80}. The Formation of Very Stable [5,6]- and [6,6]-Adducts. *J Am Chem Soc* **133:** 1563–1571.
131. Wang G-W, Liu T-X, Jiao M, *et al.* (2011) The Cycloaddition Reaction of I_h-$Sc_3N@C_{80}$ with 2-Amino-4,5-diisopropoxybenzoic Acid and Isoamyl Nitrite to Produce an Open-Cage Metallofullerene. *Angew Chem Int Ed* **50:** 4658–4662.
132. Lu X, Nikawa H, Tsuchiya T, *et al.* (2010) Nitrated benzyne derivatives of $La@C_{82}$: Addition of NO_2 and its positional directing effect on the subsequent addition of benzynes. *Angew Chem Int Ed* **49:** 594–597.
133. Iezzi E, Duchamp JC, Harich K, *et al.* (2002) A Symmetric Derivative of the Trimetallic Nitride Endohedral Metallofullerene, $Sc_3N@C_{80}$. *J Am Chem Soc* **124:** 524.

134. Stevenson S, Stephen RR, Amos TM, *et al.* (2005) Synthesis and Purification of a Metallic Nitride Fullerene BisAdduct: Exploring the Reactivity of $Gd_3N@C_{80}$. *J Am Chem Soc* **127:** 12776–12777.
135. Lee HM, Olmstead MM, Iezzi E, *et al.* (2002) Crystallographic Characterization and Structural Analysis of the First Organic Functionalization Product of the Endohedral Fullerene $Sc_3N@C_{80}$. *J Am Chem Soc* **124:** 3494–3495.
136. Maeda Y, Miyashita J, Hasegawa T, *et al.* (2005) Reversible and regioselective reaction of $La@C_{82}$ with cyclopentadiene. *J Am Chem Soc* **127:** 12190–12191.
137. Sato S, Maeda Y, Guo JD, *et al,* (2013) Mecanistic study of the Diels-Alder reaction of paramagnetic endohedral metallofullerene: Reaction of $La@C_{82}$ with 1,2,3,4,5-pentametyl cycropentadiene. *J Am Chem Soc* **135:** 5582–5587.
138. Ge Z, Duchamp JC, Cai T, *et al.* (2005) Purification of endohedral trimetallic nitride fullerenes in a single, facile step. *J Am Chem Soc* **127:** 16292–16298.
139. Stevenson S, Harich K, Yu H, *et al.* (2006) Nonchromatographic "Stir and Filter Approach" (SAFA) for Isolating $Sc_3N@C_{80}$ Metallofullerenes. *J Am Chem Soc* **128:** 8829–8835.
140. Kanbara T, Shibata K, Fujiki S, *et al.* (2003) N-channel field effect transistors with fullerene thin films and their application to a logic gate circuit. *Chem Phys Lett* **379:** 223–229.
141. Kobayashi S, Mori S, Iida S, *et al.* (2003) Conductivity and Field Effect Transistor of $La_2@C_{80}$ Metallofullerene. *J Am Chem Soc* **125:** 8116–8117.
142. Rikiishi Y, Kubozono Y, Hosokawa Y, *et al.* (2004) Structural and Electronic Characterizations of Two Isomers of $Ce@C_{82}$. *J Phys Chem B* **108:** 7580–7585.
143. Nagano T, Kuwahara E, Takayanagi T, *et al.* (2005) Fabrication and characterization of field-effect transistor device with C_{2v} isomer of $Pr@C_{82}$. *Chem Phys Lett* **409:** 187–191.
144. Sato S, Seki S, Honsho Y, *et al.* (2011) Semi-metallic single-Component crystal of soluble $La@C_{82}$ derivative with high electron mobility. *J Am Chem Soc* **133:** 2766–2771.
145. Sato S, Nikawa H, Seki S, *et al.* (2012) A co-crystal composed of the paramagnetic endohedral metallofullerene $La@C_{82}$ and a nickel porphyrin with high electron mobility. *Angew Chem Int Ed* **51:** 1589–1591.
146. Sato S, Seki S, Luo G, *et al.* (2012) Tunable charge-transport properties of I_h-C_{80} endohedral metallofullerenes: Investigation of $La_2@C_{80}$, $Sc_3N@C_{80}$, and $Sc_3C_2@C_{80}$. *J Am Chem Soc* **134:** 11681–11686.

147. Tsuchiya T, Kumashiro R, Tanigaki K, *et al.* (2008) Nanorods of endohedral metallofullerene derivative. *J Am Chem Soc* **130:** 450–451.
148. Ross RB, Cardona CM, Swain FB, *et al.* (2009) Tuning conversion efficiency in metallo endohedral fullerene-based organic photovoltaic devices **19:** 2332–2337.
149. Liedtke M, Sperlich A, Kraus H, *et al,* (2011) Triplet exciton generation in bulk-hetrojunction solar cells based on endohedral fullerenes. *J Am Chem Soc* **133:** 9088–9094.

CHAPTER 12

Computational Studies of Endohedral Fullerenes: Bonding, Isomerism, Internal Dynamics, Spectroscopy, and Chemical Reactivity

*Alexey A. Popov**

Computational studies of endohedral metallofullerenes (EMFs) advanced dramatically over the last two decades and currently allow researchers to predict most probable molecular structures of new EMFs or the products of chemical reactions with EMFs, address complex problem of the internal dynamics, or interpret spectroscopic properties of EMFs. This chapter reviews the current state of computational studies of EMFs and shows how quantum-chemical calculations can help experimental researchers in various fields of chemistry and physics of EMFs.

* Leibniz Institute for Solid State and Materials Research Dresden 01069 Dresden, Germany
E-mail: a.popov@ifw-dresden.de

1. INTRODUCTION

From the early stages of endohedral metallofullerene (EMF) research, experimental studies on the synthesis, separation, and characterization of EMFs were accompanied by quantum-chemical computations. The first studies of EMFs were performed at a time when computational facilities were rare and even semiempirical computations were possible for a limited number of professional theoreticians working in supercomputer centers. The limitation was not only in the hardware, but also in the few software codes that were available to the research community in the early 1990s. Furthermore, density functional theory (DFT), the method of choice in the majority of computational studies of EMFs nowadays, was still under development: the famous Becke's exchange functional (usually abbreviated as "B") was developed in 1988,[1] the Lee-Yang-Parr (LYP) correlation functional also appeared in 1988,[2] Perdew's correlation functionals P86 and PW91 in 1986[3] and 1991,[4] respectively, and 3-parameter hybrid schema used in B3LYP was first proposed in 1993 (although it was originally developed for B3PW91).[5] Thus, early first principle studies of EMFs employed either the Hartree–Fock (HF) approach or local density approximation (LDA), the workhorse in solid state physics since 1970s, whereas GGA and hybrid functionals became widely available in computational software since the mid-1990s. Semiempirical methods, which might be an inexpensive alternative to HF or LDA, suffered from the lack of reliable parameters for a description of metal atoms.

Fortuitously for theory-oriented EMF research, development of EMF science coincided in time with the dramatic progress in the computational hardware and software. Nowadays, with the availability of dozens of well-optimized proprietary and free codes, semiempirical computations can be performed on a standard office computer within one minute, point-energy DFT computation takes several hours, whereas optimization of the atomic coordinates of a typical EMF molecule can take few days. Thus, entry-level quantum-chemical calculations of EMFs at the DFT level are in principle accessible to virtually every interested researcher, not necessarily a computational scientist. Development of supercomputer resources and distribution of access to such resources on the basis of time grants allow fulfillment of much more demanding tasks. Computations of multiple possible isomers, detailed analysis of reaction mechanisms, or *ab initio* molecular dynamics simulations are becoming a standard praxis for theoreticians working in the field of EMFs. Methodologically, DFT in the form of either generalized gradient approximation (e.g. BP, PBE,[6] BLYP functionals) or hybrid theory with admixture of exact exchange (e.g. B3LYP, PBE0, etc.) is the most frequent choice in the studies of EMFs as providing the compromise between reasonable computational demands and good reliability of results. Less expensive semiempirical methods (such as NDDO-based approaches or tight-binding methods) can hardly be recommended for metal-containing fullerene molecules; the only exclusion is DF-TB approach,[7] which can give reasonable results, especially after careful fitting of the parameters.[8] *Ab initio* treatment of electron correlation beyond the second-order Møller–Plesset perturbation theory (MP2) is still prohibitively expensive, whereas MP2, although much more demanding than DFT, does not give better results (except for intermolecular interactions, which are the weak point of DFT) and is rarely used.

In this chapter, we will give an overview of the current state of the computational studies of EMFs and show how they help in understanding the electronic structure and bonding in EMFs, provide guidelines for the molecular structure elucidation and isomerism of EMFs, assist in interpretation of their spectroscopic properties, and are used to get a deeper insight into the chemical reactivity of EMFs and the internal dynamics of encapsulated species.

2. ELECTRONIC STRUCTURE AND BONDING IN EMFs

2.1. Metal-Cage Bonding

The electronic structure of EMFs was of high interest to the fullerene community since the isolation of first samples in the early 1990s. The ionic conjecture, which suggests metal cations are encapsulated in the negatively charged carbon cages (e.g. $La^{3+}@C_{82}{}^{3-}$), was soon proposed and received confirmation from ESR spectroscopic studies of paramagnetic EMFs such as $M@C_{82}$ (M = La, Y). Namely, small hyperfine coupling constants of the metal atoms showed that the spin density in $M@C_{82}$ is presumably localized on the carbon cage,[9,10] i.e. valence electrons of the metal atoms are almost completely transferred to the fullerene. The same conclusion also follows from the molecular orbital analysis, which shows that the frontier

MOs of the EMFs can be described as MOs of the fullerene cores with additional electrons borrowed from metal atoms.[11,12] Suggested first for monometallofullerenes such as $La^{3+}@C_{82}^{3-}$, the ionic model is also used to describe the electronic structure of endohedral fullerenes with more complex encapsulated species, such as dimetallofullerenes (e.g. $(La^{3+})_2@C_{80}^{6-}$), nitride clusterfullerenes (e.g. $(Sc^{3+})_3N^{3-}@C_{80}^{6-}$), carbide clusterfullerenes (e.g. $(Sc^{3+})_2C_2^{2-}@C_{82}^{4-}$), and all other EMFs.[13–15]

The XPS studies also favored the ionic model by revealing that the spectra in the range of metal levels correspond well to the spectra of metal oxides in the corresponding oxidation states.[16] The ionic conjecture is further supported by spectroscopic techniques and electrochemistry (spectra and redox potentials of isostructural EMFs with the same isomeric cage but different metal atoms are usually very similar if these metal atoms are in the same oxidation state).[13,14,17] Computational studies have also shown that the stability order of the isomers of EMFs corresponds well to the stability order of the isomers of empty fullerenes in the appropriate charge states (see Section 3).[18–20] Furthermore, the electrostatic potential (ESP) has large negative values inside negatively charged carbon cages,[21–23] which results in the strong stabilization of the metal cations if they are placed inside fullerene anions, thus favoring the formation of EMFs. Importantly, the minimum of the electrostatic potential in C_{82}^{3-}-$C_{2v}(9)$ was found in the position where the metal ion resides in M@C_{82}.[22,24–27]

The facts outlined above resulted in the wide acceptance of the ionic conjecture for EMFs by the scientific community. However, although the ionic model explains many spectroscopic and structural properties of EMFs, the transfer of a large integer number of electrons from an endohedral atom or cluster to the fullerene cage is only formal and should not be understood literally. Numerous studies show that the ionic model is oversimplified. For instance, the vibrational (IR, Raman) and UV-Vis-NIR spectra of EMFs with the same isomeric cage but different metals are indeed very similar and in many cases are well reproduced by DFT or TD-DFT calculations (e.g. vibrational spectra of all $M_3N@C_{80}$ NCFs are almost identical except for the cluster-based modes), but analogous calculations for the charged *empty* cages do not match experimental spectra of EMFs for the fullerene-based vibrations or excitations.[28–31] In fact, the electronic state of the cage cannot be considered as purely ionic, and the covalent contribution to the metal-cage bonding must be also taken into account. Likewise, while the analysis of the frontier MO energy levels in the EMFs and the corresponding empty cages can be indeed interpreted by an electron transfer from the metal to the cage, many experimental[32,33] and theoretical[21,24,34–36] studies clearly show that endohedral metal atoms have a substantial non-zero population of *nd*-levels. The metal contribution to individual π-MO orbitals of the fullerene is small and can remain hidden in a standard MO analysis, but the analysis of the system in its whole complexity, e.g. the analysis of the electron density (rather than individual orbitals) can reveal a significant degree of covalency in metal-cage interactions.[31,36–39] Computed atomic charges, although they strongly depend both on the method of theory used to compute wave-functions and on the electron density partitioning, are always significantly smaller than expected for the purely ionic bonding.[39,40] Analysis of the difference electron density obtained for ionic reference (e.g. for $Sc_3N@C_{80}$ it would be a difference of the electron density of the $Sc_3N@C_{80}$ itself and the densities of C_{80}^{6-} and Sc_3N^{6+} ions) also showed that the electron density in EMFs is not a simple superposition of the densities of its components in appropriate charge state.[31,35,36,39,41]

In summary, in terms of MO analysis, the mechanism of the metal (cluster)-cage interactions in EMFs can be described as the formal transfer of an appropriate number of electrons from the endohedral species to the carbon cage with subsequent coordination of the metal cations by the cage as a "ligand" and re-occupation of the metal *nd* orbitals (Fig. 1). In most cases there are no localized orbitals that could be ascribed to the metal-cage bonding, and the metal-cage interaction occurs through the overlap of many fullerene-based π-orbitals with *nd* orbitals of the metal atoms.

Since the metal-cage bonding in EMFs is best revealed in the analysis of the whole electron density of the molecule, for the quantitative description of the metal-cage interactions in EMFs it is appealing to use the methods based on the analysis of the total electron density such as Bader's quantum theory of atoms in molecules (QTAIM).[43,44] For EMFs, QTAIM was first used by Kobayashi and Nagase in 1998 in the analysis of the bonding in $Sc_2@C_{84}$, $Ca@C_{72}$, and $Sc_2@C_{66}$.[45,46] These studies have shown that, although the bond critical points (BCPs: critical points of the electron density, in

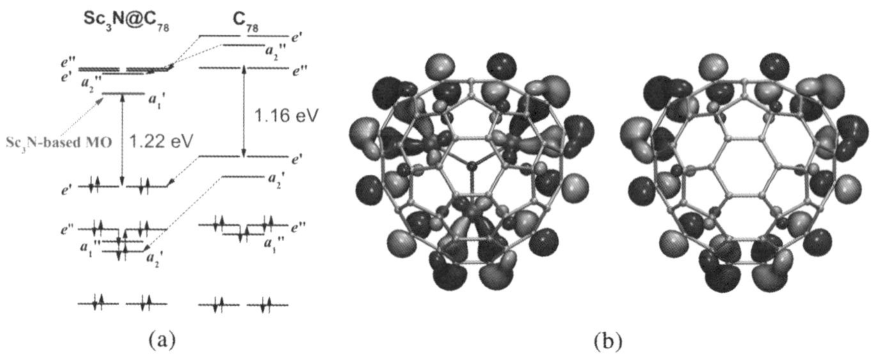

Fig. 1 (a) PBE/TZ2P Kohn–Sham MO energy levels in $Sc_3N@C_{78}$-D_{3h}(5) and C_{78}-D_{3h}(5) (based on the data from Ref. 28); (b) MO isosurfaces for a_2' orbital with considerable Sc–C bonding contribution in $Sc_3N@C_{78}$ (left), and corresponding orbital in C_{78} (right). Reproduced with permission from Ref. 42. © 2013 American Chemical Society.

which the density has a minimum along one axis and has a maximum in the plane perpendicular to this direction) could be found between Sc or Ca and certain carbon atoms of the cage, the values of the electron density at BCPs (ρ_{bcp}) were rather small (0.05 a.u. for Sc–C and 0.01–0.02 a.u. for Ca–C), and the Laplacian of the density ($\nabla^2\rho_{bcp}$) at metal-carbon BCPs was always positive. The authors concluded that such values for these descriptors are signatures of the ionic character of the metal-cage bonding. However, as summarized by Macchi and Sironi[47] and Gatti,[48] an analysis of the bonding of transition metal compounds by QTAIM (see, e.g. Ref. 49) reveals that for the bonds involving transition metals $\nabla^2\rho_{bcp}$ values are usually positive and ρ_{bcp} values are small because of the diffuse character of electron distribution[47] and hence these descriptors alone cannot properly describe the bonding situation. The analysis of the *energy density* appears more useful than the analysis of the electron density alone. In particular, Cremer and Kraka suggested that bonding between the atoms can be considered as covalent if the total energy density at BCP, H_{bcp}, is negative.[50]

The detailed study of the metal-cage and intra-cluster bonding by QTAIM/DFT was reported in 2009 for the four classes of endohedral metallofullerenes (monometallofullerenes, dimetallofullerenes, metal-nitride clusterfullerenes, metal-carbide clusterfullerenes), as well as for $Sc_3CH@C_{80}$ and $Sc_4O_x@C_{80}$ ($x = 2, 3$).[51] QTAIM studies were also reported for $Y_3@$

C_{80},[52] $M_2S@C_{82}$ (M = Sc, Y),[53] $TiSc_2N@C_{80}$,[41] and $Sc_3NC@C_{2n}$ clusterfullerenes,[54] as well as for the analysis of the metal–metal bonding in EMFs.[55]

The analysis of the BCP indicators showed that both the intra-cluster and the metal-cage interactions in EMFs were characterized by the negative total energy density H_{bcp}, which means that bonding in EMFs exhibits a high degree of covalency.[51] Furthermore, the electron density in EMF molecules was found to be a complex topological object, with many critical points, unpredictable number of the bond paths and large bond ellipticities, similar to the bonding situation in complexes of transition metals with π-carbocyclic ligands.[56] Hence, the analysis of the metal-cage bonding based only on the properties of the bond paths was found to be insufficient, and the metal-cage delocalization indices, δ(M, C), were also analyzed. By definition, δ(M, C) is the number of the electron pairs shared by the atoms M and C, which is very similar to the "bond order" in Lewis definition. In the majority of EMFs the δ(M, C) values were found to be a smooth function of M–C distances and did not exceed 0.25; however, when summed up over all metal-cage interactions, the values of δ(M, cage) close to 2–3 were obtained (Table 1).[51] While QTAIM atomic charges of the metal atoms were found to be approximately two times smaller than their formal oxidation states, the total number of the electron pairs shared by the metal atoms with the EMF molecule, Δ(M), was found to be close to the typical valence of the given element (≈3 for Sc, Y, and La, and ≈4 for Ti, see Table 1).[51] Thus, an analysis of the Δ(M) values circumvents an apparent contradiction between the ionic model and the substantial degree of covalency in metal-cage interactions revealed by more comprehensive analyses. The surplus electrons (e.g. six electrons in $M_3N@C_{2n}$ compounds) are indeed available to the carbon cage in EMFs, but these electrons are not completely transferred to the cage; instead, they are *shared* between the endohedral cluster and the carbon cage (note that *sharing* of electrons is the essence of the covalent interaction).

To conclude this section, it is useful to distinguish two definitions of the atomic (or cluster) charges in EMFs: (i) a "formal" charge and (ii) an "actual" charge. The formal charge (the same as an oxidation state or a valence state) is an integer and implies that the metal (cluster)–cage bonding is purely ionic. Formal charges do not describe the actual electron distribution in the endohedral fullerenes, but they are useful for

Table 1 QTAIM (Bader) Atomic Charges of Endohedral Metal Atoms and Cluster, and Metal-Cage and Metal-Cluster Delocalization Indices ("bond orders") in selected EMFs

EMF	q_M	$q_{cluster}$- QTAIM	$q_{cluster}$- formal	δ(M, cluster)	δ(M, cage)	Δ(M)
$Ca@C_{82}$	1.52		2		1.34	1.34
$Sc@C_{82}$	1.60		3		2.71	2.71
$Y@C_{82}$	1.80		3		2.54	2.54
$La@C_{82}$	1.82		3		2.69	2.69
$Y_2@C_{82}$	1.35	2.69	4	0.63	2.25	2.88
$Y_2@C_{79}N$	1.52	3.04	5	0.37	2.37	2.74
$La_2@C_{80}$	1.65	3.29	6	0.08	2.95	3.04
$Sc_2C_2@C_{84}$	1.72	2.01	4	0.63	2.00	2.63
$Y_2C_2@C_{82}$	1.89	2.40	4	0.66	1.90	2.56
$Ti_2C_2@C_{78}$	1.67	2.17	6	0.82	3.29	4.11
$Sc_3N@C_{80}$	1.76	3.50	6	0.82	1.77	2.59
$Y_3N@C_{80}$	1.90	3.87	6	0.85	1.77	2.62
$Y_3N@C_{88}$	1.93	3.95	6	0.79	1.65	2.44
$La_3N@C_{88}$	1.88	3.92	6	0.95	1.98	2.93
$Sc_3CH@C_{80}$	1.72	3.48	6	0.82	1.82	2.64
$Sc_3C_2@C_{80}$	1.67, 1.67	3.30	6	0.86, 0.83	1.87, 2.13	2.73, 2.76
$Sc_4O_2@C_{80}$	1.41, 1.75	3.66	6	1.21, 1.32	1.74, 1.41	2.97, 2.65
$Sc_4O_3@C_{80}$	1.87, 1.82	3.47	6	1.43, 1.11	1.07, 1.48	2.52, 2.60

Note: δ (M, cluster) is the number of the electron pairs shared by the metal atom with the cluster; δ(M, cage) is the number of the electron pairs shared by the metal atom with the carbon cage; Δ(M) is the total number of the electron pairs shared by the metal atom with all other atoms of the molecule; when EMF has non-equivalent metal atoms, corresponding values are averaged, unless they are significantly different. Based on the QTAIM-B3LYP/6–311G* calculations from Ref. 51.

understanding the spectroscopic and structural properties of endohedral fullerenes. In the next section we will show that the formal charge of the cage determines the isomeric structure of a given EMF molecule as well.

The "actual" charge should represent the real electronic structure of the EMF molecule, but unfortunately there is no unique way to partition the electron density between the atoms. Depending on the definition of the charge and on the method of theory used to compute the wavefunction, the charges can vary in a large range. However, whatever the method is used, computed charges are always considerably smaller than the formal charges, which points to the covalent contribution to the metal-cage interactions.

2.2. Intra-Cluster Interactions and Metal–Metal Bonding in EMFs

Although a large degree of covalency in metal-cage d-π bonding reduces the actual atomic charges in comparison to the formal values, metal atoms in EMFs still have a sufficiently high positive charges (Table 1) to induce strong Coulomb repulsion when two or more metals are encapsulated inside one carbon cage.[57,58] For instance, estimation of Coulomb repulsion energy between metal atoms based on the QTAIM-based interacting quantum atom (IQA) analysis yields the values of 9 and 14 eV for La_2 and Y_3 units in $La_2@C_{80}$ and $Y_3@C_{80}$, respectively,[55] which is comparable to the dissociation energy of the strongest covalent bonds (e.g. 9.8 eV in N_2[59]). Thus, the propensity of fullerenes to encapsulate clusters with non-metals has clear physical background: repulsive metal–metal interactions in clusterfullerenes are compensated by attractive interactions with negatively-charged species such as N^{3-}, C_2^{2-}, S^{2-}, O^{2-}, etc. (here formal charge states are listed). QTAIM analysis indeed shows that the non-metals in clusterfullerenes localize a substantial excess of electron density: typical charges are −1.7 to −1.9 for nitrogen in $M_3N@C_{2n}$, −1.3 to −1.5 for a C_2 unit in $M_2C_2@C_{2n}$, ca −1.2 for sulfur in $M_2S@C_{82}$, and −1.3 to −1.5 for oxygen in $Sc_4O_{2,3}@C_{80}$.[51,53,54,60] Besides, according to the QTAIM analysis, intra-cluster interactions also exhibit a significant degree of covalency, which is exemplified by negative total energy densities and considerable metal/non-metal delocalization indices.[41,51,54] For instance, $\delta(M, N)$, $\delta(M, C_2)$ and $\delta(M, S)$ values in respective clusterfullerenes span the range of 0.7–0.8, whereas $\delta(Sc, O)$ indices in oxide clusterfullerenes are 0.5–0.6.[51,53]

An interplay between ionic and covalent contributions in the intra-cluster bonding in $Sc_4O_2@C_{80}$ was recently analyzed by means of IQA-QTAIM,[55] which allows partitioning of the inter-atomic interaction energies into physically sound ionic and covalent terms.[61–63] The study showed that the net Sc–Sc and O–O interactions are strongly repulsive ($\Sigma E_{Sc-Sc} = 69$ eV, $E_{O-O} = 8$ eV), whereas net Sc–O interactions are attractive ($\Sigma E_{Sc-O} = -127$ eV), resulting for the whole Sc_4O_2 cluster in the net stabilization energy of −50 eV. Interestingly, ionic Sc–O interaction energy, −109 eV, is much larger than the covalent term, −17 eV, but the former is largely compensated by ionic Sc–Sc term, resulting in comparable covalent and ionic contributions to the intra-cluster interaction energy (−20 and −30 eV, respectively).[55]

Analysis of the formal charges in EMFs shows that group-III metals (Sc, Y, lanthanides) do not adopt their highest 3+ oxidation state in $Y_3@C_{80}$,[52] $Sc_4O_2@C_{80}$[64] as well as in a certain fraction of dimetallofullerenes, including $Lu_2@C_{76}$,[65] $M_2@C_{79}N$,[66,67] several $M_2@C_{82}$ structures (e.g. $Sc_2@C_{82}$,[68] $Y_2@C_{82}$,[69] $Er_2@C_{82}$[70–72]). The lower oxidation state implies that the outer valence electrons are not completely transferred from the metal to the carbon cage, thus yielding a possibility of the covalent metal–metal bonding. In this case, HOMO usually has pronounced M–M bonding character (Fig. 2a–c).[68,69,73] Furthermore, even when metal atoms in dimetallofullerenes are in the 3+ state (e.g. in $La_2@C_{2n}$ and $Ce_2@C_{2n}$), the LUMO of these EMFs is usually a M–M bonding orbital (Fig. 2d–e),[39,74–79] and hence metal–metal bonds can be formed in the result of the electrochemical or chemical reduction.[55] EMFs thus provide an interesting situation — on the one hand, metal atoms are strongly repulsive due to Coulomb interaction; on the other hand, they can form intermetallic covalent bonds.

Analysis of the metal–metal interactions by different approaches, including MO analysis,[51,55,66,73,80–82] QTAIM,[51,55] and ELF[51,52,55] topological analysis, consistently showed that covalent bonds between metal atoms in EMFs do exist and that their parameters are similar to those of metal–metal bonds in organometallic compounds.[83–88] Moreover, the M–M bond critical points in EMFs are characterized by negative values of the density Laplacian,[51] which is a usual situation for covalent bonds between the first row elements, but is rarely observed for metal–metal bonds. At the same time, IQA analysis showed that the covalent bonding contribution to the

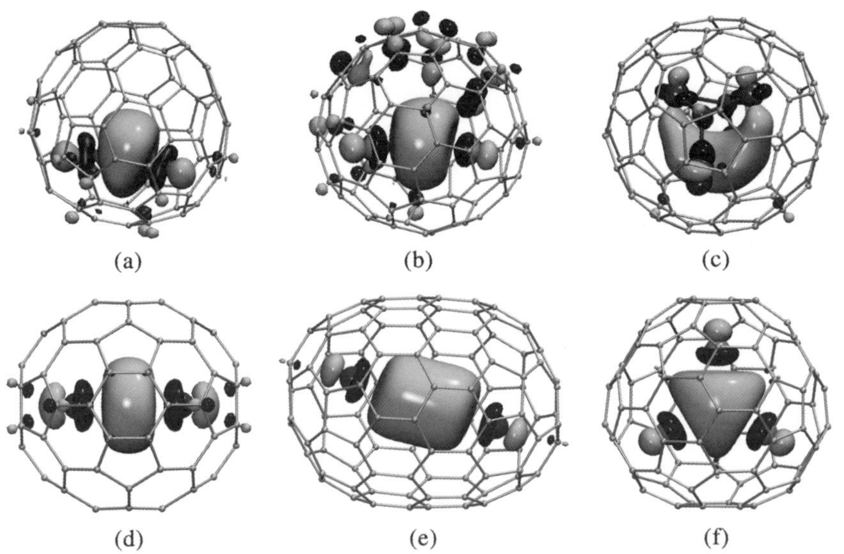

Fig. 2 Frontier molecular orbitals of metallofullerenes with metal–metal bonding character: (a) HOMO of $Sc_2@C_{82}$-C_{3v}(9); (b) HOMO of $Y_2@C_{82}$-C_{3v}(9); (c) HOMO of $Sc_4O_2@C_{80}$-I_h(7); (d) LUMO of $La_2@C_{80}$-I_h(7); (e) LUMO of $La_2@C_{100}$-D_5(450); (f) HOMO–1 of $Y_3@C_{80}$-I_h(7). Computations are performed at the PBE/TZVP level with ZORA correction, based on the data from Ref. 55.

total metal–metal interaction energy (*ca* –1 to –2 eV) is much smaller than the repulsive ionic term (*ca* 5–9 eV), which results in the net repulsion between the metal atoms.[55] Since Coulomb repulsion is the dominant term in metal–metal interactions, metal atoms in EMFs show a tend to be as far from each other as possible within the fullerene cage leading to unprecedentedly long metal–metal bonds. For instance, covalent bonding between La atoms in the anionic states of $La_2@C_{100}$ can be established at the La–La distances of *ca* 5 Å (for comparison, the single-bond covalent radius of La is 1.80 Å[89]).[55]

Whereas the valence state of metal atoms in monometallofullerenes is determined by their third ionization potentials,[90,91] comprehensive MO analysis showed that the valence state of metal atoms in dimetallofullerenes is determined by an energy matching between the carbon cage orbitals and the lowest-energy valence MO of the free metal dimer, which usually has $(ns)\sigma_g^2$ character.[55] Since the energy of the $(ns)\ \sigma_g^2$ orbital in the M_2 dimer correlates with the $ns^2(n{-}1)d^1 \cdot ns^1(n{-}1)d^2$ excitation energy of the free metal atom, this excitation energy thus determines the valence state of metal atoms in dimetallofullerenes. Namely, the $ns^2(n{-}1)d^1 \rightarrow ns^1(n{-}1)d^2$ excitation energies are increasing in the row La (0.33 eV) – Sc (1.43 eV)/Y (1.36 eV) – Lu (2.34 eV),[92] and the energies of $(ns)\ \sigma_g^{\ 2}$ MO in corresponding M_2 dimers are systematically decreasing from La to Sc/Y to Lu.[55] As a result, in dimetallofullerenes La is always trivalent, Lu prefers to adopt a divalent state, whereas the valence state of Sc and Y is more sensitive to the energy of carbon cage MOs and can be either di- or trivalent.[55]

A special M–M bonding situation is found in trimetallofullerenes such as $Y_3@C_{80}$. MO analysis showed that the highest two-fold occupied orbital of $Y_3@C_{80}$ has metal–metal bonding character and involves all three Y atoms (Fig. 2f), so that the bonding can be described as a three center–two electron bond.[52] QTAIM analysis did not show the presence of direct Y–Y bond paths in $Y_3@C_{80}$.[52] Instead, the non-nuclear attractor (i.e. the maximum of the electron density not associated with the nuclei, also referred to as "pseudoatom") was found in the center of the molecule, and each Y atom forms a bond path to the pseudoatom. Thus, pseudoatom in $Y_3@C_{80}$ mimics the nitrogen atom in $Y_3N@C_{80}$.

3. ISOMERISM OF EMFs

3.1. Formal Charge of the Carbon Cage Determines the Most Preferable Isomers

Structural studies of EMFs show that their carbon cage isomers are usually different from those of the isolated empty fullerenes, which can be explained by the metal-to-cage electron transfer as mentioned in the previous section. The fact that charging can affect relative stabilities of the fullerene isomers was recognized in early 1990s.[93,94] For instance, the $I_h(7)$ isomer of C_{80}, being the least stable structure for (+2) and (0) charge states, was found to be the lowest energy structure for (–4) and (–6) states.[94,95] In agreement with this finding, it was shown that the most stable isomer of $La_2@C_{80}$ has a $I_h(7)$ cage, as opposed to the D_2-symmetric hollow C_{80}.[96] Variation of the relative energies of C_{82} isomers in the negatively charged states was analyzed by Kobayashi and Nagase at the HF/3–21G//AM1 level to rationalize experimentally observed isomers of M@C_{82}.[22,97] The C_2(3) cage was found to be the lowest energy isomer of the neutral C_{82}, whereas the C_{2v}(9) cage is the most stable for both C_{82}^{2-} and C_{82}^{3-} as well as for M@C_{82} (M = Ca, Sc, Y, La, etc.).[22,97] Likewise, the C_{3v}(8) isomer was shown to be the most stable one for the tetraanion C_{82}^{4-},[31,40,53] in line with the finding that the most abundant isomers of Sc_2C_2@C_{82},[98,99] Y_2C_2@C_{82},[69,100] and M_2S@C_{82} (M = Sc, Y, Dy).[53,101] have this carbon cage.

Detailed studies of the relative stabilities of EMF isomers were performed for the family of nitride clusterfullerenes (NCFs) with the formal 6-fold electron transfer from the cluster to the carbon cage. The simple use of ionic conjecture in M_3N@C_{2n} to predict the most suitable cage isomers of NCFs was proposed by Poblet and coworkers.[102,103] They have supposed that only the fullerenes with a considerable gap between LUMO+2 and LUMO+3 should be considered as suitable hosts for nitride clusters since this MO gap corresponds to the HOMO-LUMO gap of M_3N@C_{2n} molecules (see Fig. 1 for an example of Sc_3N@C_{78}).[102] Among all IPR fullerenes in the C_{60}–C_{84} range, only C_{60}, C_{78}-D_{3h}(5), C_{80}-I_h(7) and C_{80}-D_{5h}(6) were found to be suitable cage isomers. Indeed, besides C_{60}, which is too small to form a nitride clusterfullerene, only these and no other IPR cage isomers were found for Sc_3N@C_{2n} clusterfullerenes. Later this work was expanded by considering also all IPR isomer in the C_{86}–C_{100} range,[103] and the isomers with the largest orbital gaps found for C_{86}, C_{88}, and C_{100} hexaanions (D_3(19), D_2(35), and D_5(450), respectively) were those proved for Tb_3N@C_{86}, Tb_3N@C_{88}, and La_2@C_{100} by single-crystal X-ray diffraction studies.[104,105] The orbital gap criterion was also applied in the studies of carbide clusterfullerenes M_2C_2@C_{2n} which have the formal four-fold electron transfer from the cluster to the cage, and the preferable formation of EMFs with C_{82}-C_{3v}(8) carbon cage was rationalized by the largest gap.[40] Obviously, the orbital gap criterion discriminates the cages based on their kinetic stability, but it does not answer the question on their thermodynamic stability.

A different approach to the search of the suitable cage isomers of NCFs is based on the thermodynamic stability of the hexaanions.[18–20,106] It was suggested that the suitable cage isomers of M_3N@C_{2n} should be found among the most stable isomers of C_{2n}^{6-}, and that both IPR and non-IPR isomers should be considered on the equal basis. The good correlation found between the relative energies of the C_{2n}^{6-} isomers computed at the DFT and semiempirical AM1 levels of theory[19] allowed the screening of many thousands of possible isomers at the relatively cheap AM1 level. Ten to twenty lowest energy C_{2n}^{6-} isomers (and corresponding $M_3N@C_{2n}$ isomers) were then studied at the DFT level of theory. By screening through the large number of the C_{70}^{6-}, $C_{76}^{6-}$and C_{78}^{6-} isomers, the molecular structures of non-IPR $Sc_3N@C_{70}$-C_{2v}(7854),[19] $DySc_2N@C_{76}$-C_s(17490),[106] and $M_3N@C_{78}$-C_2(22010) (M = Dy, Tm)[18] were proposed. The structural guesses were then confirmed by comparison of the experimental and DFT-computed vibrational spectra; besides, C_2(22010) cage isomer in $M_3N@C_{78}$ NCFs has been confirmed in 2009 by a single crystal X-ray diffraction study of $Gd_3N@C_{78}$.[107] Recently, the reports on the search for the most stable isomers of $M_3N@C_{88}$ (M = La, Gd, Lu)[108] and $La_3N@C_{92}$[109] were published.

The exhaustive search for the most favorable cage isomers of $Sc_3N@C_{2n}$ and $Y_3N@C_{2n}$ ($2n = 68 - 98$) at the PBE/TZ2P level was reported in 2007.[20] The isomers shown to exist by single-crystal X-ray studies were always found among the lowest energy isomers of C_{2n}^{6-}. For $M_3N@C_{82}$, the study has predicted the C_{82}-C_s(39663) isomer as one of the most probable structures, and later it was indeed confirmed by a single-crystal X-ray study of $Gd_3N@C_{82}$.[110] It is remarkable that C_{2n}^{6-} isomers with high thermodynamic stability (i.e. low relative energy) quite often exhibit high kinetic stability (i.e. large HOMO-LUMO gap). As a result, the IPR isomers proposed based on the orbital gap criterion[102,103] in most cases coincide with predictions based on the relative stability of the hexaanions.[20]

The broad range of fullerenes studied in Ref. 20 enabled the authors to follow the general trends in their stabilities. Figure 3a plots the absolute energies of the most stable C_{2n}^{6-} isomers normalized to the number of atoms as a function of the fullerene size. The energy decreases smoothly with the growth of the fullerene size, which can be explained by (i) the decrease of the curvature of the cage, which results in a smaller curvature-induced strain, and (ii) the decrease of the on-site Coulomb repulsions of six surplus electrons in C_{2n}^{6-} with the increase of the cage size. However, a significant deviation from the smooth curve can be seen for C_{80}-I_h(7) and C_{80}-D_{5h}(6) isomers. These isomers are 185 and 98 kJ/mol more stable than

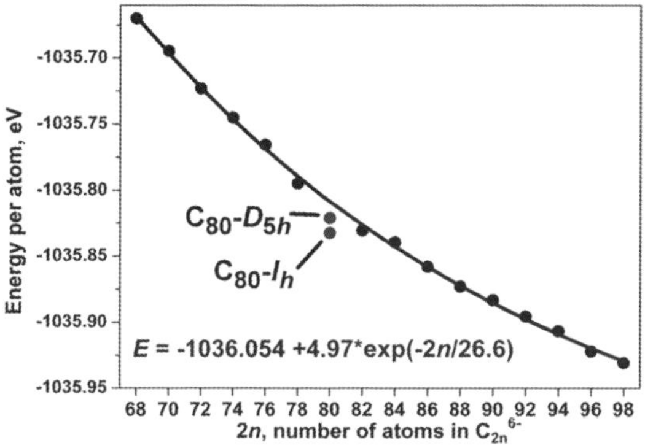

(a)

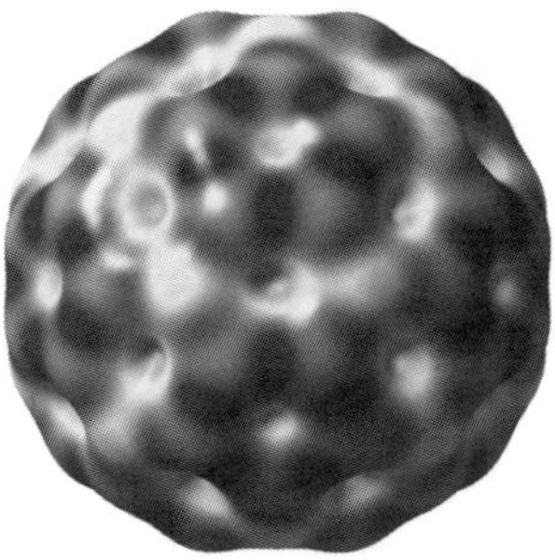

(b)

Fig. 3 (a) Normalized energies of the most stable C_{2n}^{6-} isomers (dots) and their fit with the exponential decay function (solid line). Normalized energies for C_{80}^{6-}-I_h(7) and C_{80}^{6-}-D_{5h}(6) isomers are clearly seen below the line; reproduced with permission from Ref. 20 © 2007 American Chemical Society. (b) electrostatic potential in C_{80}^{6-}-I_h(7) mapped onto the electronic density isosurface; the darkest regions correspond to the highest concentration of the negative charge.

they might be if they were like all other fullerenes (i.e. like those which obey the smooth decay in the normalized energy). Note that similar deviations of the overall smooth dependence were also found for C_{60} and C_{70} in the DFT study of the non-charged fullerenes.[111] The kinetic model developed by Curl *et al.*[111] shows that such deviations as found for C_{60} and C_{70} lead to a high yield of corresponding fullerenes. Thus, the enhanced stability of the two C_{80}^{6-} isomers explains the increased yield of $M_3N@C_{80}$ compared to all other cage sizes. Besides, it explains why the C_{80}-I_h(7) isomer is always the most preferable structure in other clusterfullerenes with formal 6-fold electron transfer to the cage, such as $La_2@C_{80}$, $Sc_4O_{2,3}@C_{80}$,[64] or $Sc_4C_2@C_{80}$.[112]

The extraordinary stability of the I_h(7) and D_{5h}(6) isomers of C_{80}^{6-} was explained in Refs. 20,31 using the concept of hexagon indices established by Raghavachari[113] to quantify the distribution of pentagon-induced curvature in fullerenes. According to Raghavachari,[113] the neighbor index of a hexagon is the number of hexagons to which it is adjacent, and every fullerene isomer can be characterized by a set of indices h_k ($k = 0 - 6$), where h_k is the number of hexagons with neighbor index k (for IPR isomers h_0, h_1, and h_2 are equal to 0, and hence only four indices (h_3, h_4, h_5, h_6) are sufficient).[113,114] The steric strain is at minimum when dispersion of the indices is minimized, i.e. the indices of all hexagons should be as close to each other as possible. Therefore, the lowest strain is expected for the isomer in which all indices are equal and pentagons are distributed in the most uniform way. For fullerenes C_{2n} with $2n < 120$ this condition is fulfilled only for C_{60}-I_h(1), C_{80}-I_h(7) and C_{80}-D_{5h}(6).[114] Thus, the exceptional stability of two C_{80}^{6-} isomers can be explained by the favorable uniform distribution of the pentagons. For other IPR fullerenes (h_3, h_4, h_5, h_6), indices minimizing the strain are ($80 - 2n$, $3n - 90$, 0, 0) for C_{2n} with $2n \leq 80$, and (0, $70 - n$, $2n - 80$, 0) for C_{2n} with $2n \geq 80$.[114] It was found that the list of the least-strained IPR isomers perfectly corresponds to the lowest energy IPR isomers of C_{76}^{6-}–C_{90}^{6-}.[20] Furthermore, the relative energies of the IPR C_{2n}^{6-} isomers follow the rationalization of the stability based on the spatial distribution of pentagons much better than the relative energies of the non-charged fullerenes (the latter usually violate the requirement of the least-strain). In 2010 this phenomenon was explained by Poblet and co-workers.[115] The authors proposed that the relative stabilities of the isomers of multiply charged fullerene anions are predominantly determined by the on-site Coulomb repulsion. Since the negative charge in anions is mostly localized on the pentagons (Fig. 3b), the lowest-energy structures should have the maximal separation of pentagons (and hence the most uniform distribution). The authors have introduced the inverse pentagon separation index and showed that its correlation with the relative stabilities of the isomers is improving with the increase of the charge. Thus, minimization of the on-site Coulomb repulsion and minimization of the steric strain are achieved by fulfillment of the same condition, the most uniform distribution of the pentagons. If for the non-charged fullerenes the steric strain factor can be outweighed by other factors such as stability of the π-system (and hence the lowest strain condition is usually violated), with the increase of the negative charge the Coulomb repulsion sooner or later outweighs all other factors. Therefore, to predict the lowest energy isomers of the EMFs with high formal charge of the cage, one can apply the same arguments as developed in early 1990s to find the isomers with the lowest steric strain.

In summary, extended search for factors determining isomerism of EMFs showed that (i) the stability of the isomers of appropriately charged empty fullerenes provides a good estimation to the relative stability of the cage isomers of EMFs, (ii) the relative energies of the fullerene isomers are dramatically affected by the charge, and (iii) the lowest energy isomers for highly charged cages are those in which pentagons are distributed in the most uniform way.

3.2. Violation of the Isolated Pentagon Rule (IPR) in EMFs

The specific feature of endohedral fullerenes is the violation of the isolated pentagon rule (IPR). This rule imposes a strict limitation on the possible isomers of *empty* fullerenes, but encapsulation of a metal or a cluster inside the cage often results in the violation of the IPR, affording thus the non-IPR endohedral fullerenes. Such possibility was first suggested by Kobayashi *et al.* in theoretical studies of Ca@C_{72} isomers,[116] and the first experimental evidence was provided in 2000 when Sc_2@C_{66}-C_{2v}(4348) [117] and Sc_3N@C_{68}-D_3(6140) [118] were isolated. Since that time, a lot of non-IPR endohedral fullerenes were reported, including Sc_2C_2@C_{68}-C_{2v}(6073),[119] Sc_3N@C_{70}-C_{2v}(7854),[19] Sc_2S@C_{70}-C_2(7892),[120] Sc_2S@C_{72}-C_s(10528),[121] La@C_{72}-C_2(10612),[122] La_2@C_{72}-D_2(10611),[75,123] $DySc_2N$@C_{76}-C_s(17490),[106] M_3N@C_{78}-C_2(22010) (M=Y, Dy, Tm, Gd),[18,107,124] Sc_3NC@C_{78}-C_2(22010),[125] Gd_3N@C_{82}-C_s(39663),[110] M_3N@C_{84}-C_s(51365) (M=Tb, Gd, Tm).[126,127] Based on computational studies, the non-IPR isomers were also proposed for Sc_2@C_{70}-C_{2v}(7854),[128] Sc_2@C_{76}-C_s(17490),[20] Ca@C_{72}-C_2(10612) or C_{2v}(11188),[116,129] the minor isomer of Yb@C_{74} (C_1(13393) or C_1(14049)),[130] M_2@C_{74} (C_2(13295) and C_2(13333), M=Sc, La),[131] M@C_{76} (C_{2v}(19138) and C_1(17459), M=Yb, Ca, Sr, Ba),[132,133] Y_2C_2@C_{84}-C_1(51383),[134] Gd_2@C_{98}-C_1(168785).[135]

The tendency to violate the IPR in endohedral fullerenes can be explained by the changes of the relative energies of the fullerene isomers with the increase of the cage charge and additional stabilization of the pentagon pairs by coordination to metal atoms. It was shown back in 1995 that the non-IPR C_{2v}(1809) isomer of C_{60} with two pairs of adjacent pentagons is substantially less stable than the sole IPR isomer in the neutral state, but with the increase of the negative charge on the cage, the difference in the relative energies is diminishing, reaching almost zero for (–4) and even inversing for (–6) states.[94] Semiempirical and DFT calculations for the whole set of 6332 isomers of C_{68} in 0, (–2), (–4), and (–6) charge states showed the penalty for a pentagon adjacency is decreasing from 92 kJ/mol in C_{68} to 70 kJ/mol in C_{68}^{6-} at the AM1 level, confirming that the increase of the charge stabilizes adjacent pentagon pairs.[136]

An explanation of the stabilization of the pentagon adjacencies in metallofullerenes was proposed by Slanina *et al.*[137] The authors noted that the pentalene (a unit of two fused pentagons) is a 8π anti-aromatic system in the neutral state, but becomes 10π aromatic in the dianionic state.[138] In a similar fashion, when the metal in the non-IPR EMF is coordinated to the pentagon pair, it donates two electrons to the pentalene unit to make it aromatic. This reasoning agrees with the fact that the maximum number of pentagon pairs found in EMFs is equal to half of the formal charge of the cage: three pairs for (6–) charge as in $Sc_3N@C_{68}$[118,139] or $Sc_3N@C_{70}$,[19] two pairs for (4–) state as in $Sc_2S@C_{70}$,[119] and one pair for (3–) and (2–) states as in $La@C_{72}$[122] and $Ca@C_{72}$.[116]

A correlation of the total number of adjacent pentagon pairs in the lowest-energy isomers of C_{2n}^{6-} ($2n$ = 68–88) with the cage size was analyzed in Ref. 20. With the increase of the cage size, the number of pentagon pairs in the most stable isomers was found to decrease: while three pairs are common for C_{68}^{6-} and C_{70}^{6-}, two pairs are preferable for C_{72}^{6-}–C_{78}^{6-} and one pair is preferable for C_{80}^{6-}–C_{86}^{6-}. Relative energies of the non-IPR isomers with respect to the IPR isomers are also increasing with the size of the fullerene, and hence formation of non-IPR fullerenes beyond C_{84} should not be common. Indeed, the largest cage reported to date for any non-IPR fullerene is C_{84}-C_s(51365).[126,127] With the increase of the cage size, a more uniform distribution of pentagons over the fullerene is possible and hence, localization of a steric strain and surplus charge in adjacent pentagon pairs is more unfavorable than for the smaller cages. At the same time, coordination of metal atoms to pentagon adjacencies results in their additional stabilization, and it cannot be excluded that stable non-IPR isomers of EMFs can be found for larger cages such as predicted recently for $Gd_2@C_{98}$-C_1(168785).[135]

3.3. Cage form Factor

In the previous chapter it was shown that the relative stability of the carbon cage in the charged state is the main factor determining isomeric distribution of the EMFs. However, this factor is not the only one which determines whether the given EMF isomer is stable or not. Another important parameter which strongly influences stability of the EMF isomers can be defined as the cage form factor. In other words, the carbon cage topology (its inner space and the distribution of the metal-bonding sites) is also an important stability factor, especially for clusterfullerenes.

An example of the decisive influence of the cage form factor is the change in the preferable cage isomer in the $M_3N@C_{78}$ system with the increase of the metal size. In particular, the most stable isomer of C_{78}^{6-} is $D_{3h}(5)$,[18] and this cage is indeed found in $Sc_3N@C_{78}$.[140] However, when the larger metal atoms are considered (such as Y or Lu), the size of the M_3N cluster is so large that the C_{78}-$D_{3h}(5)$ cage does not provide sufficient space to accommodate it. DFT calculations showed that the Y_3N cluster is pyramidal in $Y_3N@C_{78}$-$D_{3h}(5)$, which points to the large steric strain in the system (Fig. 4a).[18] For $M_3N@C_{78}$ EMFs, this results in the preference of another isomer, non-IPR C_{78}-$C_2(22010)$, whose flattened shape provides sufficient room for the large cluster (Fig. 4b).[18,54,107,124,125] Although in the empty hexaanionic form, C_{78}^{6-}-$C_2(22010)$, this isomer is less stable than C_{78}^{6-}-$D_{3h}(5)$ by 59 kJ/mol, $Y_3N@C_{78}$-$C_2(22010)$ is more stable than $Sc_3N@C_{78}$-$D_{3h}(5)$ by 84 kJ/mol (Fig. 4c).[18]

Importantly, the inner space in C_{80}-$I_h(7)$ is also rather limited, and for the nitride clusterfullerenes with large lanthanides it results in the preference for larger cages.[141,142] Even when this isomer is still preferable (such as for $Y_3N@C_{80}$), the increase of the internal strain already influences the structure (e.g. large Gd_3N cluster is pyramidalized[143]), chemical[144–148] and spectroscopic properties.[149] C_{80}^{6-}-$I_h(7)$ is more stable than any other C_{80}^{6-} isomer (except for $D_{5h}(6)$) by more than 200 kJ/mol, and almost the same situation is found for the isomers of $Sc_3N@C_{80}$.[20] However, other isomers (with more suitable cage shape) are significantly stabilized for $Y_3N@C_{80}$ with respect to the $Y_3N@C_{80}$-$I_h(7)$ structure and the lowest energy structures (e.g. non-IPR $C_1(28325)$) already compete in stability with the $Y_3N@C_{80}$-$D_{5h}(6)$ isomer.[20] It means that their formation might be, in principle,

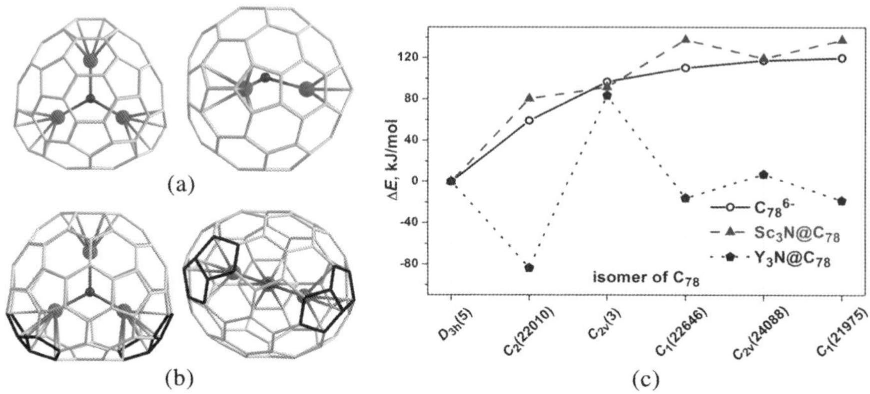

Fig. 4 (a–b) PBE/TZ2P-optimized molecular structures of $Y_3N@C_{78}$ with IPR $D_{3h}(5)$ (a) and non-IPR $C_2(22010)$ (b) cage isomers. Each structure is shown in two projections, adjacent pentagon pairs in $Y_3N@C_{78}$-$C_2(22010)$ are highlighted in black. Pyramidalization of the Y_3N cluster inside C_{78}-$D_{3h}(5)$ cage can be well seen. Reproduced with permission from Ref. 18 © 2007 American Chemical Society. (c) Relative energies of the most stable isomers of C_{78}^{6-} and corresponding $Sc_3N@C_{78}$ and $Y_3N@C_{78}$ EMFs. Based on the numerical data from Ref. 18.

possible. Note that Yang and Dunsch reported the isolation of the third isomer of $Dy_3N@C_{80}$,[150] whose structure is still not known, but $C_1(28325)$ is a likely candidate.[20]

On the other hand, if the fullerene is too large, efficient coordination of the nitride cluster is also not possible,[20] which limits the size distribution of NCFs (for Sc-NCFs, $Sc_3N@C_{80}$ is the largest isolable structure, while for La the largest NCF isolated so far is $La_3N@C_{96}$[151]).

Another way the cage form factor can influence the isomer distribution is most apparent for non-IPR cages. In all non-IPR EMFs, adjacent pentagon pairs are stabilized by coordination to the metal atoms. According to the principles discussed in the previous section, pentagon pairs should be distributed in the most uniform way. In line with these guidelines, the most stable C_{72}^{6-} isomer, $D_2(10611)$, has an elongated shape with two pentagon pairs located on the opposite poles of the molecule. This isomer is confirmed for isolated $M_2@C_{72}$ EMFs (M = La,[123] Ce[78]), but the nitride clusterfullerenes $M_3N@C_{72}$ have never been found. The reason is that the location of the pentagon pairs as well as the elongated shape of the cage are not compatible with the triangular shape of the M_3N cluster, whereas deformation of the cluster increases repulsive interaction between metal atoms.[57] M_3N cannot be coordinated to two pentagon pairs of C_{72}-D_2(10611) and preserve its regular triangular shape at once, and hence $M_3N@C_{72}$ with the D_2(10611) cage is very unstable.[20] The same reasoning explains why $M_3N@C_{74}$ NCFs have never been observed.[20] Likewise, the yield of $Sc_3N@C_{70}$, in which the cluster has to be significantly deformed, is much lower than that of other Sc-based NCFs.[19]

From the discussion above, it should be rather obvious that the cage form factor is much more important for clusterfullerene stability than for conventional EMFs. While for the former the shape and the size of the cluster play a limiting role in the cage size distribution, for mono and dimetallofullerenes these factors are not so crucial and a much broader range of structures can be obtained.[152–156]

3.4. Gibbs Energy Considerations

Stability factors described in the previous sections of this chapter determine relative energies of the EMF isomers at $T = 0$ K, whereas the synthesis of EMFs in the arc proceeds at high temperatures exceeding 2000 K. Since isolated structures roughly correlate to their stability, at least partial thermodynamic control of the products can be postulated. Hence, equilibrium composition computed for the reaction temperature may correspond to the real distribution of isomers better than their zero-temperature relative energies. Equilibrium composition of the mixture of isomers can be computed via Gibbs energies, which in due turn can be estimated using statistical thermodynamic approaches, such as rigid rotor-harmonic oscillator (RRHO) or its modification "free encapsulate model" (FEM).[157] Exhaustive analysis of the influence of the temperature on equilibrium composition of isomeric mixtures of EMFs was performed by Slanina *et al.* in a series of publications.[130,157–167] In 2010 Gibbs energy analysis showed that while $Sc_2O@C_{82}$-C_{3v}(8) is the lowest energy isomer at low temperature, increase of the temperature beyond 1000 K results in the preference of another isomer, $Sc_2O@C_{82}$-C_s(6), which was the only isomer found experimentally.[60]

It should be noted that real temperatures of the arc in EMF synthesis and the temperatures at which interconversion between different EMF isomers is still possible are not known. Besides, the relative energies of the isomers, which have crucial influence upon the temperature dependence of the isomeric composition, are usually taken from DFT calculations. Precision of such calculations remains unknown since experimental values on the relative energies of isomers of EMFs are not available so far, and hence it is not possible to validate DFT results quantitatively. When the effect of the temperature is taken into account in the analysis of the isomeric distribution, two further considerations should be kept in mind. First, the concentration of the structures unstable at $T = 0$ K should increase with the temperature as follows from the behavior of the Boltzmann term, $\exp(-\Delta E/RT)$, where ΔE is the relative energy and R is the universal gas constant. Obviously, with the increase of the temperature the role of ΔE is decreasing. Second, concentration of the low-symmetric isomers should increase with the temperature since the entropy includes a term $-R{\cdot}\ln(\sigma)$, where σ is a symmetry number. The larger the symmetry of the carbon cage (and in the MEF model only the cage symmetry counts[157]), the more negative this contribution in the entropy. Analysis of the symmetry factor for isomers of $Gd_3N@C_{82}$ as well as for some other fullerene cages is given in Ref. 168.

In addition to the composition of isomeric mixtures, thermodynamic analysis can address the question of the relative yield of EMFs with different metals. Slanina *et al.* reported such an analysis for various monometallofullerenes,[169–172] and recently also for $Sc_2S@C_{82}/Y_2S@C_{82}$.[173] These studies revealed that the relative yields of EMFs with different metal atoms can be rationalized employing saturated vapor pressures of metals and their ionization potentials.

4. DYNAMICS OF ENCAPSULATED SPECIES INSIDE THE FULLERENE CAGE

Depending on the carbon cage as well as the type of the encapsulated species (single metal atoms, two or more metal atoms, clusters of metal atoms with non-metals like nitrogen, oxygen, or sulfur, etc.), the latter can exhibit different dynamic behavior varying from the tight fixed position inside the carbon cage to nearly free rotation. Dynamics of the internal species can be also significantly altered by charging the molecule or by its chemical modification *via* the exohedral attachment of functional groups. The cluster dynamics seriously affect spectroscopic properties of EMFs and their derivatives: in case of the fast rotation of the endohedral cluster, experimentally measurable properties correspond to the time-averaged situation. For instance, high symmetry of the cage is revealed by NMR spectroscopy when the cluster rotates, whereas lower symmetry can be observed for the fixed position (an illustrative example is $Sc_2C_2@C_{80}$-C_{2v}(5) exhibiting C_s-symmetric C_{80} cage at room temperature and C_{2v}-symmetric cage at 413 K[174]). The problem of internal dynamics in EMFs can be addressed by quantum-chemical calculations in two different ways: via standard static calculations and via molecular dynamics simulations of EMF molecules.

4.1. Static Computations

The first indication on the internal dynamic behavior of EMFs can be obtained from DFT computations of different cluster orientations with respect to the carbon cage (denoted as *conformers* hereafter). When such calculations show co-existence of several almost isoenergetic minima, free rotation of the cluster can be tentatively proposed. However, this criterion cannot be definitive since the barriers to the internal rotation can still be rather high, and hence computations of transition states (TSs) are necessary to make more reliable conclusions on the internal dynamics. Unfortunately, the search for TSs is less straightforward than optimization of coordinates for the energy minima, and there are few works reporting such calculation.

The most detailed computational studies were performed for $M_2@C_{80}$ (M = La, Ce) or $M_3N@C_{80}$ (M = Sc, Y, Lu, etc.) with the I_h(7)-symmetric carbon cage. High symmetry of the cage facilitates such computations by limiting the number of possible conformers. For $La_2@C_{80}$ several unique conformers were found (note that each minimum is multiplied by the group symmetry operations of I_h group resulting thus in equivalent structures with La atoms distributed all over the carbon cage). In two lowest energy conformers metal atoms are pointing to the center of the cage hexagons (overall D_{2h} symmetry) or coordinating the carbon atoms on a triple-hexagon junction (D_{3d} symmetry).[96,175–177] The energy difference between the two structures is reported to be within few kJ/mol and depends on the method and basis set used. For instance, scalar-relativistic ZORA computations with BP functional and DZP (carbon)/TZP (La) basis set show that the D_{2h} conformer is 4.5 kJ/mol lower in energy that D_{3d} conformer, and the energy of the transition states connecting equivalent D_{2h} conformers is only 4.4 kJ/mol.[176] Room temperature ^{13}C and ^{139}La NMR studies[178] as well as synchrotron radiation powder diffraction study[179] show that La atoms are freely circulating inside the carbon cage.

Several groups reported relative energies of conformers of $M_3N@C_{80}$ (M = Sc,[35,39,180–183] Y,[39,181,183] Lu,[39] Gd[183,184]) and showed that the energies of the majority of found structures lie within the range of 10–15 kJ/mol. The C_3-symmetric conformer with metal atoms pointing the hexagons but somewhat displaced toward the pentagon/hexagon edges was found to be most stable one.[39] In Ref. 39, exhaustive search for transition states accompanied with intrinsic reaction coordinate (IRC) computations were accomplished for $Sc_3N@C_{80}$. These computations showed that at the PBE/TZ2P level the barrier to the cluster rotation also did not exceed 11.4 kJ/mol.[39]

Analogous computations were also performed for nitride clusterfullerenes with other carbon cages,[183,185] as well as $Ca@C_{74}$,[163] $Sc_2C_2@C_{82}$,[40,53] $Sc_2S@C_{82}$,[53] $Y_2C_2@C_{92}$,[186] $Sc_4O_2@C_{80}$,[82,187] etc. In certain cases, frontier MOs structure was found to be significantly dependent on the orientation of the cluster,[39,53,185] which shows that the search of the most stable conformers is a necessary prerequisite for the analysis of the electronic structure of EMFs (i.e. calculations made for one arbitrary chosen conformer can give erroneous results).

Several studies showed that changing the charge state of the EMF molecule can affect the internal dynamics of the endohedral species. For instance, the lowest-energy conformer of $Sc_3N@C_{80}^-$ has C_{3v} symmetry and corresponds to the higher-energy conformer of the neutral molecule, whereas the cluster rotation barrier is increased to 25 kJ/mol.[39] Significant changes in the relative energies of conformers and increase of the internal barriers in the anionic state was also reported for $La_2@C_{80}$,[188] $Sc_3N@C_{80}(CF_3)_2$,[189] [5,6]-pyrrolidino-$Sc_3N@C_{80}$,[190] and $Sc_4O_2@C_{80}$.[187] At the same time, computational studies show that rotation of the cluster is facilitated in the cations of $Sc_3N@C_{80}(CF_3)_2$[189] and $Sc_4O_2@C_{80}$.[187]

A special dynamic situation was found in $TiSc_2N@C_{80}$.[41] Formally, substitution of one Sc atom with Ti is equivalent to increasing the number of electron by one, and hence $TiSc_2N@C_{80}$ is isoelectronic to the anion $Sc_3N@C_{80}^-$. Indeed, relative energies of the $TiSc_2N@C_{80}$ conformers show some similarities with those of $Sc_3N@C_{80}$ anion, whereas conformers of the $TiSc_2N@C_{80}^+$ cation are similar to those of $Sc_3N@C_{80}$. Furthermore, the section of the potential energy corresponding to the rotation of the cluster around the Ti–N bond was reconstructed (Fig. 5). Rotation of the cluster can be described as a stepwise jumping between nine different conformers. Each step can be described as a rearrangement of usually one Sc atom from one pent/hex edge to another one, either in the same pentagon or in the neighboring pentagon (in the latter case transition state corresponds to Sc atom facing the center of the hexagon). Figure 5 shows that two regimes can be distinguished in the cluster rotation. There are several conformers with energies below 2 kJ/mol and barriers to the interconversion below 5 kJ/mol (rotation angles 0–30°). Further rotation requires a considerably higher barrier (16.5 kJ/mol). Once this barrier is overcome, five other conformers with relative energies in the range of 13–18 kJ/mol and barriers in the range of 15–20 kJ/mol can also be reached.[41]

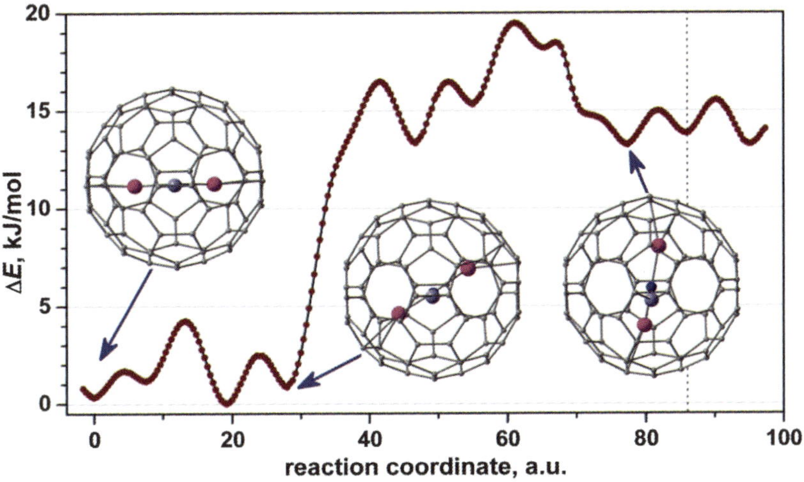

Fig. 5 The PBE/TZ2P energy profile along the rotation of the $TiSc_2N$ cluster in $TiSc_2N@C_{80}$ around the Ti–N bond. Each point of the curve corresponds to one step in intrinsic reaction coordinate calculations. Molecular structures of selected conformers are also shown projected along the Ti–N bonds to emphasize rotation of the cluster. Reproduced with permission from Ref. 41 © 2010 American Chemical Society.

Exohedral derivatization of EMFs (e.g. cycloaddition) shrinks the π-system of the fullerene cage and reduces its symmetry, thus providing a possibility for a larger number of conformers with larger variation of their relative energies. Therefore, dynamics of the endohedral species can be significantly altered in the derivatives of EMFs. The first indication that the rotational behavior can be manipulated by the cycloaddition was reported in 2003 by Kobayashi *et al.*[188] The authors found that three-dimensional random circulation of La atoms in $La_2@C_{80}$ is changed into motion in one plane in the $La_2@C_{80}(Si_2H_4CH_2)$ monocycloadduct and even more so in the bisadduct. Experimental studies of the cycloadducts of $La_2@C_{80}$ and $Ce_2@C_{80}$ proved that the motional of metal atoms was significantly changed.[191–195]

Exhaustive computations located more than 40 different conformers for $Sc_3N@C_{80}(CF_3)_2$[189] and [5,6]-pyrrolidino-$Sc_3N@C_{80}$[190] derivatives and their anions, whose relative energies spanned the range of 60–70 kJ/mol, which is several times larger than for the non-derivatized $Sc_3N@C_{80}$. In both derivatives, only two to three closely-related conformers have their energies below 10 kJ/mol with small barriers of interconversion, and few more structures are found in the 10–20 kJ/mol range. In the anions, the energy gap between few stable and all other conformers is further increased.[189,190] Thus, exohedral derivatization can significantly hinder rotation of the endohedral cluster, and multiple functionalization can completely block it as found in $Sc_3N@C_{80}(CF_3)_x$ adducts with $x \geq 4$.[196,197]

It is worth mentioning that preliminary information on the possible dynamic can be obtained from the analysis of the spatial distribution of electrostatic potential (ESP) inside the appropriately charged empty fullerenes.[21–23] For instance, ESP has well-defined minimum C_{82}^{3-}-$C_{2v}(9)$, which corresponds to the fixed metal position in the M@C_{82}-$C_{2v}(9)$ EMFs.[22,24] On the contrary, the ESP inside C_{80}^{6-} has no distinguishable minima, which agrees with almost free rotation of La_2 or Sc_3N clusters encapsulated inside the C_{80} cage. Spatial distribution of ESP was also been used to explain the positions and dynamics of the metal atoms in derivatives of EMFs.[189,191,194,195,198]

4.2. Molecular Dynamics Simulations

Molecular dynamics (MD) simulations, which allow direct "observation" of the dynamic behavior of the species on the microscopic scale, appear to be an ideal tool for addressing the problems of the cluster dynamics and their influence on spectroscopic properties. However, the wide use of such methods is still hindered by the high computational costs, especially when long propagation times are necessary. For instance, only one picosecond of the Born–Oppenheimer (BO) MD trajectory followed with the time step of 0.5–1 fs requires 1000–2000 evaluations of the energy and the energy gradient. For this reason, first principle MD studies of endohedral fullerenes are still very rare.[41,52,53,187,189,199–202] However, the continually increasing power and availability of supercomputer resources now allow longer and longer propagation times. Currently, 50–100 ps trajectories can be in principle accessible for DFT-based MD of EMFs.[187,203] Much longer dynamics (reaching the nanosecond scale) can be studied at the Density Functional based Tight Binding (DFTB) level.[202,204–207]

The first *ab initio* MD study of EMFs was reported by Andreoni and Curioni in 1996.[201] The authors used Car–Parrinello (CP) MD with local spin density approximation to study La@C_{60} and La@C_{82} on a timescale of a few picoseconds. Fluxional motion of La in C_{60} was found, whereas in La@C_{82} the metal atom was trapped in two different configurations. Unfortunately, the correct isomeric structure of La@C_{82}-C_{2v}(9) was not known yet in 1996, and calculations were performed for a C_2 cage isomer.

In 2000 Vietze *et al.* reported DFTB-BOMD studies of Eu@C_{74}-D_{3h} at different temperatures.[207] In the C_{74}-D_{3h} cage, the metal atom can occupy one of three equivalent minima. Simulations showed that already at 140 K the metal atom shows oscillations along the C_3 axis with amplitudes of ca 0.17 Å, which increased to 0.22 Å at 200 K. With further increase of temperature to 600 K, the amplitude of oscillations (0.9 Å along the axis and 0.5 Å in the transverse direction) became larger than the harmonic region of the potential, and at 900 K the Eu atom started to circulate randomly between three equivalent minima. The timescale of these simulations was not mentioned in the original paper, but experimental studies of isostructural Ca@C_{74}[208] and Yb@C_{74}[209] showed that at the NMR timescale metal atoms are circulating already at room temperature, although the pronounced temperature-dependence of the line width indicated that the fast-exchange limit was not reached. DFT calculation by Slanina *et al.* at the B3LYP/3–21G~dz level showed that the barrier of such circulation in Ca@C_{74} is in the range of 30–40 kJ/mol.[163] In 2004 Jansen *et al.* reported CPMD simulations for Ba@C_{74}-D_{3h}.[199] Over the timescale of ca 15 ps at room temperature, the authors observed one rearrangement between two different minima.

Extended DFTB-BOMD study of $Sc_2C_2@C_{84}$ at different temperatures and with the timescale reaching 1.5 ns was performed by Vietze and Seifert. They have found that positions of Sc atoms are almost fixed, whereas the C_2 unit is freely rotating around the Sc–Sc axis.[205] A year later these observations were confirmed in an experimental study which also showed that rotation of C_2 is quantized at low temperature.[204]

Taubert *et al.* reported BP86/def2-SVP based BOMD studies of $Sc_3C_2@C_{80}$ at 150 and 500 K propagating the system for 3.5–6.0 ps.[200] They showed that Sc atoms form a relatively rigid triangle which exhibits ratchet-like rotational motion between carbon rings, whereas the C_2 unit rapidly flips through the Sc_3 plane. The amplitudes of these motions were noticeably higher at 500 K than at 150 K, but the character of the dynamics was preserved.[200] BOMD studies of $Sc_2S@C_{82}$ and $Sc_2C_2@C_{82}$, both with $C_{3v}(8)$ carbon cage, were performed at the PBE/DZV(P) level.[53] The trajectories followed for 10 ps at 300 K are visualized in Fig. 6. The clusters tend to rotate around the C_3-axis of the carbon cage, but rotation appears to be faster for the Sc_2S cluster, which accomplishes ca 180° turn over 10 ps, whereas the Sc_2C_2 cluster only showed large-amplitude librations. It can be anticipated, however, that the Sc_2C_2 cluster also rotates on the longer timescale, since ^{13}C NMR studies show that the carbon cage has an effective C_{3v} symmetry.[98] The intra-cluster dynamics of the C_2 unit in $Sc_2C_2@C_{82}$-$C_{3v}(8)$ are rather chaotic (Fig. 6) and cannot be described as in-plane rotation as in $Sc_2C_2@C_{84}$ or through-plane flipping as in $Sc_3C_2@C_{80}$.[53]

Several BOMD studies were reported for nitride clusterfullerenes. In the early 2000s, Vietze and Seifert accomplished nanosecond-long dynamic studies of $Sc_3N@C_{78}$ and $Sc_3N@C_{80}$ at the DFTB-BOMD level.[202,205,206] Complex fluxional behavior of the Sc_3N cluster was found in $Sc_3N@C_{80}$.[206] On the timescale of ca 0.4 ps, each Sc atom was fluctuating around the carbon atoms of one pentagon. On the longer timescale (ca 1.5 ps), one of the Sc atoms hopped from one pentagon to another. Activation barrier of ca 0.144 eV was estimated from the hopping time, which agrees well with static DFT as well as experimental studies. In $Sc_3N@C_{78}$, the position of the Sc_3N cluster was locked inside the cage, but it exhibited slow in-plane rotational vibrations and in-plane deformations of the cluster.[205]

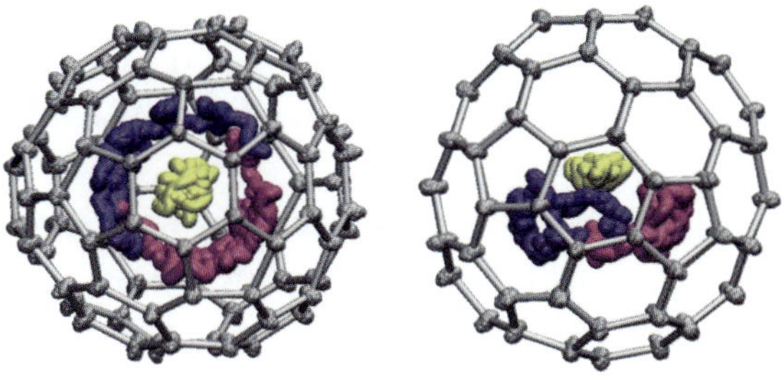

(a)

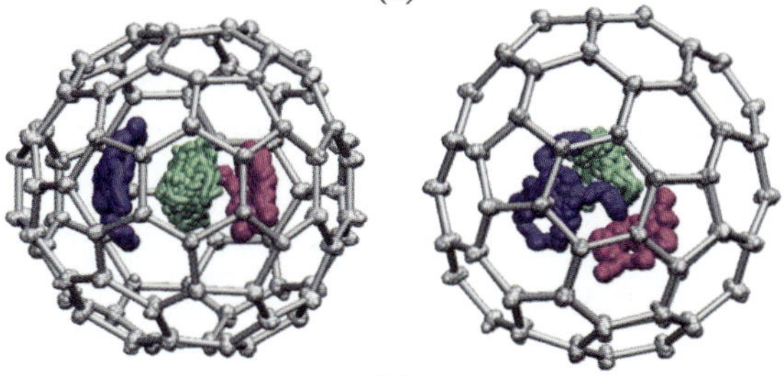

(b)

Fig. 6 (a) BOMD trajectory of $Sc_2S@C_{82}$ followed for 10 ps in the microcanonical ensemble after equilibration for 0.5 ps at 300 K; Sc atoms are blue and magenta, sulfur atom is yellow, carbon atoms are grey; (b) the same for $Sc_2C_2@C_{82}$; carbon atoms of the carbide unit are light green. For each molecule, two views are shown: one with the C_3 axis of the cage perpendicular to the paper, and another one with the axis coplanar to the paper. BOMD computations were performed at the PBE/DZV(P) level. Reproduced with permission from Ref. 53 © 2010 American Chemical Society.

DFT-based BOMD studies with the PBE functional were reported in our group. Comparative DFT-BOMD study of $Y_3@C_{80}$ and $Y_3N@C_{80}$ showed that dynamic behavior of the Y_3 cluster inside the C_{80} cage is similar to the behavior of the Y_3N cluster, both exhibiting rotation of the cluster as a whole.[52] In another work it was found that substitution of one scandium atom in $Sc_3N@C_{80}$ with a titanium atom drastically changes rotational behavior of the cluster.[41] While Sc_3N, as discussed above, is rotating freely, in $TiSc_2N@C_{80}$ the Ti atom remains fixed, at least at the 10 ps time scale, and the cluster exhibits only restricted rotations around the Ti–N bond, in line with the results of static DFT studies. DFT-based BOMD study of $Sc_3N@C_{80}(CF_3)_2$ confirmed that exohedral derivatization restricts rotation of the endohedral cluster,[189] which agrees with the results of ESR studies of the anion and trianion of $Sc_3N@C_{80}(CF_3)_2$.[210] Besides, the cluster dynamics were also strongly dependent on the charge state of the molecule in agreement with static DFT studies discussed above.[189]

Recently, 50 ps long DFT-BOMD trajectories were studied for $Sc_4O_2@C_{80}$, its cation, and anion at 300 K.[187] In agreement with ^{45}Sc NMR data, the studies revealed that the Sc_4O_2 cluster is rather rigid (i.e. there are no rearrangements of oxygen atoms between the faces of the Sc_4 tetrahedron). In the cation, the cluster exhibits large rotational displacements (i.e. rotation is virtually free), which means that not only relative energies of the conformers, but also the barriers to internal rotation are small. In general, Sc atoms prefer to occupy positions near the pentagon/hexagon (5/6) edges, and the rotation of the cluster can be described as a jumping of Sc atoms between different 5/6 edges of the symmetric C_{80}-I_h cage, similar to the behavior described for $Sc_3N@C_{80}$. Dwelling times in each conformation did not exceed a few picoseconds. For the anion, MD simulations revealed a more complex rotational behavior. The trajectory started from one of the high-energy conformers showed that the cluster rearranged rapidly between this and one other high-energy conformer during the 15 ps of equilibration and the first 12 picoseconds of the production run. But then the cluster rearranged to the position corresponding to the lowest-energy conformer in static DFT computations and remain trapped in this state for the next 27 ps till the end of the MD run showing only vibrational displacements. These data show that the rotation of the cluster in the anion is more hindered than in the cation.[187]

The BOMD studies of $TiSc_2N@C_{80}$,[41] $Sc_3N@C_{80}^-$,[189] $Sc_3N@C_{80}(CF_3)_2^{-/3-}$ (Ref. 189) and $Sc_4O_2@C_{80}^{-/+}$ (Ref. 187) also revealed important aspects of the spin density dynamics in these radicals. The spin density distribution in such molecules is very flexible and even small displacements of the metal atoms result in drastic changes of the spin populations. The effect was defined as *spin flow*,[41] which, depending on the cluster, can be either intra-cluster (net spin population of the cluster is not changing), cluster-cage spin flow (spin population of the cluster and the carbon carcass are changing) or a combination of both channels. Fourier-transformation of spin-populations resulted in spin-flow vibrational spectra, showing how spin flow is coupled to specific vibrations of the molecules.[41]

To conclude, MD studies complement static DFT well in describing the dynamic situation in endohedral fullerenes. Static DFT studies provide information on the relative energies of different conformers and, less straightforwardly, energy barriers to the rotational motions of the encapsulated clusters. Such computations can quickly become cumbersome, especially when the carbon cage has low symmetry or is chemically modified. On the other hand, MD studies reveal an overall dynamic picture but give limited information on the energetic parameters of the process. The best strategy to obtain a detailed description of the internal dynamics of EMF molecules is therefore to combine two techniques,[187,189,211] and the computational abilities provided by modern supercomputers promise that such extended studies will be more and more feasible in the near future.

5. COMPUTATIONAL STUDIES OF ELECTROCHEMICAL AND SPECTROSCOPIC PROPERTIES OF EMFs

Single energy SCF computation of the (preferably optimized) EMF molecule already provides a lot of information on the electronic structure of the molecule and can assist in interpretation of certain spectroscopic properties. Although the physical meaning of Kohn–Sham molecular orbitals is rather ambiguous, it is a common praxis to use their energies and isosurfaces in the discussion of the electronic properties. For instance, HOMO-LUMO gap gives a first indication on the kinetic stability of a given EMF. In a broader perspective, electronic density of states (DOS, which for discrete molecules is simply calculated via broadening of orbital degeneracy by a peak function, usually of a Gaussian type) is used to interpret ultraviolet photoelectron (UPS)[212,213] or scanning tunneling spectra (STS).[214,215] Summation of orbital densities in the energy range of interest is often used to simulate the scanning tunneling microscopy images.[214,216–218] However in most cases interpretation of spectroscopic data requires more dedicated computational studies, which are discussed below.

5.1. Electrochemical Properties

The energies of the frontier orbitals usually correlate well with the redox potentials, and therefore can be used for preliminary estimations of these values, whereas inspection of the shapes of the orbitals gives an information on the localization of the charge and spin in the electrochemically and chemically produced ions of EMFs. That is, localization of the LUMO (HOMO) of the EMF molecule on the encapsulated species indicates that reduction (oxidation) of this EMF is the endohedral redox process, and spin density in the electrochemically generated radical ions will be localized endohedrally.[219] A typical recent example is $Sc_4O_2@C_{80}$, which has both its HOMO and LUMO localized on the scandium-oxide cluster,[51,82] and its reduction and oxidation leads to the radicals with metal-localized spin density and rich ^{45}Sc-based hyperfine structure.[187] Some other examples of this sort are listed in Ref. 219. But localization of the frontier MOs on the carbon cage (such as in $M@C_{82}$ EMFs) indicates that the endohedral species will not participate in the redox process. Despite the fact that these simple rules are usually valid, one should be careful when using MOs computed for the neutral molecule to predict its behavior in the redox process. For instance, although Ce-4f^1 based MO in $CeLu_2N@C_{80}$ is buried 0.37 eV below the cage-based HOMO, it is the Ce-4f^1 orbital which is active in the oxidation of $CeLu_2N@C_{80}$.[220] The process can be therefore described as an endohedral Ce^{III}→Ce^{IV} oxidation rather than a cage-based redox process.

More accurate information on the electrochemical properties can be obtained by modeling the ionic states of EMFs and their solvation energies, which can be done using several variants of the polarizable continuum model[221] (such as COSMO[222] or conceptually similar C-PCM,[223] to name a few) implemented in many quantum-chemical codes. Representative examples of calculated redox potentials for different classes of clusterfullerenes are compared to the experimental values in Table 2.

BLYP/DNP studies of $Sc_3C_2@C_{80}$ in different charged states as well as DFT-COSMO calculations of its oxidation potential and the first three reduction potentials in *o*-dichlorobenzene were reported by Tan and Lu in 2006.[229] The first reduction potential of the compound is reproduced by the calculations quite well (the error of only +0.12 V), however discrepancies for the other redox potentials are rather large (up to −0.46 V for the third reduction and +0.37 V for the oxidation). Analogous study for $Sc_4C_2@C_{80}$ resulted in reasonable estimation of the first reduction and two oxidation potentials (the errors are less than 0.2 V), but the error for the

Table 2 DFT-COSMO Computed Redox Potentials of Selected EMF Compared to Experimental Values in o-dichlorobenzene (in V versus $Fe(Cp)_2^{+/0}$)

EMF		ox-II	ox-I	red-I	red-II	gap_{H-L}	gap_{EC}[a]	Ref
$Sc_3C_2@C_{80}$	exp		−0.06	−0.50	−1.64		0.44	224
	calc[b]		0.31	−0.38	−1.87	0.36	0.69	
$Sc_4C_2@C_{80}$	exp	1.10	0.40	−1.16	−1.65		1.56	112
	calc[b]	1.30	0.32	−1.33	−2.02	1.17	1.65	225
$Sc_3NC@C_{80}$	exp		0.60	−1.05	−1.68		1.65	
	calc[c]	1.72	0.97	−0.80	−1.94	1.13	1.77	54
$Sc_3N@C_{80}$-I_h	exp		0.59	−1.27			1.85	183
	calc[d]		0.53	−1.17		1.22	1.70	183
	calc[e]-ΔG		0.47	−1.05			1.48	183
$Sc_3N@C_{80}$-D_{5h}	exp		0.34	−1.33			1.67	183
	calc[d]		0.19	−1.34		1.15	1.53	183
	calc[e]-ΔG		0.18	−1.18			1.36	183
$Gd_3N@C_{80}$	exp		0.58	−1.44			2.02	226
	calc[d]		0.48	−1.51		1.55	1.99	183
$Gd_3N@C_{84}$	exp		0.32	−1.37			1.69	226
	calc[d]		0.11	−1.50		1.33	1.60	183
$Gd_3N@C_{88}$	exp		0.06	−1.43			1.49	226
	calc[d]		−0.10	−1.49		0.93	1.39	183
$Sc_2S@C_{70}$	exp	0.65	0.14	−1.44	−1.87		1.58	120
	calc[d]	0.60	0.04	−1.33	−1.82	0.95	1.37	120
	calc[c]	1.30	0.52	−1.10	−1.83	1.83[f]	1.62	227
$Sc_4O_2@C_{80}$	exp		0.00	−1.10			1.10	187,228
	calc[d]		−0.09	−0.97		0.60	0.88	82

[a] gap_{H-L} is HOMO-LUMO gap of the neutral molecule, gap_{EC} is the difference between the first oxidation and first reduction potentials; [b] BLYP/DNP; [c] PBE/DNP;
[d] BP86 with ZORA correction and Slater TZP basis set; [e] the same as [d] but calculated taken into account thermal correction at room temperature (hence ΔG); f B3LYP/6–31G(d)-Lanl2dz level

second reduction potential is more than 0.4 V.[225] COSMO calculations at the PBE/DNP level were also reported for $Sc_3NC@C_{80}$[54] and $Sc_2S@C_{70}$.[227]

A series of calculations of redox potentials of nitride clusterfullerenes,[183] $Sc_4O_2@C_{80}$,[82] and $Sc_2S@C_{70}$[120] were performed by Poblet, Rodríguez-Fortea, and coworkers using BP86 functional, Slater-type TZP basis set and COSMO model for solvation energies. Computed values show reasonable agreement with the experimental data (Table 2). The errors for the first reduction and oxidation potential are ca 0.1 V and reach 0.2 V in the worst cases. For $Sc_3N@C_{80}$, calculations correctly predicted that the isomer with $D_{5h}(6)$ cage is easier to oxidize than $I_h(7)$ isomer by ca 0.3 V.[183]

The authors also showed that the use of Gibbs functions computed using statistical thermodynamics for 300 K instead of energies (corresponding to 0 K) does not improve the agreement between computed redox potentials and experimental values. Moreover, fortuitous cancelation of error results in even better performance of 0 K values.[183] Hence, less computationally demanding energies can be used in calculations.

Importantly, experimental values are usually referred to the potential of the $Fe(Cp)_2^{+/0}$ couple, and hence computed redox potentials should be adjusted to this standard. In the BLYP/DBP and PBE/DNP calculations on $Sc_3C_2@C_{80}$,[229] $Sc_4C_2@C_{80}$,[225] $Sc_3NC@C_{80}$[54] and $Sc_2S@C_{70}$,[227] the absolute potential of ferrocene was set to −4.98 V adopted from experimental studies, whereas Poblet and coworkers used −4.68 V obtained in their own computations. Comparison of the experimental and computed redox potentials in Table 2 shows that the latter approach gives smaller errors in predicted potentials. Note that in fact, BLYP/DNP and PBE/DNP computations provide better estimation of electrochemical gaps than the BP86/TZP approach, which tends to underestimate gap_{EC} by *ca* 0.1–0.2 V.

It should be mentioned that HOMO-LUMO gaps (gap_{H-L}) computed with GGA functionals are smaller than experimental electrochemical gaps by ca 0.5 V (Table 2). At the same time, MO energies and especially the HOMO-LUMO gap are strongly dependent on the amount of exact exchange in the density functional. For instance, the 20–25% fraction of the exact exchange (as in many currently used functionals, including B3LYP, PBE0, etc.) increases the gap_{H-L} by ca 1 eV over the pure GGA-computed values (compare BP86 and B3LYP-derived gap_{H-L} values for $Sc_2S@C_{70}$ in Table 2).

In addition to prediction of redox potentials, computational studies also addressed the reason for the electrochemically irreversible reduction found for many nitride clusterfullerenes in contrast to empty fullerenes, whose reductions are electrochemically reversible.[14,146,230,231] Such irreversibility is an indication of the follow-up process accompanying the electron transfer. The first hypothesis was that the reduction induces a structural rearrangement of the endohedral cluster (e.g. pyramidalization),[230] but computational studies did not reveal any pronounced structural changes in the charged states of NCFs.[39,183] Further experimental studies suggested that dimerization of anion-radicals may be a follow-up process responsible for peculiar cathodic behavior of NCFs.[232] Indeed, computations of different dimers of $M_3N@C_{2n}$ and their exohedral derivatives showed that single-bonded dianionic dimers of NCFs are considerably more stable than analogous dimers of empty fullerenes, which indicates that the monomer↔dimer equilibrium for anion radicals of NCFs is shifted towards the dimers.[233] Interestingly, when GGA-DFT binding energies were corrected for van der Waals interactions (this correction stabilized the dimers by ca 60–80 kJ/mol), some of the dianionic dimers were found to be stable already in the gas phase (e.g. dimer of $Sc_3N@C_{68}$).

5.2. UV-Vis-NIR Absorption Spectra

UV-Vis-NIR absorption spectroscopy is one of the most frequently used spectroscopic characterization methods for EMFs. The main advantage of this technique is its high structural sensitivity. Absorption spectra of fullerenes are dominated by the $\pi \rightarrow \pi^*$ excitation of their carbon cage π-system, and hence the spectra of EMFs with the same carbon cage isomer in the same formal charge state are virtually identical irrespective of the encapsulated metal. However, computational studies of absorption spectra of EMFs are rather scarce. Semiempirical methods found to be rather efficient for empty fullerenes — ZINDO or INDO/S — can be of only limited use because of the lack of metal-related parameters. So far, the *state-of-the-art* in the computations of the electronic excitation spectra of EMFs is time-dependent (TD) DFT, whereas more advanced *ab initio* approaches are not applicable yet because of high computational costs. TD-DFT itself also has hardware-related limitations caused by the rapid growth of the required resources with the increase of the number of excited states. Besides, TD-DFT has reliability problems related to the systematic errors in predicted excitation energies, which can vary for different types of excitations in the same molecule. Yet, several works reported TD-DFT studies of absorption spectra of EMFs and showed that interpretation of the UV-vis-NIR spectra of EMFs is in principle possible.[28–30,120,121,227,232,234–236]

Several TD-DFT studies were devoted to nitride clusterfullerenes. Detailed interpretation of the absorption spectrum of $Sc_3N@C_{68}$ based on TD-PBE calculations was reported in 2006.[29] This work also showed that computed spectrum of $Sc_3N@C_{68}$ is substantially different from that of C_{68}^{6-} hexaanion, emphasizing thus the limitations of the description of the charge distribution in EMFs as purely ionic. Later, TD-DFT was also used to interpret absorption spectra of the electrochemically-generated $Sc_3N@C_{68}^{+}$ cation.[30,232] An interesting evolution of the absorption spectra was also found in $Lu_xSc_{3-x}N@C_{68}$ ($x = 0$–2).[30,237] While the lowest energy absorption was found at exactly the same energy independent of the cluster composition, the band at somewhat higher energy showed noticeable blue shift with substitution of one and two Sc atoms by Lu. With the help of TD-DFT calculations, this excitation was shown to have a significant cage-to-Sc electron transfer (0.06 *e* per Sc atom) that was not paralleled for Lu.[30] That is to say, with the increase of the number of Lu atoms in the molecule the extent of the cage-to-cluster electron transfer in the metal-dependent excitation decreases from 0.18 *e* for $Sc_3N@C_{68}$ to 0.06 *e* for $Lu_2ScN@C_{68}$. For $Sc_3N@C_{68}$ and $Sc_3N@C_{78}$, TD-DFT gradients of the excited states were used to model resonance Raman intensities.[28,29]

Recently, TD-DFT computed excitation spectra of $Sc_2S@C_{70}$[120,227] and $Sc_2S@C_{72}$[121] were reported. A good agreement with experimental data was found for both sulfide clusterfullerenes, confirming the structural assignment of $Sc_2S@C_{70}$.[120] For $Sc_2S@C_{72}$, TD-DFT computations were used to propose a detailed assignment of the absorption spectrum.[121] In 2013, Kobayashi *et al.* reported magnetic circular dichroism (MCD) spectroscopic study of $La@C_{82}$, $La_2@C_{80}$, and $Sc_3N@C_{80}$ and interpretation of their excitation spectra based on TD-DFT calculations.[234] TD-DFT computations of two-photon absorption spectra of EMFs were performed by Cheng *et al.*[238–240]

5.3. Vibrational Spectra

Vibrational spectroscopy (in particular, IR and Raman) can provide rich information on the EMF molecules under study.[31] In particular, vibrations of the fullerene cage are very structure sensitive and hence can be used for structure elucidation[18,19,106]; internal vibrations of endohedral species reveal their geometric and dynamic situation,[149,204,241,242] whereas a special class of vibrations known as "metal-cage" or "cluster-cage" modes provides information on the metal-cage bonding.[149,241,243–245] Unfortunately, the value of these methods is often underestimated, in part because the full wealth of such information can be obtained from experimental IR and Raman spectra only if reliable interpretation of the vibrational spectra is available. That is to say, it should be possible to assign the bands in the experimental spectra to specific vibrations of the studied molecules with known vibrational eigenvectors. For the medium- and large-size molecules (such as EMFs), detailed interpretation of the vibrational spectra is possible only with the help of the first principle calculations (DFT performs quite well), and preferably with the basis sets of at least DZVP quality or better. Furthermore, in view of the large vibrational density of states in large molecules, reliable prediction of IR and Raman *intensities* is also necessary for assignment of their spectra, whereas assignment based solely on the computed vibrational frequencies is inevitably ambiguous. Due to substantial computational demands, first principle vibrational calculations of EMFs were hardly possible in 1990s, but the situation has substantially improved since the mid-2000s. Nowadays, computation of the Hessian matrix (which usually automatically includes computations of IR intensities) is a feasible and highly recommended procedure for the confirmation of the true energy minimum found in the geometry optimization, and hence interpretation of IR spectra is relatively straightforward. Computations of non-resonant Raman scattering intensities is not very time consuming in comparison to the Hessian computation (in a numerical procedure it requires evaluation of energy and gradient in the presence of the electrostatic field aligned in different directions), but Raman spectra of EMFs are commonly excited by the laser lines in the visible part of the spectrum, where EMFs absorb light, and hence the spectra usually have resonant character. Prediction of resonant Raman intensities is a much more complex task and requires modeling of the electronic excitations and their coupling to the molecular vibrations, and only a few such calculations have been reported so far.[28,29,235] A detailed account of the experimental and computational vibrational spectroscopic studies of EMFs is given in Ref. 31, and here we will discuss only several representative examples.

The most detailed vibrational spectroscopic studies coupled to quantum-chemical calculations are reported for nitride clusterfullerenes. The first such report was the seminal study of $Sc_3N@C_{80}$ by Krause *et al.*,[241] in which experimental measurements were augmented with DFTB calculations of the vibrational frequencies and IR intensities. With the help of computational data, the authors could assign several Sc_3N-based vibrations, including the antisymmetric metal-nitrogen stretching mode and cluster-cage mode, which were intensely used in many subsequent studies of NCFs to analyze the state of the cluster and cluster-cage interactions. So far, the most detailed information to date on the whole set of cluster and cluster-cage vibrations in metal-nitride clusterfullerenes is available for $Sc_3N@C_{68}$[29] and $Dy_3N@C_{78}$.[18] For these molecules, almost all of their cluster modes were assigned in the experimental spectra.

The power of DFT in prediction of the vibrational spectra of EMFs is illustrated in Fig. 7, where IR spectra $Sc_3N@C_{78}$ and $Dy_3N@C_{78}$ are compared to the results of computations for $Sc_3N@C_{78}$-D_{3h}(5) and $Y_3N@C_{78}$-C_2(22010).[18,28] As already discussed in Section 3.3, an increase of the cluster size from Sc_3N to Y_3N results in the switch of the preferable cage isomer for $M_3N@C_{78}$.[18] The cage isomerism can be easily seen in the IR spectra (dominated by the cage vibrations) and is perfectly reproduced by computations (Fig. 7). Thus, the first structural elucidation of the non-IPR $M_3N@C_{78}$-C_2(22010) clusterfullerenes was done with the help of vibrational DFT calculations, and was later confirmed by single-crystal X-ray diffraction.[107] DFT-based vibrational calculations were also used for the structure elucidation and vibrational assignment of $Sc_3N@C_{68}$,[29] $Sc_3N@C_{70}$,[19] $Sc_3N@C_{78}$,[28] $DySc_2N@C_{76}$,[106] $Gd_xSc_{3-x}N@C_{2n}$ ($2n$ = 78–88),[185,242,243] $Sc_2S@C_{82}$,[53] $Sc_4O_2@C_{80}$,[187] $Sc_3NC@C_{78}$,[125] $La_2@C_{80}$,[175] $Eu@C_{74}$,[235] and $Y_2C_2@C_{92}$.[186]

Note that computational studies of EMFs containing lanthanides with a partially filled 4f-shell are severely complicated because of their intrinsically multi-reference character. However, in many instances they can be

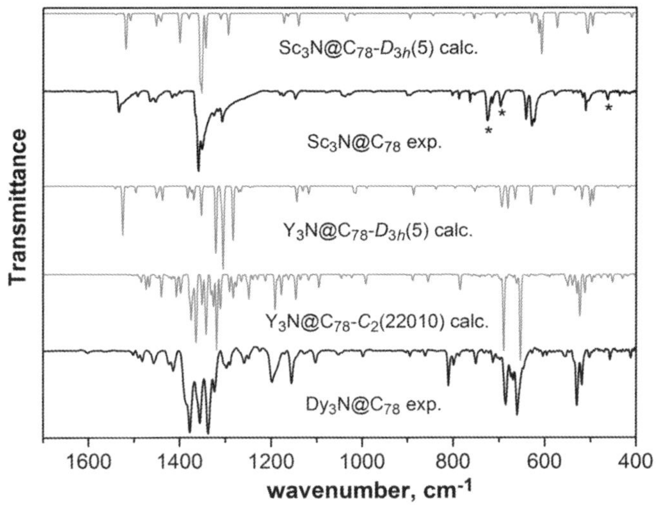

Fig. 7 Experimental FTIR spectra of $Sc_3N@C_{78}$ and $Dy_3N@C_{78}$ (thick black lines) and DFT-computed IR spectra of $Sc_3N@C_{78}$-D_{3h}(5), $Y_3N@C_{78}$-D_{3h}(5), and $Y_3N@C_{78}$-C_2(22010). Asterisks in the experimental spectrum of $Sc_3N@C_{78}$ mark absorptions of the solvent (toluene) residues in the sample. Based on the data from Ref. 18.

reasonably modeled by replacing the lanthanide with yttrium, which has similar ionic radius and valence-shell electronic structure.[18,31,106] The difference in masses of Y and lanthanides is apparent only in the low-frequency range (which is usually beyond the scope of standard IR spectroscopy), and can be taken easily into account when necessary (e.g. in the analysis of the low-frequency part of Raman spectra).[149]

The importance of the intra-cluster and cluster-cage modes stems from the fact that their frequencies can be directly related to the internal structure of the cluster or its interaction with the fullerene, and hence these modes were addressed in many reports. DFT studies show that vibrational eigenvector of the ν_{as}(M–N) mode can be described by the in-plane displacement of the nitrogen atom, while the metal atoms remain intact.[28,29,241] This mode is two-fold degenerate in homogeneous-metal NCFs (with M_3N cluster composed of the one type of metal atoms), but the degeneracy is lifted in mixed-metal nitride clusterfullerenes, and two main lines are usually observed in the IR spectra. DFT computations showed that in the optimized structures of $Y_xSc_{3-x}N@C_{80}$ NCFs, the nitrogen atom is displaced from its position on the three-fold axis (which it adopts in $Sc_3N@C_{80}$ and $Y_3N@C_{80}$) towards Sc atom(s).[242] Similar effects were observed in the DFT-optimized structures of $Gd_xSc_{3-x}N@C_{2n}$,[185,243] $Lu_xSc_{3-x}N@C_{80}$,[246] and $Lu_xSc_{3-x}N@C_{68}$[237] as well as in the X-ray structures of $MSc_2N@C_{80}$ (M = Er,[247] Ce,[248] Tb[249] and Gd[249]) and $Gd_2ScN@C_{80}$.[249] As a result, Sc–N bond are shorter than in $Sc_3N@C_{80}$ whereas M–N bonds are longer than in $M_3N@C_{80}$, and this redistribution of the bond lengths results in corresponding changes in the vibrational frequencies (shorter bonds result in higher frequencies). Thus, the shift of the vibrational frequencies of the ν_{as}(M–N) mode can be used to obtain information on the M–N bond lengths. In particular, effective metal-nitrogen force constants derived from experimental vibrational frequencies exhibit quadratic dependence on the inversed DFT-computed Sc–N bond lengths.[31] DFT calculations also helped in assignment of Sc–S and Sc–O vibrations of the encapsulated cluster in $Sc_2S@C_{82}$[53] and $Sc_4O_2@C_{80}$.[187]

While ν_{as}(M–N) mode provides information on the internal state of the cluster, the extent of the cluster-cage interaction may be studied by analysis of the frustrated rotation and translation of the cluster. Analysis of the Raman spectra of $M_3N@C_{80}$ (M = Sc, Y, Gd–Tm) revealed that one mode is particular convenient for the study of the cluster-cage interactions as it can be easily determined in the spectra of all $M_3N@C_{80}$ nitride clusterfullerenes.[149,245] DFT calculations for $Sc_3N@C_{80}$ and $Y_3N@C_{80}$ revealed that this mode can be described as a frustrated in-plane cluster translation partially mixed with the in-plane M_3N deformation.[149] It was found that the same effective force constant can describe M_3N-cage mode for almost all metals except for Sc, which has a stronger interaction with the carbon cage.[149]

It should be pointed out that a standard procedure for the computation of vibrational spectra is based on the harmonic approximation and static computations, i.e. the molecule is first optimized and then its Hessian (the matrix of second energy derivatives with respect to displacements of atomic coordinates) is computed numerically or analytically. Another approach, which is rarely used but can be potentially more precise, is based on molecular dynamics simulations. Fourier-transformation of the autocorrelation function of kinetic energy calculated along the MD trajectory gives vibrational density of states, whereas the same for dipole moment gives IR spectrum of the molecule. Kemner and Zerbetto computed vibrational density of states for $M@C_{82}$ and $Eu@C_{74}$ molecules using classical MD simulation and showed that metal motions are chaotic at temperatures above 50 K.[250] Vietze and Seifert reported vibrational spectra of several EMF molecules computed based on the DFTB-BOMD trajectories.[205–207,251] They also showed that MD-based spectra can be noticeably different from those computed in the static approximation.

5.4. Electron Spin Resonance (ESR) Spectra

ESR spectroscopy (also abbreviated as EPR, electron paramagnetic resonance) is a very sensitive tool for the study of paramagnetic states and can give a lot of information on the spin-density distribution in the radical based on the values of the g-factor and hyperfine coupling constants (hfccs). In particular, significant shift of the g-factor from its free-electron value of 2.0023 and large metal-based hfccs in EMF radicals point to the localization of spin density on the metal atoms. Of course, to be eligible for ESR spectroscopic studies, EMFs should be paramagnetic. Therefore, the EMFs studied by ESR spectroscopy can be classified into two groups: (i) EMFs, which are paramagnetic in their pristine states, and (ii) diamagnetic EMFs, which are transformed into paramagnetic forms via either chemical or electrochemical redox reactions. Since spin density distribution corresponds to the spatial distribution of single-occupied molecular orbital (SOMO), ESR studies of charged EMFs radicals provide information on the frontier orbitals of the neutral molecules and on the mechanism of the electron transfer. The majority of ESR studies of EMFs were performed in solution, and hence isotropic hfc constants and g-factors are usually discussed. However in several works low-temperature studies allowed determination of the hyperfine tensors.

In principle, computation of isotropic hyperfine constants is rather straightforward: a spin-unrestricted SCF calculation of the radical produces spin-up and spin-down MOs which can be used to construct a spin density and to determine hfccs. Namely, isotropic hfcc is proportional to the spin density at the nuclei (not to be confused with the atomic spin population, which is an integral over the whole atom either in the real or Hilbert space). Since all atomic orbitals with angular momentum equal or larger than 1 (*p*, *d*, *f*, etc.) have nodes on the nuclei, large atomic spin population does not necessarily lead to the large isotropic hfcc; the latter is determined only by contribution of *s*-orbitals to the spin density. Two major terms in the isotropic hfcc are usually distinguished[252,253]: direct term arises from the valence s-orbital contribution to SOMO and is always positive, whereas polarization term arises from the spin polarization of *s* shells and can be either positive or negative. The basis set used in the calculation should be flexible enough to be able to describe the spin polarization near the nuclei, which can bring in additional difficulties. First, hfcc calculations obviously require the use of full-electron basis sets, whereas effective core potentials often used for heavy elements are not applicable. Second, standard full-electron basis sets are not very flexible for core electrons because they are of no interest for chemical tasks. Therefore, specially tailored basis sets should be used. Finally, relativistic effects, including spin-orbit coupling, can have a significant influence on the predicted hfcc values for heavy elements and it is worth to take them into account.

The first dedicated computational studies of hfccs in monometallofullerenes appeared in the late 1990s. Seifert *et al.* reported a detailed analysis of ^{13}C hyperfine parameters in $M^{III}@C_{82}$ (M^{III} = Sc, Y, La) in 1998.[254] A well-resolved ^{13}C hyperfine structure of all three EMFs allowed identification of 7–8 individual ^{13}C hfc constant and was found to be very similar pointing to the identical geometric structure of all $M@C_{82}$ species. DFTB calculations of the spin density distribution in $La@C_{82}$ were performed to aid in assignment of ^{13}C hfc constants to the cage carbon atoms, however computations were based on the C_{82}-C_{3v}(8) carbon cage since C_{2v}(9) isomeric structure of $M@C_{82}$ was not well established at that moment. In 2005, DFT-based assignment of the ^{13}C satellite structure in the ESR spectrum of $Sc@C_{82}$-C_{2v}(9) was accomplished by Morley *et al.*[255] The authors also measured the ESR spectrum in frozen solution and determined a hfc tensor for ^{45}Sc. DFT-computed values showed good agreement with experimental data. In 2010 Misochko *et al.* studied ESR spectra of $Y@C_{82}$ at low temperature in Ar matrix and showed that relativistic GGA DFT calculations produce good agreement with the experimental data for all $M^{III}@C_{82}$ radicals (M^{III} = Sc, Y, La).[256] For $Y_2@C_{79}N$, the ^{89}Y hfc constant of 62.4 G predicted at the B3LYP/DZVP(Y)+6–31G*(C,N) level agrees with the large value found in experiment (81.23 G) but noticeably underestimates it, probably because of the limited size of the basis set and non-relativistic treatment of Y.[66]

Several ESR/computational studies were performed for radical anion of $Sc_3N@C_{80}$ and its derivatives. The first experimental study by Jakes and Dinse showed that $Sc_3N@C_{80}^-$ exhibits large $a(^{45}Sc)$ constant, 55.6 G.[257] RO-B3LYP/3–21G* and RO-HF calculations performed by these authors gave only 7.14 and 14.27 G, respectively. Such a small values are in part due to the spin-restricted formalism (which cannot properly treat polarization contribution), but strong underestimation is still quite remarkable. Extended computations of $a(^{45}Sc)$ constant in $Sc_3N@C_{80}^-$ were reported in 2008.[39] The authors used different basis sets and density functionals, analyzed the influence of scalar relativistic and spin-orbit corrections as well as geometry factors. It was shown that computed $a(^{45}Sc)$ constant is very sensitive to all of these parameters (except for spin-orbit coupling, whose contribution was less than 0.3 G). In particular, $a(^{45}Sc)$ hfcc dramatically depends on the cluster orientation inside the cage: rotation of the cluster around the C_3 axis from C_{3v} to D_3 configuration changes the value from –6 G to 25 G (B3LYP//PBE calculations). Comparison between RO-B3LYP and U-B3LYP calculations showed that polarization term remained constant during this rotation, and hence structural sensitivity of the $a(^{45}Sc)$ constants is due to the direct term. Surprisingly, even at the highest levels of theory used, with hybrid functionals, large basis sets, and relativistic corrections, predicted $a(^{45}Sc)$ hfcc values for the lowest energy conformer of $Sc_3N@C_{80}^-$ were still below 40 G.[39] In a following study, this strong underestimation was ascribed to the dynamic effects.[189] The $a(^{45}Sc)$ value averaged over the PBE-BOMD trajectory was found to be significantly larger than the constant computed at the same level of theory in the lowest energy configuration. In the same work, static and dynamic calculations of $a(^{45}Sc)$ were also performed for mono- and trianion of $Sc_3N@C_{80}(CF_3)_2$,[189] whose ESR spectra showed rich hyperfine structure caused by non-symmetric Sc_3N cluster.[210] Contribution of dynamic effects was found to be significant for these radical anions as well. Furthermore, computations of hfccs along the IRC trajectories connecting closely lying energy minima showed that the values exhibit strong variations even for very small displacements of Sc atoms.[210] On the contrary, B3LYP/6–311G*-computed hfccs in the radical anion $Sc_3N@C_{80}(CF_3)_{10}^-$ with fixed Sc_3N cluster showed good agreement to the experiment.[196] Very recently, extended experimental and computational study of the radical anion of [5,6]-pyrrolidino-$Sc_3N@C_{80}$ was reported.[190] ^{45}Sc hyperfine structure showed that rotation of the Sc_3N cluster was hindered by functionalization, similar to $Sc_3N@C_{80}(CF_3)_2^-$. DFT-predicted $a(^{45}Sc)$ constants were in qualitative agreement with experimental values and showed large variations along IRC trajectories between three energy minima. However, the values are significantly underestimated, probably due to the dynamic effects.[190]

Computations of ESR parameters were also performed for cation and anion radicals of $Sc_3N@C_{68}$.[232] For both species, experimental $a(^{45}Sc)$ hfccs did not exceed 2 G in good agreement with DFT-calculated values, which also showed that spin density in these radical ions is localized on the carbon cage with insignificant contribution from the Sc_3N cluster.

Detailed experimental and computational study of cation and anion radicals of $Sc_4O_2@C_{80}$ with mixed valence states of Sc^{II}/Sc^{III} atoms was reported in 2012.[187] The assignment of the experimentally determined $a(^{45}Sc)$ values to the two types of Sc atoms in the Sc_4O_2 cluster was accomplished by extended DFT calculations. Dynamics effects were taken into account by DFT-BOMD simulations (up to 40 ps were followed after 15 ps of equilibration, revealing large variations of the instantaneous $a(^{45}Sc)$ constants, see Fig. 8). In both ions, DFT showed that spin density is localized primarily on the Sc_4O_2 cluster, resulting in enhanced $a(^{45}Sc)$ constants. Sc atoms adopting a divalent state in the neutral $Sc_4O_2@C_{80}$ exhibited an especially large coupling constant of 150.4 G in the cation radical, which is the record high $a(^{45}Sc)$ value for Sc-based endohedral metallofullerenes. Such a high value is explained by the nature of the HOMO localized on the six-atomic Sc_4O_2 cluster. This HOMO is a Sc–Sc bonding MO and hence has large contributions from 4s atomic orbitals of Sc^{II}.[187]

The anion radical of $Sc_3NC@C_{80}$ was studied by ESR spectroscopy in 2013.[258] The hyperfine structure revealed the presence of two types of Sc atoms with hfccs of 3.89×2 and 1.95 G. Since X-ray structure was rather ambiguous in terms of the orientation of the CN unit with respect to the Sc_3 cluster,[259] the authors computed $a(^{45}Sc)$ values for two configurations with either carbon or nitrogen atoms in the middle of the Sc_3NC cluster. The values computed for the former configuration, –4.10/–3.95 and –3.33 G, are in better agreement with experiment. Calculations also showed that spin density is mainly localized on the CN unit.[258]

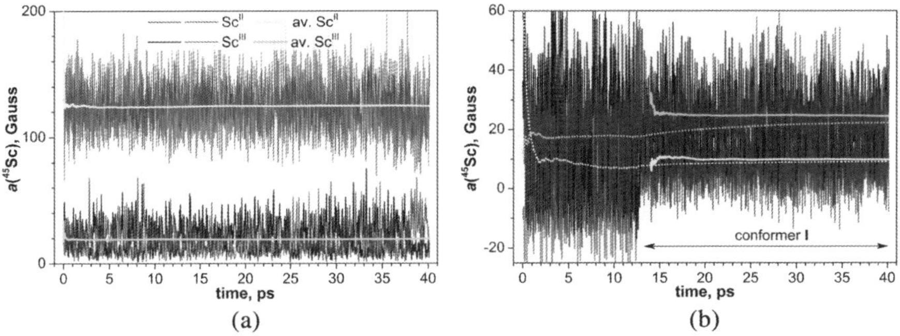

Fig. 8 Instantaneous (dark lines) and time-averaged (light lines) $a(^{45}Sc)$ values in the cation (a) and anion (b) of $Sc_4O_2@C_{80}$ computed along the BOMD trajectories. BOMD calculations were performed at the PBE/DZVP level, hfcc values were then computed at each 5 fs using SVP basis set for carbon, TZVP for oxygen, and CP3P basis set for Sc. Reproduced with permission from Ref. 187 © 2012 American Chemical Society.

Although calculation of g-factor is less computationally straightforward and requires the use of perturbation theory, such possibilities are currently implemented into several quantum-chemical codes making the computations feasible. Good agreement between experimental and computed g-factors was reported for $Sc_3N@C_{80}^-$, $Sc_3N@C_{80}(CF_3)_2^-$, $Sc_3N@C_{80}(CF_3)_2^{3-}$, $Sc_4O_2@C_{80}^+$, and $Sc_4O_2@C_{80}^-$.[39,187,189]

5.5. Nuclear Magnetic Resonance (NMR) Spectra

NMR spectroscopy is a popular technique in chemistry of EMFs for several reasons: first, ^{13}C NMR spectra give information on the carbon cage symmetry, which often helps in structure elucidation; second, multinuclear NMR spectra addressing the atoms of the endohedral species (Sc, La, Y, N, etc.) give information on the valence state, symmetry, and dynamics of the encapsulated clusters; third, when EMFs are paramagnetic but still enable NMR measurements (e.g. Ce-EMFs), paramagnetic shifts in NMR spectra and their temperature dependence give information on position of metal atoms and their dynamics. Quantum-chemical calculations of chemical shifts were found to be helpful in all three mentioned aspects. The most common technique used in such calculations is the gauge-including atomic orbital (GIAO) method, which gives reasonable good accuracy in predicting ^{13}C chemical shifts when coupled to hybrid DFT functionals such as B3LYP.

Prediction of ^{13}C chemical shifts is the most common use of computational methods in NMR spectroscopy of EMFs.[128,137,261–266] ^{13}C NMR spectra give information on the symmetry of the carbon cage, which leaves some ambiguities in the structure elucidation, especially when several isomers of a given symmetry group are plausible. In such cases, calculations of ^{13}C NMR spectra can help in sorting out the isomers. The spectra computed at the GIAO-B3LYP level with basis sets of DZVP or better quality usually provide good overall agreement with the experimental spectra in terms of the range covered by the $\delta(^{13}C)$ values as well as strong shielding or deshielding of special carbon atoms such as those in metal-bonded adjacent pentagons. Structure determination in a series of Yb@C_{2n} compounds ($2n = 80, 82, 84$) can be an illustrative example (Fig. 9).[261]

In several studies, ambiguous structural elucidations were corrected based on the results of computations of relative energies and ^{13}C NMR spectra. Kobayashi and Nagase showed[46] that Sc_2@C_{66}-C_{2v}(4059) with two pairs of threefold fused pentagons is much more stable and fits experimental ^{13}C spectrum of Sc_2@C_{66} much better than the spectrum computed

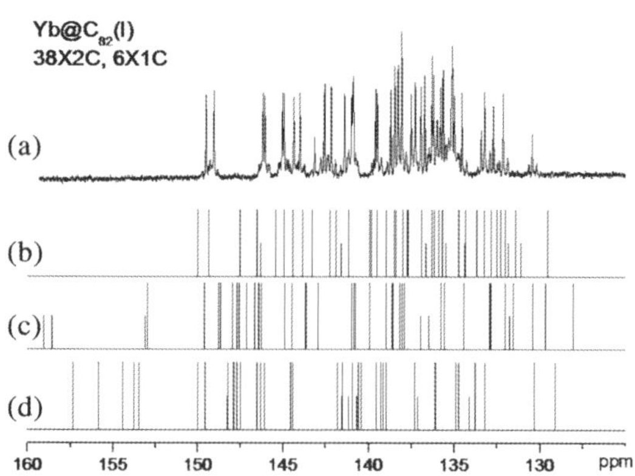

Fig. 9 (a) ^{13}C NMR spectrum of Yb@C_{82}(I) with $38 \times 2C$, $6 \times 1C$ intensity pattern. (b)-(d) calculated ^{13}C NMR spectra of Yb@C_{82} isomers with C_s-symmetric carbon cages: (b) C_s(6) (the most stable C_s isomer of Yb@C_{82}), (c) C_s(4) (52.5 kJ/mol), (d) C_s(2) (96.3 kJ/mol). Single-crystal X-ray study reported later confirmed that Yb@C_{82}(I) has C_s(6) carbon cage.[260] Computations were performed at the GIAO-B3LYP/6–311G*-SDD level. Reproduced with permission from Ref. 261 © 2010 American Chemical Society.

for $Sc_2@C_{66}$-C_{2v}(4348) isomer proposed in the original study.[117] Likewise, recent computations[128] showed that $Sc_2@C_{70}$-C_{2v}(7854) is more stable than $Sc_2C_2@C_{68}$-C_{2v}(6073) proposed in the first report[119] and its computed ^{13}C NMR spectrum agrees with the experimental data for Sc_2C_{70} better than that of $Sc_2C_2@C_{68}$. An opposite situation is found for Ti_2C_{80}: $Ti_2C_2@C_{78}$-D_{3h}(5) was found to be much more stable than $Ti_2@C_{80}$ and its computed spectrum first experimental data very well,[265] which serves as an additional argument to disprove an original conjecture of the mixture of two $Ti_2@C_{80}$ isomers.[267]

Importantly, although the level of agreement between experimental and computed ^{13}C chemical shifts of a few ppm is often good enough to confirm a structural assignment, it is usually not sufficient for peak-to-peak assignment of the spectra of low-symmetric cages (C_2, C_s, C_{2v}), where many lines in close proximity can be found. For such structures, assignment is possible only for selected salient carbon atoms. At the current state of computational NMR studies, complete DFT-based assignment of experimental spectra is possible only for high-symmetry cages (D_{3h}, D_{5h}, I_h, etc.), where the distance between observed peaks is rather large and additional restrictions are imposed by relative intensities of the peaks.

In most cases calculations of cage ^{13}C signals in EMFs are static, i.e. performed for the optimized structures. When the endohedral cluster is known to rotate and the signals of the carbon atoms are averaged in the experimental spectra, statically-computed signals are also averaged in an analogous way. This procedure cannot be considered fully satisfactory since it does not take into account the real dynamics of the cluster and may be in error when rotation energy barriers are low. More elaborate computational study of ^{13}C NMR spectra of $Sc_3N@C_{80}$ and $Sc_3N@C_{68}$ were reported by Heine *et al.*[202,266] The authors performed DFTB-based molecular dynamics simulations of the studied molecules on the nanosecond timescale and then computed NMR spectra for selected snapshots along the BOMD trajectory. For $Sc_3N@C_{80}$ with freely rotating Sc_3N cluster this procedure resulted in a better agreement of experimental and computed chemical shifts.[202]

Carbide clusterfullerenes are special objects for ^{13}C NMR spectroscopy due to the presence of endohedral C_2 units, whose ^{13}C signals are strongly deshielded and are usually found ca 100 ppm lower field than the carbon cage signals. Computational studies of carbide clusterfullerenes showed that deshielding can be predicted with rather good accuracy,[268,269] however dynamical averaging of the signals in certain carbide clusterfullerenes can bring in some difficulties.

In certain cases calculation can help in interpretation of NMR spectra even without computations of the chemical shifts. One such example is a series of $Y_xLu_{3-x}N@C_{80}$-I_h and $Lu_xSc_{3-x}N@C_{80}$-I_h nitride clusterfullerenes.[246] Two ^{13}C signals of the icosahedral carbon cage exhibited noticeable metal dependence. Computations showed that local pyramidalization of the metal-coordinated carbon atoms (quantified in terms of Haddon's π-orbital axis vector, POAV[270]) was changing with the size of the endohedral cluster and that the chemical shifts of the carbon cage showed perfect correlation with POAV angles.[246] Another groups of examples is Ce-based EMFs,[78,271–273] whose paramagnetic shifts were modeled based on the DFT-optimized atomic coordinates and Bleaney theory of paramagnetic NMR.[274]

Computational studies of multinuclear NMR spectra of EMFs for nuclei other than ^{13}C are rather scarce. In 2006 Slanina reported GIAO-B3LYP calculations of ^{139}La chemical shifts in $La_2@C_{80}$ and two isomers of $La_2@C_{72}$ with the aim to confirm the structural assignment of $La_2@C_{72}$-D_2(10611).[137] Whereas the basis sets with effective core potentials strongly underestimated the difference of ^{139}La shifts for $La_2@C_{80}$ and $La_2@C_{72}$, the use of full-electron UGBS basis overcorrected computed values. Douglas–Kroll–Hess relativistic correction improved the agreement with experiment, but the increase of the La basis set to UGBS1P again gave large error.[137] Analogous GIAO-B3LYP calculations with ZORA relativistic corrections were reported by Hao *et. al.* in 2009.[236] The good agreement with experimental $\Delta\delta(^{139}La)$ value (computed 180 ppm *versus* experimental 173 pm) was achieved when spin-orbit interactions were taken into account and full-electron Slater-type TZP basis set was used for La.

Extended analysis of ^{14}N chemical shifts in mixed-metal nitride clusterfullerenes based on Ramsey perturbation theory and DFT-GIAO and DFT-IGLO calculations was performed in 2011.[275] Analysis of the orbital contributions to the nitrogen chemical shifts in nitride clusterfullerenes and several model compounds showed that the differences in the chemical shifts can be explained simply based on the nitrogen-projected partial density of electronic states.

Recently, PBE-GIAO calculations of ^{45}Sc chemical shifts in a series of Sc-containing EMFs and DFT-based assignment of Sc signals in $Sc_4O_2@C_{80}$ were reported.[187] Computed signals were found to be strongly dependent on the position of Sc atoms and emphasized importance of the dynamics effects.

6. CHEMICAL REACTIVITY OF EMFs: COMPUTATIONAL PERSPECTIVES

Extended studies of chemical reactivity of EMFs became possible since the 2000s, and dozens of reactions of different kinds of EMFs (mostly different types of cycloaddition) have been reported by now.[276] Computational studies can be of strong help in interpretation of the reaction mechanisms and prediction of the possible products. However, this task can be very challenging from the computational point of view. First, one has to distinguish if the reaction proceeds under thermodynamic or kinetic control. In the former case, it can be sufficient to compute possible reaction products and find those with the lowest energies. However, for a kinetically-controlled process, conclusions based on the thermodynamic stability of the products can be erroneous, and the whole reaction mechanism has to be reconstructed since the products will be determined by stages with high activation barriers. Second, additional complications stem from the interaction between endohedral species and exohedral groups. It was already discussed in Section 4 that dynamics of the endohedral cluster can be controlled by modification of the π-system of the fullerene and that variation of the relative energies of different conformers can be very high. Therefore, for each position of the exohedral addend on the carbon cage it is necessary to consider different positions of the encapsulated cluster to ensure that the lowest energy product or transition state is found. The use of computations for understanding chemical reactivity of EMFs was exhaustingly reviewed by Osuna, Swart and Solà in 2011.[277] Here we will briefly discuss the main aspects and consider recent reports on the subject.

6.1. Reactivity from the Analysis of Pristine EMF Molecules

Although complete exploration of reactivity requires computations of molecular structures and relative energies of different products as well as reaction barriers, in many cases rather reliable predictions can be made based on the analysis of the molecular and electronic structure of the pristine EMF molecule. In particular, such parameters as atomic spin populations (for paramagnetic EMFs), atomic charges, local pyramidalization parameters (e.g. POAV angles), as well as the energy and spatial extension of frontier orbitals were successfully employed to predict and rationalize reaction products of mono- and dimetallofullerenes by Akasaka, Nagase and coworkers (see Ref. 276 for a detailed review).

Radical and cycloaddition reaction of monometallofullerenes provide especially illustrative examples of the direct relation between reactivity and molecular structure. It is natural to expect that addition of radicals to EMFs should proceed to the atoms with the highest spin populations. According to B3LYP calculations, such sites in La@C_{82} are C10 and C18 near the equator of the molecule and C23 on the pole (Fig. 10).[278] Indeed, these atoms are functionalized in the most abundant products of photochemical reactions of La@C_{82}-C_{2v}(9).[278] Importantly, calculations also showed that relative energies of these adducts are several kJ/mol above the most stable thermodynamic products indicating the kinetic control. Likewise, addition of the $C_6H_3Cl_2$ group to La@C_{82}-C_{3v}(7),[279] La@C_{72} (three isomers),[122] and La@C_{74} (two isomers)[280,281] proceeds to the atoms with the highest spin populations and POAV angles.

A different part of the M^{III}@C_{82}-C_{2v}(9) molecule is functionalized by the addition of photochemically generated carbene (e.g. adamantylidene, Ad). For M = Y,[282] La,[283] Ce,[284] and Gd,[285] reaction results in two dominant products whose structures were elucidated by single-crystal X-ray diffraction.

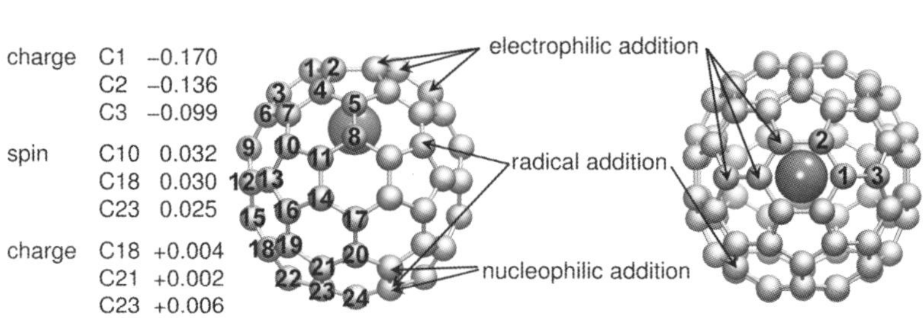

Fig. 10 Molecular structure of La@C_{82}-C_{2v}(9) with numbering system of the cage carbon atoms used in this section. The most negative and positive atomic charges and the largest spin populations computed at the B3LYP level are listed on the left (the values are from the Ref. 276). The most reactive sites in different types of reactions (according to the structures of the products) are indicated.

DFT calculations showed that these results can be explained by the analysis of POAV angles and atomic charges (the atoms with the most negative charges should be the most active ones in electrophilic addition). According to these criteria, the most active cage atom is C1 followed by C2 and C3 (Fig. 10), all on the pole of the molecule near the metal atom. Indeed, experimental results show that carbene addition functionalizes C1–C2 ad C1–C3 bonds. Sc@C_{82} has somewhat different electronic structure due to the higher electronegativity and smaller radius of Sc.[256] In particular, spin populations of C1 and C2 are larger than in other M@C_{82}.[286] As a result, Sc@C_{82} forms four Sc@C_{82}Ad adducts, and the most reactive CC bonds are likewise near the Sc coordination sites.

As discussed in Section 2, metal-cage interactions are rather local, i.e. the charge transferred to the fullerene from the metal is distributed non-homogeneously, the largest negative charges being in the vicinity of the metal coordination. Metal-coordinated carbon atoms also have large POAV pyramidalization angles. Thus, metal-coordinated fragments might be expected to be most reactive for electrophonic addition. On the other hand, metal-cage bonding has a large covalent contribution via d-π overlap and hence reduction of the π-system is not favorable. Therefore, if addition to metal-coordinated sites was resulting in the changing of the hybridization state of carbon atoms from sp^2 to sp^3, the situation with the metal-cage bonding would be unfavorably altered. Instead, electrophilic carbene cycloaddition preserves the sp^2 state of cage carbon atoms and proceeds via "opening" of the cage (i.e. breaking fullerene sigma-bonds). This way, metal-cage bonding is still efficient even in the vicinity of the addend.

Whereas the most negatively charged carbon atoms are located near the metal atom, the most positive ones are found on the opposite pole of the M@C_{82} cage (Fig. 10). It can be expected that these atoms should be attacked by nucleophiles, and analysis of the single-bonded adducts of the Bingel–Hirsch reaction are in line with this reasoning. In particular, the single-crystal X-ray structure of the La@C_{82} adduct shows that the malonate group is attached to C23, which is the most positively charged carbon atom in La@C_{82}.[287,288]

Diels–Alder addition to La@C_{82} also affects the positively charged sites of the molecule. In the X-ray structure of the La@C_{82}Cp* adduct the cyclopentadiene moiety is added across the C21–C23 bond; POAV angles are also rather large for these carbon atoms. According to computation of the relative energies of the possible cycloadducts, C21–C23 addition site appears to be both kinetically and thermodynamically favorable.[289] Direction and efficiency of Diels–Alder addition can be also correlated to the LUMO of the molecule (fullerene is a dienophile), whose energy should match HOMO energy of a diene. LUMO of La@C_{82} is computed to be 0.83 eV below that of C_{60} at the B3LYP/6-31G* level, and hence the EMF is much more reactive.[290]

To summarize, reactivity of M@C_{82} in different kinds of addition reactions can be well correlated to the DFT-computed charge and spin distribution in the pristine M@C_{82} molecule: electrophiles attack the most-negatively atoms on the metal-coordinated pole, radical addition affects the atoms at the equator with the largest spin populations, whereas nucleophiles attack the most positively charged atom on the pole which is opposite to the metal-coordination.

Similar arguments based on the electronic and structural properties of pristine EMF molecules were also used to predict and explain reactivity of other EMFs. For instance, possible reaction sites in $Sc_3N@C_{2n}$ (68, 78, 80) were predicted by Poblet *et al.* based on the analysis of bond lengths, bond orders, and POAV angles.[291] Low reactivity of dianionic $Sc_3N@C_{80}$ in comparison to reactive $[Lu_3N@C_{80}]^{2-}$ was explained by the endohedral localization of LUMO in $Sc_3N@C_{80}$, which makes its dianion less reactive in comparison to the dianion of $Lu_3N@C_{80}$ with cage-based LUMO.[292] Atomic spin populations in La@C_{82}-C_{3v}(7) show that the most reactive site is the one to which $C_6H_3Cl_2$ group is attached.[279] Reactivity of empty fullerenes and EMFs with disilirane correlates with their frontier MO energy levels.[293]

6.2. Reactivity Studies Based on Detailed Exploration of Thermodynamic and Kinetic Factors

Although predictions of reactivity and addition sites of EMFs based on the properties of pristine molecules often provide useful background for understanding of the experimental phenomena, they do not take into account many factors that can affect reaction pathways. Thus, more reliable — but more expensive — computational studies are focusing on the predictions of thermodynamic stability of the possible adducts and on the reaction barriers. The latter may become especially important for partially or fully kinetically controlled process such as Bingel–Hirsch addition (see below). So far, detailed computational studies of thermodynamics and kinetics were performed for three cycloaddition reactions to EMFs: 1,3-dipolar cycloaddition (Prato reaction), Diels–Alder cycloaddition, and Bingel–Hirsch cycloaddition.

Prato reaction. One of the first dedicated studies was the analysis of the rearrangement barrier between [5,6] and [6,6] isomers of pyrrolidino adducts of $Sc_3N@C_{80}$ and $Y_3N@C_{80}$.[145] Experimental studies showed that 1,3-dipolar cycloaddition of azomethine on $Sc_3N@C_{80}$ predominantly results in [5,6] monoadduct, whereas formation of the [6,6] adduct is more typical for $Y_3N@C_{80}$.[294] However, the latter appeared to be a kinetic product since it could be thermally rearranged into the more thermodynamically stable [5,6] adduct. Computations of Poblet and coworkers showed that [5,6] adduct is indeed a thermodynamic product for both $Sc_3N@C_{80}$ and $Y_3N@C_{80}$, but relative stability of [5,6] isomer is decreased from Sc to Y.[145] At the same time, transition state computations revealed that for $Y_3N@C_{80}$ the barrier to [6,6] addition is 15 kJ/mol smaller than that for the [5,6] addition proving that the [6,6] adduct. In contrast, for $Sc_3N@C_{80}$ the barrier to [5,6] addition is smaller than that for [6,6]. Finally, the barrier of the pirouette-like rearrangement between [6,6] and [5,6] adducts of $Y_3N@C_{80}$ was found to be 120–150 kJ/mol.[145]

The study of 1,3-dipolar cycloadditions to $Gd_xSc_{3-x}N@C_{80}$ ($x = 0–3$) reported by Wang, Lu and coworkers in 2007 showed that regioselectivity of the reaction was strongly dependent on the cluster composition.[148] In line with experimental data, DFT computations showed that with the increase of the number of Gd atoms in the cluster, the energy difference between [5,6] and [6,6] pyrrolidine adducts was gradually changing from 49 kJ/mol in favor of [5,6] isomer for $Sc_3N@C_{80}$ through intermediate values of 36 and 10 kJ/mol for $GdSc_2N@C_{80}$ and $Gd_2ScN@C_{80}$, respectively, to the slight preference (by 2 kJ/mol) of the [6,6] isomer over the [5,6] for $Gd_3N@C_{80}$.[148]

Experimental study by Dorn and coworkers showed that the $D_{5h}(6)$ isomer of $Sc_3N@C_{80}$ is more reactive than the $I_h(7)$ isomer in the dipolar cycloaddition reaction, and the authors proposed that the difference in the reactivity is due to the presence of pyracylene units in the former (i.e. addition was supposed to proceed to the [6,6] bonds connecting pyracylene pentagons).[295] However, recent DFT study by Osuna *et al.* showed that addition to the [5,6] bonds in $Sc_3N@C_{80}$-D_{5h}(6) is more energetically favored than addition to the pyracylenic [6,6] bonds by 51 kJ/mol.[296] Besides, computations of the proton chemical shifts also discouraged addition to the pyracylenic [6,6] bond and proved formation of the [5,6] adduct.

Diels–Alder addition. The most extended computational studies were performed for Diels–Alder (DA) [4+2] cycloaddition of dienes to EMFs. In 2008 Osuna *et al.* reported extended computational study of the DA reaction between 1,3-butadiene and $Sc_3N@C_{78}$-D_{3h}(5) and corresponding empty cage.[297] The study showed that encapsulation of the Sc_3N cluster strongly affected regioselectivity, e.g. the most stable adducts for $Sc_3N@C_{78}$ and C_{78} were different. Besides, $Sc_3N@C_{78}$ was found to be significantly less reactive than C_{78} in terms of both thermodynamics (less exothermic reaction energies of DA addition to EMF) and kinetics (higher reaction barriers for $Sc_3N@C_{78}$). The most energetically preferable site of DA addition to $Sc_3N@C_{78}$, one of the [6,6] bonds (reaction energy −53 kJ/mol at the BP86/TZP//PB86/DZP level), also corresponded to the lowest reaction barrier (77 kJ/mol) and agreed well with experimental data of Cai *et al.*[298] Finally, the authors concluded that results of the extended study of thermodynamics and kinetics of DA addition do not correlate well with the predictions based on the bond lengths, POAV angles, and MO shapes and energies.[297] A year later the same group performed analogous study for DA addition to two isomers of $Y_3N@C_{78}$.[299] When Y_3N cluster was encapsulated in C_{78}-D_{3h}(5) (the same as found in $Sc_3N@C_{78}$), the high strain of the structure resulted in peculiar addition pattern different from that in $Sc_3N@C_{78}$. However, the non-IPR C_2(22010) isomer is more preferable for $Y_3N@C_{78}$ as was predicted by computations[18] and then proved experimentally.[124] For this isomer, the most preferable addition site with the reaction energy of −108 kJ/mol was [5,5] bond, i.e. a bond on the junction of two pentagons.[299] Recently this group also reported a study of DA addition to Ti_2C_2@C_{78}-D_{3h}(5) and showed that this EMF exhibits substantially different regioselectivity than $Sc_3N@C_{78}$ with the same carbon cage.[300]

In 2012, an extended DFT study of DA addition to D_{5h}(6) and I_h(7) cage isomers of $M_3N@C_{80}$ with various metal clusters, including Sc_3N, Lu_3N, Gd_3N, was reported by Osuna *et al.*[301] The study revealed that [5,6] bonds are most preferable cycloaddition sites for both cages. However, whereas addition to the [5,6] bond of the I_h isomer of $Sc_3N@C_{80}$ is slightly more exothermic (reaction energy is −72 kJ/mol versus −66 kJ/mol in $Sc_3N@C_{80}$-D_{5h}), the D_{5h} isomer is more reactive kinetically with the transition state energy of 56 kJ/mol (the value found for the I_h isomer is 66 kJ/mol). The reactivity of the $Lu_3N@C_{80}$ isomers is rather similar to that of $Sc_3N@C_{80}$, but further increase of the cluster size in $Gd_3N@C_{80}$ results in the reduced regioselectivity of addition (i.e. [6,6] adducts start to compete with [5,6] adducts). Besides, kinetic and thermodynamics products of the DA addition to $Gd_3N@C_{80}$ are different.[301]

Two groups recently reported on the computational studies of the Diels-Alder reaction between $La@C_{82}$ and cyclopentadiene (Cp) or pentamethylcyclopentadiene (Cp*). Garcia-Borràs *et al.*[302] studied thermodynamics and kinetics of Cp and Cp* addition to $La@C_{82}$ at the B3LYP-D_2/TZP//BLYP-D_2/DZP level (i.e. the structures were optimized using Slater-type DZP basis set, while final point-energy computations were then performed with TZP basis, and D_2 stands for dispersion correction by Grimme *et al.*[303]). The authors found the same regioselectivity of Cp and Cp* addition (in terms of the relative stability of the monoadduct isomers), favoring the structure characterized experimentally for $La@C_{82}$Cp* (C21–C23 bond, see Fig. 10). Kinetic studies showed that reaction proceeded via concerted mechanism. At the first stage, a reactant complex between $La@C_{82}$ and Cp or Cp* was formed, which was then transformed to a cycloadduct via an energy barrier of ca 40–50 kJ/mol. The barriers ($\Delta E^{\neq}$) for the retro-DA reaction were found to be 101 kJ/mol (Cp) and 95 kJ/mol (Cp*), which compares reasonably well with experimentally determined activation energies of 100[290] and 113[304] kJ/mol, respectively. The authors also found that significant difference in the stability of $La@C_{82}$Cp and $La@C_{82}$Cp* adducts should be attributed to the different stability of the reactant complex. Higher stability of the latter in $La@C_{82}$Cp*, which is predominantly caused by dispersion interaction, makes this adduct more kinetically stable. The authors also showed that dispersion correction is of paramount importance for the reliable modeling of the reaction

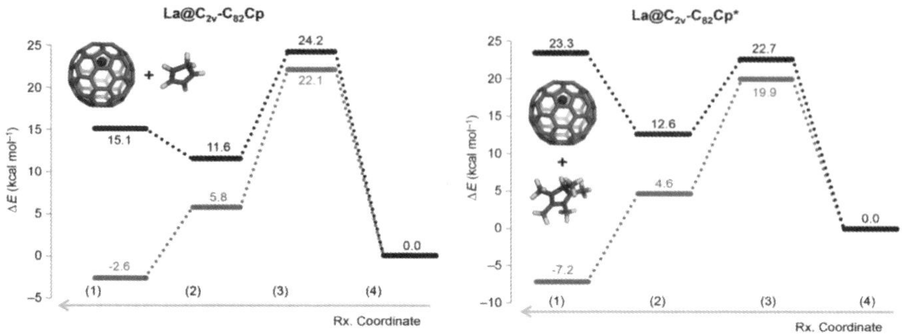

Fig. 11 Energy profiles of the Diels–Alder cycloaddition between La@C_{82} and Cp (left) and Cp* (right) for the attack of the most reactive C21–C23 bond computed at the B3LYP-D_2/TZP//BLYP-D_2/DZP level. Black lines denote computations with dispersion effects taken into account, while grey lines are for the computation without dispersion correction. Stationary points are (1) reactants, (2) reactant complex, (3) transition state, (4) cycloadduct; relative energies are in kcal/mol. Reproduced with permission from Ref. 302 © 2013 Wiley-VCH Verlag GmbH & Co. KGaA, Weinheim.

mechanism (see Fig. 11 for the comparison of the reaction profiles computed with and without D_2 correction).

Similar results were obtained by Nagase and Akasaka *et al.*,[304] who used M06-2X functional with 6-31G(d) basis set for C and H and Lanl2dz basis set with corresponding effective core potential for La. An intermediate complex between La@C_{82} and Cp* was found to be 44 kJ/mol more stable than the reactants, whereas the energy barrier between intermediate complex and the cycloadducts was computed to be 51 kJ/mol. Overall, the DA reaction was found to be exothermic with the release of 118 kJ/mol. The computed barrier for the retro-DA reaction, 125 kJ/mol, agreed reasonably well with the experimentally determined activation energy of 113 kJ/mol. The experimentally characterized isomer of the La@C_{82}Cp* adduct was found to be favorable both thermodynamically (lowest energy) and kinetically (large orbital coefficients).

To simplify the search for the most probable cycloaddition products to EMFs, in 2012 Garcia-Borràs *et al.*[305] proposed the Frozen Cage Model based on the single-point energy calculations of EMF adducts using geometries optimized for the free fullerene adducts and the cluster geometry optimized separately for underivatized EMF molecule. This way, relatively fast screening can be performed to find the most reactive bonds, and more extended thermodynamic and kinetic studies can be performed then for selected addition pathways.[305]

Bingel–Hirsch addition. Experimental studies of the Bingel-Hirsch (BH) addition to EMFs showed that they prefer to form open-cage cycloadducts (i.e. with the breaking of the sigma bonds of the cage). For $M_3N@C_{80}$-I_h, [6,6]-open adducts were formed predominantly,[144] whereas for the non-IPR $Gd_3N@C_{82}$, $Gd_3N@C_{84}$, and $Sc_3N@C_{68}$ EMFs functionalization of the bonds in the vicinity of the adjacent pentagon pairs was proposed.[306,307] Interestingly, the first study of the BH addition to $Sc_3N@C_{68}$ showed that the predominant product, whose structure was proposed based on the NMR data, was probably a kinetically favored adduct. Its relative energy was predicted to be 47 kJ/mol higher than the energy of the most stable isomer.[306]

Regioselectivity of Bingel–Hirsch addition to non-IPR $Gd_3N@C_{82}$ and $Gd_3N@C_{84}$ was studied by Poblet *et al.* based on relative energies of cycloadducts.[307] Extended computations were performed for $Y_3N@C_{82}$ and $Y_3N@C_{84}$ and then the results were verified in computations of selected $Gd_3N@C_{82}$ and $Gd_3N@C_{84}$ adducts. Computations showed that the most stable are open-cage adducts in which long C–C bonds with high pyramidalization angles are functionalized. [5,5] bonds in the pentagon junctions were not found to be very reactive. From thermodynamic point of view, the most reactive are [5,6] bonds near the pentagon junctions.

In 2013, this group also reported extended thermodynamic and kinetic study of Bingel–Hirsch addition to $Sc_3N@C_{68}$ and $Sc_3N@C_{80}$.[308] For $Sc_3N@C_{80}$ the authors found that addition across [6,6] and [5,6] gives essentially isoenergetic adducts, and hence if the reaction were proceeding under thermodynamic control, a mixture of regioisomers would be expected. This conclusion would contradict experimental data showing that only [6,6] bond of $Sc_3N@C_{80}$ is reacting.[309] In mechanistic studies the authors followed suggestion of Hirsch[310] that at the first step bromomalonate carbanion reacts with fullerene forming a single-bonded negatively charged intermediate, and at the second stage the bromide anion is leaving and the second bond between malonate carbon atom and the carbon cage is formed. DFT calculations showed that the energies of three intermediates (**1a**, **1b**, and **2b**, see Fig. 12) lie within 10 kJ/mol. However, the reaction

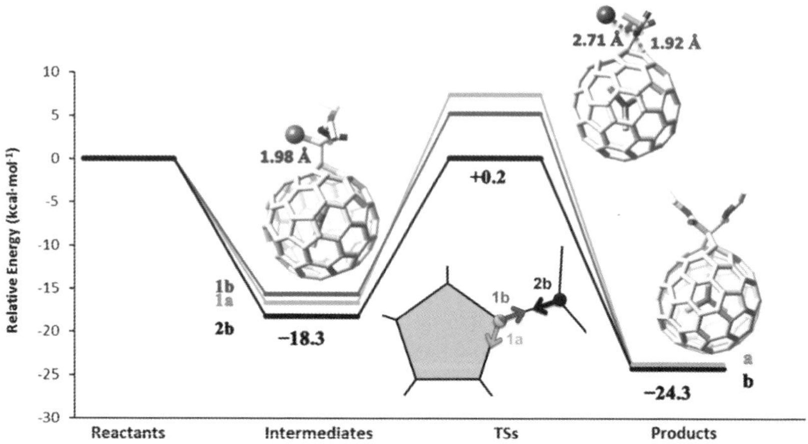

Fig. 12 Energy profiles of the Bingel–Hirsch addition of bromomalonate to $Sc_3N@C_{80}$. Reproduced with permission from Ref. 308 © 2013 Wiley-VCH Verlag GmbH & Co. KGaA, Weinheim.

barrier for the second step is considerably higher for the process forming [5,6] adduct from the intermediate **1a** (100 kJ/mol) than for transformation of intermediates **1b** and **2b** into the [6,6] adduct (87 and 77 kJ/mol, respectively). These results proved that Bingel–Hirsch addition to $Sc_3N@C_{80}$ proceeds under kinetic control. Calculations for $Sc_3N@C_{68}$ showed much larger spread of the relative energies of regioisomers. More importantly, it was found that reaction barrier for the most thermodynamically stable adduct is a few kJ/mol higher than that of the less stable adduct proposed earlier by Dorn *et al.*[306] Thus, it was concluded that kinetic rather than thermodynamic factors determine regioselectivity of the Bingel–Hirsch addition to EMFs.[308]

6.3. Trifluoromethylation of EMFs: Thermodynamic Control

Radical trifluoromethylation is a rare example of fullerene functionalization techniques proceeding at high temperature and hence ensuring predominant thermodynamic control of the products (CF_3 radicals are produced by thermal decomposition of either solid $AgCF_3CO_2$ or gaseous CF_3I at temperatures 300–600°C). However, reaction usually results in multiply functionalized EMFs (derivatives with up to 16 were isolated and up to 22 groups were detected in mass spectra of crude products). Only a few such derivatives were characterized by single-crystal X-ray diffraction.[196,197,311,312] For many others, structure elucidation is based on the ^{19}F NMR, which enables one to deduce the relative position of CF_3 groups based on ^{19}F multiplet structure and 2D COSY correlations caused by through-space F···F interactions of neighboring groups.[196,197,313,314] However, exact positions of CF_3 groups cannot be obtained by this technique, and computational search for the most stable isomers is necessary. Such studies are complicated by the dramatic increase in the number of possible isomers with the increase of the number of added groups.

The first high-temperature trifluoromethylation of EMF was reported in 2005 for Y@C_{82}.[314] Two isomers of the derivative with five CF_3 groups were isolated, and NMR studies showed that CF_3 groups form a chain of four 1,4-$C_6(CF_3)_2$ hexagons. DFT computations of all 56 isomers of Y@$C_{82}(CF_3)_5$ with NMR-compatible addition pattern revealed two particularly stable structures, which were followed by the third isomer at 14 kJ/mol (Fig. 13). All other structures were less stable by at least 50 kJ/mol.[314]

Extended studies of trifluoromethylation were reported for Sc_3N@C_{80}.[196,197,311–313] In a search for the most stable isomers of Sc_3N@$C_{80}(CF_3)_n$ for a given number of n, the authors of Ref. 196 supposed that (i) they are

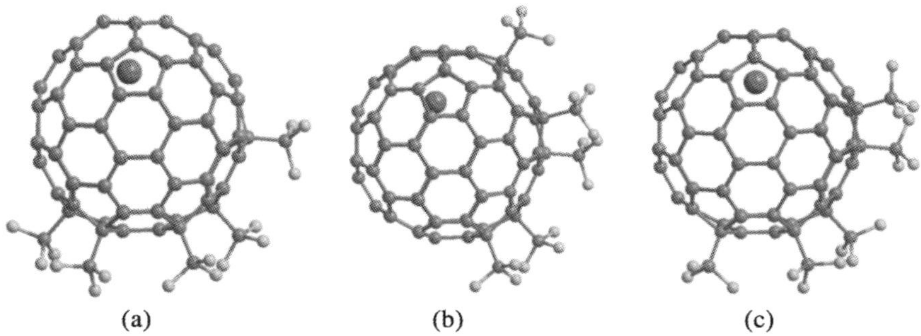

Fig. 13 Three lowest energy isomer of Y@$C_{82}(CF_3)_5$ with ribbons of 1,4-$C_6(CF_3)_2$ edge-sharing hexagons calculated at the DFT level of theory. Structures (a) and (b) were found to be 14 kJ/mol more stable than structure (c) and at least 50 kJ/mol more stable than the other 53 calculated isomers. Reproduced with permission from Ref. 314 © 2005 Wiley-VCH Verlag GmbH & Co. KGaA, Weinheim.

based on the most stable isomers of $C_{80}(CF_3)_n^{6-}$ and (ii) have suitable arrangement of CF_3 groups so that efficient coordination of the Sc_3N cluster to the carbon cage π-system can be realized. Therefore the search for the lowest energy isomers of $Sc_3N@C_{80}(CF_3)_n$ was performed in two steps: (i) the most stable $C_{80}(CF_3)_n^{6-}$ isomers were identified, followed by (ii) the studies of $Sc_3N@C_{80}(CF_3)_n$ isomers with the same CF_3 addition patterns. However, for large number of added CF_3 groups (eight or more) even these and NMR-based restrictions still allowed tens of thousands of isomers. Hence, the following multistep strategy was developed by the authors.

First, AM1 point energy calculations of all possible isomers of $C_{80}H_n^{6-}$ were performed to exclude the structures with unstable π-system. The isomers of $C_{80}H_n^{6-}$ were then sorted according to their energy and AM1 computations of the corresponding $C_{80}(CF_3)_n^{6-}$ isomers were performed with the full optimization (starting from the most stable isomers of $C_{80}H_n^{6-}$). The computations were stopped when it became clear that all stable isomers of $C_{80}(CF_3)_n^{6-}$ with the relative energies less than 50–70 kJ/mol were found. At the third stage, point energy DFT/AM1 calculations were performed for selected structure to reveal the most stable structures with the largest HOMO-LUMO gaps.

After the most stable isomers of $C_{80}(CF_3)_n^{6-}$ had been found, the calculations with the Sc_3N cluster placed inside the carbon cage were performed. At the first step, the less computationally intensive DFTB approach was used. At this level, suitable pre-optimized coordinates for the following DFT calculations could be found and the isomers could be sorted by their energy beforehand. However, subsequent DFT calculations showed that the DFTB method predicted high stability for the isomers in which Sc was coordinated to CF_3-bearing C-sp^3 atoms of the cage, contrary to the results of the DFT method that showed that such isomers are not very stable. At the same time, the isomers with the DFTB-calculated relative energies higher than 50 kJ/mol were also found to be unstable at the DFT level. Therefore these higher-energy structures were discarded from the final stage of the calculations that relied on the full DFT optimization.

This complicated approach was used to find the most stable isomer of $Sc_3N@C_{80}(CF_3)_n$ for $n = 8 - 14$; for $n = 10$ the proposed structure was then confirmed by single-crystal X-ray diffraction study.[196] The study has also provided a background for some methodological conclusions. First of all, judicious choice of computational approaches in terms of the balance between reliability and computational demands allows one to screen many thousands of isomers and provide reliable predictions of the most stable structures even for multiply functionalized EMFs. However, a word of caution is necessary since the correspondence between the relative energies of $Sc_3N@C_{80}(CF_3)_n$ and $C_{80}(CF_3)_n^{6-}$ isomers is not as good as that for the non-derivatized cage isomer (see Section 3).[197] Besides, it was revealed that although DFTB may be a promising approach for reducing computational demands, reliable predictions of EMF derivatives at this level are hardly possible now and hence tailoring of parameters may be necessary for more dedicated studies.

7. CONCLUDING REMARKS

In this chapter we have given an overview of the state of the art in computational studies of the molecular and electronic structure, stability, molecular dynamics, spectroscopic properties, and chemical reactivity of EMFs. We have shown how modern computational approaches allow one to address and — importantly — provide reliable description of many experimentally observed phenomena, predict molecular structure and reveal invaluable information on reaction mechanisms of EMFs. Importantly, many of these computations can be performed nowadays with comparably modest computational resources (i.e. with standard office computers) and hence we would like to encourage fullerene scientists toward a broader use of computational approaches in their everyday work. Furthermore, continual development in computational hardware and software allow us to hope that even more complex tasks and application of post-Hartree-Fock *ab initio* approaches in the studies of EMFs will become possible in the near future.

ACKNOWLEDGMENTS

This work was supported by DFG (project PO 1602/1–1). The author would like to thank his colleagues at IFW Dresden, Colorado State University, Hefei National Laboratory for Physical Sciences at Microscale

(USTC), and Moscow State University for fruitful discussions. Continuous support of computational studies by the Research Computing Center of the Moscow State University (supercomputer "Chebyshev"), Research Centre Jülich (supercomputer JUROPA), and the Supercomputing Center at the Technical University of Dresden are highly appreciated.

REFERENCES

1. Becke AD. (1988) Density-functional exchange-energy approximation with correct asymptotic behavior. *Phys Rev A* **38**: 3098–3100.
2. Lee C, Yang W, Parr RG. (1988) Development of the Colle-Salvetti correlation-energy formula into a functional of the electron density. *Phys Rev B* **37**: 785–789.
3. Perdew JP. (1986) Density-Functional Approximation for the Correlation Energy of the Inhomogeneous Electron-Gas (P86). *Phys Rev B* **33**: 8822–8824.
4. Perdew JP (1991). PW91. Ziesche P, Eschrig H (eds). *Electronic Structure of Solids '91*. Akademie Verlag: 11.
5. Becke AD. (1993) Density-functional thermochemistry. III. The role of exact exchange. *J Chem Phys* **98**: 5648–5652.
6. Perdew JP, Burke K, Ernzerhof M. (1996) Generalized gradient approximation made simple. *Phys Rev Lett* **77**: 3865–3868.
7. Porezag D, Frauenheim T, Köhler T, *et al.* (1995) DFTB. *Phys Rev B* **51**: 12947–12960.
8. Nishimoto Y, Wang Z, Morokuma K, *et al.* (2011) Molecular and electronic structures of endohedral fullerenes, $Sc_2C_2@C_{3v}$-C_{82} and $Sc_2@C_{3v}$-C_{82}: Benchmark for SCC-DFTB and proposal of new inner cluster structures. *Phys Status Solidi B* **249**: 324–334.
9. Johnson RD, Devries MS, Salem J, *et al.* (1992) Electron-Paramagnetic Resonance Studies of Lanthanum-Containing C_{82}. *Nature* **355**: 239–240.
10. Shinohara H, Sato H, Saito Y, *et al.* (1992) Mass Spectroscopic and ESR Characterization of Soluble Yttrium-Containing Metallofullerenes $Y@C_{82}$ and $Y_2@C_{82}$. *J Phys Chem* **96**: 3571–3573.
11. Rosen A, Wastberg B. (1988) 1st-Principle Calculations of the Ionization-Potentials and Electron-Affinities of the Spheroidal Molecules C_{60} and LaC_{60}. *J Am Chem Soc* **110**: 8701–8703.
12. Rosen A, Wastberg B. (1989) Electronic-Structure of Spheroidal Metal Containing Carbon Shells — Study of the LaC_{60} and C_{60} Clusters and Their

Ions within the Local Density Approximation. *Z Phys D-Atoms Mol Clusters* **12**: 387–390.
13. Shinohara H. (2000) Endohedral metallofullerenes. *Rep Prog Phys* **63**: 843–892.
14. Chaur MN, Melin F, Ortiz AL, *et al.* (2009) Chemical, Electrochemical, and Structural Properties of Endohedral Metallofullerenes. *Angew Chem-Int Edit* **48**: 7514–7538.
15. Dunsch L, Yang S. (2007) Metal Nitride Cluster Fullerenes: Their Current State and Future Prospects. *Small* **3**: 1298–1320.
16. Weaver JH, Chai Y, Kroll GH, *et al.* (1992) XPS Probes of Carbon-Caged Metals. *Chem Phys Lett* **190**: 460–464.
17. Suzuki T, Kikuchi K, Oguri F, *et al.* (1996) Electrochemical properties of fullerenolanthanides. *Tetrahedron* **52**: 4973–4982.
18. Popov AA, Krause M, Yang SF, *et al.* (2007) C_{78} cage isomerism defined by trimetallic nitride cluster size: A computational and vibrational spectroscopic study. *J Phys Chem B* **111**: 3363–3369.
19. Yang SF, Popov AA, Dunsch L. (2007) Violating the Isolated Pentagon Rule (IPR): The endohedral Non-IPR cage of $Sc_3N@C_{70}$. *Angew Chem-Int Edit* **46**: 1256–1259.
20. Popov AA, Dunsch L. (2007) Structure, Stability, and Cluster-Cage Interactions in Nitride Clusterfullerenes $M_3N@C_{2n}$ (M = Sc, Y; 2n = 68–98): a Density Functional Theory Study. *J Am Chem Soc* **129**: 11835–11849.
21. Nagase S, Kobayashi K, Akasaka T. (1997) Recent progress in endohedral dimetallofullerenes. *J Mol Struct (Theochem)* **398**: 221–227.
22. Kobayashi K, Nagase S. (1998) Structures and electronic states of $M@C_{82}$ (M = Sc, Y, La and lanthanides). *Chem Phys Lett* **282**: 325–329.
23. Kobayashi K, Nagase S, Akasaka T. (1996) Endohedral dimetallofullerenes $Sc_2@C_{84}$ and $La_2@C_{80}$. Are the metal atoms still inside the fullerene cages? *Chem Phys Lett* **261**: 502–506.
24. Lu J, Zhang XW, Zhao XG, *et al.* (2000) Strong metal-cage hybridization in endohedral $La@C_{82}$, $Y@C_{82}$ and $Sc@C_{82}$. *Chem Phys Lett* **332**: 219–224.
25. Takata M, Nishibori E, Sakata M, *et al.* (2003) Synchrotron radiation for structural chemistry — Endohedral natures of metallofullerenes found by synchrotron radiation powder method. *Struct Chem* **14**: 23–38.
26. Suzuki M, Lu X, Sato S, *et al.* (2012) Where Does the Metal Cation Stay in $Gd@C_{2v}(9)$-C_{82}? A Single-Crystal X-ray Diffraction Study. *Inorg Chem* **51**: 5270–5273.
27. Sato S, Nikawa H, Seki S, *et al.* (2012) A Co-Crystal Composed of the Paramagnetic Endohedral Metallofullerene $La@C_{82}$ and a Nickel Porphyrin with High Electron Mobility. *Angew Chem-Int Edit Engl* **51**: 1589–1591.

28. Krause M, Popov A, Dunsch L. (2006) Vibrational structure of endohedral fullerene $Sc_3N@C_{78}(D_{3h}')$: Evidence for a strong coupling between the Sc_3N cluster and C_{78} cage. *ChemPhysChem* **7**: 1734–1740.
29. Yang SF, Kalbac M, Popov A, *et al.* (2006) A facile route to the non-IPR fullerene $Sc_3N@C_{68}$: Synthesis, spectroscopic characterization, and density functional theory computations (IPR =isolated pentagon rule). *Chem-Eur J* **12**: 7856–7863.
30. Popov AA, Yang SF, Kalbac M, *et al.* (2009). Electronic Structure of $Sc_3N@C_{68}$ in Neutral and Charged States: An Experimental and TD-DFT Study. Simos TE, Maroulis G (eds). *Computational Methods in Science and Engineering, Vol 2 — Advances in Computational Science.* Amer Inst Physics. **1148:** 712–715.
31. Popov AA. (2009) Metal-Cage Bonding, Molecular Structures and Vibrational Spectra of Endohedral Fullerenes: Bridging Experiment and Theory. *J Comput Theor Nanosci* **6**: 292–317.
32. Kessler B, Bringer A, Cramm S, *et al.* (1997) Evidence for incomplete charge transfer and la-derived states in the valence bands of endohedrally doped $La@C_{82}$. *Phys Rev Lett* **79**: 2289–2292.
33. Alvarez L, Pichler T, Georgi P, *et al.* (2002) Electronic structure of pristine and intercalated $Sc_3N@C_{80}$ metallofullerene. *Phys Rev B* **66**: 035107.
34. Muthukumar K, Larsson JA. (2008) A density functional study of Ce@C-82: Explanation of the Ce preferential bonding site. *J Phys Chem A* **112**: 1071–1075.
35. Campanera JM, Bo C, Olmstead MM, *et al.* (2002) Bonding within the Endohedral Fullerenes $Sc_3N@C_{78}$ and $Sc_3N@C_{80}$ as Determined by Density Functional Calculations and Reexamination of the Crystal Structure of $Sc_3N@C_{78}$•Co(OEP)•1.5(C_6H_6)•0.3($CHCl_3$). *J Phys Chem A* **106**: 12356–12364.
36. Yang S, Yoon M, Hicke C, *et al.* (2008) Electron transfer and localization in endohedral metallofullerenes: Ab initio density functional theory calculations. *Phys Rev B* **78**: 115435.
37. Liu D, Hagelberg F, Park SS. (2006) Charge transfer and electron backdonation in metallofullerenes encapsulating NSc3. *Chem Phys* **330**: 380–386.
38. Wu J, Hagelberg F. (2008) Computational Study on C_{80} Enclosing Mixed Trimetallic Nitride Clusters of the Form $Gd_xM_{3-x}N$ (M = Sc, Sm, Lu) *J Phys Chem C* **112**: 5770–5777.
39. Popov AA, Dunsch L. (2008) Hindered Cluster Rotation and ^{45}Sc Hyperfine Splitting Constant in Distonoid Anion Radical $Sc_3N@C_{80}$, and Spatial Spin Charge Separation as a General Principle for Anions of Endohedral

Fullerenes with Metal-Localized Lowest Unoccupied Molecular Orbitals. *J Am Chem Soc* **130**: 17726–17742.

40. Valencia R, Rodríguez-Fortea A, Poblet JM. (2008) Understanding the Stabilization of Metal Carbide Endohedral Fullerenes $M_2C_2@C_{82}$ and Related Systems. *J Phys Chem A* **112**: 4550–4555.
41. Popov AA, Chen C, Yang S, *et al.* (2010) Spin-Flow Vibrational Spectroscopy of Molecules with Flexible Spin Density: Electrochemistry, ESR, Cluster and Spin Dynamics, and Bonding in $TiSc_2N@C_{80}$. *ACS Nano* **4**: 4857–4871.
42. Popov AA, Yang S, Dunsch L. (2013) Endohedral Fullerenes. *Chem Rev* **113**: 5989–6113.
43. Bader RFW (1990). *Atoms in Molecules — A Quantum Theory*, Oxford University Press.
44. Matta CF, Boyd RJ (eds). (2007) *The Quantum Theory of Atoms in Molecules. From Solid State to DNA and Drug Design.*, WILEY-VCH Verlag.
45. Kobayashi K, Nagase S. (1999) Bonding features in endohedral metallofullerenes. Topological analysis of the electron density distribution. *Chem Phys Lett* **302**: 312–316.
46. Kobayashi K, Nagase S. (2002) A stable unconventional structure of $Sc_2@C_{66}$ found by density functional calculations. *Chem Phys Lett* **362**: 373–379.
47. Macchi P, Sironi A. (2003) Chemical bonding in transition metal carbonyl clusters: complementary analysis of theoretical and experimental electron densities. *Coord Chem Rev* **238–239**: 383–412.
48. Gatti C. (2005) Chemical bonding in crystals: new directions. *Z Kristall* **220**: 399–457.
49. Cortés-Guzmán F, Bader RFW. (2005) Complementarity of QTAIM and MO theory in the study of bonding in donor–acceptor complexes. *Coord Chem Rev* **249**: 633–662.
50. Cremer D, Kraka E. (1984) Chemical Bonds without Bonding Eectron Density — Does the Difference Electron-Density Analysis Suffice for a Description of the Chemical Bond? *Angew Chem-Int Edit Engl* **23**: 627–628.
51. Popov AA, Dunsch L. (2009) The Bonding Situation in Endohedral Metallofullerenes as Studied by Quantum Theory of Atoms in Molecules (QTAIM). *Chem-Eur J* **15**: 9707–9729.
52. Popov AA, Zhang L, Dunsch L. (2010) A Pseudoatom in a Cage: Trimetallofullerene $Y_3@C_{80}$ Mimics $Y_3N@C_{80}$ with Nitrogen Substituted by a Pseudoatom. *ACS Nano* **4**: 795–802.
53. Dunsch L, Yang S, Zhang L, *et al.* (2010) Metal Sulfide in a C_{82} Fullerene Cage: A New Form of Endohedral Clusterfullerenes. *J Am Chem Soc* **132**: 5413–5421.

54. Jin P, Zhou Z, Hao C, *et al.* (2010) NC unit trapped by fullerenes: a density functional theory study on $Sc_3NC@C_{2n}$ (2n = 68, 78 and 80). *Phys Chem Chem Phys* **12**: 12442–12449.
55. Popov AA, Avdoshenko SM, Pendás AM, *et al.* (2012) Bonding between strongly repulsive metal atoms: an oxymoron made real in a confined space of endohedral metallofullerenes. *Chem Commun* **48**: 8031–8050.
56. Farrugia LJ, Evans C, Lentz D, *et al.* (2009) The QTAIM Approach to Chemical Bonding Between Transition Metals and Carbocyclic Rings: A Combined Experimental and Theoretical Study of $(\eta^5\text{-}C_5H_5)Mn(CO)_3$, $(\eta^6\text{-}C_6H_6)Cr(CO)_3$, and (E)-$\{(\eta^5\text{-}C_5H_4)CF{=}CF(\eta^5\text{-}C_5H_4)\}(\eta^5\text{-}C_5H_5)_2Fe_2$. *J Am Chem Soc* **131**: 1251–1268.
57. Gan L-H, An J, Pan F-S, *et al.* (2011) Geometrical and Electronic Rules in Fullerene-Based Compounds. *Chem-Asian J* **6**: 1304–1314.
58. Gan LH. (2006) Two contrary effects of electron transfer on the structures and stability of dimetallofullerenes. *Chem Phys Lett* **429**: 185–188.
59. Glockler G. (1961) The heat of dissociation of nitrogen. *Pure Appl Chem* **2**: 49–54.
60. Mercado BQ, Sruart MA, Mackey MA, *et al.* (2010) $Sc_2(\mu_2\text{-}O)$ Trapped in a Fullerene Cage: The Isolation and Structural Characterization of $Sc_2(\mu_2\text{-}O)@C_s(6)\text{-}C_{82}$ and the Relevance of the Thermal and Entropic Effects in Fullerene Isomer Selection. *J Am Chem Soc* **132**: 12098–12105.
61. Martín Pendás A, Francisco E, Blanco MA. (2006) Binding Energies of First Row Diatomics in the Light of the Interacting Quantum Atoms Approach. *J Phys Chem A* **110**: 12864–12869.
62. Francisco E, Martín Pendás A, Blanco MA. (2005) A Molecular Energy Decomposition Scheme for Atoms in Molecules. *J Chem Theory Comput* **2**: 90–102.
63. Blanco MA, Martín Pendás A, Francisco E. (2005) Interacting Quantum Atoms: A Correlated Energy Decomposition Scheme Based on the Quantum Theory of Atoms in Molecules. *J Chem Theory Comput* **1**: 1096–1109.
64. Stevenson S, Mackey MA, Stuart MA, *et al.* (2008) A Distorted Tetrahedral Metal Oxide Cluster inside an Icosahedral Carbon Cage. Synthesis, Isolation, and Structural Characterization of $Sc_4(\mu_3\text{-}O)_2@I_h\text{-}C_{80}$. *J Am Chem Soc* **130**: 11844–11845.
65. Umemoto H, Ohashi K, Inoue T, *et al.* (2010) Synthesis and UHV-STM observation of the T_d-symmetric Lu metallofullerene: $Lu_2@C_{76}(T_d)$. *Chem Commun* **46**: 5653–5655.
66. Zuo T, Xu L, Beavers CM, *et al.* (2008) $M_2@C_{79}N$ (M = Y, Tb): Isolation and Characterization of Stable Endohedral Metallofullerenes Exhibiting M ... M

Bonding Interactions inside Aza[80]fullerene Cages. *J Am Chem Soc* **130**: 12992–12997.

67. Fu W, Zhang J, Fuhrer T, *et al.* (2011) $Gd_2@C_{79}N$: Isolation, Characterization, and Monoadduct Formation of a Very Stable Heterofullerene with a Magnetic Spin State of S = 15/2. *J Am Chem Soc* **133**: 9741–9750.
68. Kurihara H, Lu X, Iiduka Y, *et al.* (2012) $Sc_2@C_{3v}(8)$-C_{82} *vs.* $Sc_2C_2@C_{3v}(8)$-C_{82}: drastic effect of C_2 capture on the redox properties of scandium metallofullerenes. *Chem Commun* **48**: 1290–1292.
69. Inoue T, Tomiyama T, Sugai T, *et al.* (2004) Trapping a C_2 radical in endohedral metallofullerenes: Synthesis and structures of $(Y_2C_2)@C_{82}$ (Isomers I, II, and III). *J Phys Chem B* **108**: 7573–7579.
70. Olmstead MM, de Bettencourt-Dias A, Stevenson S, *et al.* (2002) Crystallographic characterization of the structure of the endohedral fullerene {$Er_2@C_{82}$ Isomer I} with *Cs* cage symmetry and multiple sites for erbium along a band of ten contiguous hexagons. *J Am Chem Soc* **124**: 4172–4173.
71. Olmstead MM, Lee HM, Stevenson S, *et al.* (2002) Crystallographic characterization of Isomer 2 of $Er_2@C_{82}$ and comparison with Isomer 1 of $Er_2@C_{82}$. *Chem Commun*: 2688–2689.
72. Ito Y, Okazaki T, Okubo S, *et al.* (2007) Enhanced 1520 nm photoluminescence from Er^{3+} ions in di-erbium-carbide metallofullerenes $(Er_2C_2)@C_{82}$ (isomers I, II, and III). *ACS Nano* **1**: 456–462.
73. Yang T, Zhao X, Osawa E. (2011) Can a Metal–Metal Bond Hop in the Fullerene Cage? *Chem-Eur J* **17**: 10230–10234.
74. Suzuki T, Maruyama Y, Kato T, *et al.* (1995) Electrochemistry and Ab-Initio Study of the Dimetallofullerene $La_2@C_{80}$. *Angew Chem-Int Edit Engl* **34**: 1094–1096.
75. Lu X, Nikawa H, Nakahodo T, *et al.* (2008) Chemical Understanding of a Non-IPR Metallofullerene: Stabilization of Encaged Metals on Fused-Pentagon Bonds in $La_2@C_{72}$. *J Am Chem Soc* **130**: 9129–9136.
76. Iiduka Y, Ikenaga O, Sakuraba A, *et al.* (2005) Chemical reactivity of $Sc_3N@C_{80}$ and $La_2@C_{80}$. *J Am Chem Soc* **127**: 9956–9957.
77. Cao BP, Wakahara T, Tsuchiya T, *et al.* (2004) Isolation, characterization, and theoretical study of $La_2@C_{78}$. *J Am Chem Soc* **126**: 9164–9165.
78. Yamada M, Wakahara T, Tsuchiya T, *et al.* (2008) Spectroscopic and Theoretical Study of Endohedral Dimetallofullerene Having a Non-IPR Fullerene Cage: $Ce_2@C_{72}$. *J Phys Chem A* **112**: 7627–7631.
79. Muthukumar K, Larsson JA. (2010) The role of ellipticity on the preferential binding site of Ce and La in C_{78}-D_{3h} — A density functional theory study. *Nanoscale* **2**: 1250–1255.

80. Wang J, Irle S. (2011) Erbium Photoluminescence in $Er_2C_2@C_{82}$ and $Er_2@C_{82}$ Elucidated by Density Functional Theory. *ECS Meeting Abstracts* **1101**: 1782.
81. Wu X, Lu X. (2007) Dimetalloendofullerene U-2@C-60 has a U-U multiple bond consisting of sixfold one-electron-two-center bonds. *J Am Chem Soc* **129**: 2171–2177.
82. Valencia R, Rodriguez-Fortea A, Stevenson S, *et al.* (2009) Electronic Structures of Scandium Oxide Endohedral Metallofullerenes, $Sc_4(\mu_3\text{-}O)_n@I_h\text{-}C_{80}$ (n = 2, 3). *Inorg Chem* **48**: 5957–5961.
83. Farrugia L, Macchi P (2012). Bond Orders in Metal–Metal Interactions Through Electron Density Analysis. Stalke D (eds). *Electron Density and Chemical Bonding I: Experimental Charge Density Studies.* Springer Berlin/Heidelberg. **146:** 127–158.
84. Gatti C, Lasi D. (2007) Source function description of metal–metal bonding in d-block organometallic compounds. *Faraday Discuss* **135**: 55–78.
85. Gervasio G, Bianchi R, Marabello D. (2004) About the topological classification of the metal–metal bond. *Chem Phys Lett* **387**: 481–484.
86. Macchi P, Garlaschelli L, Martinengo S, *et al.* (1999) Charge Density in Transition Metal Clusters: Supported vs Unsupported Metal-Metal Interactions. *J Am Chem Soc* **121**: 10428–10429.
87. Farrugia LJ, Evans C, Senn HM, *et al.* (2012) QTAIM View of Metal–Metal Bonding in Di- and Trinuclear Disulfido Carbonyl Clusters. *Organometallics* **31**: 2559–2570.
88. Chevreau H, Martinsky C, Sevin A, *et al.* (2003) The nature of the chemical bonding in the D_{3h} and C_{2v} isomers of $Fe_3(CO)_{12}$. *New J Chem* **27**: 1049–1053.
89. Pyykkö P, Atsumi M. (2009) Molecular Single-Bond Covalent Radii for Elements 1–118. *Chem-Eur J* **15**: 186–197.
90. Okazaki T, Suenaga K, Lian YF, *et al.* (2001) Intrafullerene electron transfers in Sm-containing metallofullerenes: $Sm@C_{2n}$ ($74 <= 2n <= 84$). *J Mol Graph* **19**: 244–251.
91. Okazaki T, Suenaga K, Lian YF, *et al.* (2000) Direct EELS observation of the oxidation states of Sm atoms in $Sm@C_{2n}$ metallofullerenes ($74 <= 2n <= 84$). *J Chem Phys* **113**: 9593–9597.
92. Ralchenko Y, Kramida AE, Reader J (2011). NIST Atomic Spectra Database (ver. 4.1.0), [Online]. *National Institute of Standards and Technology.* URL: http://physics.nist.gov/asd.
93. Fowler PW, Manolopoulos DE. (1992) Magic Numbers and Stable Structures for Fullerenes, Fullerides and Fullerenium Ions. *Nature* **355**: 428–430.
94. Fowler PW, Zerbetto F. (1995) Charging and Equilibration of Fullerene Isomers. *Chem Phys Lett* **243**: 36–41.

95. Nakao K, Kurita N, Fujita M. (1994) Ab-Initio Molecular-Orbital Calculation for C_{70} and Seven Isomers of C_{80}. *Phys Rev B* **49**: 11415–11420.
96. Kobayashi K, Nagase S, Akasaka T. (1995) A Theoretical Study of C_{80} and $La_2@C_{80}$. *Chem Phys Lett* **245**: 230–236.
97. Kobayashi K, Nagase S. (1997) Structures of the $Ca@C_{82}$ isomers: A theoretical prediction. *Chem Phys Lett* **274**: 226–230.
98. Iiduka Y, Wakahara T, Nakajima K, *et al.* (2006) ^{13}C NMR spectroscopic study of scandium dimetallofullerene, $Sc_2@C_{84}$ *vs.* $Sc_2C_2@C_{82}$. *Chem Commun*: 2057–2059.
99. Nishibori E, Ishihara M, Takata M, *et al.* (2006) Bent (metal)$_2C_2$ clusters encapsulated in $(Sc_2C_2)@C_{82}$(III) and $(Y_2C_2)@C_{82}$(III) metallofullerenes. *Chem Phys Lett* **433**: 120–124.
100. Nishibori E, Narioka S, Takata M, *et al.* (2006) A C_2 molecule entrapped in the pentagonal-dodecahedral Y_2 cage in $Y_2C_2@C_{82}$(III). *ChemPhysChem* **7**: 345–348.
101. Chen N, Chaur MN, Moore C, *et al.* (2010) Synthesis of a new endohedral fullerene family, $Sc_2S@C_{2n}$ (n = 40–50) by the introduction of SO_2. *Chem Commun* **46**: 4818–4820.
102. Campanera JM, Bo C, Poblet JM. (2005) General rule for the stabilization of fullerene cages encapsulating trimetallic nitride templates. *Angew Chem-Int Edit* **44**: 7230–7233.
103. Valencia R, Rodriguez-Fortea A, Poblet JM. (2007) Large fullerenes stabilized by encapsulation of metallic clusters. *Chem Commun*: 4161–4163.
104. Zuo TM, Beavers CM, Duchamp JC, *et al.* (2007) Isolation and structural characterization of a family of endohedral fullerenes including the large, chiral cage fullerenes $Tb_3N@C_{88}$ and $Tb_3N@C_{86}$ as well as the I_h and D_{5h} isomers of $Tb_3N@C_{80}$. *J Am Chem Soc* **129**: 2035–2043.
105. Beavers CM, Jin H, Yang H, *et al.* (2011) Very Large, Soluble Endohedral Fullerenes in the Series La_2C_{90} to La_2C_{138}: Isolation and Crystallographic Characterization of $La_2@D_5(450)$-C_{100}. *J Am Chem Soc* **133**: 15338–15341.
106. Yang S, Popov AA, Dunsch L. (2007) The role of an asymmetric nitride cluster on a fullerene cage: The Non-IPR endohedral $DySc_2N@C_{76}$. *J Phys Chem B* **111**: 13659–13663.
107. Beavers CM, Chaur MN, Olmstead MM, *et al.* (2009) Large Metal Ions in a Relatively Small Fullerene Cage: The Structure of $Gd_3N@C_2(22010)$-C_{78} Departs from the Isolated Pentagon Rule. *J Am Chem Soc* **131**: 11519–11524.
108. Xu L, Li S-F, Gan L-H, *et al.* (2012) The structures of trimetallic nitride fullerenes $M_3N@C_{88}$: Theoretical evidence of corporation between electron transfer interaction and size effect. *Chem Phys Lett* **521**: 81–85.

109. Zheng J, Zhao X, Dang J, *et al.* (2011) Density functional theory characterization of lanthanum nitride endohedral fullerene: $La_3N@C_{92}$. *Chem Phys Lett* **514**: 104–108.
110. Mercado BQ, Beavers CM, Olmstead MM, *et al.* (2008) Is the Isolated Pentagon Rule Merely a Suggestion for Endohedral Fullerenes? The Structure of a Second Egg-Shaped Endohedral Fullerene — $Gd_3N@C_s(39663)$-C_{82}. *J Am Chem Soc* **130**: 7854–7855.
111. Curl RF, Lee MK, Scuseria GE. (2008) C_{60} Buckminsterfullerene High Yields Unraveled. *J Phys Chem A* **112**: 11951–11955.
112. Wang T-S, Chen N, Xiang J-F, *et al.* (2009) Russian-Doll-Type Metal Carbide Endofullerene: Synthesis, Isolation, and Characterization of $Sc_4C_2@C_{80}$. *J Am Chem Soc* **131**: 16646–16647.
113. Raghavachari K. (1992) Ground state of C84: two almost isoenergetic isomers. *Chem Phys Lett* **190**: 397–400.
114. Fowler PW, Manolopoulos DE (1995). *An Atlas of Fullerenes*, Claredron Press.
115. Rodriguez-Fortea A, Alegret N, Balch AL, *et al.* (2010) The maximum pentagon separation rule provides a guideline for the structures of endohedral metallofullerenes. *Nat Chem* **2**: 955–961.
116. Kobayashi K, Nagase S, Yoshida M, *et al.* (1997) Endohedral metallofullerenes. Are the isolated pentagon rule and fullerene structures always satisfied? *J Am Chem Soc* **119**: 12693–12694.
117. Wang CR, Kai T, Tomiyama T, *et al.* (2000) C_{66} fullerene encaging a scandium dimer. *Nature* **408**: 426–427.
118. Stevenson S, Fowler PW, Heine T, *et al.* (2000) Materials science — A stable non-classical metallofullerene family. *Nature* **408**: 427–428.
119. Shi ZQ, Wu X, Wang CR, *et al.* (2006) Isolation and characterization of $Sc_2C_2@C_{68}$: A metal-carbide endofullerene with a non-IPR carbon cage. *Angew Chem-Int Edit* **45**: 2107–2111.
120. Chen N, Mulet-Gas M, Li Y-Y, *et al.* (2013) $Sc_2S@C_2(7892)$-C_{70}: a metallic sulfide cluster inside a non-IPR C_{70} cage. *Chem Sci* **4**: 180–186.
121. Chen N, Beavers CM, Mulet-Gas M, *et al.* (2012) $Sc_2S@C_s(10528)$-C_{72}: A Dimetallic Sulfide Endohedral Fullerene with a Non-IPR Cage. *J Am Chem Soc* **134**: 7851–7860.
122. Wakahara T, Nikawa H, Kikuchi T, *et al.* (2006) $La@C_{72}$ having a non-IPR carbon cage. *J Am Chem Soc* **128**: 14228–14229.
123. Kato H, Taninaka A, Sugai T, *et al.* (2003) Structure of a missing-caged metallofullerene: $La_2@C_{72}$. *J Am Chem Soc* **125**: 7782–7783.

124. Ma Y, Wang T, Wu J, *et al.* (2011) Size effect of endohedral cluster on fullerene cage: Preparation and structural studies of $Y_3N@C_{78}$-C_2. *Nanoscale* **3**: 4955–4957.
125. Wu J, Wang T, Ma Y, *et al.* (2011) Synthesis, Isolation, Characterization, and Theoretical Studies of $Sc_3NC@C_{78}$-C_2. *J Phys Chem C* **115**: 23755–23759.
126. Beavers CM, Zuo TM, Duchamp JC, *et al.* (2006) $Tb_3N@C_{84}$: An improbable, egg-shaped endohedral fullerene that violates the isolated pentagon rule. *J Am Chem Soc* **128**: 11352–11353.
127. Zuo T, Walker K, Olmstead MM, *et al.* (2008) New egg-shaped fullerenes: non-isolated pentagon structures of $Tm_3N@C_s(51365)$-C_{84} and $Gd_3N@C_s(51365)$-C_{84}. *Chem Commun*: 1067–1069.
128. Zheng H, Zhao X, Wang W-W, *et al.* (2012) $Sc_2@C_{70}$ rather than $Sc_2C_2@C_{68}$: Density functional theory characterization of metallofullerene Sc_2C_{70}. *J Chem Phys* **137**: 014308.
129. Nagase S, Kobayashi K, Akasaka T. (1999) Unconventional cage structures of endohedral metallofullerenes. *Theochem-J Mol Struct* **462**: 97–104.
130. Slanina Z, Uhlik F, Nagase S. (2006) Computed structures of two known $Yb@C_{74}$ isomers. *J Phys Chem A* **110**: 12860–12863.
131. Zheng H, Zhao X, Ren T, *et al.* (2012) C_{74} Endohedral Metallofullerenes Violating the Isolated Pentagon Rule: A Density Functional Theory Study. *Nanoscale* **4**: 4530–4536.
132. Yang T, Zhao X, Xu Q, *et al.* (2011) Non-IPR endohedral fullerene $Yb@C_{76}$: density functional theory characterization. *J Mater Chem* **21**: 12206–12209.
133. Yang T, Zhao X, Xu Q, *et al.* (2012) Probing the role of encapsulated alkaline earth metal atoms in endohedral metallofullerenes $M@C_{76}$ (M = Ca, Sr, and Ba) by first-principles calculations. *Dalton Trans* **41**: 5294–5300.
134. Yang T, Zhao X, Li S-T, *et al.* (2012) Is the Isolated Pentagon Rule Always Satisfied for Metallic Carbide Endohedral Fullerenes? *Inorg Chem* **51**: 11223–11225.
135. Zhao X, Gao W-Y, Yang T, *et al.* (2012) Violating the Isolated Pentagon Rule (IPR): Endohedral Non-IPR C_{98} Cages of $Gd_2@C_{98}$. *Inorg Chem* **51**: 2039–2045.
136. Chen DL, Q. TW, Feng JK, *et al.* (2008) C68 Fullerene Isomers, Anions, and their Metallofullerenes: Charge-Stabilizing Different Isomers. *ChemPhysChem* **9**: 454–461.
137. Slanina Z, Chen ZF, Schleyer PV, *et al.* (2006) $La_2@C_{72}$ and $Sc_2@C_{72}$: Computational characterizations. *J Phys Chem A* **110**: 2231–2234.

138. Zywietz TK, Jiao H, Schleyer R, *et al.* (1998) Aromaticity and Antiaromaticity in Oligocyclic Annelated Five-Membered Ring Systems. *J Org Chem* **63**: 3417–3422.
139. Olmstead MM, Lee HM, Duchamp JC, *et al.* (2003) $Sc_3N@C_{68}$: Folded pentalene coordination in an endohedral fullerene that does not obey the isolated pentagon rule. *Angew Chem-Int Edit* **42**: 900–903.
140. Olmstead MM, de Bettencourt-Dias A, Duchamp JC, *et al.* (2001) Isolation and structural characterization of the endohedral fullerene $Sc_3N@C_{78}$. *Angew Chem-Int Edit* **40**: 1223–1225.
141. Melin F, Chaur MN, Engmann S, *et al.* (2007) The Large $Nd_3N@C_{2n}$ (40< = 2*n*< = 49) Cluster Fullerene Family: Preferential Templating of a C_{88} Cage by a Trimetallic Nitride Cluster. *Angew Chem-Int Edit* **46**: 9032–9035.
142. Chaur MN, Melin F, Elliott B, *et al.* (2008) New $M_3N@C_{2n}$ Endohedral Metallofullerene Families (M=Nd, Pr, Ce; n = 40–53): Expanding the Preferential Templating of the C_{88} Cage and Approaching the C_{96} Cage. *Chem-Eur J* **14**: 4594–4599.
143. Stevenson S, Phillips JP, Reid JE, *et al.* (2004) Pyramidalization of Gd_3N inside a C_{80} cage. The synthesis and structure of $Gd_3N@C_{80}$. *Chem Commun*: 2814–2815.
144. Lukoyanova O, Cardona CM, Rivera J, *et al.* (2007) "Open Rather than Closed" Malonate Methano-Fullerene Derivatives. The Formation of Methanofulleroid Adducts of $Y_3N@C_{80}$. *J Am Chem Soc* **129**: 10423–10430.
145. Rodriguez-Fortea A, Campanera JM, Cardona CM, *et al.* (2006) Dancing on a fullerene surface: Isomerization of $Y_3N@$(N-ethylpyrrolidino-C_{80}) from the 6,6 to the 5,6 regioisomer. *Angew Chem-Int Edit* **45**: 8176–8180.
146. Cardona CM, Elliott B, Echegoyen L. (2006) Unexpected chemical and electrochemical properties of $M_3N@C_{80}$ (M = Sc, Y, Er). *J Am Chem Soc* **128**: 6480–6485.
147. Chen N, Fan LZ, Tan K, *et al.* (2007) Comparative Spectroscopic and Reactivity Studies of $Sc_{3-x}Y_xN@C_{80}$ (x = 0–3). *J Phys Chem C* **111**: 11823–11828.
148. Chen N, Zhang EY, Tan K, *et al.* (2007) Size effect of encaged clusters on the exohedral chemistry of endohedral fullerenes: A case study on the pyrrolidino reaction of $Sc_xGd_{3-x}N@C_{80}$ (x = 0–3). *Org Lett* **9**: 2011–2013.
149. Yang SF, Troyanov SI, Popov AA, *et al.* (2006) Deviation from the planarity — a large Dy_3N cluster encapsulated in an I_h-C_{80} cage: An X-ray crystallographic and vibrational spectroscopic study. *J Am Chem Soc* **128**: 16733–16739.

150. Yang SF, Dunsch L. (2006) Expanding the number of stable isomeric structures of the C_{80} cage: A new fullerene $Dy_3N@C_{80}$. *Chem-Eur J* **12**: 413–419.
151. Chaur MN, Melin F, Ashby J, *et al.* (2008) Lanthanum Nitride Endohedral Fullerenes $La_3N@C_{2n}$ (43 < n < 55): Preferential Formation of $La_3N@C_{96}$. *Chem-Eur J* **14**: 8213–8219.
152. Yang H, Jin H, Zhen H, *et al.* (2011) Isolation and Crystallographic Identification of Four Isomers of $Sm@C_{90}$. *J Am Chem Soc* **133**: 6299–6306.
153. Yang H, Jin H, Hong B, *et al.* (2011) Large Endohedral Fullerenes Containing Two Metal Ions, $Sm_2@D_2(35)$-C_{88}, $Sm_2@C_1(21)$-C_{90}, and $Sm_2@D_3(85)$-C_{92}, and Their Relationship to Endohedral Fullerenes Containing Two Gadolinium Ions. *J Am Chem Soc* **133**: 16911–16919.
154. Mercado BQ, Jiang A, Yang H, *et al.* (2009) Isolation and Structural Characterization of the Molecular Nanocapsule $Sm_2@D_{3d}(822)$-C_{104}. *Angew Chem-Int Edit Engl* **48**: 9114–9116.
155. Yang H, Lu C, Liu Z, *et al.* (2008) Detection of a Family of Gadolinium-Containing Endohedral Fullerenes and the Isolation and Crystallographic Characterization of One Member as a Metal-Carbide Encapsulated inside a Large Fullerene Cage. *J Am Chem Soc* **130**: 17296–17300.
156. Che Y, Yang H, Wang Z, *et al.* (2009) Isolation and Structural Characterization of Two Very Large, and Largely Empty, Endohedral Fullerenes: $Tm@C_{3v}$-C_{94} and $Ca@C_{3v}$-C_{94}. *Inorg Chem* **48**: 6004–6010.
157. Slanina Z, Lee SL, Uhlik F, *et al.* (2007) Computing relative stabilities of metallofullerenes by Gibbs energy treatments. *Theor Chem Acc* **117**: 315–322.
158. Slanina Z, Uhlik F, Lee SL, *et al.* (2007) Computed structures and relative stabilities of $Be@C_{74}$. *Int J Quantum Chem* **107**: 2494–2498.
159. Uhlik F, Slanina Z, Nagase S. (2007) $Mg@C_{74}$ isomers: Calculated relative concentrations and comparison with $Ca@C_{74}$. *Phys Status Solidi A-Appl Mat* **204**: 1905–1910.
160. Slanina Z, Nagase S. (2006) Stability computations for $Ba@C_{74}$ isomers. *Chem Phys Lett* **422**: 133–136.
161. Slanina Z, Nagase S. (2005) $Sc_3N@C_{80}$: Computations on the two-isomer equilibrium at high temperatures. *ChemPhysChem* **6**: 2060–2063.
162. Slanina Z, Adamowicz L, Kobayashi K, *et al.* (2005) Gibbs energy-based treatment of metallofullerenes: $Ca@C_{72}$, $Ca@C_{74}$, $Ca@C_{82}$, and $La@C_{82}$. *Mol Simul* **31**: 71–77.
163. Slanina Z, Kobayashi K, Nagase S. (2004) $Ca@C_{74}$ isomers: relative concentrations at higher temperatures. *Chem Phys* **301**: 153–157.

164. Slanina Z, Kobayashi K, Nagase S. (2004) Ca@C_{82} isomers: Computed temperature dependency of relative concentrations. *J Chem Phys* **120**: 3397–3400.
165. Slanina Z, Kobayashi K, Nagase S. (2004) Computed temperature development of the relative stabilities of La@C_{82} isomers. *Chem Phys Lett* **388**: 74–78.
166. Slanina Z, Kobayashi K, Nagase S. (2003) Ca@C_{72} IPR and non-IPR structures: computed temperature development of their relative concentrations. *Chem Phys Lett* **372**: 810–814.
167. Slanina Z, Zhao X, Grabuleda X, *et al.* (2001) Mg@C_{72} MNDO/d evaluation of the isomeric composition. *J Mol Graph* **19**: 252–255.
168. Mulet-Gas M, Rodríguez-Fortea A, Echegoyen L, *et al.* (2013) Relevance of Thermal Effects in the Formation of Endohedral Metallofullerenes: The Case of $Gd_3N@C_s(39663)$-C_{82} and Other Related Systems. *Inorg Chem* **52**: 1954–1959.
169. Slanina Z, Uhlik F, Nagase S (2009). Computations on Encapsulations of Lanthanides into C_{74}. Laudon M, Romanowicz B (eds). *Nanotech Conference & Expo 2009, Vol 3, Technical Proceedings — Nanotechnology 2009: Biofuels, Renewable Energy, Coatings Fluidics and Compact Modeling.* Crc Press-Taylor & Francis Group: 312–315.
170. Slanina Z, Uhlik H, Nagase S. (2007) Computational evaluation of the relative production yields in the X@C_{74} series (X = Ca, Sr, Ba). *Chem Phys Lett* **440**: 259–262.
171. Slanina Z, Uhlik F, Lee SL, *et al.* (2007) Computations of production yields for Ba@C_{74} and Yb@C_{74}. *Mol Simul* **33**: 563–568.
172. Slanina Z, Uhlik F, Lee S-L, *et al.* (2011) Computed Stabilities in Metallofullerene Series: Al@C_{82}, Sc@C_{82}, Y@C_{82}, and La@C_{82}. *Int J Quantum Chem* **111**: 2712–2718.
173. Slanina Z, Uhlík F, Lee S-L, *et al.* (2011) Calculated relative yields for $Sc_2S@C_{82}$ and $Y_2S@C_{82}$. *Theor Chem Acc* **130**: 549–554.
174. Kurihara H, Lu X, Iiduka Y, *et al.* (2011) $Sc_2C_2@C_{80}$ Rather than $Sc_2@C_{82}$: Templated Formation of Unexpected $C_{2v}(5)$-C_{80} and Temperature-Dependent Dynamic Motion of Internal Sc_2C_2 Cluster. *J Am Chem Soc* **133**: 2382–2385.
175. Shimotani H, Ito T, Iwasa Y, *et al.* (2004) Quantum chemical study on the configurations of encapsulated metal ions and the molecular vibration modes in endohedral dimetallofullerene $La_2@C_{80}$. *J Am Chem Soc* **126**: 364–369.

176. Zhang J, Hao C, Li SM, *et al.* (2007) Which configuration is more stable for $La_2@C_{80}$, D_{3d} or D_{2h}? Recomputation with ZORA methods within ADF. *J Phys Chem C* **111**: 7862–7867.
177. Muthukumar K, Larsson JA. (2008) Explanation of the different preferential binding sites for Ce and La in $M_2@C_{80}$ (M = Ce, La) — a density functional theory prediction. *J Mater Chem* **18**: 3347–3351.
178. Akasaka T, Nagase S, Kobayashi K, *et al.* (1997) ^{13}C and ^{139}La NMR studies of $La_2@C_{80}$: First evidence for circular motion of metal atoms in endohedral dimetallofullerenes. *Angew Chem-Int Edit Engl* **36**: 1643–1645.
179. Nishibori E, Takata M, Sakata M, *et al.* (2001) Pentagonal-dodecahedral La_2 charge density in [80-I_h]fullerene: $La_2@C_{80}$. *Angew Chem-Int Edit* **40**: 2998–2999.
180. Kobayashi K, Sano Y, Nagase S. (2001) Theoretical study of endohedral metallofullerenes: $Sc_{3-n}La_nN@C_{80}$ (n = 0–3). *J Comput Chem* **22**: 1353–1358.
181. Gan LH, Yuan R. (2006) Influence of cluster size on the structures and stability of trimetallic nitride fullerenes $M_3N@C_{80}$. *ChemPhysChem* **7**: 1306–1310.
182. Yanov I, Kholod Y, Simeon T, *et al.* (2006) Local minima conformations of the $Sc_3N@C_{80}$ endohedral complex: Ab initio quantum chemical study and suggestions for experimental verification. *Int J Quantum Chem* **106**: 2975–2980.
183. Valencia R, Rodriguez-Fortea A, Clotet A, *et al.* (2009) Electronic Structure and Redox Properties of Metal Nitride Endohedral Fullerenes $M_3N@C_{2n}$ (M = Sc, Y, La, and Gd; 2n = 80, 84, 88, 92, 96). *Chem-Eur J* **15**: 10997–11009.
184. Lu J, Sabirianov RF, Mei WN, *et al.* (2006) Structural and magnetic properties of $Gd_3N@C_{80}$. *J Phys Chem B* **110**: 23637–23640.
185. Svitova AL, Popov AA, Dunsch L. (2013) Gd/Sc-based Mixed Metal Nitride Cluster Fullerenes: The Mutual Influence of the Cage and Cluster Size and the Role of Sc in the Electronic Structure. *Inorg Chem* **52**: 3368–3380.
186. Burke BG, Chan J, Williams KA, *et al.* (2011) Vibrational spectrum of the endohedral $Y_2C_2@C_{92}$ fullerene by Raman spectroscopy: Evidence for tunneling of the diatomic C_2 molecule. *Phys Rev B* **83**: 115457.
187. Popov AA, Chen N, Pinzón JR, *et al.* (2012) Redox-Active Scandium Oxide Cluster inside a Fullerene Cage: Spectroscopic, Voltammetric, Electron Spin Resonance Spectroelectrochemical, and Extended Density Functional Theory Study of $Sc_4O_2@C_{80}$ and Its Ion Radicals. *J Am Chem Soc* **134**: 19607–19618.

188. Kobayashi K, Nagase S, Maeda Y, *et al.* (2003) $La_2@C_{80}$: is the circular motion of two La atoms controllable by exohedral addition? *Chem Phys Lett* **374**: 562–566.
189. Popov AA, Dunsch L. (2011) Charge controlled changes in the cluster and spin dynamics of $Sc_3N@C_{80}(CF_3)_2$: the flexible spin density distribution and its impact on ESR spectra. *Phys Chem Chem Phys* **13**: 8977–8984.
190. Elliott B, Pykhova AD, Rivera J, *et al.* (2013) Spin Density and Cluster Dynamics in $Sc_3N@C_{80}^-$ upon [5,6] Exohedral Functionalization: An ESR and DFT Study. *J Phys Chem C* **117**: 2344–2348.
191. Yamada M, Wakahara T, Nakahodo T, *et al.* (2006) Synthesis and structural characterization of endohedral pyrrolidinodimetallofullerene: $La_2@C_{80}(CH_2)_2NTrt$. *J Am Chem Soc* **128**: 1402–1403.
192. Wakahara T, Yamada M, Takahashi S, *et al.* (2007) Two-dimensional hopping motion of encapsulated La atoms in silylated $La_2@C_{80}$. *Chem Commun*: 2680–2682.
193. Yamada M, Nakahodo T, Wakahara T, *et al.* (2005) Positional control of encapsulated atoms inside a fullerene cage by exohedral addition. *J Am Chem Soc* **127**: 14570–14571.
194. Ishitsuka MO, Sano S, Enoki H, *et al.* (2011) Regioselective Bis-functionalization of Endohedral Dimetallofullerene, $La_2@C_{80}$: Extremal La-La Distance. *J Am Chem Soc* **133**: 7128–7134.
195. Yamada M, Akasaka T, Nagase S. (2010) Endohedral Metal Atoms in Pristine and Functionalized Fullerene Cages. *Acc Chem Res* **43**: 92–102.
196. Shustova NB, Peryshkov DV, Kuvychko IV, *et al.* (2011) Poly (perfluoroalkylation) of Metallic Nitride Fullerenes Reveals Addition-Pattern Guidelines: Synthesis and Characterization of a Family of $Sc_3N@C_{80}(CF_3)_n$ (n = 2–16) and Their Radical Anions. *J Am Chem Soc* **133**: 2672–2690.
197. Shustova NB, Chen Y-S, Mackey MA, *et al.* (2009) $Sc_3N@(C_{80}\text{-}I_h(7))(CF_3)_{14}$ and $Sc_3N@(C_{80}\text{-}I_h(7))(CF_3)_{16}$. Endohedral Metallofullerene Derivatives with Exohedral Addends on Four and Eight Triple-Hexagon Junctions. Does the Sc_3N Cluster Control the Addition Pattern or Vice Versa? *J Am Chem Soc* **131**: 17630–17637.
198. Yamada M, Okamura M, Sato S, *et al.* (2009) Two Regioisomers of Endohedral Pyrrolidinodimetallofullerenes $M_2@I_h\text{-}C_{80}(CH_2)_2NTrt$ (M=La, Ce; Trt=trityl): Control of Metal Atom Positions by Addition Positions. *Chem-Eur J* **15**: 10533–10542.
199. Reich A, Panthofer M, Modrow H, *et al.* (2004) The structure of $Ba@C_{74}$. *J Am Chem Soc* **126**: 14428–14434.

200. Taubert S, Straka M, Pennanen TO, *et al.* (2008) Dynamics and magnetic resonance properties of $Sc_3C_2@C_{80}$ and its monoanion. *Phys Chem Chem Phys* **10**: 7158–7168.
201. Andreoni W, Curioni A. (1996) Freedom and constraints of a metal atom encapsulated in fullerene cages. *Phys Rev Lett* **77**: 834–837.
202. Heine T, Vietze K, Seifert G. (2004) ^{13}C NMR fingerprint characterizes long time-scale structure of $Sc_3N@C_{80}$ endohedral fullerene. *Magn Reson Chem* **42**: S199–S201.
203. Popov AA (2012). Internal Dynamics in Endohedral Metallofullerenes as Studied by *ab initio* Born-Oppenheimer Molecular Dynamics. Binder K, Münster G, Kremer M (eds). *NIC Symposium 2012: 25 Years HLRZ/NIC,* Forschungszentrum Jülich GmbH: 123–130.
204. Krause M, Hulman M, Kuzmany H, *et al.* (2004) Fullerene quantum gyroscope. *Phys Rev Lett* **93**: 137403.
205. Vietze K, Seifert G. (2003) Intra-cage Dynamics in Endohedral Fullerenes. *AIP Conference Proceedings* **685**: 7–10.
206. Vietze K, Seifert G. (2002) Fluxional behaviour of Sc_3N in endohedral $Sc_3N@C_{80}$. *AIP Conference Proceedings* **633**: 39–42.
207. Vietze K, Seifert G, Fowler PW. (2000) Structure and dynamics of endohedral fullerenes. *AIP Conference Proceedings* **544**: 131–134.
208. Kodama T, Fujii R, Miyake Y, *et al.* (2004) ^{13}C NMR study of $Ca@C_{74}$: the cage structure and the site-hopping motion of a Ca atom inside the cage. *Chem Phys Lett* **399**: 94–97.
209. Xu JX, Tsuchiya T, Hao C, *et al.* (2006) Structure determination of a missing-caged metallofullerene: $Yb@C_{74}$ (II) and the dynamic motion of the encaged ytterbium ion. *Chem Phys Lett* **419**: 44–47.
210. Popov AA, Shustova NB, Svitova AL, *et al.* (2010) Redox-Tuning Endohedral Fullerene Spin States: From the Dication to the Trianion Radical of $Sc_3N@C_{80}(CF_3)_2$ in Five Reversible Single-Electron Steps. *Chem-Eur J* **16**: 4721–4724.
211. Chen C, Liu F, Li S, *et al.* (2012) Titanium/Yttrium Mixed Metal Nitride Clusterfullerene $TiY_2N@C_{80}$: Synthesis, Isolation, and Effect of the Group-III Metal. *Inorg Chem* **51**: 3039–3045.
212. Hino S, Zenki M, Zaima T, *et al.* (2012) Photoelectron Spectroscopy of $Sc_3N@C_{78}$. *J Phys Chem C* **116**: 165–170.
213. Hino S, Kato M, Yoshimura D, *et al.* (2007) Effect of encapsulated atoms on the electronic structure of the fullerene cage: A case study on $La_2@C_{78}$ and $Ti_2C_2@C_{78}$ via ultraviolet photoelectron spectroscopy. *Phys Rev B* **75**: 125418.

214. Muthukumar K, Stróżecka A, Mysliveček J, *et al.* (2013) Endohedral Fullerene Ce@C_{82} on Cu(111): Orientation, Electronic Structure, and Electron-Vibration Coupling. *J Phys Chem C* **117**: 1656–1662.
215. Stróżecka A, Muthukumar K, Dybek A, *et al.* (2009) Modification of the conductance of single fullerene molecules by endohedral doping. *Appl Phys Lett* **95**: 133118.
216. Huang T, Zhao J, Feng M, *et al.* (2010) Superatom orbitals of Sc_3N@C_{80} and their intermolecular hybridization on Cu(110)-(2x1)-O surface. *Phys Rev B* **81**: 085434.
217. Iwamoto M, Ogawa D, Yasutake Y, *et al.* (2010) Molecular Orientation of Individual Lu@C_{82} Molecules Demonstrated by Scanning Tunneling Microscopy. *J Phys Chem C* **114**: 14704–14709.
218. Leigh DF, Owen JHG, Lee SM, *et al.* (2005) Distinguishing two isomers of Nd@C-82 by scanning tunneling microscopy and density functional theory. *Chem Phys Lett* **414**: 307–310.
219. Popov AA, Dunsch L. (2011) Electrochemistry in Cavea: Endohedral Redox Reactions of Encaged Species in Fullerenes. *J Phys Chem Lett* **2**: 786–794.
220. Zhang L, Popov AA, Yang S, *et al.* (2010) An endohedral redox system in a fullerene cage: The Ce based mixed cluster fullerene Lu_2CeN@C_{80}. *Phys Chem Chem Phys* **12**: 7840–7847.
221. Tomasi J, Mennucci B, Cammi R. (2005) Quantum Mechanical Continuum Solvation Models. *Chem Rev* **105**: 2999–3094.
222. Klamt A. (2011) The COSMO and COSMO-RS solvation models. *WIREs Comput Mol Sci* **1**: 699–709.
223. Cossi M, Rega N, Scalmani G, *et al.* (2003) Energies, structures, and electronic properties of molecules in solution with the C-PCM solvation model. *J Comput Chem* **24**: 669–681.
224. Wakahara T, Sakuraba A, Iiduka Y, *et al.* (2004) Chemical reactivity and redox property of Sc_3@C_{82}. *Chem Phys Lett* **398**: 553–556.
225. Tan K, Lu X, Wang CR. (2006) Unprecedented mu^4-C_2^{6-} anion in Sc_4C_2@C_{80}. *J Phys Chem B* **110**: 11098–11102.
226. Chaur MN, Melin F, Elliott B, *et al.* (2007) Gd_3N@C_{2n} (n = 40, 42, and 44): Remarkably Low HOMO-LUMO Gap and Unusual Electrochemical Reversibility of Gd_3N@C_{88}. *J Am Chem Soc* **129**: 14826–14829.
227. Yang T, Zhao X, Nagase S. (2013) Quantum Chemical Insight of the Dimetallic Sulfide Endohedral Fullerene Sc_2S@C_{70}: Does It Possess the Conventional D_{5h} Cage? *Chem-Eur J* **19**: 2649–2654.
228. Rivera-Nazario DM, Pinzón JR, Stevenson S, *et al.* (2013) Buckyball Maracas: Exploring the Inside and Outside Properties of Endohedral Fullerenes. *J Phys Org Chem* **26**: 194–205.

229. Tan K, Lu X. (2006) Electronic structure and redox properties of the open-shell metal-carbide endofullerene $Sc_3C_2@C_{80}$: A density functional theory investigation. *J Phys Chem A* **110**: 1171–1176.
230. Yang SF, Zalibera M, Rapta P, *et al.* (2006) Charge-induced reversible rearrangement of endohedral fullerenes: Electrochemistry of tridysprosium nitride clusterfullerenes $Dy_3N@C_{2n}$ (2n=78, 80). *Chem-Eur J* **12**: 7848–7855.
231. Elliott B, Yu L, Echegoyen L. (2005) A Simple Isomeric Separation of D_{5h} and I_h $Sc_3N@C_{80}$ by Selective Chemical Oxidation. *J Am Chem Soc* **127**: 10885–10888.
232. Rapta P, Popov AA, Yang SF, *et al.* (2008) The Charged States of $Sc_3N@C_{68}$: An in situ Spectroelectrochemical Study of the Radical Cation and Radical Anion of a Non-IPR Fullerene. *J Phys Chem A* **112**: 5858–5865.
233. Popov AA, Avdoshenko SM, Cuniberti G, *et al.* (2011) Dimerization of Radical-Anions: Nitride Clusterfullerenes versus Empty Fullerenes. *J Phys Chem Lett* **2**: 1592–1600.
234. Yamada M, Slanina Z, Mizorogi N, *et al.* (2013) Application of MCD spectroscopy and TD-DFT to endohedral metallofullerenes for characterization of their electronic transitions. *Phys Chem Chem Phys* **15**: 3593–3601.
235. Rappoport D, Furche F. (2009) Structure of endohedral fullerene $Eu@C_{74}$. *Phys Chem Chem Phys* **11**: 6353–6358.
236. Jin P, Mi W, Hao C, *et al.* (2009) Relativistic Density Functional Theory Study on the ^{139}La Chemical Shifts and Electronic Spectra of La_2C_{72}. *J Comput Theor Nanosci* **6**: 434–438.
237. Yang SF, Popov AA, Dunsch L. (2008) Large mixed metal nitride clusters encapsulated in a small cage: the confinement of the C_{68}-based clusterfullerenes. *Chem Commun*: 2885–2887.
238. Hu H, Cheng WD, Huang SP, *et al.* (2008) Size effect of encased atom on absorption and nonlinear optical properties of embedded metallofullerenes $M@C_{82}$ (M = Sc, Y, La). *J Theor Comput Chem* **7**: 737–749.
239. Wang J-Y, Lin C-S, Zhang M-Y, *et al.* (2012) Effect of cage size on the third-order optical properties of endohedral metallofullerenes $Sc_3N@C_{2n}$ ($2n = 68$, 70, 78, and 80): A theoretical study. *Int J Quantum Chem* **112**: 759–769.
240. Cheng W-D, Hu H, Wu D-S, *et al.* (2009) Effect of Cage Charges on Multiphoton Absorptions: First-Principles Study on Metallofullerenes $Sc_2C_2@C_{68}$ and $Sc_3N@C_{68}$. *J Phys Chem A* **113**: 5966–5971.
241. Krause M, Kuzmany H, Georgi P, *et al.* (2001) Structure and stability of endohedral fullerene $Sc_3N@C_{80}$: A Raman, infrared, and theoretical analysis. *J Chem Phys* **115**: 6596–6605.
242. Yang SF, Kalbac M, Popov A, *et al.* (2006) Gadolinium-based mixed metal nitride clusterfullerenes $Gd_xSc_{3-x}N@C_{80}$ (x = 1, 2). *ChemPhysChem* **7**: 1990–1995.

243. Yang SF, Popov AA, Kalbac M, *et al.* (2008) The Isomers of Gadolinium Scandium Nitride Clusterfullerenes $Gd_xSc_{3-x}N@C_{80}$ (x = 1, 2) and Their Influence on Cluster Structure. *Chem-Eur J* **14**: 2084–2092.
244. Lebedkin S, Renker B, Heid R, *et al.* (1998) A spectroscopic study of $M@C_{82}$ metallofullerenes: Raman, far-infrared, and neutron scattering results. *Appl Phys A-Mater Sci Process* **66**: 273–280.
245. Krause M, Liu XJ, Wong J, *et al.* (2005) The electronic and vibrational structure of endohedral $Tm_3N@C_{80}$(I) fullerene — Proof of an encaged Tm^{3+}. *J Phys Chem A* **109**: 7088–7093.
246. Yang S, Popov AA, Dunsch L. (2008) Carbon Pyramidalization in Fullerene Cages Induced by the Endohedral Cluster: Non-Scandium Mixed Metal Nitride Clusterfullerenes. *Angew Chem-Int Edit Engl* **47**: 8196–8200.
247. Olmstead MM, de Bettencourt-Dias A, Duchamp JC, *et al.* (2000) Isolation and crystallographic characterization of $ErSc_2N@C_{80}$: an endohedral fullerene which crystallizes with remarkable internal order. *J Am Chem Soc* **122**: 12220–12226.
248. Wang XL, Zuo TM, Olmstead MM, *et al.* (2006) Preparation and structure of $CeSc_2N@C_{80}$: An icosahedral carbon cage enclosing an acentric $CeSc_2N$ unit with buried f electron spin. *J Am Chem Soc* **128**: 8884–8889.
249. Stevenson S, Chancellor C, Lee HM, *et al.* (2008) Internal and External Factors in the Structural Organization in Cocrystals of the Mixed-Metal Endohedrals ($GdSc_2N@I_h$-C_{80}, $Gd_2ScN@I_h$-C_{80}, and $TbSc_2N@I_h$-C_{80}) and Nickel(II) Octaethylporphyrin. *Inorg Chem* **47**: 1420–1427.
250. Kemner E, Zerbetto F. (2005) Guest dynamics in endohedrally doped fullerenes. *J Phys Chem B* **109**: 15048–15051.
251. Vietze K, Seifert G, Richter M, *et al.* (1999) Endohedral rare earth fullerenes — electronic and dynamic properties. *AIP Conference Proceedings* **486**: 128–131.
252. Gerson F, Huber W (2003). *Electron Spin Resonance Spectroscopy of Organic Radicals*, WILEY-VCH Verlag GmbH & Co. KGaA.
253. Kaupp M, Buhl M, Malkin VG (eds). (2004) *Calculation of NMR and EPR Parameters. Theory and Applications*, WILEY-VCH Verlag GmbH & Co. KGaA.
254. Seifert G, Bartl A, Dunsch L, *et al.* (1998) Electron spin resonance spectra: geometrical and electronic structure of endohedral fullerenes. *Appl Phys A-Mater Sci Process* **66**: 265–271.
255. Morley GW, Herbert BJ, Lee SM, *et al.* (2005) Hyperfine structure of $Sc@C_{82}$ from ESR and DFT. *Nanotechnology* **16**: 2469–2473.
256. Misochko EY, Akimov AV, Belov VA, *et al.* (2010) EPR spectrum of the $Y@C_{82}$ metallofullerene isolated in solid argon matrix: hyperfine structure from EPR spectroscopy and relativistic DFT calculations. *Phys Chem Chem Phys* **12**: 8863–8869.

257. Jakes P, Dinse KP. (2001) Chemically induced spin transfer to an encased molecular cluster: an EPR study of $Sc_3N@C_{80}$ radical anions. *J Am Chem Soc* **123**: 8854–8855.
258. Feng Y, Wang T, Wu J-Y, *et al.* (2013) Spin-Active Metallofullerene Stabilized by the Core of NC Moiety. *Chem Commun* **49**: 2148–2150.
259. Wang T-S, Feng L, Wu J-Y, *et al.* (2010) Planar Quinary Cluster inside a Fullerene Cage: Synthesis and Structural Characterizations of $Sc_3NC@C_{80}$-I_h. *J Am Chem Soc* **132**: 16362–16364.
260. Suzuki M, Slanina Z, Mizorogi N, *et al.* (2012) Single-Crystal X-ray Diffraction Study of Three $Yb@C_{82}$ Isomers Cocrystallized with Ni^{II}(octaethylporphyrin). *J Am Chem Soc* **134**: 18772–18778.
261. Lu X, Slanina Z, Akasaka T, *et al.* (2010) $Yb@C_{2n}$ (n = 40, 41, 42): New Fullerene Allotropes with Unexplored Electrochemical Properties. *J Am Chem Soc* **132**: 5896–5905.
262. Fu W, Xu L, Azurmendi H, *et al.* (2009) ^{89}Y and ^{13}C NMR Cluster and Carbon Cage Studies of an Yttrium Metallofullerene Family, $Y_3N@C_{2n}$ (n = 40–43). *J Am Chem Soc* **131**: 11762–11769.
263. Duchamp JC, Demortier A, Fletcher KR, *et al.* (2003) An isomer of the endohedral metallofullerene $Sc3N@C_{80}$ with D_{5h} symmetry. *Chem Phys Lett* **375**: 655–659.
264. Hao C, Li H, Jia G, *et al.* (2012) Electron Transfer Makes D_{3h} (78:5) Cage Easy To Form $M_2@C_{78}$ (M = La, Ce): A Relativistic Density-Functional Theory Study. *J Theor Comput Chem* **11**: 197–207.
265. Tan K, Lu X. (2005) Ti_2C_{80} is more likely a titanium carbide endohedral metallofullerene $(Ti_2C_2)@C_{78}$. *Chem Commun*: 4444–4446.
266. Reveles JU, Heine T, Koster AM. (2005) ^{13}C NMR pattern of $Sc_3N@C_{68}$. Structural assignment of the first fullerene with adjacent pentagons. *J Phys Chem A* **109**: 7068–7072.
267. Cao BP, Hasegawa M, Okada K, *et al.* (2001) EELS and ^{13}C NMR characterization of pure $Ti_2@C_{80}$ metallofullerene. *J Am Chem Soc* **123**: 9679–9680.
268. Zhang J, Fuhrer T, Fu W, *et al.* (2012) Nanoscale Fullerene Compression of a Yttrium Carbide Cluster. *J Am Chem Soc* **134**: 8487–8493.
269. Yamazaki Y, Nakajima K, Wakahara T, *et al.* (2008) Observation of ^{13}C NMR Chemical Shifts of Metal Carbides Encapsulated in Fullerenes: $Sc_2C_2@C_{82}$, $Sc_2C_2@C_{84}$, and $Sc_3C_2@C_{80}$. *Angew Chem-Int Edit Engl* **47**: 7905 –7908.
270. Haddon RC, Scott LT. (1986) Pi-Orbital Conjugation and Rehybridization in Bridged Annulenes and Deformed Molecules in General — Pi-Orbital Axis Vector Analysis. *Pure Appl Chem* **58**: 137–142.
271. Yamada M, Wakahara T, Tsuchiya T, *et al.* (2008) Location of the metal atoms in $Ce_2@C_{78}$ and its bis-silylated derivative. *Chem Commun*: 558–560.

272. Yamada M, Wakahara T, Lian Y, *et al.* (2006) Analysis of Lanthanide-Induced NMR Shifts of the Ce@C_{82} Anion. *J Am Chem Soc* **128**: 1400–1401.
273. Yamada M, Mizorogi N, Tsuchiya T, *et al.* (2009) Synthesis and Characterization of the D_{5h} Isomer of the Endohedral Dimetallofullerene Ce_2@C_{80}: Two-Dimensional Circulation of Encapsulated Metal Atoms Inside a Fullerene Cage. *Chem-Eur J* **15**: 9486–9493.
274. Bleaney B. (1972) Nuclear Magnetic Resonance Shifts in Solution Due to Lanthanide Ions. *J Magn Reson* **8**: 91–100.
275. Popov AA, Schiemenz S, Avdoshenko SM, *et al.* (2011) The State of Asymmetric Nitride Clusters in Endohedral Fullerenes as Studied by ^{14}N NMR Spectroscopy: Experiment and Theory. *J Phys Chem C* **115**: 15257–15265.
276. Lu X, Akasaka T, Nagase S. (2011) Chemistry of endohedral metallofullerenes: the role of metals. *Chem Commun* **47**: 5942–5957.
277. Osuna S, Swart M, Solà M. (2011) The reactivity of endohedral fullerenes. What can be learnt from computational studies? *Phys Chem Chem Phys* **13**: 3585–3603.
278. Takano Y, Yomogida A, Nikawa H, *et al.* (2008) Radical Coupling Reaction of Paramagnetic Endohedral Metallofullerene La@C_{82}. *J Am Chem Soc* **130**: 16224–16230.
279. Akasaka T, Lu X, Kuga H, *et al.* (2010) Dichlorophenyl Derivatives of La@C_{3v}(7)-C_{82}: Endohedral Metal Induced Localization of Pyramidalization and Spin on a Triple-Hexagon Junction. *Angew Chem-Int Edit Engl* **49**: 9715–9719.
280. Nikawa H, Kikuchi T, Wakahara T, *et al.* (2005) Missing metallofullerene La@C_{74}. *J Am Chem Soc* **127**: 9684–9685.
281. Lu X, Nikawa H, Kikuchi T, *et al.* (2011) Radical Derivatives of Insoluble La@C_{74}: X-ray Structures, Metal Positions, and Isomerization. *Angew Chem-Int Edit Engl* **50**: 6356–6359.
282. Lu X, Nikawa H, Feng L, *et al.* (2009) Location of the Yttrium Atom in Y@C_{82} and Its Influence on the Reactivity of Cage Carbons. *J Am Chem Soc* **131**: 12066–12067.
283. Maeda Y, Matsunaga Y, Wakahara T, *et al.* (2004) Isolation and characterization of a carbene derivative of La@C_{82}. *J Am Chem Soc* **126**: 6858–6859.
284. Takano Y, Aoyagi M, Yamada M, *et al.* (2009) Anisotropic Magnetic Behavior of Anionic Ce@C_{82} Carbene Adducts. *J Am Chem Soc* **131**: 9340–9346.
285. Akasaka T, Kono T, Takematsu Y, *et al.* (2008) Does Gd@C_{82} Have an Anomalous Endohedral Structure? Synthesis and Single Crystal X-ray Structure of the Carbene Adduct. *J Am Chem Soc* **130**: 12840–12841.

286. Hachiya M, Nikawa H, Mizorogi N, *et al.* (2012) Exceptional Chemical Properties of Sc@C_{2v}(9)-C_{82} Probed with Adamantylidene Carbene. *J Am Chem Soc* **134**: 15550–15555.
287. Feng L, Wakahara T, Nakahodo T, *et al.* (2006) The bingel monoadducts of La@C_{82}: Synthesis, characterization, and electrochemistry. *Chem-Eur J* **12**: 5578–5586.
288. Feng L, Nakahodo T, Wakahara T, *et al.* (2005) A singly bonded derivative of endohedral metallofullerene: La@C_{82}CBr($COOC_2H_5$)$_2$. *J Am Chem Soc* **127**: 17136–17137.
289. Maeda Y, Sato S, Inada K, *et al.* (2010) Regioselective Exohedral Functionalization of La@C_{82} and its 1,2,3,4,5-Pentamethylcyclopentadiene and Adamantylidene Adducts. *Chem-Eur J* **16**: 2193–2197.
290. Maeda Y, Miyashita J, Hasegawa T, *et al.* (2005) Reversible and regioselective reaction of La@C_{82} with cyclopentadiene. *J Am Chem Soc* **127**: 12190–12191.
291. Campanera JM, Bo C, Poblet JM. (2006) Exohedral reactivity of trimetallic nitride template (TNT) endohedral metallofullerenes. *J Org Chem* **71**: 46–54.
292. Li FF, Rodriguez-Fortea A, Poblet JM, *et al.* (2011) Reactivity of metallic nitride endohedral metallofullerene anions: Electrochemical synthesis of a Lu_3N@I_h-C_{80} derivative. *J Am Chem Soc* **133**: 2760–2765.
293. Maeda Y, Miyashita J, Hasegawa T, *et al.* (2005) Chemical reactivities of the cation and anion of M@C_{82} (M = Y, La, and Ce). *J Am Chem Soc* **127**: 2143–2146.
294. Cardona CM, Kitaygorodskiy A, Echegoyen L. (2005) Trimetallic nitride endohedral metallofullerenes: Reactivity dictated by the encapsulated metal cluster. *J Am Chem Soc* **127**: 10448–10453.
295. Cai T, Xu LS, Anderson MR, *et al.* (2006) Structure and enhanced reactivity rates of the D_{5h} Sc_3N@C_{80} and Lu_3N@C_{80} metallofullerene isomers: The importance of the pyracylene motif. *J Am Chem Soc* **128**: 8581–8589.
296. Osuna S, Rodriguez-Fortea A, Poblet JM, *et al.* (2012) Product formation in the Prato reaction on Sc_3N@D_{5h}-C_{80}: preference for [5,6]-bonds, and not pyracylenic bonds. *Chem Commun* **48**: 2486–2488.
297. Osuna S, Swart M, Campanera JM, *et al.* (2008) Chemical Reactivity of D_{3h}-C_{78} (Metallo)Fullerene: regioselectivity Changes Induced by Sc_3N Encapsulation. *J Am Chem Soc* **130**: 6206–6214.
298. Cai T, Xu L, Gibson HW, *et al.* (2007) Sc_3N@C_{78}: Encapsulated Cluster Regiocontrol of Adduct Docking on an Ellipsoidal Metallofullerene Sphere. *J Am Chem Soc* **129**: 10795–10800.

299. Osuna S, Swart M, Sola M. (2009) The Diels–Alder Reaction on Endohedral Y3N@C78: The Importance of the Fullerene Strain Energy. *J Am Chem Soc* **131**: 129–139.
300. Garcia-Borràs M, Osuna S, Luis JM, *et al.* (2012) The Exohedral Diels–Alder Reactivity of the Titanium Carbide Endohedral Metallofullerene $Ti_2C_2@D_{3h}$-C_{78}: Comparison with D_{3h}-C_{78} and M3N@D_{3h}-C_{78} (M=Sc and Y) Reactivity. *Chem-Eur J* **18**: 7141–7154.
301. Osuna S, Valencia R, Rodríguez-Fortea A, *et al.* (2012) Full Exploration of the Diels–Alder Cycloaddition on Metallofullerenes $M_3N@C_{80}$ (M=Sc, Lu, Gd): The D_{5h} versus Ih Isomer and the Influence of the Metal Cluster. *Chem-Eur J* **18**: 8944–8956.
302. Garcia-Borràs M, Luis JM, Swart M, *et al.* (2013) Diels–Alder and Retro-Diels–Alder Cycloadditions of (1,2,3,4,5-Pentamethyl)cyclopentadiene to La@C_{2v}-C_{82}: Regioselectivity and Product Stability. *Chem-Eur J* **19**: 4468–4479.
303. Grimme S. (2006) Semiempirical GGA-type density functional constructed with a long-range dispersion correction. *J Comput Chem* **27**: 1787–1799.
304. Sato S, Maeda Y, Guo J-D, *et al.* (2013) Mechanistic Study of the Diels–Alder Reaction of Paramagnetic Endohedral Metallofullerene: Reaction of La@C_{82} with 1,2,3,4,5–Pentamethylcyclopentadiene. *J Am Chem Soc* **135**: 5582–5587.
305. Garcia-Borràs M, Romero-Rivera A, Osuna S, *et al.* (2012) The Frozen Cage Model: A Computationally Low-Cost Tool for Predicting the Exohedral Regioselectivity of Cycloaddition Reactions Involving Endohedral Metallofullerenes. *J Chem Theory Comput* **8**: 1671–1683.
306. Cai T, Xu L, Shu C, *et al.* (2008) Synthesis and Characterization of a Non-IPR Fullerene Derivative: $Sc_3N@C_{68}[C(COOC_2H_5)_2]$. *J Phys Chem C* **112**: 19203–19208.
307. Alegret N, Chaur MN, Santos E, *et al.* (2010) Bingel-Hirsch Reactions on Non-IPR $Gd_3N@C_{2n}$ ($2n = 82$ and 84). *J Org Chem* **75**: 8299–8302.
308. Núria Alegret, Rodríguez-Fortea A, Poblet JM. (2013) Bingel–Hirsch Addition on Endohedral Metallofullerenes: Kinetic Versus Thermodynamic Control. *Chem-Eur J* **19**: 5061–5069.
309. Pinzon JR, Zuo TM, Echegoyen L. (2010) Synthesis and Electrochemical Studies of Bingel-Hirsch Derivatives of $M_3N@I_h$-C_{80} (M = Sc, Lu). *Chem-Eur J* **16**: 4864–4869.
310. Hirsch A, Brettreich M (2005). *Fullerenes. Chemistry and Reactions.*, Wiley-VCH Verlag GmbH & Co. KGaA.

311. Yang S, Chen C, Lanskikh MA, *et al.* (2011) New Isomers of Trifluoromethylated Derivatives of Metal Nitride Cluster Fullerene: $Sc_3N@C_{80}(CF_3)_n$ (n = 14 and 16). *Chem-Asian J* **6**: 505–509.
312. Yang S, Chen C, Jiao M, *et al.* (2011) Synthesis, Isolation, and Addition Patterns of Trifluoromethylated D_{5h} and I_h Isomers of $Sc_3N@C_{80}$: $Sc_3N@D_{5h}$-$C_{80}(CF_3)_{18}$ and $Sc_3N@I_h$-$C_{80}(CF_3)_{14}$. *Inorg Chem* **50**: 3766–3771.
313. Shustova NB, Popov AA, Mackey MA, *et al.* (2007) Radical Trifluoromethylation of $Sc_3N@C_{80}$. *J Am Chem Soc* **129**: 11676–11677.
314. Kareev IE, Lebedkin SF, Bubnov VP, *et al.* (2005) Trifluoromethylated endohedral metallofullerenes: Synthesis and characterization of $Y@C_{82}(CF_3)_5$. *Angew Chem-Int Edit* **44**: 1846–1849.

CHAPTER 13

Biomedical Applications of Trimetallic Nitride Endohedral Metallofullerenes

Jianyuan Zhang,[b,c] *Boris M. Kiselev,*[a] *Youqing Ye,*[b,c] *and Harry C. Dorn*[b,c,*]

It is now well established that nanoparticles offer promising new opportunities for biomedical applications. Nanoparticles represent a new toolset with a substantial multi-functional capability. It is well recognized that endohedral metallofullerenes (EMFs) are especially attractive nanoparticles because of their shape, capacity for different encapsulants and excellent isolation from the bio-environment, representing an ideal nanoplatform on which to engineer next-generation diagnostic and therapeutic biomedical agents. As illustrated in Fig. 1, the idealized nanoparticle can have multifunctional capability with signaling, targeting and therapeutic components. This is an important consideration allowing the same biodistribution for each component. In addition, nanoparticles can easily undergo

[a] Virginia Tech Carilion School of Medicine, Roanoke, VA, 24016, United States.
[b] Virginia Tech Carilion Research Institute, Roanoke, VA, 24016, United States.
[c] Department of Chemistry, Virginia Polytechnic Institute and State University, Blacksburg, VA, 24061, United States.
*E-mail: hdorn@vt.edu

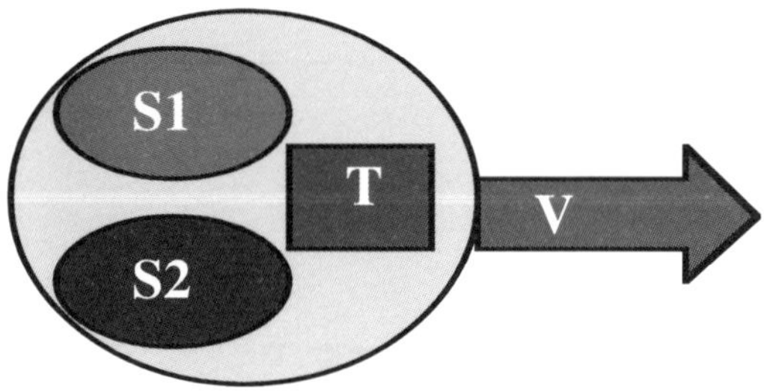

Fig. 1 Conceptual Multimodality Nanoparticle: S1, S2 = signaling or diagnostic component (MRI, fluorescence, radiolabel) ; V = vector component for targeted delivery (antibody, targeting protein); T = therapeutic unit (anti-tumor drug, radiotherapeutic) . In this conceptual approach S1, S2 and T are *endo* components, but they could be *exo*.

extravasation from the blood pool into tumor tissues and be retained because of poor lymphatic drainage.[1] This phenomenon of selective accumulation of nano-sized particles near tumor tissues is termed the enhanced permeability and retention (or EPR) effect, and leads to the accumulation of nanoparticles near tumor tissues. A special class of EMFs are the trimetallic nitride template EMFs (TNT EMFs, $A_{3-x}B_xN@C_{2y}$, x = 0–3, A,B = metal, y = 34, 39–44), which were first reported in 1999 and are one of the most promising classes of nanoparticle EMFs.[2, 3] The first member of the TNT EMF family, $Sc_3N@C_{80}$, was determined to be $Sc_3N@I_h$-C_{80}, with stabilization of the icosahedral I_h cage by transfer of six electrons to the fullerene cage $(C_{80})^{6-}$ (Fig. 2).[4] However, it was later discovered by ^{13}C NMR[5] that a second isomer, $Sc_3N@D_{5h}$-C_{80}, was present in the Krätschmer–Huffman reaction mixture. This isomer was later isolated and structurally characterized using single crystal X-ray diffraction analysis by Balch and coworkers.[6] As illustrated in Fig. 2, with the smaller scandium ions present in the trimetallic cluster $(Sc_3N)^{6+}$, smaller cage structures are possible, and $Sc_3N@D_3$-C_{68} was one of the first examples that violates the isolated pentagon rule (IPR) for fullerenes and metallofullerenes.[7] This non-IPR structure, $Sc_3N@D_3$-C_{68}, was first identified by ^{13}C NMR and, later, structural characterization by single crystal X-ray diffraction analysis.[7, 8] A second cage smaller than 80 carbons, $Sc_3N@D_{3h}$-C_{78}, was also characterized by both ^{13}C NMR and single crystal X-ray studies.[9] Larger lanthanide trimetallic cluster $(A_3N)^{6+}$ TNT EMFs have been produced and characterized in the laboratories of Balch, Dunsch, Echegoyen and others.[3]

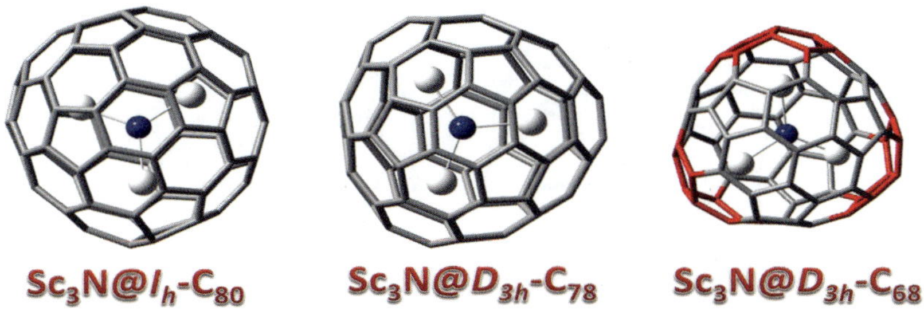

Fig. 2 Representative members of the scandium trimetallic nitride template endohedral metallofullerene (TNT EMF) family. The three non-IPR pentalene motifs are shown in red for $Sc_3N@C_{68}$.

There are several advantages of the TNT EMFs for biomedical applications. First, TNT EMFs can be produced in relatively large quantities — $Sc_3N@I_h$-C_{80} is the third most abundant fullerene, with only C_{60} and C_{70} produced in greater quantity. Second, an important trait of TNT EMFs is their high thermal and chemical reactivity stability. For biomedical applications, the focus of this report is on the highly stable lutetium and gadolinium lanthanide EMFs, $Gd_3N@I_h$-C_{80} and $Lu_3N@I_h$-C_{80} (Fig. 3). For the $Gd_3N@I_h$-C_{80}, the Gd_3N cluster is bound to the I_h-C_{80} cage by a strong binding energy of 13.63 eV, and is extremely unlikely to break and release the encapsulated Gd atoms.[10] The I_h-C_{80} fullerene cage has inherent advantages because of its high stability and characteristic resistance to any potential biological metabolic cage-opening processes, preventing the release of toxic metal ions from the TNT EMF into surrounding tissue, serum, and other biological components. In comparison with empty-cage fullerenes C_{60}, C_{70}, and C_{84}, the higher stability of the TNT EMFs has enabled selective chemical reactions and separation schemes based on the corresponding lower chemical reactivity of the $M_3N@C_{2n}$ families.[11–13] The TNT EMFs still exhibit sufficient chemical reactivity for a variety of different chemical reactions including Diels–Alder,[14, 15] Bingel,[16, 17] Prato,[18–20] and free radical reactions.[21] Third, the ability of the TNT EMF's internal A_3N cluster allows encapsulation of a wide range of lanthanide and Group III elements, as well as different elements in the same cluster (e.g. AB_2N), allowing additional flexibility for multi-modal diagnostic and therapeutic applications *vide infra.*

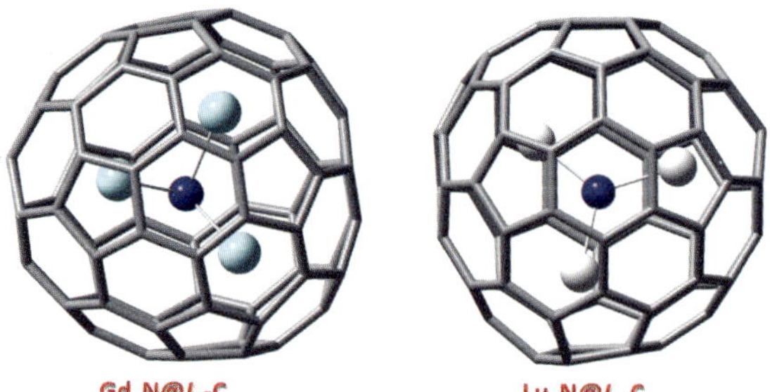

Fig. 3 Representative members of the scandium trimetallic nitride template endohedral metallofullerene (TNT EMF) family. The three non-IPR pentalene motifs are shown in red for $Sc_3N@C_{68}$.

1. TNT EMF DIAGNOSTIC IMAGING

TNT EMFs, like other fullerenes and metallofullerenes, have a hydrophobic sp^2 hybridized carbon cage surface that must be converted to a hydrophilic surface for most biomedical applications. The first water-soluble TNT EMF derivative was made by Iezzi *et al.* shortly after the discovery of TNT EMFs.[22] In this early preparation, a toluene solution of $Sc_3N@C_{80}$ was refluxed with sodium metal in argon to yield black precipitate, which converted to a water-soluble metallofullerene derivative by addition of water and exposure to air. This method introduces both hydroxyl groups and highly oxygenated carbon (carbonyl or hemi-ketal groups) to the cage, with an average formula of $Sc_3N@C_{80}O_{10}(OH)_{10}$. In a revised synthetic approach, the $Gd_3N@C_{80}$ TNT EMF was first functionalized with poly(ethylene glycol) and the carbon cage was subsequently hydroxylated in a second reaction step (Fig. 4).[23, 24] The ^{1}H magnetic resonance imaging (MRI) relaxivities (r_1) for $Gd_3N@C_{80}$[DiPEG5000(OH)$_x$] were approximately 40 times greater than conventional gadolinium-containing MRI contrast agents (e.g. gadodiamide, Omniscan™).[23, 24] This increased relaxivity allows for the use of significantly lower concentrations of $Gd_3N@C_{80}$ (10–100-fold) in this new nanoparticle MR contrast agent. The *in vivo* MRI results illustrated in Fig. 5 demonstrate different distribution characteristics of a conventional commercial contrast agent (gadodiamide, 0.5 mmol/L, left hemisphere) and $Gd_3N@C_{80}$[DiPEG5000(OH)$_x$] (0.013 mmol/L, right hemisphere) infused into a normal rat brain by means of a concurrent

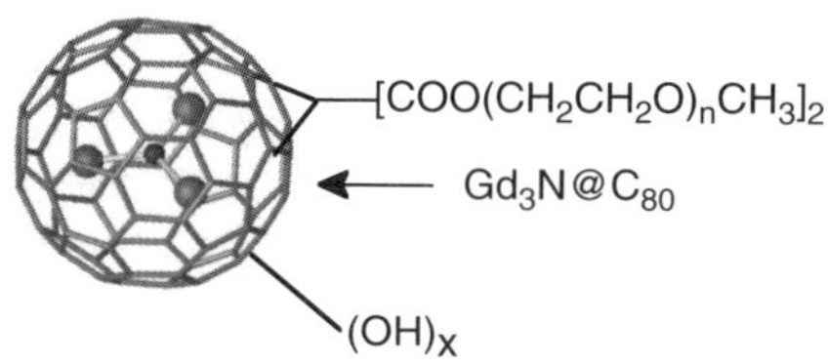

Fig. 4 $Gd_3N@C_{80}[DiPEG5000(OH)_x]$, a pegylated and hydroxylated $Gd_3N@C_{80}$ MR imaging nanoparticle. The polyethylene glycol (PEG) chain has an average molecular weight of 5000.

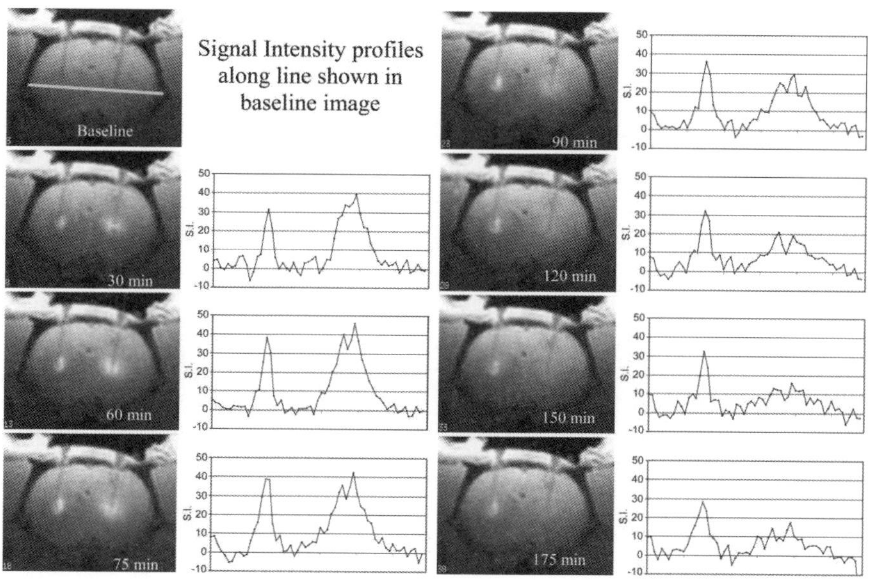

Fig. 5 The T1-weighted 2.4 T MR images (700/10) of bilateral infusion into a normal rat brain of 0.0131 mmol/L pegylated and hydroxylated $Gd_3N@C_{80}$ (see Fig. 4) and 0.5 mmol/L gadodiamide at a 0.2• L/min infusion rate, left and right side of image, respectively. Displayed times are in minutes from the start of bilateral infusion. Positive pressure infusion was performed for 75 minutes, with an additional 100 minutes follow-up for diffusion. The line across the baseline image indicates the location of the displayed profiles. The profiles represent plots of image signal intensity (S.I.) against distance.

bilateral infusion, shown in successive T1-weighted MR images. As both the visual inspection (with over 40 times difference in concentration) and signal intensity profiles through the infusion sites demonstrate, the gadodiamide completely disappeared within three hours from the start of the infusion, compared with the $Gd_3N@C_{80}[DiPEG5000(OH)_x]$, which diffuses very slowly and, for the normal brain, appears to remain in the vicinity of the infusion catheter tip. A more facile method to functionalize EMFs in a shorter time period has been reported by Shu and coworkers.[25] In this elaboration, $Gd_3N@C_{80}$ reacts with a succinyl peroxide to produce a highly water-soluble derivative, $Gd_3N@C_{80}(OH)_{\sim 26}(CH2CH2COOM)_{\sim 16}$ (M = H, Na), in high yield (Fig. 6). This free radical method allows for the chemical attachment of a variety of functional groups to the carbon cage through covalent carbon–carbon bonds without destroying the integrity of the cage. In addition, carbon-centered radicals generated from acyl peroxides have terminal functional groups, –COOH, that provide sites for further elaboration by reaction with other compounds. The method was extended for magnetic resonance imaging (MRI) contrast agent development by Wang and coworkers on Sc-Gd mixed TNT EMFs, $Sc_2GdN@C_{80}$ and $ScGd_2N@C_{80}$ [26] and higher derivatives, $Sc_2GdN@C_{80}O_{12}(OH)_{26}$ and $ScGd_2N@C_{80}O_{12}(OH)_{26}$. In a recent study on a different carbonaceous nanomaterial, Dorn and coworkers have reported successful synthesis of single-wall nanohorn (SWNH)-peapod structures (insertion of TNT EMFs $Gd_3N@C_{80}$, $Sc_3N@C_{80}$, and $Lu_3N@C_{80}$ into SWNHs) that were subsequently exohedrally functionalized utilizing a peroxide and high-speed

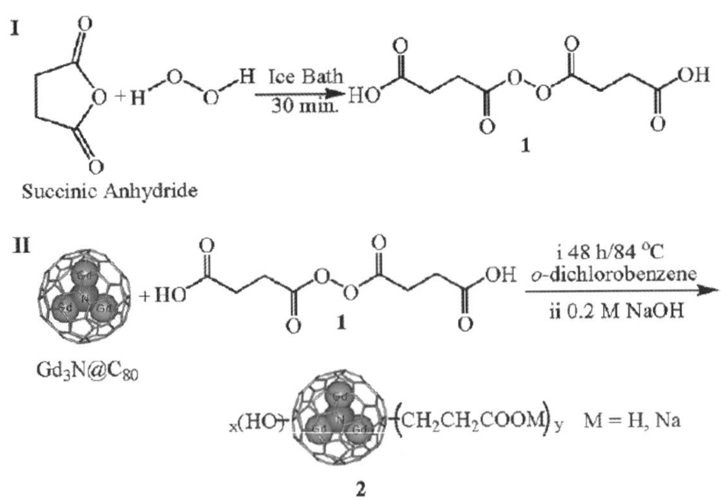

Fig. 6 Reaction scheme for facile preparation MR imaging nanoparticle $Gd_3N@C_{80}(OH)_{\sim 26}(CH2CH2COOM)_{\sim 16}$ (M = H, Na).

vibration treatment.[27] These nanoparticles were also externally functionalized by attaching quantum dots, and additional fluorescent diagnostic and *in vivo* studies of this platform have been reported. In addition, we have obtained data for synthesis of SWNHs that can be assembled as supported liposome nanoparticles.[28] In 2010, Zhang *et al.* employed a similar approach to attach monodisperse PEG chains with different molecular weight[24] to study the influence of chain length on TNT EMF derivatives' ^{1}H relaxivity. Braun and coworkers have reported the synthesis and application of a $Gd_xSc_{3-x}N@C_{80}$-bioshuttle-conjugate referred which features very high ^{1}H relaxivity in comparison to the commonly used commercial contrast agents at very low Gd concentrations.[29] This modularly designed contrast agent represents a new tool for improved monitoring and evaluation of interventions at the gene transcription level. Widespread tracking of individual cells and sensory of microenvironments is possible. In another approach, modified oligoethylene glycol groups were used by Macfarland *et al.* at Luna Innovations to make water-soluble $Gd_3N@C_{80}$ derivatives.[30] More recently, the Luna Innovations research team has reported synthesizing an atherosclerotic-targeting contrast agent (ATCA), in which the gadolinium (Gd)-containing EMF $Gd_3N@C_{80}$ was functionalized and formulated into liposomes, with CD36 ligands intercalated into the lipid bilayer.[31] In this diagnostic study, the research group followed the accumulation of plaque in vessel walls that leads to atherosclerosis, where vulnerable plaque can dislodge, enter the blood stream, and result in acute myocardial infarction and stroke. Mice that were given ATCA exhibited time-dependent accumulation of ATCA in atherosclerotic plaque lesions of ApoE – /– mice as determined by MRI. This provides a new, non-invasive MRI strategy to detect atherosclerotic plaque lesions in patients with chronic plaque accumulation.

The design of a nanoparticle TNT EMF with dual-modality imaging agent for combination positron emission tomography (PET) and magnetic resonance imaging (MRI) has been reported.[39] The dual-modal signaling agent is based on an exohedral ^{124}I (PET) radiolabeled on the surface of the TNT EMF, $Gd_3N@C_{80}$, described above with endohedral Gd labeling for MRI. Specifically, the $Gd_3N@C_{80}$ metallofullerenes were surface functionalized with carboxyl and hydroxyl groups (f-$Gd_3N@C_{80}$) using previously developed procedures, and subsequently iodinated with ^{124}I to produce

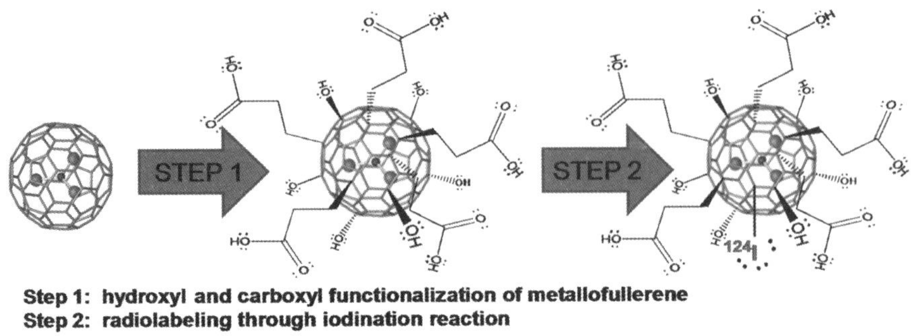

Fig. 7 Illustration of the reaction scheme starting with the $Gd_3N@C_{80}$ and resulting in the aqueous soluble $Gd_3N@C_{80}(OH)_{\sim 26}(CH2CH2COOM)_{\sim 16}$ and final iodination to provide the dual-modality (MR/PET) imaging probe.

^{124}I-f-$Gd_3N@C_{80}$ (Fig. 7). Orthotopic tumor-bearing rats were infused intratumorally by convection-enhanced delivery (CED) with the ^{124}I-f-$Gd_3N@C_{80}$ agent and imaged by MRI or micro PET (Fig. 8). As shown, the anatomical positioning and distribution of the ^{124}I-f-$Gd_3N@C_{80}$ agent were comparable between the MRI and PET scans. The ^{124}I-f-$Gd_3N@C_{80}$ dual-agent distribution and infusion site within the tumor is clearly evident in both T_1- and T_2-weighted MR images. These results demonstrate the successful preparation of a dual-modality imaging agent, ^{124}I-f-$Gd_3N@C_{80}$, designed for simultaneous PET/MR imaging. It should be noted that Siemens has recently introduced a dual PET/MRI instrument for clinical diagnostic applications. When compared with computed tomography (CT) approaches, the PET/MRI dual modality has a distinct clinical advantage because a lower level of ionizing radiation is utilized. In addition, the PET/MRI has the added advantage of combining biochemical soft tissue information (MRI) while the PET approach provides higher sensitivity with specific targeting dependent on the spatial location of the radiolabel.

2. TNT EMF THERAPEUTIC

The therapeutic potential of EMFs has been an area of great interest. Numerous studies have demonstrated the free radical scavenging capabilities of fullerenes, to such a degree that some authors have described them as "free radical sponges."[32,33] In addition, the antineoplastic and radical

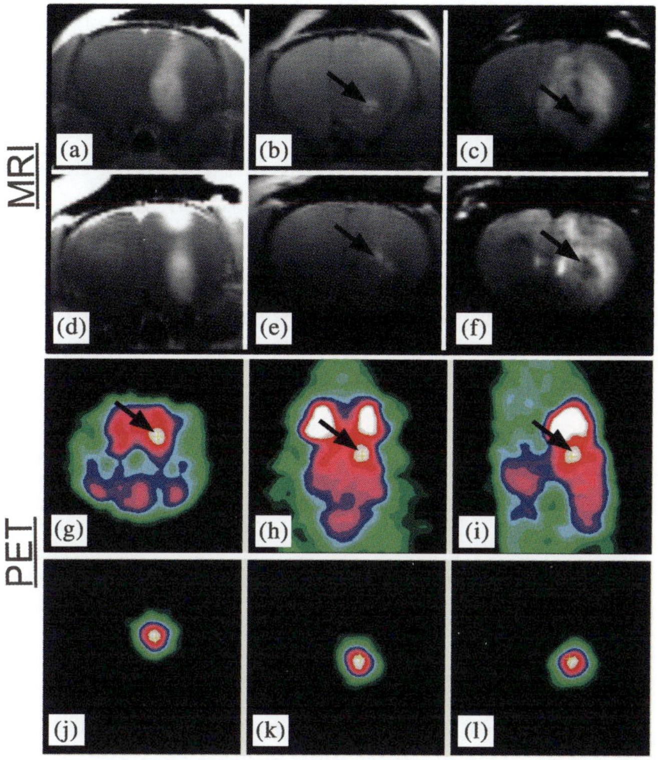

Fig. 8 2.4 T MR images from two representative orthotopic tumor-bearing rat brains and microPET images from one representative rat infused with the ^{124}I-f-$Gd_3N@C_{80}$ product. (A, D) T_1-weighted images with gadodiamide contrast. (B, E) T_1-weighted images showing bright contrast at the infusion site due to the ^{124}I-f-$Gd_3N@C_{80}$. (C, F) T_2-weighted images with dark contrast due to the ^{124}I-f-$Gd_3N@C_{80}$. (J–L) microPET images showing signal from the ^{124}I-f-$Gd_3N@C_{80}$. (G) Coronal, (H) axial, and (I) sagittal micro-PET images following ^{18}F-FDG injection with the additive image signal allowing for localization of the ^{124}I-f-$Gd_3N@C_{80}$ within the right hemisphere of the rat brain (arrows point to infusion sites).

scavenging properties for the metallofullerene, $Gd@C_{82}$ platform have been reported.[40] These desirable anti-free radical effects also hold true for TNT EMFs. In a 2010 study by Ehrich *et al.*, the cytoprotective effect of TNT EMFs was reported to be equal to or even greater than that of empty fullerenes.[34] In this study, a xanthine/xanthine oxidase system was used to generate free radicals. $Gd_3N@C_{80}$ was found to have significant free radical scavenging ability, at 1 μM decreasing free radical activity to less than 50%

of control, and at 10 μM decreasing free radical activity to less than 20% of control. Another TNT EMF, $Sc_3N@C_{80}$-OH, was found to have radical scavenging capability at a concentration as low as 1 μM. At concentrations of 1 μM, $Gd_3N@C_{80}$-OH and $SC_3N@C_{80}$-OH were also found to counteract the anti-Acetylcholinesterase (anti-AChE) activity of the organophosphates paraoxon and diisopropylphosphorofluoridate (DFP) in both human neuroblastoma cells and hen brain homogenates. $Gd_3N@C_{80}$-OH was found to counteract anti-AChE effects at concentrations as low as 0.001 nM. Corresponding cytotoxicity studies found that $Gd_3N@C_{80}$-OH had no effect on primary mouse neuronal cells or human neuroblastoma cells at a concentration of 10 μM using three different assays of cell viability. 100 μM concentrations of the $Gd_3N@C_{80}$-OH were found to decrease ATP production in both cell lines by as much as 40% when measured by cellular luminescence, but this concentration was far greater than that necessary to achieve a therapeutic effect.

The most rigorous *in vivo* study to date of the therapeutic potential of TNT EMFs was conducted by Shultz *et al.* in 2010.[35] In this study, researchers examined the potential of functionalized $Gd_3N@C_{80}$ chelated to lutetium-177, a β-emitting radionuclide with a tissue penetration of 0.7 mm and 500 keV maximum energy range, for tumor imaging and brachytherapy. The researchers used a nude athymic mouse with xenografted human U87GBM (glioblastoma multiforme) as a model. After waiting eight days post-tumor implantation, a technique known as convection-enhanced delivery (CED) was used to directly infuse the nanoparticle into the tumor. Researchers then conducted studies that included MRI, biodistribution, survival times, and histopathology.

Using both MRI data and biodistribution studies, the researchers found that approximately one-quarter of the infused metallofullerenes were retained within the tumor after 52 days. In comparison, when not conjugated to a TNT EMF, lanthanide chelates such as ^{177}Lu-DOTA otherwise clear the brain at much higher rate, with 99% clearance after only 24 hours. This long retention time inside the tumor was credited for the increased survival of treated mice. The mean survival time for control mice was 20.7 days after treatment, while the treated mice continued to thrive to day 52 after treatment. This was an increased survival of more than 2.5 times. Post-mortem histological examination of the treated mice revealed findings consistent with radiation-induced damage to the tumor. Additionally, post-mortem biodistribution studies found that the EMF did not accumulate in the lungs, liver, spleen, or kidneys, meaning that although the nanoparticle was highly retained in the tumor and surrounding tissue, once it entered the blood stream it was quickly cleared from the body. Thus, the authors were able to demonstrate a true "theranostic" nanoparticle that could be used to both longitudinally image the tumor via MRI and, simultaneously, deliver brachytherapy that extended the survival time of treated animals by a significant margin. More recently, Wilson and coworkers have reported a similar longer term combined therapeutic and diagnostic EMF study with ^{177}Lu-DOTA-f-$Gd_3N@C_{80}$ that consists of ^{177}Lu brachytherapy and an MRI diagnostic ($Gd_3N@C_{80}$) in a murine glioblastoma multiforme (GBM) model that shows remarkable therapeutic efficacy out to 120 days, as illustrated in Fig. 9.[36] However, a significant drawback of these studies is that the radionuclide ^{177}Lu is bound to an external DOTA ligand, and therefore the possibility of loss of free radioactive $^{177}Lu^{+3}$ ion is still a possibility for long-term therapeutic applications.

Several other studies on the potential of TNT EMFs with respect to the treatment of GBM have been conducted. In a 2011 study, researchers

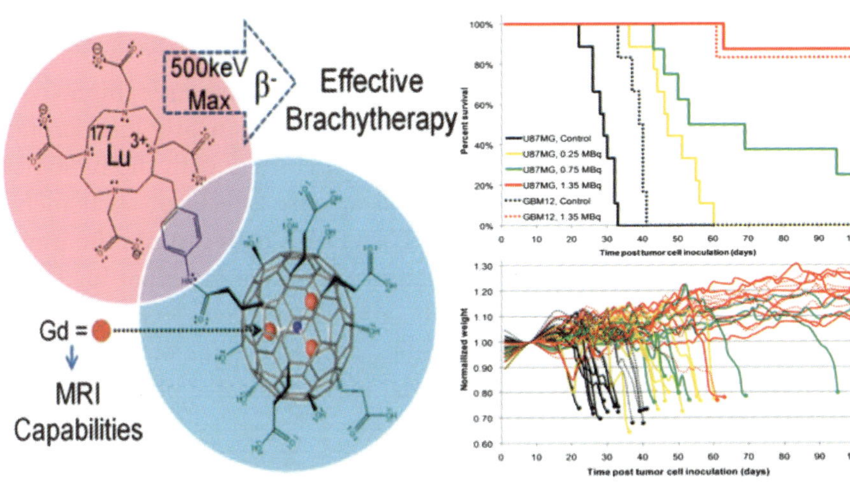

Fig. 9 EMF Theranostic Agent ^{177}Lu-DOTA-f-$Gd_3N@C_{80}$ (left); Suvival Studies with U87MG and GBM ortthotopic tumor-bearing mice (top right); Mice weight normalized to day of treatment (bottom right).

created a metallofullerene nanoplatform designed to selectively target as well as image GBM cells.[37] Targeting was accomplished by attaching an IL-13 related peptide to the carboxyl functionalized metallofullerene $Gd_3N@C_{80}(OH)_{26}(CH_2CH_2COOH)_{16}$. *In vitro* cell binding studies showed a significantly increased affinity of the peptide-conjugated EMF versus a non-conjugated control. *In vivo* experiments in a murine orthotopic model using human U87 GBM cells were also performed. By attaching the IL-13 peptide to a fluorescent label, the authors were able to demonstrate a high level of agreement between areas on the MRI that demonstrated increased contrast and the actual location of the conjugated EMFs on post-mortem histological analysis.

A nanoparticle related to the functionalized $Gd_3N@C_{80}$-IL-13 nanoparticle but with an endohedral radioactive ^{177}Lu was also synthesized. This particle consisted of the same IL-13-related peptide for tumor-targeting conjugated to a functionalized EMF; however, this EMF contained $^{177}Lu_3N$ encapsulated within the cage to give the nanoparticle brachytherapeutic capabilities.[38] The authors demonstrated that the cage was able to contain the encapsulated lutetium ions for 6.7 days without any evidence of release. The authors also argued that this EMF could be used for imaging, as ^{177}Lu is a • emitter that could be identified via Single Photon Emission Computer Tomography (SPECT).

In conclusion, metallofullerene nanoparticles show great promise as important platforms for both diagnostic and therapeutic biomedical applications. As illustrated, the carbon cage provides nearly ideal protection against contact of the endohedral metal clusters with the biological environment and thus against potential adverse effects of such metal-based toxicity. The limited studies of TNT EMFs demonstrate the multi-modality capability of this unique platform and the potential for future biomedical diagnostic and therapeutic applications. However, additional biodistribution and cytotoxicity studies are needed before TNT EMF nanoparticle agents can undergo translation to clinical applications.

REFERENCES

1. Kelkar SS, Reineke TM (2011) Theranostics: Combining Imaging and Therapy. *Bioconjugate Chem* **22:** 1879–1903.

2. Stevenson S, Rice G, Glass T, *et al.* (1999) Small–bandgap endohedral metallofullerenes in high yield and purity. *Nature* **401:** 55–57.
3. Zhang J, Stevenson S, Dorn HC (2013) Trimetallic Nitride Template Endohedral Metallofullerenes: Discovery, Structural Characterization, Reactivity, and Applications. *Acc of Chem Res* **46:** 1548–1557.
4. Kobayashi K, Sano Y, Nagase S (2001) Theoretical study of endohedral metallofullerenes: $Sc_{3-n}La_nN@C_{80}$ (n=0-3). *J Comput Chem* **22:** 1353–1358.
5. Duchamp JC, Demortier A, Fletcher KR, *et al.* (2003) An isomer of the endohedral metallofullerene $Sc_3N@C_{80}$ with D_{5h} symmetry. *Chem Phys Lett* **375:** 655–659.
6. Cai T, Xu LS, Anderson MR, *et al.* (2006) Structure and enhanced reactivity rates of the D_{5h} $Sc_3N@C_{80}$ and $Lu_3N@C_{80}$ metallofullerene isomers: The importance of the pyracylene motif. *J Am Chem Soc* **128:** 8581–8589.
7. Stevenson S, Fowler PW, Heine T, *et al.* (2000) Materials science-A stable non-classical metallofullerene family. *Nature* **408:** 427–428.
8. Olmstead MM, Lee HM, Duchamp JC, *et al.* (2003) $Sc_3N@C_{68}$: Folded Pentalene Coordination in an Endohedral Fullerene that Does Not Obey the Isolated Pentagon Rule. *Angew Chem Int Ed* **42:** 900–903.
9. Olmstead MM, Duchamp JC, *et al.* (2001) Isolation and Structural Characterization of the Endohedral Fullerene $Sc_3N@C_{78}$. *Angew Chem Int Ed* **40:** 1223–1225.
10. Qian MC, Ong SV, Khanna SN, *et al.* (2007) Magnetic endohedral metallofullerenes with floppy interiors. *Phys Rev B* **75:** 104424–104430.
11. Ge ZX, Duchamp JC, Cai T, *et al.* (2005) Purification of endohedral trimetallic nitride fullerenes in a single, facile step. *J Am Chem Soc* **127:** 16292–16298.
12. Stevenson S, Harich K, Yu H, *et al.* (2006) Nonchromatographic "stir and filter approach" (SAFA) for isolating $Sc_3N@C_{80}$ metallofullerenes. *J Am Chem Soc* **128:** 8829–8835.
13. Stevenson S, Mackey MA, Coumbe CE, *et al.* (2007) Rapid removal of D_{5h} isomer using the "stir and filter approach" and isolation of large quantities of isomerically pure $Sc_3N@C_{80}$ metallic nitride fullerenes. *J Am Chem Soc* **129:** 6072–6073.
14. Iezzi EB, Duchamp JC, Harich K, *et al.* (2002) A symmetric derivative of the trimetallic nitride endohedral metallofullerene, $Sc_3N@C_{80}$. *J Am Chem Soc* **124:** 524–525.
15. Stevenson S, Stephen RR, Amos TM, *et al.* (2005) Synthesis and purification of a metallic nitride fullerene BisAdduct: Exploring the reactivity of $Gd_3N@C_{80}$. *J Am Chem Soc* **127:** 12776–12777.

16. Lukoyanova O, Cardona CM, Rivera J, *et al.* (2007) "Open rather than closed" malonate methano fullerene derivatives. The formation of methanofulleroid adducts of $Y_3N@C_{80}$. *J Am Chem Soc* **129:** 10423–10430.
17. Pinzon JR, Zuo TM, Echegoyen L (2010) Synthesis and Electrochemical Studies of Bingel-Hirsch Derivatives of $M_3N@I_hC_{80}$ (M = Sc, Lu). *Chem Eur J* **16:** 4864–4869.
18. Cardona CM, Kitaygorodskiy A, Echegoyen L (2005) Trimetallic nitride endohedral metallofullerenes: Reactivity dictated by the encapsulated metal cluster. *J Am Chem Soc* **127:** 10448–10453.
19. Cai T, Slebodnick C, Xu L, *et al.* (2006) A pirouette on a metallofullerene sphere: Interconversion of isomers of N-tritylpyrrolidino I_h-$Sc_3N@C_{80}$. *J Am Chem Soc* **128:** 6486–6492.
20. Chen N, Zhang EY, Tan K, *et al.* (2007) Size effect of encaged clusters on the exohedral chemistry of endohedral fullerenes: A case study on the pyrrolidino reaction of $Sc_xGd_{3-x}N@C_{80}$ (x=0-3). *Org Lett* **9:** 2011–2013.
21. Shu CY, Cai T, Xu LS, *et al.* (2007) Manganese(III)-catalyzed free radical reactions on trimetallic nitride endohedral metallofullerenes. *J Am Chem Soc* **129:** 15710–15717.
22. Iezzi EB, Cromer F, Stevenson P, *et al.* (2002) Synthesis of the first water-soluble trimetallic nitride endohedral metallofullerols. *Synth Met* **128:** 289–291.
23. Fatouros PP, Corwin FD, Chen ZJ, *et al.* (2006) In vitro and in vivo imaging studies of a new endohedral metallofullerene nanoparticle. *Radiology* **240:** 756–764.
24. Zhang JF, Fatouros PP, Shu CY, *et al.* (2010) High Relaxivity Trimetallic Nitride (Gd_3N) Metallofullerene MRI Contrast Agents with Optimized Functionality. *Bioconjugate Chem* **21:** 610–615.
25. Shu CY, Corwin FD, Zhang JF, *et al.* (2009) Facile Preparation of a New Gadofullerene-Based Magnetic Resonance Imaging Contrast Agent with High H^+ Relaxivity. *Bioconjugate Chem* **20:** 1186–1193.
26. Zhang EY, Shu CY, Feng L, *et al.* (2007) Preparation and characterization of two new water-soluble endohedral metallofullerenes as magnetic resonance imaging contrast agents. *J Phys Chem B* **111:** 14223–14226.
27. Zhang JF, Ge JC, Shultz MD, *et al.* (2010) In Vitro and in Vivo Studies of Single-Walled Carbon Nanohorns with Encapsulated Metallofullerenes and Exohedrally Functionalized Quantum Dots. *Nano Lett* **10:** 2843–2848.
28. Huang W, Zhang JF, Dorn HC, *et al.* (2011) Assembly of Single-Walled Carbon Nanohorn Supported Liposome Particles. *Bioconjugate Chem* **22:** 1012–1016.

29. Braun K, Dunsch L, Pipkorn R, *et al.* (2010) Gain of a 500 fold sensitivity on an intravital MR Contrast Agent based on an endohedral Gadolinium Cluster Fullerene Conjugate: A new chance in cancer diagnostics. *Int J Med Sci* **7:** 136–146.
30. MacFarland DK, Walker KL, Lenk RP, *et al.* (2008) Hydrochalarones: A novel endohedral metallofullerene platform for enhancing magnetic resonance imaging contrast. *J Med Chem* **51:** 3681–3683.
31. Dellinger A, Olson J, Link K, *et al.* (2013) Functionalization of gadolinium metallofullerenes for detecting atherosclerotic plaque lesions by cardiovascular magnetic resonance. *J Cardiovasc Magn Reson* **15 ASAP**
32. Xiao L, Takada H, Gan XH, *et al.* (2006) The water-soluble fullerene derivative 'Radical Sponge' exerts cytoprotective action against UVA irradiation but not visible-light-catalyzed cytotoxicity in human skin keratinocytes. *Bioorg & Med Chem Lett* **16:** 1590–1595.
33. Yin JJ, Lao F, Fu PP, *et al.* (2009) The scavenging of reactive oxygen species and the potential for cell protection by functionalized fullerene materials. *Biomaterials* **30:** 611–621.
34. Ehrich M, Van Tassell R, Li Y, *et al.* (2011) Fullerene antioxidants decrease organophosphate-induced acetylcholinesterase inhibition in vitro. *Toxicol in Vitro* **25:** 301–307.
35. Shultz MD, Wilson JD, Fuller CE, *et al.* (2011) Metallofullerene-based Nanoplatform for Brain Tumor Brachytherapy and Longitudinal Imaging in a Murine Orthotopic Xenograft Model. *Radiology* **261:** 136–143.

CHAPTER 14

Higher LUMO Level Endohedral Fullerene and Fullerene Bisadduct Acceptors for Polymer Solar Cells

*Yongfang Li**

1. INTRODUCTION

Bulk heterojunction (BHJ) polymer solar cells (PSCs) are commonly composed of a blend active layer of conjugated polymer donor and fullerene derivative acceptor sandwiched between a transparent ITO positive electrode and a low workfunction negative electrode.[1] When light irradiates on the PSCs through the transparent ITO electrode, the semiconducting donor and acceptor materials in the photoactive layer absorb photons to produce excitons (bounded electron-hole pairs); the excitons then diffuse to the donor/acceptor interfaces where they dissociate into electrons on the LUMO (the lowest unoccupied molecular orbital) level of the acceptor and holes on the HOMO (the highest occupied molecular orbital) level of the donor. Then, under the driving force of the built-in electric field of the device from the workfunction difference between the positive and negative electrodes, the photo-induced electrons and holes move to and are

* Institute of Chemistry, Chinese Academy of Sciences, Beijing 100190, China, and College of Chemistry, Chemical Engineering and Materials Science, Soochow University, Suzhou, Jiangsu 215123, China. E-mail: liyf@iccas.ac.cn

collected by the negative and positive electrodes respectively to realize the optoelectronic conversion. In comparison with traditional inorganic semiconductor solar cells, PSCs possess the advantages of simple device structure, low-cost fabrication by solution-processing, light weight, and possibility to be fabricated into flexible or transparent devices. Therefore, PSCs have attracted great attention in recent years, and the power conversion efficiency of the PSCs has reached 8~9%.[2-4]

Present studies on PSCs are mainly focused on increasing power conversion efficiency (PCE). For the solar cells, the PCE is proportional to the short circuit current density (J_{sc}), open circuit voltage (V_{oc}) and fill factor (FF) of the devices. The key points to increase PCE are high-efficiency donor and acceptor photovoltaic materials[5,6] and new device structures[2-4,7] to achieve higher values of J_{sc}, V_{oc} and FF.

As for the photovoltaic materials, the most representative polymer donor and fullerene derivative acceptor used in the PSCs are regioregular poly(3-hexylthiophene) (P3HT) and [6,6]-phenyl-C_{61}- butyric acid methyl ester ($PC_{60}BM$) or [6,6]-phenyl-C_{71}- butyric acid methyl ester ($PC_{70}BM$), respectively (molecular structures of P3HT, $PC_{60}BM$ and $PC_{70}BM$ are shown in Fig. 1), and the PCE of the PSCs based on P3HT/$PC_{60}BM$ reached reproducibly over 4% by solvent annealing and thermal annealing.[8,9] But further improvement of PCE of the PSCs is limited by the mismatch of the absorption spectra of the P3HT/$PC_{60}BM$ active layer with the solar spectrum (the absorption edge of P3HT film is only 650 nm and the visible absorption of $PC_{60}BM$ is very weak) and the mismatch of the electronic energy levels of P3HT and $PC_{60}BM$. The LUMO and HOMO energy levels of P3HT are −2.74 and −4.76 eV, respectively,[10] while the LUMO and HOMO energy levels of $PC_{60}BM$ are −3.91 and −5.93 eV, respectively.[6]. We need the LUMO level difference ($\Delta \boldsymbol{E}_1$) between the polymer donor and the fullerene acceptor to be larger than 0.3 eV for efficient exciton dissociation to overcome binding energy of the excitons. But $\Delta \boldsymbol{E}_1$ of 1.17 eV for P3HT and $PC_{60}BM$ is too big, which results in low V_{oc} of the devices, because the V_{oc} of the PSCs is proportional to the difference between the LUMO of the fullerene acceptor and the HOMO of the polymer donor. For increasing efficiency of the PSCs, various new donor and acceptor photovoltaic materials have been developed in recent years,[5] in pursuing broad and strong absorption in visible-near infrared

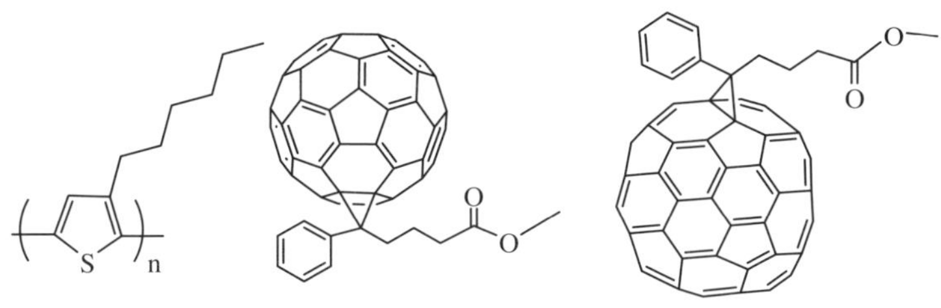

Fig. 1 Molecular structures of P3HT, $PC_{60}BM$ and $PC_{70}BM$.

region, higher charge carrier mobility and suitable electronic energy levels (relatively lower HOMO energy level for the donor materials with PCBM as acceptor and relatively higher LUMO energy level for the acceptor materials with P3HT as donor). In this chapter, I will mainly focus on fullerene acceptors for improving PCE of the PSCs, especially focusing on the new fullerene acceptors with LUMO energy level up-shifted than that of PCBM.

2. $PC_{60}BM$ AND $PC_{70}BM$

$PC_{60}BM$ was firstly synthesized by Hummelen and Wudl *et al.*[11] and used as acceptor in the first BHJ PSC reported in 1995. $PC_{60}BM$ is a crystalline dark brown powder, and has good solubility in common organic solvents such as chloroform, toluene and o-dichlorobenzene. The solubility of $PC_{60}BM$ in toluene is ca. 35 mg/mL, reported by American Dye Source, Inc. Electron mobility of $PC_{60}BM$ is 2×10^{-3} cm^2/Vs, measured by space-charge-limited-current (SCLC) method.[12] The LUMO and HOMO energy levels of $PC_{60}BM$ are −3.91 and −5.93 eV, respectively, measured by electrochemical cyclic voltammetry.[6]

A weak point of $PC_{60}BM$ as a photovoltaic material is its weak absorption in the visible region, as shown in Fig. 2, due to the high degree of symmetry of C_{60}, which makes the lowest-energy transitions formally dipole forbidden. The corresponding C_{70} derivative $PC_{70}BM$[13] possesses much stronger visible absorption than $PC_{60}BM$ (see Fig. 2) with similar solubility and LUMO energy level with $PC_{60}BM$ (the LUMO and HOMO energy levels of $PC_{70}BM$ are −3.91 and −5.87 eV, respectively[6]). Therefore,

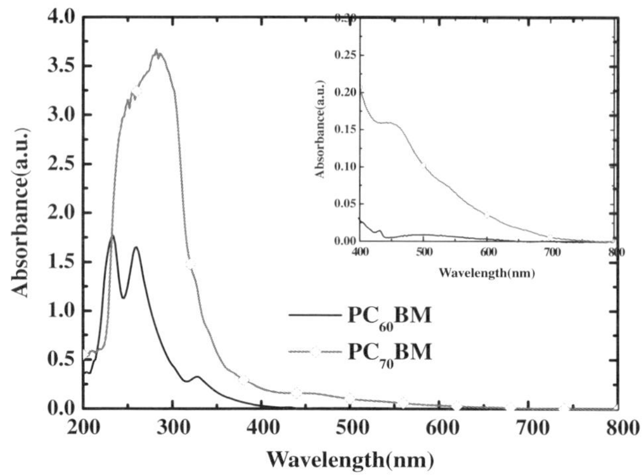

Fig. 2 Absorption spectra of $PC_{60}BM$ and $PC_{70}BM$ in THF solutions (10^{-5} mol/L); the inset is the enlarged absorption spectra in the visible region.

$PC_{70}BM$ has been widely used instead of $PC_{60}BM$ in the PSCs based on narrow bandgap polymer donors for enhancing the solar light harvest in the wavelength region of 350~600 nm.

3. ENDOHEDRALL FULLERENES

As mentioned in the introduction, fullerene derivatives with higher LUMO energy levels are desirable for increasing the V_{oc} of the PSCs. Ross *et al.*[14] noticed that the LUMO energy levels of trimetallic nitride endohedral fullerenes (TNEFs)[15] are much higher than those of their corresponding empty-cage fullerenes.[16,17] In order to utilize TNEFs in the BHJ PSCs, they synthesized a series of soluble PCBM-like trilutetium nitride endohedral C_{80} ($Lu_3N@C_{80}$) derivatives, $Lu_3N@C_{80}$-PCBX (X=M for methyl, B for butyl, H for hexyl and O for octyl end groups on the substituents). Figure 3 shows the molecular structures of $Lu_3N@C_{80}$-PCBM and $Lu_3N@C_{80}$-PCBH, and Fig. 4 displays the synthetic route of $Lu_3N@C_{80}$-PCBH.[14] It should be mentioned that the synthesis conditions for $Lu_3N@C_{80}$-PCBH are quite different from those for PCBM due to different chemical reactivity of the endohedral fullerenes with the empty-cage fullerenes.[14]

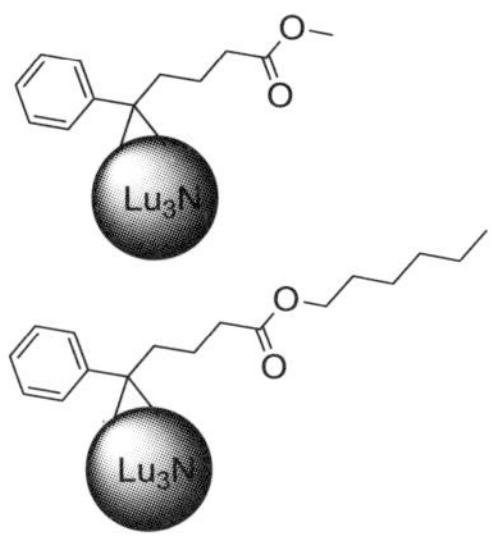

Fig. 3 Molecular structures of trilutetium nitride endohedral C_{80} ($Lu_3N@C_{80}$) derivatives.

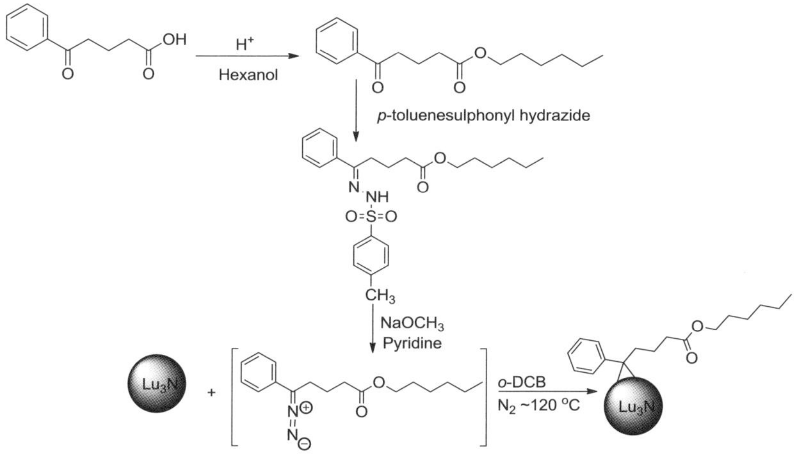

Fig. 4 Synthetic route of $Lu_3N@C_{80}$-PCBH.

$Lu_3N@C_{80}$-PCBH possesses similar solubility and miscibility to $PC_{60}BM$ when it is used as an acceptor blended with P3HT. The LUMO energy level of $Lu_3N@C_{80}$-PCBH is 0.28 eV higher than that of $PC_{60}BM$.[14] The electron mobility of $Lu_3N@C_{80}$-PCBH is 4.0×10^{-4} cm^2/V.s in comparison with 1.4×10^{-3} cm^2/V.s for $PC_{60}BM$.

The photovoltaic properties of $Lu_3N@C_{80}$-PCBH were investigated by fabricating the PSCs with P3HT as donor and $Lu_3N@C_{80}$-PCBH as acceptor, and compared with the PSC based on P3HT/$PC_{60}BM$. Figure 5 shows the current density (J) — voltage (V) curves of the devices. The power conversion efficiency (PCE) of the PSC based on P3HT/$Lu_3N@C_{80}$-PCBH

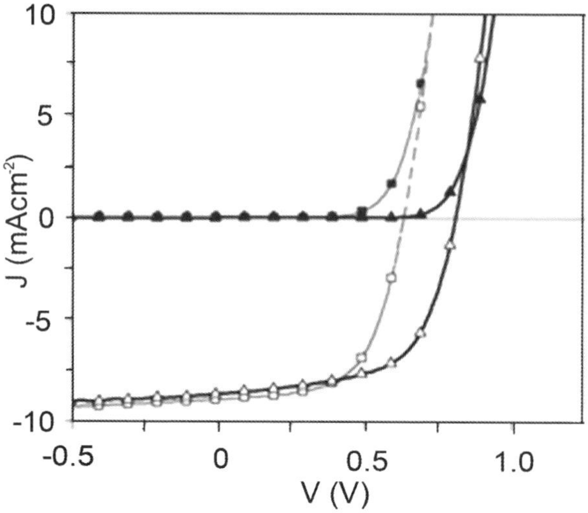

Fig. 5 *J-V* curves of the PSCs based on P_3HT/PCBM (filled and open square) and P_3HT/$Lu_3N@C_{80}$-PCBH (filled and open triangle) at dark and under the illumination of AM1.5, 100 mW/cm^2.

reached 4.2% with a higher V_{oc} of 0.81 V, a short circuit current density (J_{sc}) of 8.64 mA/cm^2 and a fill factor (FF) of 0.61 under the illumination of AM1.5, 100 mW/cm^2, while the PCE of the PSC based on P3HT/$PC_{60}BM$ was only 3.4% with a V_{oc} of 0.63 V, a J_{sc} of 8.9 mA/cm^2 and a FF of 0.61 under the same experimental conditions.[14] The V_{oc} of the device with $Lu_3N@C_{80}$-PCBH as acceptor is 0.18 V higher than that of the PSC with $PC_{60}BM$ as acceptor, which is benefited from the up-shifted LUMO energy level of $Lu_3N@C_{80}$-PCBH. The result indicates that the endohedrall fullerene derivatives could be a type of high V_{oc} acceptor for the application in PSCs.

4. FULLERENE BISADDUCTS

In recent years, fullerene bisadducts have attracted great attention due to their higher-lying LUMO energy levels than PCBM and good photovoltaic performance in the P3HT-based PSCs. The fullerene bisadducts reported in the literature include PCBM bisadducts, indene-fullerene bisadducts, dihydronaphthyl fullerene bisadducts, etc.

4.1. PCBM Bisadducts

Figure 6 shows the molecular structure of the PCBM-like fullerene bisadducts. The PCBM-bisadduct (bisPCBM) can be separated by silicon gel chromatography from the product mixture during the synthesis of PCBM.[18,19] Lenes *et al.* studied the physical and photovoltaic properties of bisPC$_{60}$BM in detail.[19] bisPC$_{60}$BM has very good solubility in common organic solvents as PC$_{60}$BM; the LUMO energy level of bisPC$_{60}$BM is 0.1 eV higher than that of PC$_{60}$BM, and its electron mobility is 7×10^{-4} cm^2/Vs,[18] a little lower than that of PC$_{60}$BM (2×10^{-3} cm^2/Vs).[12] Choi *et al.* synthesized a series of thienyl-substituted methanofullerenes, including ThC$_{60}$BM and ThC$_{60}$BM-bisadduct (bisThC$_{60}$BM, see Fig. 6) for improving the miscibility of the polymer donor and the fullerene acceptor and increasing the V_{oc} of the PSCs.[20] They found that the first one-electron reduction potentials of the fullerene derivatives are shifted toward more negative values from −0.57 V for ThC$_{60}$BM to −0.66 V for bisThC$_{60}$BM, indicating that the LUMO energy level of bisThC$_{60}$BM raised by 0.09 eV more than ThC$_{60}$BM.[20] Obviously, the LUMO energy levels of the PCBM-like bisadducts are ca. 0.1 eV up-shifted than those of the corresponding PCBM monoadducts.

The photovoltaic properties of the fullerene derivatives were studied by fabricating the PSCs based on P3HT as donor and the fullerene derivatives as acceptor, and the results are listed in Table 1. The V_{oc}, J_{sc} and PCE of the

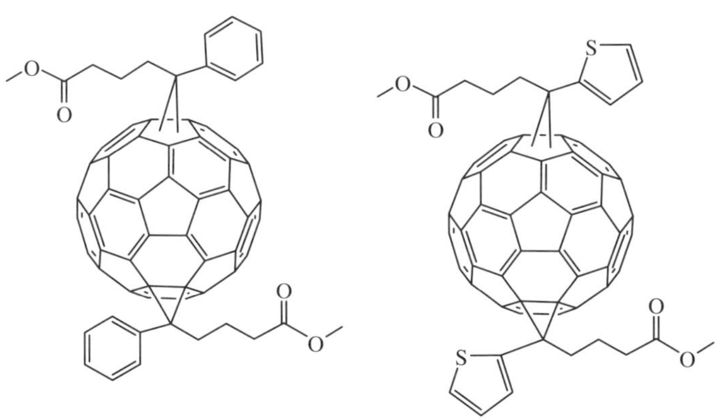

bisPC$_{60}$BM **bisTh C$_{60}$BM**

Fig. 6 Molecular structures of PCBM-like fullerene bisadducts.

Table 1 Photovoltaic Performance of the Fullerene Multiadduct Acceptors

Donor/Acceptor (weight ratio)	J_{sc}(mA/cm^2)	V_{oc}(V)	FF	PCE (%)	Ref.
P3HT/PC$_{60}$BM(1:0.8)	10.9	0.62	0.62	4.18	20
P3HT/ThC$_{60}$BM(1:1)	10.9	0.60	0.61	3.97	20
P3HT/bisThC$_{60}$BM(1:1.2)	5.91	0.72	0.41	1.73	20
P3HT/bisPC$_{60}$BM(1:1.2)	9.14	0.724	0.68	4.5	18
P3HT/PC$_{60}$BM(1:1)	10.8	0.58	0.62	3.88	21
P3HT/IC$_{60}$MA(1:1)	9.66	0.63	0.64	3.89	21
P3HT/IC$_{60}$BA(1:1)	9.67	0.84	0.67	5.44	21
P3HT/IC$_{60}$BA(1:1)	10.61	0.84	0.73	6.48	22
P3HT/IC$_{70}$BA(1:1)	9.73	0.84	0.69	5.64	23
P3HT/IC$_{70}$BA(1:1) with 1-chloronaphthalene additive	11.35	0.87	0.75	7.40	24
P3HT/NC$_{60}$BA(1:1)	9.88	0.82	0.67	5.37	25
P3HT/NC$_{70}$BA(1:1)	10.71	0.83	0.67	5.95	26

devices with bisPC$_{60}$BM as acceptor were 0.724 V, 9.14 mA/cm^2 and 4.5%.[18] The V_{oc} of the device with bisPC$_{60}$BM as acceptor was 0.15 V higher than that of the device based on PC$_{60}$BM, and the J_{sc} and fill factor (FF) of the two devices were similar. The higher V_{oc} benefited from the higher LUMO energy level of bisPC$_{60}$BM. The PCE of the device based on bisPC$_{60}$BM was 18% higher than that of the device based on PC$_{60}$BM, mainly resulting from the higher V_{oc}.[18] The photovoltaic properties of bisThC$_{60}$BM and corresponding monoadduct ThC$_{60}$BM are also listed in Table 1. It can be seen that the photovoltaic performance of ThC$_{60}$BM is similar to that of PC$_{60}$BM. But its bisadduct bisThC$_{60}$BM showed poorer PCE due to lower J_{sc} and FF values, although the V_{oc} is increased by ca. 0.12 V benefiting from the higher LUMO energy level of bisThC$_{60}$BM.[20] The lower J_{sc} and FF values could be due to the lower electron mobility of bisThC$_{60}$BM.

4.2. Indene-C_{60} Bisadduct and Indene-C_{70} Bisadduct

For further up-shifting the LUMO energy level of the fullerene derivatives, He and Li *et al.* designed and synthesized a new kind of bisadduct of fullerenes with electron-donating indene as substituent, indene-C_{60} bisadduct ($IC_{60}BA$)[21,22] and indene-C_{70} bisadduct ($IC_{70}BA$).[23] Figure 7 shows the molecular structures and synthetic routes of $IC_{60}BA$ and $IC_{70}BA$, together with indene-C_{60} monoadduct ($IC_{60}MA$) and indene-C_{70} monoadduct ($IC_{70}MA$). The indene-fullerene adducts can be obtained by a one-pot reaction of indene and fullerene (C_{60} or C_{70}) in o-dichlorobenzene or trichlorobenzene. But the product is a mixture of unreacted fullerene, indene-fullerene monoadduct, indene-fullerene bisadduct and indene-fullerene multiadduct. The mixture products can be separated and purified by silicon gel chromatrography.[21,23]

$IC_{60}BA$ possesses good solubility, which is even better than that of $PC_{60}BM$. The solubility of $IC_{60}MA$ and $IC_{70}BA$ is a little poorer than that of $IC_{60}BA$, while the solubility of $IC_{70}MA$ is poor and it is not suitable for

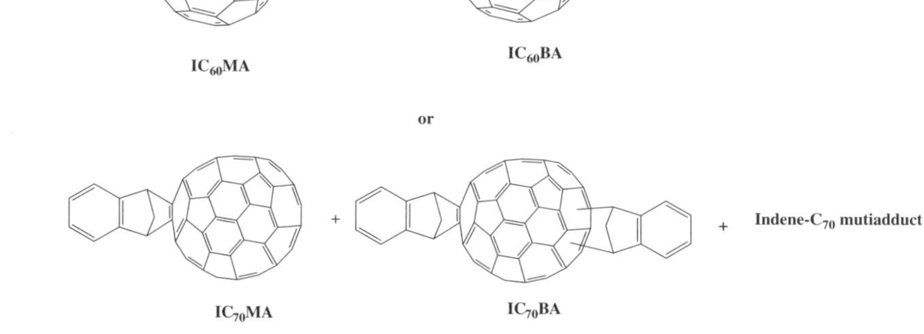

Fig. 7 Molecular structures and synthetic routes of indene-fullerene adducts.

the use in PSCs. The LUMO energy levels of $IC_{60}BA$ and $IC_{70}BA$ are 0.17 eV and 0.19 eV higher than that of $PC_{60}BM$, respectively. The LUMO energy levels of the indene monoadduct $IC_{60}MA$ and $IC_{70}MA$ are ca. 0.05 eV higher than that of $PC_{60}BM$.[21,23]

The photovoltaic properties of the indene-fullerene bisadducts acceptors with P3HT as donor were also listed in Table 1. The V_{oc} of the PSCs with $IC_{60}BA$ or $IC_{70}BA$ as acceptor reached 0.84 V,[21,23] which is 0.26 V higher than that of the PSCs with $PC_{60}BM$ as acceptor. Obviously, the great increase on V_{oc} should be benefited from the higher LUMO energy level of the two bisadduct. The PCE values of the PSCs based on P3HT/$IC_{60}BA$ and P3HT/$IC_{70}BA$ reached 5.44% and 5.64% respectively,[21,23] which is more than 40% higher in comparison with that of the PSC based on P3HT/$PC_{60}BM$. After further device optimization, the PSC based on P3HT/$IC_{60}BA$ displayed a V_{oc} of 0.84 V, a J_{sc} of 10.61 mA/cm^2, a FF of 72.7% and a PCE of 6.48%.[22] For the device based on P3HT/$IC_{70}BA$ with 3 vol% 1-chloronaphthalene additive and pre-thermal annealing at 150 °C for 10 min, the PCE of the device reached 7.40% with *Voc* of 0.87 V, *Jsc* of 11.35 mA/cm^2 and *FF* of 75.0%, under the illumination of AM 1.5G, 100 mW/cm^2.[24]

4.3. Hydronaphthyl-C_{60} Bisadduct and Hydronaphthyl-C_{70} Bisadduct

Recently, Wang *et al.* synthesized a new series of fullerene bisadducts, including hydronaphthyl-C_{60} bisadduct ($NC_{60}BA$)[25] and hydronaphthyl-C_{70} bisadduct ($NC_{70}BA$).[26] $NC_{60}BA$ was synthesized from Diels–Alder reaction of C_{60} and the *in-situ* generated *o*-quinodimethane at 80°C in 72% yield (see Fig. 8),[25] and $NC_{70}BA$ was synthesized by the same synthetic route using C_{70} instead of C_{60}.[26] The LUMO energy levels of $NC_{60}BA$ and $NC_{70}BA$ are 0.16 eV and 0.20 eV higher than that of $PC_{60}BM$, respectively.

The photovoltaic performance of the PSCs based on P3HT as donor and $NC_{60}BA$ or $NC_{70}BA$ as acceptor was also listed in Table 1 for a clear comparison. The device based on P3HT/ $NC_{60}BA$ demonstrated a PCE of 5.37% with a high V_{oc} of 0.82 V, a J_{sc} of 9.88 mA/cm^2 and a FF of 67%.[25] The PCE of the device based on P3HT/ $NC_{70}BA$ further increased to 5.95% with V_{oc} of 0.83 V, a J_{sc} of 10.71 mA/cm^2 and a FF of 67%.[26] The photovoltaic performance of $NC_{60}BA$ and $NC_{70}BA$ is similar to $IC_{60}BA$ and

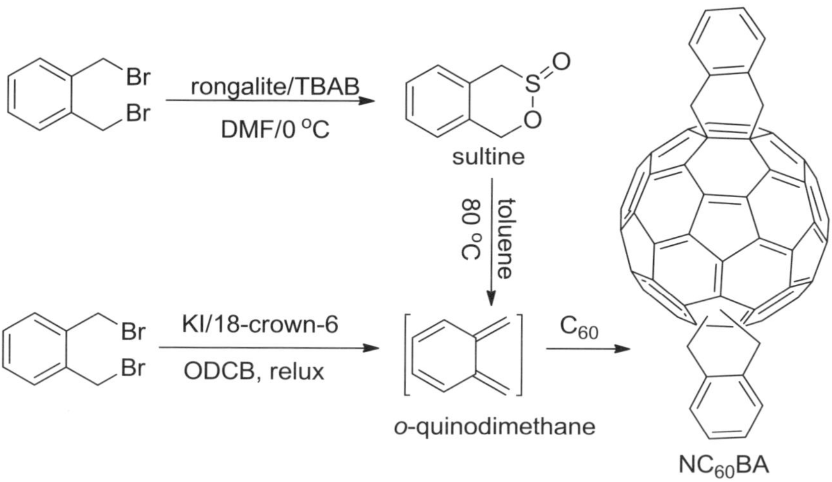

Fig. 8 Molecular structure and synthetic route of $NC_{60}BA$.

$IC_{70}BA$, which is greatly improved in comparison with that of PCBM, benefiting from the higher V_{oc} due to the higher-lying LUMO energy levels of the fullerene bisadducts.

It should be mentioned that the fullerene bisadducts are amorphous. The amorphous character of the fullerene bisadduct acceptors leads to better thermal stability of the PSCs with P3HT as donor in comparison with the crystalline PCBM acceptor which results in poor thermal stability due to the morphology unstability of PCBM under higher temperature. But the amorphous nature of the fullerene bisadductors is a disadvantage when blending with the amorphous narrow bandgap polymer donors, which limits their application. In addition, the fullerene bisadducts mentioned in this section are mixtures of their regioisomers. If we can separate the mixture into pure isomers, the photovoltaic performance of the fullerene bisadductor could be further improved.

5. SUMMARY

The bulk-heterojunction PSCs have developed quickly in recent years. The key issue is the design and synthesis of new conjugated polymer donors and new fullerene derivative acceptors for further improving the photovoltaic performance. In comparison with numerous new polymer donor materials, there are much fewer new fullerene derivative acceptors showing better photovoltaic properties than $PC_{60}BM$. $PC_{60}BM$ is still the most important acceptor material used in the studies of PSCs.

Important research progress on the new acceptors has been made with fullerene derivatives with higher LUMO energy levels. In this respect, endohedral fullerenes (such as $Lu_3N@C_{80}$-PCBH) and fullerene bisadducts (bisPCBM, ICBA and NCBA) are successful acceptors used in the P3HT-based PSCs. The V_{oc} and PCE of the PSCs increased significantly, benefiting from the higher LUMO energy levels of the acceptors. Actually, from the viewpoint of energy level matching, the PCE of over 8% for the P3HT-based PSCs can be expected if we can find a new fullerene derivative acceptor with LUMO energy level ca. 0.5 eV higher than that of $PC_{60}BM$.[27] $Lu_3N@C_{80}$-PCBH and the fullerene bisadducts are successful examples towards achieving this goal. For the fullerene derivatives, tuning the LUMO energy level upwards and improving electron mobility while keeping its high solubility should be pursued in the future for high-performance PSCs.

LIST OF ABBREVIATIONS

AM	air mass
BHJ	bulk heterojunction
FF	fill factor
HOMO	highest occupied molecular orbital
J_{sc}	short circuit current density
$IC_{60}BA$	indene-C_{60} bisadduct
$IC_{60}MA$	indene-C_{60} monoadduct
$IC_{70}BA$	indene-C_{70} bisadduct
$IC_{70}MA$	indene-C_{70} monoadduct
ITO	indium tin oxide
LUMO	lowest unoccupied molecular orbital
$NC_{60}BA$	hydronaphthyl -C_{60} bisadduct
$NC_{70}BA$	hydronaphthyl -C_{70} bisadduct
PSCs	polymer solar cells
$PC_{60}BM$	[6,6]-phenyl-C_{61}- butyric acid methyl ester
$PC_{70}BM$	[6,6]-phenyl-C_{71}- butyric acid methyl ester

P3HT	poly(3-hexylthiophene)
PCE	power conversion efficiency
SCLC	space-charge-limited-current
V_{oc}	open circuit voltage

REFERENCES

1. Yu G, Gao J, Hummelen JC, *et al.* (1995) Polymer photovoltaic cells: enhanced efficiencies via a network of internal donor-acceptor heterojunctions. *Science* **270:** 1789.
2. He ZC, Zhong CM, Su SJ, *et al.* (2012) Enhanced power-conversion efficiency in polymer solar cells using an inverted device structure. *Nature Photonics* **6:** 591.
3. Dou LT, You JB, Yang J, *et al.* (2012) Tandem polymer solar cells featuring a spectrally matched low-bandgap polymer. *Nature Photonics* **6:** 180.
4. Li XH, Choy WCH, Huo LJ, *et al.* (2012) Dual plasmonic nanostructures for high performance inverted organic solar cells. *Adv Mater* **24:** 3046.
5. Li YF. (2012) Molecular design of photovoltaic materials for polymer solar cells: towards suitable electronic energy levels and broad absorption. *Acc Chem Res* **45:** 723.
6. He YJ, Li YF. (2011) Fullerene derivative acceptors for high performance polymer solar cells. *Phys Chem Chem Phys* **13:** 2011.
7. Tan ZA, Zhang WQ, Zhang ZG, *et al.* (2012) High performance inverted polymer solar cells with solution-processed titanium chelate as electron-collecting layer on ITO electrode. *Adv Mater* **24:** 1476.
8. Li G, Shrotriya V, Huang JS, *et al.* (2005) High-efficiency solution processable polymer photovoltaic cells by self-organization of polymer blends. *Nat Mater* **4:** 864.
9. Ma W, Yang C, Gong X, *et al.* (2005) Thermally stable, efficiency polymer solar cells with nanoscale control of the interpenetrating network morphology. *Adv Funct Mater* **15:** 1617.
10. Hou JH, Tan ZA, Yan Y, *et al.* (2006) Synthesis and photovoltaic properties of two-dimensional conjugated polythiophenes with bi(thienylenevinylene) side chains. *J Am Chem Soc* **128:** 4911.
11. Hummelen JC, Knight BW, Lepeq F, Wudl F. (1995) Preparation and characterization of fulleroid and methanofullerene derivatives. *J Org Chem* **60:** 532.
12. Mihailetchi VD, van Duren JKJ, Blom PWM, *et al.* (2003) Electron transport in a methanofullerene. *Adv Funct Mater* **13:** 43.

13. Wienk MM, Kroon MJ, Verhees WJH, *et al.* (2003) Efficient methano[70]fullerene/MDMO-PPV bulk heterojunction photovoltaic cells. *Angew Chem Int Ed* **42:** 3371.
14. Ross RB, Cardona CM, Guldi DM, *et al.* (2009) Endohedral fullerenes for organic photovoltaic devices. *Nat Mater* **8:** 208.
15. Stevenson S, Rice G, Glass T, *et al.* (1999) Small-bandgap endohedral metallofullerenes in high yield and purity. *Nature* **401:** 55.
16. Campanera JM, Bo C, Olmstead MM, *et al.* (2002) "http://apps.webofknowledge.com/full_record.do?product=WOS&search_mode=GeneralSearch&qid=10&SID=Z2CVtEcT8Yo267LS8Yf&page=1&doc=1" Bonding within the endohedral fullerenes Sc3N@C-78 and Sc3N@C-80 as determined by density functional calculations and reexamination of the crystal structure of {Sc3N@C-78}center dot Co(OEP)center dot 1.5(C6H6)center dot 0.3(CHCl3)". *J Phys Chem A* **106:** 12356.
17. Cardona CM, Elliott B, Echegoyen L. (2006) Unexpected chemical and electrochemical properties of M3N@C-80 (M = Sc, Y, Er). *J Am Chem Soc* **128:** 6480.
18. Lenes M, Wetzelaer GJAH, Kooist FB, *et al.* (2008) Fullerene bisadducts for enhanced open-circuit voltages and efficiencies in polymer solar cells. *Adv Mater* **20:** 2116.
19. Lenes M, Shelton SW, Sieval AB, *et al.* (2009) Electron trapping in higher adduct fullerene-based solar cells. *Adv Funct Mater* **19:** 3002.
20. Choi JH, Son KI, Kim T, *et al.* (2010) Thienyl-substituted methanofullerene derivatives for organic photovoltaic cells. *J Mater Chem* **20:** 475.
21. He YJ, Chen HY, Hou JH, Li YF. (2010) Indene-C_{60} bisadduct: a new acceptor for high performance polymer solar cells. *J Am Chem Soc* **132:** 5532.
22. Zhao GJ, He YJ, Li YF. (2010) 6.5% Efficiency of the polymer solar cells based on poly(3-hexylthiophene) and indene-C_{60} bisadduct by device optimization. *Adv Mater* **22:** 4355.
23. He YJ, Zhao GJ, Peng B, Li YF. (2010) High yield synthesis, electrochemical and photovoltaic properties of indene-C_{70} bisadduct. *Adv Funct Mater* **20:** 3383.
24. Guo X, Cui CH, Zhang MJ, *et al.* (2012) High efficiency polymer solar cells based on poly(3-hexylthiophene)/indene-C_{70} bisadduct with solution additive. *Energy Environ Sci* **5:** 7943.
25. Meng XY, Zhang WQ, Tan ZA, *et al.* (2012) Dihydronaphthyl-based [60] fullerene bisadducts for efficient and stable polymer solar cells. *Chem Commun* **48:** 425.

26. Meng XY, Zhang WQ, Tan ZA, et al. (2012) Highly efficient and thermally stable polymer solar cells with dihydronaphthyl-based [70]fullerene bisadduct derivative as acceptor. *Adv Funct Mater* **22:** 2187.
27. Koster LJA, Mihailetchi VD, Blom PWM. (2006) Ultimate efficiency of polymer/fullerene bulk heterojunction solar cells. *Appl Phys Lett* **88:** 093511.

Index